D1157016

Methods in Enzymology

Volume 401
GLUTHIONE TRANSFERASES AND
GAMMA-GLUTAMYL TRANSPEPTIDASES

METHODS IN ENZYMOLOGY

EDITORS-IN-CHIEF

John N. Abelson   Melvin I. Simon

DIVISION OF BIOLOGY
CALIFORNIA INSTITUTE OF TECHNOLOGY
PASADENA, CALIFORNIA

FOUNDING EDITORS

Sidney P. Colowick and Nathan O. Kaplan

Methods in Enzymology

Volume 401

Gluthione Transferases and Gamma-Glutamyl Transpeptidases

EDITED BY

Helmut Sies

INSTITUTE FOR BIOCHEMISTRY AND MOLECULAR BIOLOGY I,
HEINRICH HEINE UNIVERSITY, DÜSSELDORF,
GERMANY

Lester Packer

MOLECULAR PHARMACOLOGY AND TOXICOLOGY
SCHOOL OF PHARMACY
UNIVERSITY OF SOUTHERN CALIFORNIA
LOS ANGELES, CALIFORNIA

AMSTERDAM • BOSTON • HEIDELBERG • LONDON
NEW YORK • OXFORD • PARIS • SAN DIEGO
SAN FRANCISCO • SINGAPORE • SYDNEY • TOKYO
Academic Press is an imprint of Elsevier

QP 601
.C733
vol. 401

0 623584809

Elsevier Academic Press
525 B Street, Suite 1900, San Diego, California 92101-4495, USA
84 Theobald's Road, London WC1X 8RR, UK

This book is printed on acid-free paper. ∞

Copyright © 2005, Elsevier Inc. All Rights Reserved.

No part of this publication may be reproduced or transmitted in any form or by any
means, electronic or mechanical, including photocopy, recording, or any information
storage and retrieval system, without permission in writing from the Publisher.

The appearance of the code at the bottom of the first page of a chapter in this book
indicates the Publisher's consent that copies of the chapter may be made for
personal or internal use of specific clients. This consent is given on the condition,
however, that the copier pay the stated per copy fee through the Copyright Clearance
Center, Inc. (www.copyright.com), for copying beyond that permitted by
Sections 107 or 108 of the U.S. Copyright Law. This consent does not extend to
other kinds of copying, such as copying for general distribution, for advertising
or promotional purposes, for creating new collective works, or for resale.
Copy fees for pre-2005 chapters are as shown on the title pages. If no fee code
appears on the title page, the copy fee is the same as for current chapters.
0076-6879/2005 $35.00

Permissions may be sought directly from Elsevier's Science & Technology Rights
Department in Oxford, UK: phone: (+44) 1865 843830, fax: (+44) 1865 853333,
E-mail: permissions@elsevier.com. You may also complete your request on-line
via the Elsevier homepage (http://elsevier.com), by selecting
"Support & Contact" then "Copyright and Permission" and then "Obtaining Permissions."

For all information on all Elsevier Academic Press publications
visit our Web site at www.books.elsevier.com

ISBN-13: 978-0-12-182806-6
ISBN-10: 0-12-182806-9

PRINTED IN THE UNITED STATES OF AMERICA
05 06 07 08 09 9 8 7 6 5 4 3 2 1

Working together to grow
libraries in developing countries

www.elsevier.com | www.bookaid.org | www.sabre.org

ELSEVIER   BOOK AID International   Sabre Foundation

Table of Contents

Contributors to Volume 401

Article numbers are in parantheses following the names of Contributors.
Affiliations listed are current.

BÜNYAMIN AKGÜL (13), *Department of Biochemistry and Molecular Biology, Pennsylvania State University, University Park, Pennsylvania*

M. W. ANDERS (4), *Department of Pharmacology and Physiology, University of Rochester Medical Center, Rochester, New York*

G. A. S. ANSARI (24), *Department of Human Biological Chemistry and Genetics, University of Texas Medical Branch, Galveston, Texas*

RICHARD N. ARMSTRONG (23), *Department of Biochemistry, Vanderbilt University, Nashville, Tennessee*

SANJAY AWASTHI (24), *Department of Chemistry and Biochemistry, University of Texas at Arlington, Arlington, Texas*

YOGESH C. AWASTHI (24), *Department of Human Biological Chemistry and Genetics, University of Texas Medical Branch, Galveston, Texas*

KATJA BECKER (15), *Interdisciplinary Research Center, Giessen University, Giessen, Germany*

LAUREN A. BEIHOFFER (23), *Department of Biochemistry, Vanderbilt University, Nashville, Tennessee*

PRIYARANJAN BHAKAT (10), *Department of Biosciences at Novum, Karolinska Institute, Huddinge, Sweden*

PHILIP G. BOARD (1, 4, 5), *Molecular Genetics Group, John Curtin School of Medical Research, Australian National University, Canberra, Australia*

JAMES B. BRUNS (26), *Pittsburgh, Department of Medicine, Renal-Electrolyte Division, University of Pittsburgh, Pittsburgh, Pennsylvania*

ROSELYNE CASTONGUAY (27), *Department of Chemistry, Université de Montréal, Montréal, Canada*

JINAH CHOI (28), *School of Natural Sciences, University of California, Merced, California*

MARJORIE COGGAN (5), *John Curtin School of Medical Research, Australian National University, Canberra, Australia*

BRIAN F. COLES (2), *Division of Pharmacogenomics and Molecular Epidemiology, National Center for Toxicological Research, Jefferson, Arkansas*

ALESSANDRO CORTI (29), *Department of Experimental Pathology, University of Pisa Medical School, Pisa, Italy*

MARCEL DEPONTE (15), *Interdisciplinary Research Center, Giessen University, Giessen, Germany*

DAVID P. DIXON (11), *Centre for Bioactive Chemistry, School of Biological and Biomedical Sciences, Durham University Durham, United Kingdom*

SILVIA DOMINICI (29), *Department of Experimental Pathology, University of Pisa Medical School, Pisa, Italy*

MARYAM H. EDALAT (22), *Department of Biochemistry, Uppsala University Biomedical Center, Uppsala, Sweden*

ROBERT EDWARDS (11), *Centre for Bioactive Chemistry, School of Biological and Biomedical Sciences, Durham University, United Kingdom*

KERRY L. FILLGROVE (23), *Department of Biochemistry, Vanderbilt University, Nashville, Tennessee*

VICTORIA L. FINDLAY (19), *Medical University of South Carolina, Department of Cell and Molecular Pharmacology and Experimental Therapeutics, Charleston, South Carolina*

HENRY JAY FORMAN (28), *School of Natural Sciences, University of California, Merced, California*

F. PETER GUENGERICH (21), *Department of Biochemistry and Center in Molecular Toxicology, Vanderbilt University School of Medicine, Nashville, Tennessee*

JOHN D. HAYES (1), *Biomedical Research Centre, Ninewells Hospital and Medical School, University of Dundee, Dundee, Scotland, United Kingdom*

HANS HEBERT (10), *Department of Biosciences at Novum, Karolinska Institute, Huddinge, Sweden*

JANET HEMINGWAY (14), *Liverpool School of Tropical Medicine, Liverpool L3 5QA, United Kingdom*

COLIN J. HENDERSON (7), *Cancer Research UK, Molecular Pharmacology Unit, Biomedical Research Centre, Ninewells Hospital and Medical School, Dundee, United Kingdom*

PETER J. HOLM (10), *Molecular Biophysics Center for Chemistry and Chemical Engineering, Lund University, Lund, Sweden*

WOLFGANG W. HUBER (20), *Institut für Krebsforschung, Department of Toxicology, Medical University of Vienna, Vienna, Austria*

REBECCA P. HUGHEY (26), *Department of Medicine, Renal-Electrolyte Division, University of Pittsburgh, Pittsburgh, Pennsylvania*

YOSHITAKA IKEDA (25), *Division of Molecular Cell Biology, Department of Biomolecular Sciences, Saga University Faculty of Medicine, Saga, Japan*

PER-JOHAN JAKOBSSON (9), *Department of Medical Biochemistry and Biophysics, The Karolinska Institute Department of Medicine, The Karolinska University Hospital, Stockholm, Sweden*

CAROLINE JEGERSCHÖLD (10), *Department of Biosciences at Novum, Karolinska Institute, Huddinge, Sweden*

FRED F. KADLUBAR (2), *Division of Pharmacogenomics and Molecular Epidemiology, National Center for Toxicological Research, Jefferson, Arkansas*

JEFFREY W. KEILLOR (27), *Department of Chemistry, Université de Montréal, Montréal, Canada*

ALBERT J. KETTERMAN (6), *Institute of Molecular Biology and Genetics, Mahidol University, Salaya, Nakhon Pathom, Thailand*

CAROL L. KINLOUGH (26), *Department of Medicine, Renal-Electrolyte Division, University of Pittsburgh, Pittsburgh, Pennsylvania*

CHRISTIAN LHERBET (27), *Department of Chemistry, Université de Montréal, Montréal, Canada*

IRVING LISTOWSKY (1, 18), *Department of Biochemistry, Albert-Einstein College of Medicine, Bronx, New York*

EMILIA MAELLARO (29), *Department of Pathophysiology, Experimental Medicine and Public Health, Univerisity of Sienna, Sienna, Italy*

BENGT MANNERVIK (1, 16, 17, 22), Department of Biochemistry, Uppsala University Biomedical Center, Uppsala, Sweden

AMIR MASOUMI (5), Office of the Gene Technology Regulator, Canberra, Australia

RALF MORGENSTERN (8), Division of Biochemical Toxicology, Institute of Environmental Medicine, Karolinska Institute, Stockholm, Sweden

MASAMI MURAMATSU (3), Research Center for Genomic Medicine, Saitama Medical School, Saitama, Japan

ALDO PAOLICCHI (29), Department of Experimental Pathology, University of Pisa Medical School, Pisa, Italy

WOLFRAM PARZEFALL (20), Institut für Krebsforschung, Department of Toxicology, Medical University of Vienna, Vienna, Austria

WILLIAM R. PEARSON (1, 12), Department of Biochemistry and Molecular Genetics, University of Virginia, Charlottesville, Virginia

PÄR L. PETTERSSON (9), Department of Medical Biochemistry and Biophysics, Karolinska Institute, Stockholm, Sweden

PAUL A. POLAND (26), Department of Medicine, Renal-Electrolyte Division, University of Pittsburgh, Pittsburgh, Pennsylvania

ALFONSO POMPELLA (29), Department of Experimental Pathology BMIE, University of Pisa Medical School, Pisa, Italy

FRANÇOISE RAFFALLI-MATHIEU (17), Department of Biochemistry, Uppsala University Biomedical Center, Uppsala, Sweden

HILARY RANSON (14), Liverpool School of Tropical Medicine, Liverpool, United Kingdom

RACHEL E. RIGSBY (23), Department of Biochemistry, Vanderbilt University, Nashville, Tennessee

MASAHARU SAKAI* (3), Department of Biochemistry, Graduate School of Medicine, The University of Tokyo, Tokyo, Japan

ERICA SCHMUCK (5), Deceased

NAOYUKI TANIGUCHI (25), Department of Biochemistry, Osaka University Graduate School of Medicine, Osaka, Japan

NATASHA TETLOW (5), Department of Physiology and Pharmacology, University of Queensland, Brisbane, Australia

KENNETH D. TEW (19), Medical University of South Carolina, Department of Cell and Molecular Pharmacology and Experimental Therapeutics, Charleston, South Carolina

STAFFAN THORÉN (9), Department of Medical Biochemistry and Biophysics, Karolinska Institute, Stockholm, Sweden

DANYELLE M. TOWNSEND (19), Medical University of South Carolina, Department of Cell and Molecular Pharmacology and Experimental Therapeutics, Charleston, South Carolina

CHEN-PEI D. TU (13), Department of Biochemistry and Molecular Biology, Pennsylvania State University, University Park, Pennsylvania

ASTRID K. WHITBREAD (5), School of Life Sciences, Queensland University of Technology, Brisbane, Australia

C. ROLAND WOLF (7), Cancer Research UK Molecular Pharmacology Unit, Biomedical Research Centre, Ninewells Hospital and Medical School, Dundee, United Kingdom

*Current address: Research Center for Genomic Medicine, Saitama Medical School, Saitama, Japan.

JANTANA WONGSANTICHON (6), *Institute of Molecular Biology and Genetics, Mahidol University, Salaya, Nakhon Pathom, Thailand*

HONGQIAO ZHANG (28), *School of Natural Sciences, University of California, Merced, California*

Preface

This volume on glutathione transferases and gamma-glutamyl transpeptidases serves to bring together current methods and concepts in an interesting, important and rapidly developing field of cell and systems biology. It focuses on particular aspects of the so-called Phase II of drug detoxication, which has important ramifications for endogenous metabolism and nutrition. This volume of *Methods in Enzymology* presents current knowledge in the field of research. Together with the volumes on *Quinones and Quinone Enzymes (volumes 378 and 382)*, and on *Phase II Conjugation Enzymes and Transport Systems (volume 400)*, the state of knowledge on proteomics and metabolomics of many pathways of (waste) product elimination, enzyme protein induction and gene regulation and feedback control is provided. We trust that this volume will help stimulate future investigations and speed the advance of knowledge in systems biology.

The editors thank the members of the Advisory Committee, in particular Bengt Mannervik, Uppsala, for their valuable suggestions and wisdom in selecting contributions for this volume. We also thank Marlies Scholtes and Cindy Minor for their valuable help.

HELMUT SIES
LESTER PACKER

METHODS IN ENZYMOLOGY

VOLUME 72. Lipids (Part D)
Edited by JOHN M. LOWENSTEIN

VOLUME 73. Immunochemical Techniques (Part B)
Edited by JOHN J. LANGONE AND HELEN VAN VUNAKIS

VOLUME 74. Immunochemical Techniques (Part C)
Edited by JOHN J. LANGONE AND HELEN VAN VUNAKIS

VOLUME 75. Cumulative Subject Index Volumes XXXI, XXXII, XXXIV–LX
Edited by EDWARD A. DENNIS AND MARTHA G. DENNIS

VOLUME 76. Hemoglobins
Edited by ERALDO ANTONINI, LUIGI ROSSI-BERNARDI, AND EMILIA CHIANCONE

VOLUME 77. Detoxication and Drug Metabolism
Edited by WILLIAM B. JAKOBY

VOLUME 78. Interferons (Part A)
Edited by SIDNEY PESTKA

VOLUME 79. Interferons (Part B)
Edited by SIDNEY PESTKA

VOLUME 80. Proteolytic Enzymes (Part C)
Edited by LASZLO LORAND

VOLUME 81. Biomembranes (Part H: Visual Pigments and Purple Membranes, I)
Edited by LESTER PACKER

VOLUME 82. Structural and Contractile Proteins (Part A: Extracellular Matrix)
Edited by LEON W. CUNNINGHAM AND DIXIE W. FREDERIKSEN

VOLUME 83. Complex Carbohydrates (Part D)
Edited by VICTOR GINSBURG

VOLUME 84. Immunochemical Techniques (Part D: Selected Immunoassays)
Edited by JOHN J. LANGONE AND HELEN VAN VUNAKIS

VOLUME 85. Structural and Contractile Proteins (Part B: The Contractile Apparatus and the Cytoskeleton)
Edited by DIXIE W. FREDERIKSEN AND LEON W. CUNNINGHAM

VOLUME 86. Prostaglandins and Arachidonate Metabolites
Edited by WILLIAM E. M. LANDS AND WILLIAM L. SMITH

VOLUME 87. Enzyme Kinetics and Mechanism (Part C: Intermediates, Stereo-chemistry, and Rate Studies)
Edited by DANIEL L. PURICH

VOLUME 88. Biomembranes (Part I: Visual Pigments and Purple Membranes, II)
Edited by LESTER PACKER

VOLUME 89. Carbohydrate Metabolism (Part D)
Edited by WILLIS A. WOOD

VOLUME 176. Nuclear Magnetic Resonance (Part A: Spectral Techniques and Dynamics)
Edited by NORMAN J. OPPENHEIMER AND THOMAS L. JAMES

VOLUME 177. Nuclear Magnetic Resonance (Part B: Structure and Mechanism)
Edited by NORMAN J. OPPENHEIMER AND THOMAS L. JAMES

VOLUME 178. Antibodies, Antigens, and Molecular Mimicry
Edited by JOHN J. LANGONE

VOLUME 179. Complex Carbohydrates (Part F)
Edited by VICTOR GINSBURG

VOLUME 180. RNA Processing (Part A: General Methods)
Edited by JAMES E. DAHLBERG AND JOHN N. ABELSON

VOLUME 181. RNA Processing (Part B: Specific Methods)
Edited by JAMES E. DAHLBERG AND JOHN N. ABELSON

VOLUME 182. Guide to Protein Purification
Edited by MURRAY P. DEUTSCHER

VOLUME 183. Molecular Evolution: Computer Analysis of Protein and Nucleic Acid Sequences
Edited by RUSSELL F. DOOLITTLE

VOLUME 184. Avidin-Biotin Technology
Edited by MEIR WILCHEK AND EDWARD A. BAYER

VOLUME 185. Gene Expression Technology
Edited by DAVID V. GOEDDEL

VOLUME 186. Oxygen Radicals in Biological Systems (Part B: Oxygen Radicals and Antioxidants)
Edited by LESTER PACKER AND ALEXANDER N. GLAZER

VOLUME 187. Arachidonate Related Lipid Mediators
Edited by ROBERT C. MURPHY AND FRANK A. FITZPATRICK

VOLUME 188. Hydrocarbons and Methylotrophy
Edited by MARY E. LIDSTROM

VOLUME 189. Retinoids (Part A: Molecular and Metabolic Aspects)
Edited by LESTER PACKER

VOLUME 190. Retinoids (Part B: Cell Differentiation and Clinical Applications)
Edited by LESTER PACKER

VOLUME 191. Biomembranes (Part V: Cellular and Subcellular Transport: Epithelial Cells)
Edited by SIDNEY FLEISCHER AND BECCA FLEISCHER

VOLUME 192. Biomembranes (Part W: Cellular and Subcellular Transport: Epithelial Cells)
Edited by SIDNEY FLEISCHER AND BECCA FLEISCHER

VOLUME 210. Numerical Computer Methods
Edited by LUDWIG BRAND AND MICHAEL L. JOHNSON

VOLUME 211. DNA Structures (Part A: Synthesis and Physical Analysis of DNA)
Edited by DAVID M. J. LILLEY AND JAMES E. DAHLBERG

VOLUME 212. DNA Structures (Part B: Chemical and Electrophoretic Analysis of DNA)
Edited by DAVID M. J. LILLEY AND JAMES E. DAHLBERG

VOLUME 213. Carotenoids (Part A: Chemistry, Separation, Quantitation, and Antioxidation)
Edited by LESTER PACKER

VOLUME 214. Carotenoids (Part B: Metabolism, Genetics, and Biosynthesis)
Edited by LESTER PACKER

VOLUME 215. Platelets: Receptors, Adhesion, Secretion (Part B)
Edited by JACEK J. HAWIGER

VOLUME 216. Recombinant DNA (Part G)
Edited by RAY WU

VOLUME 217. Recombinant DNA (Part H)
Edited by RAY WU

VOLUME 218. Recombinant DNA (Part I)
Edited by RAY WU

VOLUME 219. Reconstitution of Intracellular Transport
Edited by JAMES E. ROTHMAN

VOLUME 220. Membrane Fusion Techniques (Part A)
Edited by NEJAT DÜZGÜNEŞ

VOLUME 221. Membrane Fusion Techniques (Part B)
Edited by NEJAT DÜZGÜNEŞ

VOLUME 222. Proteolytic Enzymes in Coagulation, Fibrinolysis, and Complement Activation (Part A: Mammalian Blood Coagulation Factors and Inhibitors)
Edited by LASZLO LORAND AND KENNETH G. MANN

VOLUME 223. Proteolytic Enzymes in Coagulation, Fibrinolysis, and Complement Activation (Part B: Complement Activation, Fibrinolysis, and Nonmammalian Blood Coagulation Factors)
Edited by LASZLO LORAND AND KENNETH G. MANN

VOLUME 224. Molecular Evolution: Producing the Biochemical Data
Edited by ELIZABETH ANNE ZIMMER, THOMAS J. WHITE, REBECCA L. CANN, AND ALLAN C. WILSON

VOLUME 225. Guide to Techniques in Mouse Development
Edited by PAUL M. WASSARMAN AND MELVIN L. DEPAMPHILIS

VOLUME 299. Oxidants and Antioxidants (Part A)
Edited by LESTER PACKER

VOLUME 300. Oxidants and Antioxidants (Part B)
Edited by LESTER PACKER

VOLUME 301. Nitric Oxide: Biological and Antioxidant Activities (Part C)
Edited by LESTER PACKER

VOLUME 302. Green Fluorescent Protein
Edited by P. MICHAEL CONN

VOLUME 303. cDNA Preparation and Display
Edited by SHERMAN M. WEISSMAN

VOLUME 304. Chromatin
Edited by PAUL M. WASSARMAN AND ALAN P. WOLFFE

VOLUME 305. Bioluminescence and Chemiluminescence (Part C)
Edited by THOMAS O. BALDWIN AND MIRIAM M. ZIEGLER

VOLUME 306. Expression of Recombinant Genes in
Eukaryotic Systems
Edited by JOSEPH C. GLORIOSO AND MARTIN C. SCHMIDT

VOLUME 307. Confocal Microscopy
Edited by P. MICHAEL CONN

VOLUME 308. Enzyme Kinetics and Mechanism (Part E: Energetics of
Enzyme Catalysis)
Edited by DANIEL L. PURICH AND VERN L. SCHRAMM

VOLUME 309. Amyloid, Prions, and Other Protein Aggregates
Edited by RONALD WETZEL

VOLUME 310. Biofilms
Edited by RON J. DOYLE

VOLUME 311. Sphingolipid Metabolism and Cell Signaling (Part A)
Edited by ALFRED H. MERRILL, JR., AND YUSUF A. HANNUN

VOLUME 312. Sphingolipid Metabolism and Cell Signaling (Part B)
Edited by ALFRED H. MERRILL, JR., AND YUSUF A. HANNUN

VOLUME 313. Antisense Technology (Part A: General Methods, Methods of
Delivery, and RNA Studies)
Edited by M. IAN PHILLIPS

VOLUME 314. Antisense Technology (Part B: Applications)
Edited by M. IAN PHILLIPS

VOLUME 315. Vertebrate Phototransduction and the Visual Cycle (Part A)
Edited by KRZYSZTOF PALCZEWSKI

VOLUME 316. Vertebrate Phototransduction and the Visual Cycle (Part B)
Edited by KRZYSZTOF PALCZEWSKI

VOLUME 335. Flavonoids and Other Polyphenols
Edited by LESTER PACKER

VOLUME 336. Microbial Growth in Biofilms (Part A: Developmental and
Molecular Biological Aspects)
Edited by RON J. DOYLE

VOLUME 337. Microbial Growth in Biofilms (Part B: Special Environments and
Physicochemical Aspects)
Edited by RON J. DOYLE

VOLUME 338. Nuclear Magnetic Resonance of Biological
Macromolecules (Part A)
Edited by THOMAS L. JAMES, VOLKER DÖTSCH, AND ULI SCHMITZ

VOLUME 339. Nuclear Magnetic Resonance of Biological
Macromolecules (Part B)
Edited by THOMAS L. JAMES, VOLKER DÖTSCH, AND ULI SCHMITZ

VOLUME 340. Drug–Nucleic Acid Interactions
Edited by JONATHAN B. CHAIRES AND MICHAEL J. WARING

VOLUME 341. Ribonucleases (Part A)
Edited by ALLEN W. NICHOLSON

VOLUME 342. Ribonucleases (Part B)
Edited by ALLEN W. NICHOLSON

VOLUME 343. G Protein Pathways (Part A: Receptors)
Edited by RAVI IYENGAR AND JOHN D. HILDEBRANDT

VOLUME 344. G Protein Pathways (Part B: G Proteins and Their Regulators)
Edited by RAVI IYENGAR AND JOHN D. HILDEBRANDT

VOLUME 345. G Protein Pathways (Part C: Effector Mechanisms)
Edited by RAVI IYENGAR AND JOHN D. HILDEBRANDT

VOLUME 346. Gene Therapy Methods
Edited by M. IAN PHILLIPS

VOLUME 347. Protein Sensors and Reactive Oxygen Species (Part A:
Selenoproteins and Thioredoxin)
Edited by HELMUT SIES AND LESTER PACKER

VOLUME 348. Protein Sensors and Reactive Oxygen Species (Part B: Thiol
Enzymes and Proteins)
Edited by HELMUT SIES AND LESTER PACKER

VOLUME 349. Superoxide Dismutase
Edited by LESTER PACKER

VOLUME 350. Guide to Yeast Genetics and Molecular and Cell Biology (Part B)
Edited by CHRISTINE GUTHRIE AND GERALD R. FINK

VOLUME 351. Guide to Yeast Genetics and Molecular and Cell Biology (Part C)
Edited by CHRISTINE GUTHRIE AND GERALD R. FINK

VOLUME 352. Redox Cell Biology and Genetics (Part A)
Edited by CHANDAN K. SEN AND LESTER PACKER

VOLUME 353. Redox Cell Biology and Genetics (Part B)
Edited by CHANDAN K. SEN AND LESTER PACKER

VOLUME 354. Enzyme Kinetics and Mechanisms (Part F: Detection and Characterization of Enzyme Reaction Intermediates)
Edited by DANIEL L. PURICH

VOLUME 355. Cumulative Subject Index Volumes 321–354

VOLUME 356. Laser Capture Microscopy and Microdissection
Edited by P. MICHAEL CONN

VOLUME 357. Cytochrome P450, Part C
Edited by ERIC F. JOHNSON AND MICHAEL R. WATERMAN

VOLUME 358. Bacterial Pathogenesis (Part C: Identification, Regulation, and Function of Virulence Factors)
Edited by VIRGINIA L. CLARK AND PATRIK M. BAVOIL

VOLUME 359. Nitric Oxide (Part D)
Edited by ENRIQUE CADENAS AND LESTER PACKER

VOLUME 360. Biophotonics (Part A)
Edited by GERARD MARRIOTT AND IAN PARKER

VOLUME 361. Biophotonics (Part B)
Edited by GERARD MARRIOTT AND IAN PARKER

VOLUME 362. Recognition of Carbohydrates in Biological Systems (Part A)
Edited by YUAN C. LEE AND REIKO T. LEE

VOLUME 363. Recognition of Carbohydrates in Biological Systems (Part B)
Edited by YUAN C. LEE AND REIKO T. LEE

VOLUME 364. Nuclear Receptors
Edited by DAVID W. RUSSELL AND DAVID J. MANGELSDORF

VOLUME 365. Differentiation of Embryonic Stem Cells
Edited by PAUL M. WASSAUMAN AND GORDON M. KELLER

VOLUME 366. Protein Phosphatases
Edited by SUSANNE KLUMPP AND JOSEF KRIEGLSTEIN

VOLUME 367. Liposomes (Part A)
Edited by NEJAT DÜZGÜNEŞ

VOLUME 368. Macromolecular Crystallography (Part C)
Edited by CHARLES W. CARTER, JR., AND ROBERT M. SWEET

VOLUME 369. Combinational Chemistry (Part B)
Edited by GUILLERMO A. MORALES AND BARRY A. BUNIN

VOLUME 370. RNA Polymerases and Associated Factors (Part C)
Edited by SANKAR L. ADHYA AND SUSAN GARGES

[1] Nomenclature for Mammalian Soluble Glutathione Transferases

By BENGT MANNERVIK, PHILIP G. BOARD, JOHN D. HAYES,
IRVING LISTOWSKY, and WILLIAM R. PEARSON

Abstract

The nomenclature for human soluble glutathione transferases (GSTs) is extended to include new members of the GST superfamily that have been discovered, sequenced, and shown to be expressed. The GST nomenclature is based on primary structure similarities and the division of GSTs into classes of more closely related sequences. The classes are designated by the *names* of the Greek letters: Alpha, Mu, Pi, etc., abbreviated in Roman capitals: A, M, P, and so on. (The Greek characters should not be used.) Class members are distinguished by Arabic numerals and the native dimeric protein structures are named according to their subunit composition (e.g., GST A1–2 is the enzyme composed of subunits 1 and 2 in the Alpha class). Soluble GSTs from other mammalian species can be classified in the same manner as the human enzymes, and this chapter presents the application of the nomenclature to the rat and mouse GSTs.

Introduction

Advances in genomics, proteomics, and bioinformatics have facilitated identification of, and distinctions among, multiple forms of glutathione transferase (GST). Most investigations have been performed on human, rat, and mouse GSTs. However, many different forms of soluble GSTs exist in each mammalian species, causing inconsistencies in nomenclature and confusion in the interpretation of studies of the proteins and their genes. For instance, several different designations for a single GST appear in the literature. Accordingly, a major purpose of this chapter is to update the record of human GSTs and provide accurate designations for the known mouse and rat GSTs and their corresponding genes with references to database accessions.

GSTs are prominent contributors to the cellular biotransformation of electrophilic compounds. They provide protection against genotoxic and carcinogenic effects of numerous substances of both xenobiotic and endogenous origins. The GST activity was discovered as an enzyme-catalyzed conjugation of glutathione (GSH) with halogenated aromatic compounds

METHODS IN ENZYMOLOGY, VOL. 401
Copyright 2005, Elsevier Inc. All rights reserved.
0076-6879/05 $35.00
DOI: 10.1016/S0076-6879(05)01001-3

such as bromosulfophthalein (Combes and Stakelum, 1961) and chloronitrobenzenes (Booth et al., 1961). The enzyme was named GSH S-aryltransferase, and other GSTs acting on epoxides and alkyl halides were designated S-epoxide transferase and S-alkyl transferase, etc. The subsequent finding that a "general" GST substrate, 1-chloro-2,4-dinitrobenzene (CDNB), is a substrate for several dissimilar GSTs (Clark et al., 1973) demonstrated the inaccuracy of the nomenclature. In fact, many GSTs display overlapping substrate selectivities. The original names, based on assumed substrate specificities, were therefore abandoned and replaced by letters (Habig et al., 1974a).

The Enzyme Commission of the International Union of Biochemistry and Molecular Biology has provided the systematic name "RX: glutathione R-transferase" (E.C. 2.5.1.18) and recommends the trivial name "glutathione transferase" (without the prefix "S"). The commonly used term "glutathione S-transferase" is a misnomer, because the sulfur atom per se is not transferred, but the glutathionyl group, GS– . Actually, most substrates of biological origin undergo addition reactions (or, considered in the reverse direction, elimination reactions), and the GSTs could consequently be classified as lyases rather than as transferases. Furthermore, some of the GSTs act as isomerases (Benson et al., 1977; Fernández-Cañón and Peñalva, 1998; Johansson and Mannervik, 2001). Despite the preceding deliberations, the commonly used abbreviation GST should be retained. (In the context of glutathione conjugations, GST could be considered to stand for "glutathionyl" (i.e., GS–) transferase.

Another issue is the concept isoenzyme, which originally designated enzymes present in multiple forms and catalyzing the same reaction (Markert and Møller, 1959; Cahn et al., 1962). GSTs generally promote reactions between glutathione and electrophiles, but individual transferases have distinct substrate specificities and catalyze different chemical transformations. The term "isoenzyme" should therefore be used with care and is best suited for variant GSTs derived from a particular gene (e.g., by alternative splicing of the pre-mRNA) (Ranson et al., 1998). Isoenzymes can also arise by binary combination of subunits from different GST genes (Mannervik and Jensson, 1982).

Originally, GSTs were found in the soluble cell fraction, and the enzymes occurring in the cytoplasm are usually referred to as soluble or cytosolic GSTs, even if they may also be present in organelles such as the nucleus, mitochondria, and peroxisomes.

An enzyme traditionally designated Kappa GST, found in mitochondria and peroxisomes, is structurally distinct from the GSTs in Tables I–III. Like the soluble GST proteins, it forms a dimer, but the secondary structural elements of its subunit have been permuted (Ladner et al., 2004).

TABLE I
HUMAN (*HOMO SAPIENS*) SOLUBLE GLUTATHIONE TRANSFERASES, CLASSES, AND CHROMOSOMAL
LOCATIONS OF THEIR GENES

Enzyme designation	Class	Gene	Chromosome band	Accession no.
GST A1-1	Alpha	*GSTA1*	6p12	NP_665683
GST A2-2	Alpha	*GSTA2*	6p12	NP_000837
GST A3-3	Alpha	*GSTA3*	6p12	NP_000838
GST A4-4	Alpha	*GSTA4*	6p12	NP_001503
GST A5-5[a]	Alpha	*GSTA5*	6p12	NP_714543
GST M1-1	Mu	*GSTM1*	1p13	NP_666533
GST M2-2	Mu	*GSTM2*	1p13	NP_000839
GST M3-3	Mu	*GSTM3*	1p13	NP_000840
GST M4-4	Mu	*GSTM4*	1p13	NP_671489
GST M5-5	Mu	*GSTM5*	1p13	NP_000842
GST P1-1	Pi	*GSTP1*	11q13	NP_000843
GST T1-1	Theta	*GSTT1*	22q11.2	NP_000844
GST T2-2	Theta	*GSTT2*	22q11.2	NP_000845
GST Z1-1	Zeta	*GSTZ1*	14q24.3	NP_665877
GST O1-1	Omega	*GSTO1*	10q24.3	NP_004823
GST O2-2	Omega	*GSTO2*	10q24.3	NP_899062
PGD2/GST S1-1[b]	Sigma	*PGD2*	4q22.3	NP_055300

[a] A protein expressed from the Alpha class gene *GSTA5* has yet to be demonstrated.
[b] The Sigma class enzyme is known as the glutathione-dependent prostaglandin D_2 synthase.

Instead of the canonical N-terminal α/β-domain followed by an all α-helical domain, the structure has an α-helical domain inserted in the α/β-domain. This topology defines a separate family of proteins including glutaredoxin-2. The human (NP_057001), rat (NP_852036, P24473), and mouse sequences (NP_083831, Q9DCM2) have been annotated.

An independent group of proteins, called "membrane-associated proteins in eicosanoid and glutathione metabolism" (MAPEG), are integral membrane components in microsomal and mitochondrial cell fractions (Jakobsson *et al.*, 1999). Several of the MAPEG members have GSH conjugating activities similar to the soluble GSTs, but the microsomal GST and its related membrane-bound proteins are structurally unrelated to the soluble GSTs and are not included in this nomenclature overview.

The soluble GSTs have been rediscovered as binding proteins. In liver, an abundant protein with affinity for various ligands, called "ligandin" (Litwack *et al.*, 1971), was recognized as a GST (Habig *et al.*, 1974b). A specific macular zeaxanthin-binding protein in the human retina is identical to GST P1-1 (Bhosale *et al.*, 2004). GSTs also bind to proteins active in cellular signaling pathways. The first reported interaction entailed GST P1-1, which can form a complex with c-Jun N-terminal kinase1 and

TABLE II
RAT (*RATTUS NORVEGICUS*) SOLUBLE GLUTATHIONE TRANSFERASES

Enzyme designation	Class	Gene	Accession number	Other designations
GST A1-1	Alpha	*GSTA1*	P00502 NP_058709	GST B, GST 1-1, ligandin, Ya$_1$, 1a-1a SP:GSTA2
GST A2-2	Alpha	*GSTA2*	P04903 NP_001010921	Ya$_2$, 1b-1b SP:GSTA1
GST A3-3	Alpha	*GSTA3*	P04904 NP_113697	GST AA, GST 2-2, Yc$_1$, 2 SP:GSTC1_RAT
GST A4-4	Alpha	*GSTA4*	P14942 XP_217195	GST K, GST 8-8, Yk SP:GSTA3_RAT
GST A5-5	Alpha	*GSTA5*	P46418 NP_001009920	Yc$_2$ SP:GSTC2_RAT
GST M1-1	Mu	*GSTM1*	P04905 NP_058710	GST A, GST 3-3, Yb$_1$
GST M2-2	Mu	*GSTM2*	P08010 NP_803175	GST D, GST 4-4, Yb$_2$
GST M3-3	Mu	*GSTM3*	P08009 NP_112416	GST 6-6, Yb$_3$, Yn
GST M4-4	Mu	*GSTM4*	B29231 NP_065415	Yb$_4$
GST M5-5	Mu	*GSTM5*	NP_742035	GST 11-11, Yo
GST M6-6	Mu	*GSTM6*	XP_215682	GST 9-9, Yn$_2$
GST P1-1	Pi	*GSTP1*	P04906 NP_036709	GST P, GST 7-7, Yf, Yp
GST T1-1	Theta	*GSTT1*	Q01579 NP_445745	GST E, GST 5-5, Yrs'
GST T2-2	Theta	*GSTT2*	P30713 NP_036928	GST M, GST 12-12, Yrs
GST Z1-1	Zeta	*GSTZ1*	CO387161[a]	MAAI[b]
GST O1-1	Omega	*GSTO1*	BAA34217	DHAR[c]
GST O2-2		*GSTO2*	CK603665[a]	
PTGDS2/ GST S1-1[d]	Sigma	*PTGDS2*	O35543 NP_113832	

[a] cDNA sequence.
[b] MAAI, maleylacetoacetate isomerase.
[c] DHAR, dehydroascorbate reductase.
[d] The Sigma class enzyme is known as the glutathione-dependent prostaglandin D$_2$ synthase.

thereby inhibit its kinase activity (Adler *et al.*, 1999). Some of the proteins, such as the cephalopod eye lens crystallins (Tomarev *et al.*, 1995), which are clearly homologous to other members of the GST superfamily, may not serve as detoxication enzymes. CLIC represents another group of proteins that have functions (chloride intracellular channels) distinct from the conventional GSTs (Cromer *et al.*, 2002). Despite this structural

TABLE III
MOUSE (*MUS MUSCULUS*) SOLUBLE GLUTATHIONE TRANSFERASES

Enzyme designation	Class	Gene	Accession number	Other designations
GST A1-1	Alpha	*Gsta1*	P13745 NP_032207	Ya$_1$
GST A2-2	Alpha	*Gsta2*	P10648 NP_032208	Yc$_2$
GST A3-3	Alpha	*Gsta3*	P30115 NP_034486	GT10.6, Ya$_3$, Yc
GST A4-4	Alpha	*Gsta4*	P24472 NP_034487	Yk, GST5.7
GST A5-5	Alpha	Gsta5		A5
GST M1-1	Mu[a]	*Gstm1*	P10649 NP_034488	GT8.7, Yb$_1$
GST M2-2	Mu[a]	*Gstm2*	P15626 NP_032209	Yb$_2$
GST M3-3	Mu[a]	*Gstm3*	P19639 NP_034489	GT9.3, μ4
GST M4-4	Mu[a]	*Gstm4*	NP_081040	Yb$_5$, μ7
GST M5-5	Mu[a]	*Gstm5*	P48774 NP_034490	Fsc2, mGSTM5
GST M6-6	Mu[a]	*Gstm6*	NP_032210	(also called mGSTM5)
GST M7–7	Mu[a]	*Gstm7*	XP_289885	μ3
GST P1-1	Pi	*Gstp1*	P19157 NP_038569	Yf, piB
GST P2-2	Pi	*Gstp2*	P46425 NP_861461	Yf, piA
GST T1-1	Theta	*Gstt1*	Q64471 NP_032211	5
GST T2-2	Theta	*Gstt2*	NP_034491	Yrs
GST T3-3	Theta	*Gstt3*	NP_598755	
GST Z1-1	Zeta	*Gstz1*	Q9WVL0 NP_034493	MAAI[b]
GST O1-1	Omega	*Gsto1*	O09131 NP_034492	p28
GST O2-2	Omega	*Gsto2*	NP_080895	
PTGDS2/GST S1-1[c]	Sigma	*Ptgds2*	Q9JHF7 NP_062328	

[a] Some designations for mu-class GSTs differ from those of Andorfer *et al.* (2004): their subunit μ3 here is GSTM7, their subunit μ4 is GSTM3, and their subunit μ7 is GSTM4.
[b] MAAI, maleylacetoacetate isomerase.
[c] The Sigma class enzyme is known as the glutathione-dependent prostaglandin D$_2$ synthase.

divergence, it may still be appropriate that some of these proteins receive a designation showing their common phylogeny with conventional GSTs. Glutathione-dependent prostaglandin D_2 synthase (PGD_2 or $PTGDS_2$) is an example of a protein interchangeably designated as GST S1-1 because of its similarities to Sigma class GSTs in nonmammalian organisms (Meyer and Thomas, 1995; Kanaoka *et al.*, 1997).

A rational nomenclature for human GSTs has been adopted (Mannervik *et al.*, 1992), and new members of the superfamily have been added as they were discovered, sequenced, and shown to be expressed. The GST nomenclature is based on sequence similarities and the division of GSTs into classes of more closely related sequences. Members of a given class may have more than 90% sequence identity. A tentative limit of 50% sequence identity was set as a criterion for membership of a given class of the mammalian GSTs (Mannervik *et al.*, 1992). By this criterion, all known expressed human GST genes from a given class are clustered on the same chromosome. This gene distribution lends support to the GST classification. Furthermore, the exon/intron structures of GST genes belonging to the same class are similar but differ from genes of other classes.

The classes are designated by the *names* of the Greek letters: Alpha, Mu, Pi, etc., abbreviated in Roman capitals: A, M, P, and so on. (The Greek characters should not be used, because they are not well matched with computational bioinformatics tools.) A given class may contain one or several protein sequences (protein subunits); within the class, the subunit sequences are numbered using Arabic numerals.

For example, the gene for the Mu class subunit 1 is written *GSTM1* (*italicized*). GSTs are dimeric proteins, and the homodimeric protein composed of two copies of subunit 1 of the Mu class is called GST M1-1 (the protein name is *not* italicized). Heterodimeric structures within the same class can also occur naturally in tissues; for example, the enzyme composed of subunits 1 and 2 in the Alpha class is called GST A1-2. When there is a need to distinguish GSTs from different biological species, a prefix could be added; mGST A1-1 and rGST A1-1 are enzymes from the mouse and the rat, respectively. This distinction may be significant, because there is not a strict one-to-one correspondence between the subunits from different species, and orthologs are not always readily identified. When many organisms are considered, it may be necessary to have them represented by a three-letter prefix based on their Latin name: Hsa for *Homo sapiens*, Rno for *Rattus norvegicus*, Mmu for *Mus musculus*, etc. In distinction from human and rat genes, mouse genes, by convention, are designated by lower case italicized letters; human and rat genes are placed in italicized upper case letters.

In the development of a rational system for naming the soluble GSTs, it was anticipated that other mammalian species could be classified in the same manner as the human enzymes (Mannervik *et al.*, 1985, 1992). This chapter presents the current application of the nomenclature to the rat and mouse GSTs, as presented in Tables II and III. It should be noted that GSTs, like other proteins, have sequence variations caused by allelic polymorphisms or differences between animal strains.

Further details about previous GST designations have been published (Hayes and Pulford, 1995; Hayes *et al.*, 2005; Jakoby *et al.*, 1984; Mannervik and Danielson, 1988); a web site for the mouse GSTs is available at URL: www.people.virginia.edu/~wrp/gst_mouse.html.

References

Adler, V., Yin, Z., Fuchs, S. Y., Benezra, M., Rosario, L., Tew, K. D., Pincus, M. R., Sardana, M., Henderson, C. J., Wolf, C. R., Davis, R. J., and Ronai, Z. (1999). Regulation of JNK signaling by GSTp. *EMBO J.* **18,** 1321–1334.

Andorfer, J. H, Tchaikovskaya, T., and Listowsky, I. (2004). Selective expression of glutathione S-transferase genes in the murine gastrointestinal tract in response to dietary organosulfur compounds. *Carcinogenesis* **25,** 359–367.

Benson, A. M., Talalay, P., Keen, J. H., and Jakoby, W. B. (1977). Relationship between the soluble glutathione-dependent Δ^5-3-ketosteroid isomerase and the glutathione S-transferases of the liver. *Proc. Natl. Acad. Sci. USA* **74,** 158–162.

Bhosale, P., Larson, A. J., Frederick, J. M., Southwick, K., Thulin, C. D., and Bernstein, P. S. (2004). Identification and characterization of a Pi isoform of glutathione S-transferase (GSTP1) as a zeaxanthin-binding protein in the macula of the human eye. *J. Biol. Chem.* **279,** 49447–49454.

Booth, J., Boyland, E., and Sims, P. (1961). An enzyme from rat liver catalyzing conjugations with glutathione. *Biochem. J.* **79,** 516–524.

Cahn, R. D., Kaplan, N. O., Levine, L., and Zwilling, E. (1962). Nature and development of lactic dehydrogenases. *Science* **136,** 962–969.

Clark, A. G., Smith, J. N., and Speir, T. W. (1973). Cross-specificity in some vertebrate and insect glutathione-transferases with methyl parathion (dimethyl *p*-nitrophenyl phosphorothionate), 1-chloro-2,4-dinitro-benzene and *S*-crotonyl-*N*-acetylcysteamine as substrates. *Biochem. J.* **135,** 385–392.

Combes, B., and Stakelum, G. S. (1961). A liver enzyme that conjugates sulfobromophthalein sodium with glutathione. *J. Clin. Invest.* **40,** 981–988.

Cromer, B. A., Morton, C. J., Board, P. G., and Parker, M. W. (2002). From glutathione transferase to pore in a CLIC. *Eur. Biophys. J.* **31,** 356–364.

Fernández-Cañón, J. M., and Peñalva, M. A. (1998). Characterization of a fungal maleylacetoacetate isomerase gene and identification of its human homologue. *J. Biol. Chem.* **273,** 329–337.

Habig, W. H., Pabst, M. J., and Jakoby, W. B. (1974a). Glutathione S-transferases. The first enzymatic step in mercapturic acid formation. *J. Biol. Chem.* **249,** 7130–7139.

Habig, W. H., Pabst, M. J., Fleischner, G., Gatmaitan, Z., Arias, I. M., and Jakoby, W. B. (1974b). The identity of glutathione S-transferase B with ligandin, a major binding protein of liver. *Proc. Natl. Acad. Sci. USA* **71,** 3879–3882.

Hayes, J. D., and Pulford, D. J. (1995). The glutathione S-transferase supergene family: Regulation of GST and the contribution of the isoenzymes to cancer chemoprotection and drug resistance. *Crit. Rev. Biochem. Mol. Biol.* **30,** 445–600.

Hayes, J. D., Flanagan, J. U., and Jowsey, I. R. (2005). Glutathione transferases. *Annu. Rev. Pharmocol. Toxicol.* **45,** 51–88.

Jakobsson, P. J., Morgenstern, R., Mancini, J., Ford-Hutchinson, A., and Persson, B. (1999). Common structural features of MAPEG—A widespread superfamily of membrane associated proteins with highly divergent functions in eicosanoid and glutathione metabolism. *Protein Sci.* **8,** 689–692.

Jakoby, W. B., Ketterer, B., and Mannervik, B. (1984). Glutathione transferases: Nomenclature. *Biochem. Pharmacol.* **33,** 2539–2540.

Johansson, A.-S., and Mannervik, B. (2001). Human glutathione transferase A3-3, a highly efficient catalyst of double-bond isomerization in the biosynthetic pathway of steroid hormones. *J. Biol. Chem.* **276,** 33061–33065.

Kanaoka, Y., Ago, H., Inagaki, E., Nanayama, T., Miyano, M., Kikuno, R., Fujii, Y., Eguchi, N., Toh, H., Urade, Y., and Hayaishi, O. (1997). Cloning and crystal structure of hematopoietic prostaglandin D synthase. *Cell* **90,** 1085–1095.

Ladner, J. E., Parsons, J. F., Rife, C. F., Gilliland, G. L., and Armstrong, R. N. (2004). Parallel evolutionary pathways for glutathione transferases: Structure and mechanism of the mitochondrial class Kappa enzyme rGSTK1-1. *Biochemistry* **43,** 352–361.

Litwack, G., Ketterer, B., and Arias, I. M. (1971). Ligandin: A hepatic protein which binds steroids, bilirubin, carcinogens and a number of exogenous organic anions. *Nature* **234,** 466–467.

Mannervik, B., and Danielson, U. H. (1988). Glutathione transferases—structure and catalytic activity. *CRC Crit. Rev. Biochem.* **23,** 283–337.

Mannervik, B., and Jensson, H. (1982). Binary combinations of four protein subunits with different catalytic specificities explain the relationship between six basic glutathione S-transferases in rat liver cytosol. *J. Biol. Chem.* **257,** 9909–9912.

Mannervik, B., Ålin, P., Guthenberg, C., Jensson, H., Tahir, M. K., Warholm, M., and Jörnvall, H. (1985). Identification of three classes of cytosolic glutathione transferase common to several mammalian species: Correlation between structural data and enzymatic properties. *Proc. Natl. Acad. Sci. USA* **82,** 7202–7206.

Mannervik, B., Awasthi, Y. C., Board, P. G., Hayes, J. D., Di Ilio, C., Ketterer, B., Listowsky, I., Morgenstern, R., Muramatsu, M., Pearson, W. R., Pickett, C. B., Sato, K., Widersten, M., and Wolf, C. R. (1992). Nomenclature for human glutathione transferases. *Biochem. J.* **282,** 305–306.

Markert, C. L., and Møller, F. (1959). Multiple forms of enzymes: Tissue, ontogenetic, and species specific patterns. *Proc. Natl. Acad. Sci. USA* **45,** 753–763.

Meyer, D. J., and Thomas, M. (1995). Characterization of rat spleen prostaglandin H D-isomerase as a sigma-class GSH transferase. *Biochem. J.* **311,** 739–742.

Ranson, H., Collins, F., and Hemingway, J. (1998). The role of alternative mRNA splicing in generating heterogeneity within the *Anopheles gambiae* class I glutathione S-transferase family. *Proc. Natl. Acad. Sci. USA* **95,** 14284–14289.

Tomarev, S. I., Chung, S., and Piatigorsky, J. (1995). Glutathione S-transferase and S-crystallins of cephalopods: Evolution from active enzyme to lens-refractive proteins. *J. Mol. Evol.* **41,** 1048–1056.

[2] Human Alpha Class Glutathione S-Transferases: Genetic Polymorphism, Expression, and Susceptibility to Disease

By BRIAN F. COLES and FRED F. KADLUBAR

Abstract

The human alpha class glutathione S-transferases (GSTs) consist of 5 genes, *hGSTA1–hGSTA5,* and 7 pseudogenes on chromosome 6p12.1–6p12.2. hGSTA1–hGSTA4 have been well characterized as proteins, but *hGSTA5* has not been detected as a gene product. hGSTA1-1 (and to a lesser extent hGSTA2-2) catalyzes the GSH-dependent detoxification of carcinogenic metabolites of environmental pollutants and tobacco smoke (e.g., polycyclic aromatic hydrocarbon diolepoxides) and several alkylating chemotherapeutic agents and has peroxidase activity toward fatty acid hydroperoxides (FA-OOH) and phosphatidyl FA-OOH. hGSTA3-3 has high activity for the GSH-dependent Δ^5–Δ^4 isomerization of steroids, and hGSTA4-4 has high activity for the GSH conjugation of 4-hydroxyno-nenal. hGSTA4 is expressed in many tissues; hGSTA1-1 and hGSTA2-2 are expressed at high levels in liver, intestine, kidney, adrenal gland, and testis; and hGSTA3 is expressed in steroidogenic tissues. Functional, allelic, single nucleotide polymorphisms occur in an SP1-binding element of *hGSTA1* and in the coding regions of *hGSTA2* and *hGSTA3.* The main effects of these polymorphisms are the low hepatic expression of hGSTA1 in individuals homozygous for *hGSTA1*B* and the low specific activity of the hGSTA2E-2E variant toward FA-OOH. These properties suggest that alpha class GSTs will be involved in susceptibility to diseases with an environmental component (such as cancer, asthma, and cardiovascular disease) and in response to chemotherapy. Although *hGSTM1,* *hGSTT1,* and *hGSTP1* have been associated with such diseases (on the basis of genetic polymorphisms as indicators of expression), alpha class GSTs have been little studied in this respect. Nevertheless, *hGSTA1*B* has been associated with increased susceptibility to colorectal cancer and with increased efficacy of chemotherapy for breast cancer. Methods for identification and quantitation of human alpha class GST protein, mRNA, and genotype are reviewed, and the potential for GST-alpha in plasma to be used as a marker for hepatic expression and induction is discussed.

METHODS IN ENZYMOLOGY, VOL. 401

0076-6879/05 $35.00
DOI: 10.1016/S0076-6879(05)01002-5

TABLE I

HUMAN ALPHA CLASS GSTs: BIOLOGICALLY RELEVANT SUBSTRATES AND
NONSUBSTRATE LIGANDS[a]

Class of substrate or activity	Protein	Specific substrates
Carcinogens/mutagens and other xenobiotics (Phase I–activated derivatives of environmental pollutants, or dietary components)	A1-1	Polycyclic aromatic hydrocarbon epoxides: benzo[a]pyrene diolepoxide, benzo[c]chrysene diolepoxide, benzo[g] chrysene diolepoxide, benzo[c]phenanthrene diolepoxide, dibenz[a,h] anthracene diolepoxide, dibenzo[a,l]pyrene diolepoxide; N-acetoxy-PhIP
	A2-2	Dibenzo[a,l]pyrene diolepoxide
	A3-3	Limited study: dibenzo[a,l]pyrene diolepoxide, phenethylisothiocyanate, sulforaphane
	A4-4	Limited study: ethacrynic acid
Alkylating chemotherapeutic agents	A1-1	Busulphan, chlorambucil, melphalan, phosphoramide mustard, 4-OH-cyclophosphamide, thiotepa
	A2-2	Busulphan[b]
	A3-3	Not studied
	A4-4	Not studied
Steroids (Δ^5–Δ^4 isomerase activity)	A1-1	Δ^5-androstene-3,17-dione, Δ^5-pregnene-3,20-dione
	A2-2	Very low activity toward Δ^5-androstene-3,17-dione (~0.001 of hGSTA3 at 0.1 mM steroid)
	A3-3	Δ^5-androstene-3,17-dione, Δ^5-pregnene-3,20-dione
	A4-4	Very low activity toward Δ^5-androstene-3,17-dione (~0.0001 of hGSTA3 at 0.1 mM steroid)
Peroxidase activity and products of oxidative stress	A1-1	Cumene hydroperoxide. Fatty acid hydroperoxides: arachidonic acid hydroperoxide, linoleic acid hydroperoxide. Phospholipid hydroperoxides: phosphatidylethanolamine hydroperoxide, phosphatidylcholine hydroperoxide (also specifically 1-palmitoyl-2-(13-hydroperoxy-cis-9, trans-11-octadecadienoyl)-l-3-phosphatidylcholine)

(continued)

TABLE I *(continued)*

Class of substrate or activity	Protein	Specific substrates
	A2-2	Cumene hydroperoxide. Fatty acid hydroperoxides: arachidonic acid hydroperoxide, linoleic acid hydroperoxide. Phospholipid hydroperoxides: phosphatidylethanolamine hydroperoxide, phosphatidylcholine hydroperoxide
	A3-3	Cumene hydroperoxide
	A4-4	Nonenal, 4-hydroxynonenal, 4-hydroxydecenal
Non-catalytic binding[c]	A1-1	Haem, bilirubin, azo dyes, steroids
	A2-2	Haem, bilirubin, azo dyes, steroids
	A3-3	Not studied
	A4-4	Not studied

[a] Data are taken from references cited in the text; see also the following: Dreij, K., Sundberg, K., Johansson, A.-F., Nordling, E., Seidel, A., Persson B., Mannervik, B., and Jernström, B. (2002). Catalytic activities of human alpha class glutathione transferases toward carcinogenic dibenzo[*a,l*]pyrene diolepoxides. *Chem. Res. Toxicol.* **15**, 825–631; Hurst, R, Bao, Y., Jemth, P., Mannervik, B., and Williamson, G. (1998). Phospholipid hydroperoxide glutathione peroxidase activity of human glutathione transferases. *Biochem. J.* **332**, 97–100; Jernström, B., Funk M., Frank, H., Mannervik, B, and Seidel, A. (1996). Glutathione S-transferase A1-1-catalysed conjugation of bay and fjord region diol epoxides or polycyclic aromatic hydrocarbon epoxides with glutathione. *Carcinogenesis* **17**, 1491–1498; and Zhao, T., Singhal, S. S., Piper, J. T., Cheng, J., Pandya, U., Clark-Wronski, J., Awasthi, S., and Awasthi, Y. C. (1999). The role of human glutathione S-transferases hGSTA1-1 and hGSTA2-2 in protection against oxidative stress. *Arch. Biochem. Biophys.* **367**, 216–224.
[b] Recombinant hGSTA2-2 showed ∼ 50% activity of hGSTA1-1, but enzyme prepared from tissue showed poor catalytic activity (Czerwinski *et al.*, 1996).
[c] For non-catalytic binding of MAP kinases see text.

Introduction

The alpha class glutathione *S*-transferases (GSTs) consist, in humans, of at least four distinct proteins, hGSTA1–hGSTA4. Two members of the class, hGSTA1 and hGSTA2 (which have been known for many years), catalyze the glutathione conjugation of a wide range of electrophiles, possess glutathione-dependent steroid isomerase activity, and glutathione-dependent (selenium-independent) peroxidase activity (Table I). hGSTA3 and hGSTA4 have been characterized relatively recently; hGSTA3 has high specific activity for gluta-thione-dependent Δ^5–Δ^4 isomerization of steroids, and hGSTA4 has high activity for the glutathione-dependent detoxification of alkenyl products of lipid peroxidation (notably 4-hydroxynonenal) (Board, 1998; Hubatsch *et al.,* 1998, Johansson and Mannervik, 2001, Liu *et al.,* 1998).

Several recent reviews have dealt with the GSTs, including those of the alpha class, particularly the rat and mouse genes and proteins (Hayes and Pulford, 1995; Hayes *et al.,* 2005). This review focuses on the human alpha class GSTs, particularly on novel aspects concerning genetic polymorphism, expression, susceptibility to human disease, and the methods that are available to examine these interactions. The apparently specific functions of hGSTA3 and hGSTA4 are dealt with in more detail in Chapters 17 and 24, respectively.

Genes

The human alpha class GST genes form a cluster on chromosome 6p12.1–6p12.2, spanning approximately 300 kb and consisting of five genes (shown in bold text) and seven pseudogenes: 5'-*hGSTA4, hGSTAP4, hGSTA3,* *hGSTAP2, hGSTAP3,* **hGSTA5,** *HGSTAP1,* **hGSTA1,** *hGSTAP5,* **hGSTA2,** *hGSTAP6, hGSTAP7*-3' (Morel *et al.,* 2002). *hGSTA1–hGSTA4* have been well characterized as genes, cDNA, and protein (Board, 1998; Desmots *et al.,* 1998; Hayes and Pulford, 1995; Hayes *et al.,* 2005; Hubatsch *et al.,* 1998; Johansson and Mannervik, 2001; Morel *et al.,* 1994, 2002; Suzuki *et al.,* 1993, 1994). *hGSTA4* shows the least homology with the other alpha GST genes (Morel *et al.,* 2002), and it has been suggested that *GSTA4* forms a subclass within the alpha class (Board, 1998). *hGSTA5* seems to contain a complete coding sequence closely related to *hGSTA1/hGSTA2* with no obvious defect to indicate that it is nonfunctional. However, no mRNA corresponding to this gene has been identified in tissues, and sequence corresponding to *hGSTA5* has not been found in the expressed sequence tag (EST) database (Morel *et al.,* 2002). The seven pseudogenes are characterized by the absence of exons, single nucleotide deletions in exons, the presence of stop codons in the open reading frame, and/or irregular splice signals; ESTs corresponding to these pseudogene sequences have not been found (Morel *et al.,* 2002).

Expression

Alpha class GSTs are widely expressed in human tissues. hGSTA4 is present in most (possibly all) tissues on the basis of mRNA (Morel *et al.,* 2002) and has been found as protein in several cell types of the liver, kidney, skin, muscle, and brain (Desmots *et al.,* 2001). Reports differ as to the localization of the protein, predominantly mitochondrial (Gardner and Gallagher, 2001) or cytosolic (Desmots *et al.*, 2001). hGSTA1 and hGSTA2 are more restricted in their distribution (although usually coexpressed) and are present at high levels (as mRNA and protein) in liver, small intestine, testis, kidney, adrenal gland, and pancreas and at low levels in a wide range of tissues (Coles *et al.*, 2000, 2001a, 2002; Hayes and Pulford, 1995; Morel *et al.*, 2002; Mulder *et al.*, 1999; Rowe *et al.*, 1997). hGSTA3 is apparently the most restricted in distribution, being present at low levels (as mRNA) in steroidogenic tissues: ovary, mammary gland, placenta, testis, and adrenal gland but also detected in lung, stomach, and trachea (Johanssen and Mannervik, 2001, Morel *et al.*, 2002).

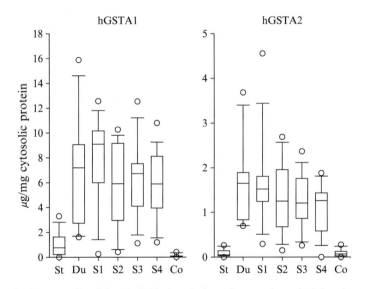

FIG. 1. An example of interindividual variation of expression of alpha class GSTs: hGSTA1 and hGSTA2 in the gastrointestinal tract. GSTs of the alpha, mu, and pi classes were isolated from cytosols by GSH-affinity chromatography and analyzed by HPLC using the method described herein. The samples were from 16 individuals for whom the entire (or substantial portions of) GI tract (stomach-colon) was available. The range of expression in each region of the GI tract is represented by "box plots." The boxes represent the 25th–75th percentiles of the data, and the "error bars" the 10th–90th percentiles; values outside of these ranges are represented as circles; the median is indicated within each box. St, stomach; Du, duodenum; S1–S4, sequentially distal portions of the small intestine; Co, colon. Reprinted from *Archives of Biochemistry and Biophysics*, Vol. 403; Coles, B. F., Chen, G., Kadlubar, F. F., and Radominska-Pandya, A., Interindividual variation and organ-specific patterns of glutathione *S*-transferase alpha mu and pi expression in gastrointestinal tract mucosa of normal individuals, pp. 270–276, copyright (2002); with permission from Elsevier.

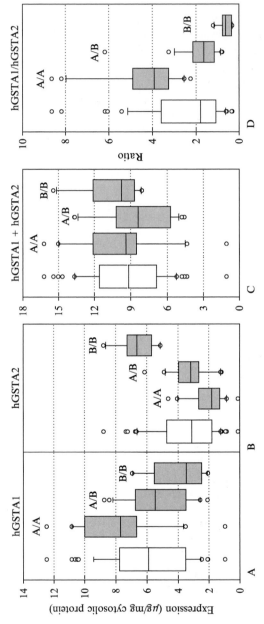

FIG. 2. Variation of hepatic expression of hGSTA1, hGSTA2, and the ratio of their expression in liver according to hGSTA1 genotype. Hepatic expression of GSTs in 52 human liver samples from Caucasian tissue donors was determined by glutathione affinity column chromatography and HPLC and *hGSTA1* genotyped using genomic DNA isolated from the same tissue samples (both using methods described herein). The range of expression according to genotype is represented by "box plots." The boxes represent the 25th–75th percentiles of the data and the "error bars" the 10th–90th percentiles; values outside of these ranges are represented as circles; the median is indicated within each box. *A/A*, homozygous hGSTA1*A; *A/B*, heterozygous hGSTA1*A/*B samples; *B/B*, homozygous hGSTA1*B samples; data for all samples are given in unshaded boxes. Note the individual variation in expression and the differences in mean expression according to genotype. From Coles, B. F., Morel, F., Rauch, C., Huber, W. W., Yang, M., Teitel, C. H., Green, B., Lang, N. P., and Kadlubar, F. F. (2001a). Effect of polymorphism in the human glutathione S-transferase A1 (*hGSTA1*) promoter on hepatic GSTA1 and GSTA2 expression, *Pharmacogenetics* **11**, 663–669; with permission.

GST expression is modified by genetic polymorphism, induction, disease, and (presumed) individual differences in the efficiency of transcriptional or posttranscriptional events (see Figs. 1 and 2 for examples of individual variation of expression). hGSTA3 and hGSTA4 have been little studied in these respects, although the genes are polymorphic (see later). hGSTA4 seems to be induced under conditions of oxidative stress (cirrhosis of the liver, ultraviolet (UV)-irradiated skin, and myocardial infarction (Desmots *et al.*, 2001).

Induction of hGSTA1/A2 is known to occur in response to diet, therapeutic drugs, and other xenobiotics; however, the range of inducers explicitly studied in humans is very limited compared with those of the mouse and rat enzymes (Morel *et al.*, 1994; Hayes and Pulford, 1995). By use of human hepatocytes in primary culture, induction of hGSTA1/A2 mRNA and protein has been shown to occur by treatment with oltipraz, dithiolethione, phenobarbitone, and 3-methylcholanthrene (Morel *et al.*, 1993). Changes in blood plasma (or serum) GST levels have been used as an indication of induction of hepatic hGSTA1/A2 protein after experimental diets high in cruciferous vegetables (see later). Similarly, induction of hGSTA1/A2 has been shown in the saliva of individuals who continually ingested large quantities of broccoli or coffee (Sreerama *et al.*, 1995), and hGSTA1/A2 is high in the duodenal mucosa of individuals who had a normal diet high in vegetables compared with mucosa from individuals who had a low vegetable diet (Hoensch *et al.*, 2002).

The elements responsible for induction of human alpha GSTs are not known. Both positive and negative regulatory regions are present in the 5' noncoding region of *hGSTA1* and *hGSTA2* (Lörper *et al.*, 1996), including a polymorphic SP1-binding site within the proximal promoter of *hGSTA1* (see later). Binding of the transcription factor AP1 has been suggested as a common mechanism for up-regulation of GSTs (Hayes and Pulford, 1995), and *hGSTA1* and/or *hGSTA2* are up-regulated in human hepatocytes in response to interleukin 4 and the associated increase in AP1-binding activity (Langouët *et al.*, 1995). However, unlike the rat and mouse alpha class genes, antioxidant responsive elements (AREs) and xenobiotic responsive elements (XREs) have not been identified in *hGSTA1* or *hGSTA2* (Hayes and Pulford, 1995; Hayes *et al.*, 2005; Morel *et al.*, 1994; Suzuki *et al.*, 1993, 1994). *hGSTA4* contains putative binding sites for the transcription factors AP1, STAT, GATA1, and NF-ebb, although it does not possess an efficient transcription-initiation site (i.e., lacking TATA or CCAAT boxes) (Desmots *et al.*, 1998).

Several diseases are known to affect hGSTA1/A2 expression. For example, expression is low (compared with normal tissue) for mucosa of the stomach of individuals infected with *Helicobacter pylori* (Verhulst *et al.*,

2000) and the small intestine of patients with untreated celiac disease (Wahab *et al.*, 2001). Decrease in alpha class GSTs has been observed in stomach and liver tumors (Howie *et al.*, 1990); conversely, increased expression has been observed in colorectal cancer (Hengstler *et al.*, 1998) and lung cancer (Carmichael *et al.*, 1988).

Genetic Polymorphism

Functional, allelic, single nucleotide polymorphisms (SNPs) are known in *hGSTA1*, *hGSTA2*, and *hGSTA3*. *hGSTA1* has a functional SNP, G-52A, in an SP1-responsive element within the proximal promoter (Morel *et al.*, 2002), plus at least four SNPs further upstream and a silent SNP A375G (Bredschneider *et al.*, 2002; Guy *et al.*, 2004; Morel *et al.*, 2002; Tetlow *et al.*, 2001). Two variants, *hGSTA1*A* and *hGSTA1*B*, have been named according to the linked SNPs G-52A, C-69T, and T-567G (Table II) (Coles *et al.*, 2001a). Other haplotypes within this nomenclature but including SNPs C-115T, T-631G, and C-1142G have been proposed (Bredschneider *et al.*, 2002, Guy *et al.*, 2004). The G-52 variant promoter caused approximately fourfold higher expression of luciferase reporter constructs transfected into HepG2, GLC4, and Caco-2 cells (Morel *et al.*, 2002). Similarly, mean expression of hGSTA1 in liver samples from homozygous *hGSTA1*A* donors was approximately fourfold higher than that of homozygous *hGSTA1*B* donors (Fig. 2) (Coles *et al.*, 2001a). It is not clear whether the polymorphism is functional in other tissues that express *h*GSTA1 at high levels, although it is known that *hGSTA1* genotype does not correlate with hGSTA1 expression in the pancreas (Coles and Kadlubar, 2003). Polymorphisms upstream of G-52C seem to have little effect on hGSTA1 expression (Morel *et al.*, 2002), although the SNP C-115T modified the effect of the G-52C on expression of reporter constructs in a medulloblastoma cell line (Guy *et al.*, 2004). The frequencies of occurrence of *hGSTA1*A* and *hGSTA1*B* are similar in African, Caucasian, and Pacific Islander populations, but *hGSTA1*B* is rarer (on the basis of C-69T) in Japanese (Table II) (Matsuno *et al.*, 2004; Ning *et al.*, 2004; Tetlow *et al.*, 2001).

The proximal promoter of *hGSTA2* does not seem to be polymorphic (although there is disagreement between workers; Guy *et al.*, 2004). However, the gene has three validated SNPs in the coding region (exons 5 and 7) that result in three amino acid substitutions and four variant proteins, hGSTA2A, hGSTA2B, hGSTA2C, and hGSTA2E (Table II). The polymorphisms are widespread, being found in African, Caucasian, and Chinese populations (Tetlow *et al.*, 2001, Ning *et al.*, 2004, Tetlow and Board, 2004). The fifth named variant, *hGSTA2*D*, has not been observed in population studies (Tetlow *et al.*, 2001). The variant (recombinant)

TABLE II
ALLELIC POLYMORPHISM OF HUMAN ALPHA CLASS GSTs

Allele	DNA	Protein	Racial distribution			Consequences of polymorphism
			African	Asian[c]	Caucasian	
hGSTA1*A[a]	T-567, C-69, G-52[b]	(5'-noncoding sequence)	0.65	0.84	0.58–0.60	Referent allele
hGSTA1*B[a]	G-567, T-69, A-52[b]	(5'-noncoding sequence)	0.35	0.16	0.40–0.42	G-52A in SP1-binding element of proximal promoter; causes 4× lower mean hepatic expression. Other SNPs (and named haplotypes) seem to have little effect on expression (see text).
hGSTA2*A	C328, G335, G588, A629	P110, S112, K196, E210	~0.3[d]	—	~0.3[d]	Referent allele
hGSTA2*B	C328, G335, G588, C629	P110, S112, K196, A210	0.28–0.40	0.18	0.08	Little effect on known activities
hGSTA2*C	C328, C335, G588, A629	P110, T112, K196, E210	0.30	—	0.57	Little effect on known activities; associated with reduced hepatic expression (in Caucasians).
hGSTA2*D	C328, G335, T588, C629	P110, S112, N196, A210	0	0	0	(Not observed in human populations) Little effect on known activities
hGSTA2*E	T328, G335, G588, A629	S110, S112, K196, E210	0.01–0.1	0.11	0.05–0.06	Reduced activity toward CDNB and organic hydroperoxides; increased activity toward 4-nitrophenyl acetate.

(continued)

TABLE II (continued)

Allele	DNA	Protein	Racial distribution			Consequences of polymorphism
			African	Asian[c]	Caucasian	
hGSTA3*A	A211	I71	0.85	1.00	1.00	Referent allele
hGSTA3*B	C211	L71	0.15	0.00	0.00	Reduced activity to model substrates, no change for Δ^5-androstene-3, 17-dione; reduced stability at 45°.
hGSTA4	A-227G, A-226G, G-194A, A-158G	(5'-non coding sequence)	—	—	—	Not well characterized, but nonfunctional.
hGSTA5	Not known					Not confirmed as a functional gene.

[a] Few populations have been studied explicitly for the haplotype; however, data available indicate consistency of the linkage of SNPs at −69 and −59 nucleotides.

[b] Polymorphisms upstream of −567 bp are not considered here (see text).

[c] Japanese.

[d] The approximate frequency of occurrence of hGSTA2*A has been calculated from the occurrence of the G335 variant (hGSTA2*A + hGST2*B + hGSTA2*E) as reported by Ning et al. (2004) and the C629 variant (hGSTA2*B) as reported by Tetlow et al.(2001).

enzymes hGSTA2A-2A, 2B-2B, 2C-2C, and 2D-2D show little difference in specific activities examined. However, hGSTA2E-2E, in which the conserved Pro110 residue is replaced by Ser, shows reduced specific activity toward the "classic" GST substrate 1-chloro-2,4-dinitrobenzene (CDNB) and organic hydroperoxides (including ~fourfold reduction of activity toward arachidonic acid hydroperoxide) but increased specific activity toward 4-nitrophenylacetate (Ning *et al.*, 2004, Tetlow and Board, 2004). The hGSTA2 Thr112 and Ser112 variants have been shown to be differentially expressed in the liver (Ning *et al.*, 2004). This is illustrated for tissue samples heterozygous for hGSTA2 Thr112 (hGSTA2C) and a Ser variant (hGSTA2A, B, or E), where the variants are coexpressed in a ratio of approximately 1 (Thr):4 (Ser) (Fig. 3A–C). However, hGSTA2 Thr and Ser variants were expressed at approximately equal amounts in the pancreas (Fig. 3D-E). In Caucasians, the hGSTA1/A2 polymorphisms are in linkage disequilibrium: *hGSTA1*A/hGSTA2*C335 (Thr112) *hGSTA1*B/ hGSTA2*G335 (Ser112) (Ning *et al.*, 2004). It seems that the higher hepatic expression of hGSTA1 in homozygous *hGSTA1*A* individuals, the lower the hepatic expression of hGSTA2 in *hGSTA2*C335 (Thr112) individuals, and their linkage in Caucasians causes the ratio of hepatic hGSTA1/ GSTA2 to be reasonably accurately predicted by the hGSTA1*A/*B or hGSTA2G335C (Ser112Thr) polymorphisms (Fig. 2D) (Coles *et al.*, 2001a, Ning *et al.*, 2004). The mechanism of these gene interactions is not known, but possibly occurs by means of linkage of the known polymorphisms in *hGSTA1* or *hGSTA2* with unidentified regulatory elements in *hGSTA2*. Alternately, there could be transacting elements in *hGSTA2* that respond (inversely) to hGSTA1 expression (Ning *et al.*, 2004). However, it should also be borne in mind that these studies of hepatic expression relate to a single population of Caucasian origin, and the effects on expression in African or Asian populations have not been confirmed. The *hGSTA1/ hGSTA2* polymorphisms were found to be in equilibrium in an African-American population (Ning *et al.*, 2004).

The polymorphism in *hGSTA3*, A211C, is in the coding region of exon 4 and results in an amino acid change Ile71Leu. This conservative substitution alters the specific activity of the protein toward several model substrates but not toward steroid isomerase activity (for Δ^5-androstene-3,17-dione), although the Leu71 variant shows reduced stability at 45° *in vitro*. The polymorphism was detected in African populations at 15% but was not found in Australian Caucasian or southern Chinese populations (Tetlow *et al.*, 2004).

Four SNPs were reported by Guy *et al.* (2004) in the 5' noncoding region of *hGSTA4*: A-227G, A-226G, G-194A, and A-158G (renumbered according to the transcription-initiation site determined by Desmots *et al.*,

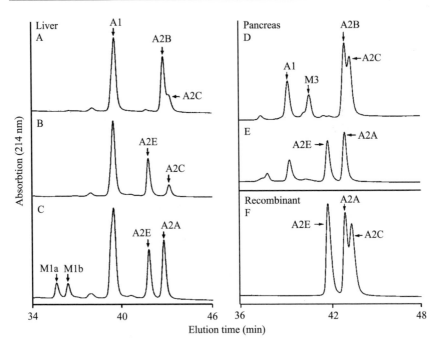

Fᴵɢ. 3. Separation of hGSTA1 and hGSTA2 variants by HPLC. Cytosolic GSTS of the alpha mu and pi classes were isolated by glutathione affinity chromatography and quantitated by HPLC as described herein. hGSTA2 variants were determined by PCR-RFLP for the G335C SNP (as described herein) and/or cloning and sequencing of the cDNA; phenotype was confirmed by determination of accurate mass (for details, see Ning *et al.*, 2004). The tissue samples are liver (A–C) and pancreas (D, E); (F) shows the separation of recombinant enzymes. Note the separation of hGSTA2A; (C and E) the approximate equal expression of the Thr112 and Ser112 variants (A2C *cf* A2A, A2B, and A2E) in the pancreas and their differential expression (i.e., a ratio of approximately 1:4) in the liver. Reprinted from Ning, B., Wang, C., Morel, F., Nowell, S., Ratnasinghe, D. L., Carter, W., Kadlubar, F. F., and Coles, B. (2004). Human glutathione *S*-transferase A2 polymorphisms: variant expression, distribution in prostate cancer cases/controls and a novel form. Pharmacogenetics **14**, 35–44.

1998). These SNPs are not in the putative transcription-factor binding sites of the gene (see previously) and do not seem to be functional (Guy *et al.*, 2004). A polymorphism in intron 7 of *hGSTA4* was examined by McGlynn *et al.* (2003) during a study of susceptibility to hepatocellular carcinoma (see later), but it is not known whether the polymorphism has any functional significance.

In addition to the polymorphisms discussed previously, many SNPs for human alpha class GSTs can be found in gene databases (e.g., the National Cancer Institute SNP 500 Cancer Database—http://snp500cancer.nci.nih. gov/snp.cfm, the Applied Biosystems Celera site—http://myscience. appliedbiosystems.com/genotype/servlet/com.celera.web.cdsentry.servlets. GetCdsEntryListServlet, and the NCBI SNP and related databases—http:// www.ncbi.nlm.nih.gov/entrez/query/Snp/entrezSNP/datamodel.gif). Most of these SNPs are in the intergenic, intronic or 3′ noncoding regions and do not seem to have been validated with respect to any functionality. Nevertheless, several have been validated for their occurrence in Caucasian, African, or Asian populations, and the Applied Biosystems Celera site (see URL previously) lists (to date) validated population frequencies and validated proprietary genotyping methods for two noncoding region SNPs in *hGSTA2*, 2 in *hGSTA3*, and 16 in *hGSTA4*.

GSTs and Disease

It has long been hypothesized that variation of GST expression will influence susceptibility to human diseases that have an environmental component such as exposure to tobacco smoke, diesel exhausts, or other environmental pollutants (Hayes and Pulford, 1995). Because of their range of catalytic properties, human alpha class GSTs would be expected to be important in this respect, particularly considering the high level of expression of hGSTA1 and hGSTA2 in the liver, the prime site of metabolism of xenobiotics. However, these GSTs have been little studied with respect to disease, primarily because genetic polymorphisms that can be used in epidemiological studies as a measure of deduced "activity" are not known (hGSTA3, hGSTA4) or have only recently been identified (hGSTA1, hGSTA2).

Susceptibility to Cancer

GSTs are of particular interest with respect to susceptibility to cancer and the outcome of chemotherapy, because many genotoxic and cytotoxic electrophiles are accepted as substrates. Variation of GST "activity" *in vivo* and susceptibility to human cancers has been studied primarily by use of homozygous *hGSTM1* null and *hGSTT1* null genotypes (*hGSTM1*0* and *hGSTT1*0*) as indicators of systemic lack of expression of the corresponding proteins or, in the case of *hGSTP1* alleles, alteration of specific activity to environmental carcinogens. A feature of these studies is the moderate association of cancer risk with GST polymorphism, the variability of the result among different study populations, and the dependence of the degree of risk on other population characteristics (Coles and Kadlubar, 2003). Nevertheless, polymorphism of hGSTM1, hGSTT1, and hGSTP1 has been associated with incidence of more than 20 types of cancer, with consistent associations (established by meta-analyses of several studies) for hGSTM1 null and increased risk of bladder, laryngeal, and lung cancer; hGSTT1 null and increased risk of astrocytoma and meningioma; and hGSTP1*B and increased risk of bladder cancer in smokers (Habdous et al., 2004). (For earlier reviews see Rebbeck, 1997; Hayes and Strange, 2000; Landi, 2000; Coles and Kadlubar, 2003.)

The most extensive studies of human alpha class GSTs concern hGSTA1 and susceptibility to colorectal cancer (CRC). hGSTA1/A2 expression in the organs of the gastrointestinal (GI) tract shows a striking inverse relationship to the frequency of occurrence of cancers. hGSTA1 in the colon is particularly low (e.g., see Fig. 1), and colon cancer is the most frequently diagnosed cancer of the GI tract (Peters et al., 1993). The risk of CRC was found to increase ~twofold for homozygous hGSTA1*B (low hepatic expression) individuals of an Arkansas (United States) population (Coles et al., 2001b) and ~threefold for individuals homozygous for hGSTA1*B who also had a preference for well-done red meat (Sweeney et al., 2002). The effect is thought to occur by means of the hepatic detoxification of N-acetoxy-PhIP (a mutagenic derivative of the predominant heterocyclic amine carcinogen found in cooked meats and a hGSTA1 substrate); the efficiency of this process affects the amount of N-acetoxy-PhIP reaching the colon (where it cannot be detoxified efficiently). PhIP-DNA adduct levels in peripheral blood lymphocytes from patients with CRC who were homozygous hGSTA1*B were found to be significantly higher than levels for hGSTA1*A or heterozygous patients, although these differences were found only in patients of less than the median age of the total population (Magagnotti et al., 2003). However, an effect of hGSTA1 genotype on risk of CRC was not observed for a Dutch population (van der Logt et al., 2004).

Polymorphism in hGSTA1 has also been studied with respect to prostate cancer using an Arkansas population of African-Americans and Caucasians. Evidence exists that PhIP is involved in the etiology of prostate cancer (but less compelling than for CRC); neither the hGSTA1*A/*B nor the hGSTA2 G335C polymorphism, which was also studied in this cohort, were associated with the risk of prostate cancer (Ning et al., 2004). Similarly, the hGSTA1 polymorphism (C-69T) was not associated with the incidence of hepatocellular carcinoma (HCC) for a Chinese population subject to a high degree of hepatitis B infection and aflatoxin B1 exposure (McGlynn et al., 2003). However, a polymorphism in intron 7 of hGSTA4 was associated with ~1.5-fold increased risk of HCC in men of the same population. The significance of this polymorphism for hGSTA4 expression is not known.

Response to Chemotherapy

Research on GSTs and response to chemotherapy has been dominated by GSTP1 overexpression in tumors and the concomitant development of the multidrug-resistant phenotype (Hayes and Pulford, 1995; Tew, 1994; Townsend and Tew, 2003). Alpha class GSTs have hardly been explored with respect to response to chemotherapy, even though a number of alkylating chemotherapeutic agents in current use are known to be substrates (e.g., busulphan, thiotepa, and the therapeutic metabolites of cyclophosphamide; Czerwinski *et al.*, 1996; Dirven *et al.*, 1996). It has also been argued that the noncatalytic binding properties of alpha class GSTs (Hayes and Pulford, 1995; Hayes *et al.*, 2005) could be important for response by means of transport of nonsubstrate therapeutic agents (Tew, 1994). In addition, hGSTP1 and hGSTM1 have been shown to bind the mitogen-activated protein (MAP) kinases JNK1, ASK1, and MEKK1 (Townsend and Tew, 2003; Ryoo et al., 2004), although the details of binding and other aspects of the interactions are not yet clear. Binding of MAP kinases is thought to affect the outcome of chemotherapy through the apoptotic cascade. Alpha class GSTs have also been shown to bind MAP kinases, but binding affinity is low (Adler *et al.*, 1999). Nevertheless, the interactions seem to be biologically significant. For example, ASK1-induced apoptosis was inhibited when rat hepatocytes transfected with ASK1 were cotransfected with hGSTA1 (Gilot *et al.*, 2002).

Several studies implicate human alpha class GSTs in response to chemotherapy. hGSTA1/A2 protein was increased in blast cells (derived from acute myeloid leukemia patients) showing resistance to doxorubicin *in vitro* (Sargent *et al.*, 1999), and a weak correlation was observed between GST alpha in gastric cancer tissues and cisplatin resistance (*in vitro*) (Kodera *et al.*, 1994). Conversely, hGSTA1/A2 protein in tumor did not predict chemoresistance in ovarian carcinoma (Germain *et al.*, 1996), nor did hGSTA1/A2 in breast tumor correlate with disease-free survival or overall survival of breast cancer (Alpert *et al.*, 1997). However, more recently, it was shown that homozygous *hGSTA1*B* breast cancer patients treated with cyclophosphamide (plus other chemotherapeutic drugs) had a reduced hazard of death during the first 5 years after diagnosis compared with homozygous *hGSTA1*A* individuals (hazard ratio, 0.3) (Sweeney *et al.*, 2003). This observation was interpreted as a role of hepatic hGSTA1 in detoxifying the therapeutic metabolites of cyclophosphamide (the hGSTA1 substrates 4-hydroxy-cyclophosphamide and phosphoramide mustard [Dirven et al., 1996] leading to higher therapeutic exposure in the hGSTA1*B [lower hepatic expression] individuals).

Asthma

GST polymorphisms have been associated with the incidence or severity of childhood or early-onset asthma and related phenotypes. The hGSTP1 "Val105" allele (hGSTP1*B plus hGSTA1*C) seems to be protective of asthma (Aynacioglu et al., 2004; Fryer et al., 2000; Habdous et al., 2004), and the null alleles of hGSTM1 and hGSTT1 seem to confer susceptibility to asthma. In addition, the effects of hGSTM1 and hGSTT1 interact with exposure to environmental tobacco smoke (ETS) (Ivaschenko et al., 2002; Kabesch et al., 2004) and (for hGSTM1) ozone (Romieu et al., 2004). Similar associations have been observed for hGSTP1, hGSTM1, and the incidence of asthma associated with extended industrial exposure to toluene diisothiocyanate (TDI) (Piirilä et al., 2001; Mapp et al., 2002). hGSTM1 null was also associated with lack of TDI-specific IgE, an effect compounded by hGSTM3*A/*A (Piirilä et al., 2001). hGSTM1 and hGSTP1 also seem to modify the adjuvant effect of diesel exhaust particles on allergic inflammation (Gilliland et al., 2004). Evidence exists that some of these effects occur in utero. The incidence of asthma in a population of children and young adults was found to correlate with hGSTP1 genotype of the mother but not the father (Child et al., 2003), and the in utero effects of maternal smoking on childhood asthma and wheezing were largely restricted to hGSTM1 null children (Gilliland et al., 2002a).

It has been suggested that GSTs affect asthma incidence (or severity) by means of the detoxification of (activated) organic components of ETS (or other environmental pollutants) and by modification of response to oxidative stress (see references previously). hGSTA1 and hGSTA2 could be involved in a similar way by means of detoxification of electrophilic xenobiotics and products of oxidative stress in the maternal or fetal liver. However, no studies have examined (e.g., hGSTA1 genotype) smoking and asthma incidence. It should be noted that the hGSTT1, hGSTM1, and hGSTP1 polymorphisms have also been associated with decrements in lung function in children (Gilliland et al., 2002b; Kabesch et al., 2004), observations that suggest that the mechanisms behind GST polymorphism and asthma are probably complex.

Cardiovascular and Other Diseases

hGSTM1 and hGSTT1 have been associated with the risk of coronary arterial disease. Like cancer risk, the associations have proved to be variable among studies. *hGSTM1* null has, in general, been associated with increased risk, primarily in smokers (although several studies show no association). However, for *hGSTT1*, the risk has variably been associated with the functional allele, null allele, or no association has been found (Li *et al.*, 2000; Masetti *et al.*, 2003; Olshan *et al.*, 2003; Tamer *et al.*, 2004; Wang *et al.*, 2002; Wilson *et al.*, 2003). It has been suggested that GSTs protect the heart from the formation of smoking-related DNA-adducts, and although aromatic adducts in heart muscle correlated with smoking and with severe coronary arterial disease (versus mild disease or no disease), there was no relationship between *hGSTM1*, *hGSTT1*, and adducts or risk of disease (Van Schooten *et al.*, 1998). hGSTA1 and hGSTA2 would be expected to be involved in similar detoxification; however, alpha class GSTs have not been studied with respect to cardiovascular disease. It has also been suggested that hGSTA4 expression could influence susceptibility to disease by means of detoxification of the lipid peroxidation product 4-hydroxynonenal, which is possibly involved in atherogenesis (Palinski, 1989) and the neuro-degenerative conditions Parkinson's disease (Yoritaka *et al.*, 1996) and Alzheimer's disease (Mark *et al.*, 1997).

Methods for Quantitation of Expression and Genotype of Human Alpha Class GSTs

Immunochemistry

The use of antibodies to distinguish GST proteins within the same class is limited by sequence homology. Polyclonal antibodies raised against hGSTA1 or hGSTA2 cross-react with each other (Mulder *et al.*, 1999) and with hGSTA3 (Johanssen *et al.*, 2001), although a hGSTA1-specific antibody has been described (Mulder *et al.*, 1999). Conversely, the high degree of sequence homology among the proteins of homologous GST families of human, rat, and mouse frequently allows the use of small mammal GSTs to be used to raise antibodies that are also specific for human GST classes. hGSTA4 is an exception, being immunologically distinct. Details of development of a peptide antibody to hGSTA4 (which was used to show that the protein is preferentially distributed in the mitochondrion) are given by Gardner and Gallagher (2001). Similarly, a peptide of mGSTA4 sequence has been used to generate a GSTA4-specific polyclonal antibody that was used to detect hGSTA4 protein, although cross-reactivity with hGSTA4 was weak (Desmots *et al.*, 2001). Conversely, an antibody raised against the entire mGSTA4 protein did not cross-react with hGSTA4 (Board, 1998).

HPLC Analysis

GST proteins of the alpha, mu, and pi classes have been quantitated by glutathione affinity chromatography and subsequent analysis of the GST pool by wide-pore reverse-phase high-performance liquid chromatography (HPLC) using water-acetonitrile gradients at low pH and UV detection at 214 nm (Coles, 2000; Meyer and Ketterer, 1995). GSTs elute from GSH affinity media as native dimers that are denatured during HPLC analysis and elute as subunits. Several authors have published methods demonstrating "improved" HPLC resolution of GSTs; however, we have not observed examples of resolution that exceed that of the Phenomenex "Jupiter" column. Therefore, this is the column used in the method given in the Appendix. For an example of a fully automated method, see Wheatley et al. (1994).

A considerable advantage of HPLC analysis of GSTs is that proteins of closely related amino acid sequence can be separated (and quantitated). This is particularly useful for the hGSTA1 and hGSTA2 subunits that, because of their high sequence homology, cannot be readily distinguished immunologically. In fact, modern HPLC media can separate GSTs that differ by a single amino acid (although the degree of separation depends on the amino acid substitution involved). For example, using the method described later, the variant subunits hGSTA2A and hGSTA2E show baseline separation even though the proteins differ by only the Pro110Ser substitution (Fig. 3C, E). Similarly, tissues heterogeneous for hGSTA2A (or B) and hGSTA2C (i.e., differing by the Ser112Thr substitution) are indicated by peak splitting or peak broadening (Fig. 3A, D), although the substitution Glu210Asn (hGSTA2A vs. hGSTA2B) does not result in any further peak separation. These patterns of separation are observed for both the native (N-terminally acetylated) proteins and the recombinant proteins (Fig. 3F). hGSTA3 and hGSTA4 have not been analyzed by HPLC, and it is not clear how methods should be modified to account for the localization of hGSTA4 to the mitochondrion (Gardner and Gallagher, 2001).

A further advantage of HPLC analysis is that all GSTs of the alpha, mu, and pi classes are quantitated simultaneously. This results in an accurate ($\pm 2\%$) relative quantitation of subunits, and in this way the differential hepatic expression of hGSTA1 and hGSTA2 according to the hGSTA1 and hGSTA2 polymorphisms (see previously) was determined. Absolute quantitation normalized to total soluble protein is (in our experience) $\pm 8\%$.

"GST-α" in Blood Plasma and Serum

"GST-α" (i.e., hGSTA1 plus hGSTA2) in blood plasma or serum has been used as a marker for hepatocellular damage. GST-α is thought to offer a better assessment of rapid changes in liver damage than aspartate aminotransferase (AST) or alanine aminotransferase (ALT) because of the short plasma half-life of GST (Mulder *et al.*, 1999). Sandwich ELISA kits for the detection of GST-α are available commercially (Biotrin International: www.biotrin.com, product BIO60HEPAS).

Within the range observed in normal individuals, plasma GST-α is thought to reflect hepatic expression by normal hepatocyte turnover (Mulder *et al.*, 1999). An explicit correlation of hepatic hGSTA1, hGSTA2, and plasma GST-α has not been undertaken; however, by using patients with beta thalassemia major, it was shown that activity of hepatic cytosols toward CDNB correlated with GST-α in plasma: $r^2 = 0.57$ (although for these patients, plasma GST-α levels were higher than the normal range) (Poonkuzhali *et al.*, 2001). Because hGSTA1 and hGSTA2 are the dominant GSTs of liver, CDNB activity is a reasonably specific measure of their expression in this case. Mulder *et al.* (1999) determined the range of plasma GST-α for 350 normal blood donors as 0.2–20.4 $\mu g/l$ (median, 2.6 $\mu g/l$). Using the same cohort, these authors showed a linear correlation of hGSTA1 and plasma GST-α ($r = 0.87$, $p < 0.0001$). A similar correlation was found for liver cytosolic hGSTA1 and GST-α.

Plasma GST-α has been shown to increase in individuals given a diet of *Brassica* vegetables. An increase of ~1.5-fold was found for male volunteers who ate 300 g of Brussels sprouts/day for 1 or 3 weeks but not for female participants (Bogaards *et al.*, 1994; Nijhoff *et al.*, 1995). Conversely, a 6-day diet of radish sprouts, cauliflower, broccoli, and cabbage increased serum GST-α by ~26% in *hGSTM1* null women (Lampe *et al.*, 2000) (*hGSTM1* genotype was not taken into account in the two earlier studies). It is thought that GSTM1 negates the effect of the isothiocyanate inducers present in brassicaceous vegetables by catalysis of GSH-conjugation (Kolm *et al.*, 1995).

The small extent of these effects and their differences by gender in different populations pose questions as to the reliability of plasma GST-α as a marker of hepatic GST induction. Nevertheless, the potential exists that a combination of plasma GST-α plus *hGSTA1* genotype will reasonably accurately reflect hGSTA1 and hGSTA2 expression in the liver. This potential is illustrated by Fig. 4, in which hGSTA1 expression is plotted against that of hGSTA1 + hGSTA2 (determined by HPLC) for the set of Caucasian liver samples discussed previously. It seems that (for this set of samples) hepatic hGSTA1 expression can be calculated from hepatic GST-α using the equations derived for the three *hGSTA1* genotypes.

Urinary GST-α (a potential marker for kidney hGSTA1 and hGSTA2 expression) did not change with a *Brassica* diet (Nijhoff *et al.*, 1995).

Isoenzyme-Specific Substrates

GSTs show overlap of substrate acceptance, and there are no truly family-specific or isoenzyme-specific substrates for alpha class GSTs. Busulphan (1,4-butanediol dimethylsulphonate) has been used as a

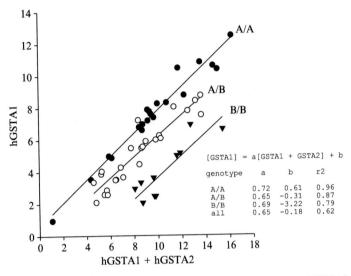

FIG. 4. Correlation of hepatic expression of hGSTA1 and hGSTA1 + hGSTA2 (μg/mg cytosolic protein) according to hGSTA1 genotype. Hepatic expression of GSTs was determined by glutathione affinity column chromatography and HPLC and *hGSTA1* genotyped using genomic DNA isolated from the same tissue samples (both using methods described herein). The lines show the result of linear regression according to each genotype. Note that the equations of the lines predict the amount of hGSTA1 from the sum (hGSTA1 + hGSTA2) with an apparent accuracy of approximately ±1, ±1.5, and ±2 μg/mg cytosolic protein for hGSTA1*A/*A, hGSTA1*A/*B, and hGSTA1*B/*B genotypes, respectively. The data are from the set of liver samples from Fig. 3

selective substrate for hGSTA1-1; however, hGSTA2-2 (recombinant) and hGSTM1-1 show almost 50% of the rate of conjugation of hGSTA1-1 (Czerwinski *et al.*, 1996). (The activity of hGSTA3-3 and hGSTA4-4 toward busulphan is not known.) The pharmacokinetic parameters of busulphan (determined for patients undergoing bone marrow transplant) were found to correlate with plasma GST-α (Poonkuzhali *et al.*, 2001). Busulphan activity (*in vitro*) was used as a "specific" measure of hGSTA1 in an examination of hGSTA1 expression in relationship to *hGSTA1* genotype (Bredschneider *et al.*, 2002). The correlation between hGSTA1 and hGSTA1 + hGSTA2 (determined immunologically), $r^2 = 0.49$, was independent of *hGSTA1* genotype and, when compared with the correlations of Fig. 4, shows the limitations of accuracy of this method compared with HPLC quantitation. The use of the busulphan-GSH product (the *S*-glutathionyl thiophenium ion) is not associated with any convenient UV absorption, and its formation is usually determined by mass spectroscopy (for a recent method, see Ritter *et al.*, 1999). Busulphan can only be used in humans under therapeutic circumstances.

Isomerization of Δ^5-androstene-3,17-dione to Δ^4-androstene-3,17-dione has potential for quantitation of hGSTA3-3, because hGSTA2-2 and hGSTA4-4 have rates <1/1000th that of hGSTA3-3 (at 0.1 mM substrate), and 3-β-hydroxysteroid dehydrogenase/isomerase has <1/200th the activity of hGSTA3-3 (Johansson and Mannervik, 2003). However, hGSTA1-1 has a rate ~1/20th of hGSTA3-3 and because hGSTA1 is more highly expressed than hGSTA3, the assay would have to be accompanied by a measure of total alpha GST. The assay is spectrophotometric, with a ΔE_{248} of 16.3 mM^{-1}cm^{-1} (Benson *et al.*, 1977), but because this is in a region where protein absorption occurs, the usefulness of the assay could be limited in some instances by high background absorption.

In addition, it should be noted that hGSTA4-4 has low activity toward the "universal" GST substrate CDNB; specific activities of recombinant hGSTA1-1, hGSTA2-2, hGSTA3-3, and hGSTA4-4 (1 mM CDNB, 1 mM GSH, 30°) were 80, 80, 23, and 7.5 μmol mg^{-1} min^{-1}, respectively (Johansson and Mannervik, 2003).

mRNA

Specific primers and conditions for reverse transcriptase polymerase chain reaction (RT-PCR) amplification have been published for all the human GST alpha cDNAs (Table III) (Morel *et al.*, 2002). These primers have been used with conventional ethidium bromide staining after agarose gel electrophoresis. Validated TaqMan assays for amplification of hGSTA1, hGSTA2, hGSTA3, and hGSTA4 cDNAs are available commercially using Applied Biosystems reagents (http://myscience.appliedbiosystems.com).

Genetic Polymorphisms

Validated polymerase chain reaction-restriction fragment length polymorphism (PCR-RFLP) methods that use genomic DNA have been published for the functional SNPs of *hGSTA1, hGSTA2,* and *hGSTA3* (Table IV). Alternatives to these methods include high-throughput direct sequencing, allelic discrimination using fluorescently labeled, allele-specific, minor groove-binding oligonucleotides (e.g., the Applied Biosystems TaqMan probes), and allele-specific methods that rely on single base-extension of gene-specific primers and determination of mass using high-throughput mass spectroscopy. However, we are not aware of published methods that use other than direct sequencing or PCR-RFLP methods for human alpha class GSTs (although it is our understanding that methods are in development commercially, at least for the *hGSTA1* G-52A

TABLE III
PCR Amplification of Human Alpha Class cDNAs

cDNA	Forward primer	Reverse primer	Annealing temperature (°)	Product size (bp)
hGSTA1	5'-AGCCAGGACGGTGACAGCG	5'-GACTGGAGTCAAGCTCCTCG	58	578
hGSTA2	5'-AGCCACAAAGGTGACAGCA	5'-AGGCTAGAGTCAAGCTCTTC	56	579
hGSTA3	5'-CGGAGACCGGCTAGACTTTA	5'-TGGAGTCAAGCTCTTCCACA	56	554
hGSTA4	5'-CGCTGACCTGGCGCTTTGTG	5'-TGGCCTAAAGATGTTGTAGACGG	56	746
hGSTA5	5'-CCCAGCCACGACAGTGACAGA	5'-TCCTCTGGTTGACATATGA	55	408

METHOD

Use a standard PCR thermocycler using the following conditions (Morel et al., 2002).

1. Reaction mix (~20 μl): 5 μl cDNA (1 μg), 12.5 μl PCR Promega buffer mix (includes dNTPs, Taq polymerase, MgCl$_2$ and buffer; Promega Corporation, Madison WI); and 400 nM of each primer (0.8 μl of 10 μM; 8 pmol).

2. 28–35 cycles - Denaturation: 30 s at 94°; anneal: 30 s (at appropriate temperature listed above); extension: 45 s at 72°. Cool to 4°.

3. Visualize by electrophoresis on a 1% agarose gel, staining with ethidium bromide, and UV illumination.

polymorphism); several validated TaqMan methods for SNPs in the non-coding (or intergenic) regions of *hGSTA2, hGSTA3,* and *hGSTA4* are available from Applied Biosystems (see earlier).

The PCR-RFLP method for the *hGSTA1* polymorphism in the proximal promoter relies on distinguishing *hGSTA1* and *hGSTA2* using the reverse primer. The upstream (forward) primer is common to both genes. The method has been validated for specificity by sequencing of the PCR products (Coles *et al.*, 2001a). In this method, it is the polymorphism at C-69T that is detected. This has been found to be linked to that in the SP1-binding sequence (G-52A) for most populations studied (i.e., African, Caucasian, and Pacific Islander, but not yet confirmed for Asians; Coles *et al.*, 2001, Matsuno *et al.*, 2004, http://snp500cancer.nci.nih.gov). The dependence of gene specificity on the reverse primer should be borne in mind when designing other methods such as "TaqMan" or mass spectroscopic methods, although for these methods the G-52A polymorphism can be examined directly. It should be noted that if hGSTA2 is co-amplified, no homozygous *hGSTA1*B* samples will be found!

Two sets of primers have been published for PCR-RFLP determination of the *hGSTA2* G335C polymorphism. The method of Ning *et al.* (2004) is given in Table IV, because it has been validated by direct sequencing of the exon 5 PCR products and by correlation with protein variant as determined by HPLC. The method of Tetlow *et al.*, (2001) seems to suffer from co-amplification of exons 5 of *hGSTA2* and *hGSTA5,* resulting in a dramatic underestimation of the frequency of *hGSTA2*C* (*hGSTA5* was not known at the time of the design of this assay). Tetlow and colleagues (Tetlow *et al.*, 2001; Tetlow and Board, 2004) have described methods for the C328T, G558T, and A629C polymorphisms of *hGSTA2* (although the G558T polymorphism has not been found in population studies). The specificity of the primers for *hGSTA2* exon 7 (G558T, A629C) has been validated by sequencing of the PCR product (Ning *et al.*, 2004). Tetlow *et al.* (2004) also give a PCR-RFLP method for the A211C polymorphism in *hGSTA3*.

Concluding Remarks

This summary has highlighted several properties of human alpha class GSTs that we believe are of particular relevance to response to environmental insult. hGSTA1 is the most highly expressed GST of the liver and could, therefore, be critical for "systemic" detoxification of electrophilic xenobiotics including carcinogens and drugs. In addition, hGSTA1 is induced by diets high in green vegetables, this being of considerable interest in that diets high in cruciferous vegetables are anticarcinogenic (and induction could also influence drug response). Although its known substrate

TABLE IV
PCR-RFLP FOR HUMAN ALPHA CLASS GSTs[a]

Gene	Polymorphism	Forward primer	Reverse primer	Annealing temperature (°)	Restriction enzyme[b]	Product sizes (PCR product) allele-specific fragments	Method
hGSTA1	C-69T	5'-TGTTGATTGTTT-GCCTGAAATT	5'-TTTGTTAAACGC-TGTCACCGTCCT	62	Ear I	(481) C:481; T:96, 385	A
hGSTA2	C328T[c]	5'-ATTTGGGTGAAA-TGATGCTT[d]	5'-CAGCTTCACTTA-CTTTTTCA	53	Mwo I	(126) C:23, 37, 66; T: 60, 66	B
	G335C	5'-GTCTTTCAGGATT-GATATGTATAT	5'-AAGTAGCGA-TTTTTTGTTTT-CTCTTG	58	Hinc II	(135) C: 135; G: 61, 74	A
	G558T	5'-TGTGCTTTGTGG-ATTACAGG	5'-CTAAGTGGG-TGAATAGGAGT	58	Apo I	(302) G: 302; T: 58, 244[e]	B
	A629C	5'-TGTGCTTTGTGG-ATTACAGG	5'-AAGGAAGCC-TCCCATGCATG[f]	55	Sph I	(221) A: 221; C: 201, 20	B
hGSTA3	A211C	5'-CATTTTATAACC-TCAGTCA-TTTCAACCATC	5'-CCTGGTCATGA-TGCCCTGTCAT-GGTCT	57	Bsp1286 I	(327) A: 60,99,168; C: 99,228	B

METHODS

A. (Coles et al., 2001; Ning et al., 2004).

1. Reaction mix (~20 μl): 0.1–0.5 μg genomic DNA (~1 μl), 14 μl water, 2 μl 10× buffer, 1.6 μl dNTP mix (2.5 mM each dNTP, final concentration = 0.2 mM), 1.2 μl 25 mM MgCl$_2$ (final concentration = 1.5 mM); 0.4 μl (0.8 U) Taq DNA polymerase (all reagents Perkin Elmer-Applied Biosystems, Foster City, CA), 0.4 μl of 10 μM each primer (4 pmol each).

2. Start with preheated block at 95°, followed by 1 min denaturation at 95°; 28–35 cycles: denaturation: 60 s at 94°; anneal: 60 s (at appropriate temperature listed above); extension: 60 s at 72°. Final extension 7 min at 72° then cool to 4°.

3. Visualize 5 μl of PCR product by electrophoresis on a 1% agarose gel and staining with ethidium bromide.

4. Digestion mix (20 μl): add 8 μl PCR product to a mix of 10 μl water, 2 μl 10× buffer (as supplied with restriction enzyme), 1 unit (1.0 μl) restriction enzyme; incubate at 37° for 2 h (or overnight).

5. Visualize 10 μl PCR product by electrophoresis on an agarose gel (1.1 % for hGSTA1; 1.8 % for hGSTA2), staining with ethidium bromide and UV illumination.

B. (Tetlow et al., 2001, 2004, 2004b)

1. Reaction mix (~20 μl): 25 ng genomic DNA, 14 μl water, 2 μl 10× buffer, 1.6 μl dNTP mix (250 mM each dNTP, final concentration = 0.2 mM), 1.4 μl 25 mM MgCl$_2$ (final concentration = 1.75 mM) (1.2 μl/1.50 mM for hGSTA3); 0.4 μl Taq DNA polymerase (all reagents Perkin Elmer-Applied Biosystems), 0.6 μl 10 μM each primer (6 pmol).

2. Denaturation for 2 min at 95°, followed by 35 cycles: denaturation: 20 s at 95°; anneal: 20 s (at appropriate temperature listed above); extension: 30 s at 72°. Final extension 3 min at 72°, cool to 4°.

3. Digestion mix (20 μl): add 8 μl PCR product to a mix of 10 μl water, 2 μl 10× buffer (as supplied with restriction enzyme), 1 unit (1.0 μl) restriction enzyme; incubate at 37° for 2 h (or overnight). (N.B., details of digestion were not given by Tetlow et al.)

4. Visualize PCR product by electrophoresis on a 6–12 % polyacrylamide gel, staining with ethidium bromide and UV illumination.

[a] Methods for the hGSTA4 polymorphisms (McGlynn et al., 2003; Guy et al., 2004) were not given in detail and are omitted from the Table.

[b] All these restriction enzymes (with 10× buffers) can be purchased from New England Biolabs (www.neb.com).

[c] The primers were incorrectly stated in Tetlow et al., 2004 (Tetlow and Board, personal communication, 2005).

[d] The underlined base is a C > G substitution to create a Mwo I restriction enzyme site.

[e] The digest pattern has not been validated because this variant has not been found in populations.

[f] The underlined base is a G > C substitution to create a Sph I restriction enzyme site.

specificity is limited, hGSTA2 would be expected to exhibit "systemic" effects because of its relative abundance in the liver and its compensatory regulation (*vis á vis* hGSTA1). hGSTA4 could also have a "systemic" effect on products of oxidative stress by its constitutive expression in most tissues.

To study these aspects further, it will be necessary to expand the range of markers, genetic or phenotypic, that can be used in epidemiological studies for example. In those cases in which tissue is available, notably tissue from surgery, use can be made of quantitative RT-PCR methods. However, such tissue, although of considerable interest, is not able to address the role of GSTs in the systemic (hepatic) metabolism of drugs, carcinogens, and response to disease (normal liver tissue being of course the exception). In this respect, the validity of plasma GST-α as a surrogate marker for hepatic alpha class GST needs to be explored in more detail, and there is also the potential that hGSTA4 expression in the leukocyte (or other blood cell type), for example, could provide a marker for systemic expression.

Finally, there is still much to be learned concerning gene regulation of human alpha class GSTs. The mechanism of GST alpha induction in humans remains enigmatic. The genetic mechanism behind variation of hGSTA2 expression in the liver is not clear. Are the little-studied SNPs in noncoding (intronic or intragenic) regions of *hGSTA2*, *hGSTA3*, and *hGSTA4* of any functional significance? It should also be remembered that genotype is not an accurate predictor of phenotype. For example, the variation of hepatic hGSTA1 expression between individuals of the same genotype is >fourfold (Fig. 2) (see Coles and Kadlubar (2003) for a discussion of this aspect).

Nevertheless, there is already evidence that at least one member of the alpha class family of GSTs, hGSTA1, is involved in susceptibility to human disease and response to chemotherapeutic drugs. The challenge of future research is to explore such interactions further and, in this way, to expand our knowledge of the role of xenobiotics (including environmental pollutants) in human disease and drug response.

Acknowledgments

We are grateful to Dr. Fabrice Morel (INSERM UMR620, University of Rennes, France) for discussion and reference to sources of information on the "web" concerning GST polymorphism, and Drs. Philip Board and Natasha Tetlow (John Curtin School of Medical Research, Australian National University, Canberra, Australia) for an update of certain aspects of PCR-RFLP assays for *hGSTA2* polymorphisms.

Appendix: GSH Affinity Chromatography and HPLC Analysis of GSTs of the Alpha, Mu, and Pi Classes

The following method gives a "low-tech" approach to GST analysis that has been found to be highly effective for small samples of tissue cytosols (e.g., 0.5–20 mg of total cytosolic protein).

A. *Reagents*

1. *S*-linked glutathione Agarose (Sigma Aldrich: www.sigmaaldrich.com product G4510; hydrated in water according to manufacturer's instructions).
2. Buffer A: 0.04 M sodium phosphate pH 7.0, 1 mM dithiothreitol (e.g., 0.4 M stock sodium phosphate made 1 mM in DTT immediately before use).
3. 5 M sodium chloride.
4. Buffer A made 0.15 M in sodium chloride.
5. 6 M guanidinium hydrochloride in water.
6. 50 mM GSH in 0.1 M Tris pH 9.6 (make up as 180 mg/15 ml Tris base, pH with 6 drops of 6 M NaOH and check pH with pH paper).
7. Acetonitrile containing 0.4 % trifluoroacetic acid.
8. Water (high-purity) containing 0.6 % trifluoroacetic acid (TFA). (The different concentrations of TFA in the water and acetonitrile are to help maintain a flat baseline during gradient separation; these values can be finely tuned empirically if desired).

B. *Equipment*

1. Dialysis tubing, MW cut-off 15,000 D (e.g., Spectrum Laboratories: www.spectrapor.com product 129115). Pasteur pipets (dropping pipets) ~150 mm × 6 mm id, loosely plugged at tapered end with cotton wool.
2. HPLC equipment (providing at least UV monitor at 214 nm, two solvent gradient system, 1–2 ml injection loop, and integrating capability).
3. HPLC column: Phenomenex: http://www.phenomenex.com C18, 300Å pore "Jupiter" column; 6-mm internal diameter, 250-mm length, 5 μM particle size (product number OOG-4053-EO).

C. *GSH-affinity chromatography (all steps at 4°)*

1. Load GSH agarose suspension into plugged pipet to depth of ~13 mm when drained.

2. Wash with 6 M guanidinium hydrochloride, ½ pipet-full.
3. Wash with buffer A, at least 5 pipets full; drain column.
4. Set aside >20 μl dialyzed cytosol for protein determination.
5. Load 100–500 μl cytosol (previously dialyzed for 12 h against ~1l of buffer A); allow to soak onto column; note volume loaded.
6. Wash cytosol onto column with ~1 ml of buffer A.
7. Wash with 6 pipets full of 0.15 M in sodium chloride in buffer A; drain.
8. Elute with 1.3–1.5 ml 50 mM GSH, 0.1 M Tris, pH 9.6; collect all but first 5 drops (in 1.5-ml "snapcap" tube).
9. Mix eluate and store at 4°, for not more than 2 days; measure volume.
10. Wash column(s) with ½ pipet-full guanidinium hydrochloride, drain.
11. Wash with 6 pipets-full of buffer A if to be used immediately OR
12. Wash with 1 pipet-full of 5 M NaCl; drain; cover gel with 5 M NaCl, cap and store at 4°.

D. *HPLC*

1. Set up method using a 60-min linear gradient of 35–65% acetonitrile in water (+TFA), 1 ml/min flow and monitor at 214 nm.
2. Inject 0.5–1 ml of column eluate onto HPLC column; note amount injected.
3. Run HPLC method.
4. Quantitate peaks (as areas at 214 nm). Use standards for quantitation (e.g., recombinant hGSTA1-1, hGSTP1-1, and hGSTM1b-1b from Invitrogen: http://www.invitrogen.com).
5. Normalize to μg/mg of total cytosolic protein.

References

Adler, V., Yin, Z., Fuchs, S. Y., Benezra, M., Rosario, L., Tew, K. D., Pincus, M. R., Sardana, M., Henderson, C. J., Wolf, C. R., Davis, R. J., and Ronai, Z. (1999). Regulation of JNK signaling by GSTp. *EMBO J.* **18,** 1321–1334.

Alpert, L. C., Schecter, R. L., Berry, D. A., Melnychuk, D., Peters, W. P., Caruso, J. A., Townsend, A. J., and Batist, G. (1997). Relationship of glutathione *S*-transferase α and μ isoforms to response to therapy in breast cancer. *Clin. Cancer. Res.* **3,** 661–667.

Aynacioglu, A. S., Nacak, M., Filiz, A., Ekinci, E., and Roots, I. (2004). Protective role of glutathione *S*-transferase P1 (GSTP1) Val105Val genotype in patients with bronchial asthma. *Br. J. Clin. Pharmacol.* **57,** 213–217.

Benson, A. M., Talalay, P., Keen, J. H., and Jakoby, W. B. (1977). Relationship between the soluble glutathione-dependent delta 5-3-ketosteroid isomerase and the glutathione S-transferases of the liver. *Proc. Natl. Acad. Sci. USA* **74,** 158–162.

Board, P. G. (1998). Identification of cDNAs encoding two human alpha class glutathione transferases (GSTA3 and GSTA4) and the heterologous expression of GSTA4-4. *Biochem. J.* **330,** 827–831.

Bogaards, J. J. P., Verhagen, H., Willems, M. I., van Poppel, G., and van Bladeren, P. J. (1994). Consumption of Brussels sprouts results in elevated α-class glutathione S-transferase levels in human blood plasma. *Carcinogenesis* **15,** 1073–1075.

Bredschneider, M., Klein, K., Thomas, E., Mürdter, T. E., Marx, C., Eichelbaum, M., Nüssler, A. K., Neuhaus, P., Zanger, U. M., and Schwab, M. (2002). Genetic polymorphisms of glutathione S-transferase A1, the major glutathione S-transferase in human liver: Consequences for enzyme expression and busulphan conjugation. *Clin. Pharmacol. Ther.* **71,** 479–487.

Carmichael, J., Forrester, L. M., Lewis, A. D., Hayes, J. D., Hayes, P. C., and Wolf, C. R. (1988). Glutathione S-transferase isoenzymes and glutathione peroxidase activity in normal and tumour samples from human lung. *Carcinogenesis* **9,** 1617–1621.

Child, F., Lenney, W., Clayton, S., Davies, S., Jones, P. W., Alldersea, J. E., Strange, R. C., and Fryer, A. A. (2003). The association of maternal but not paternal genetic variation in GSTP1 with asthma phenotypes in children. *Respir. Med.* **97,** 1247–1256.

Coles, B. F., and Kadlubar, F. F. (2003). Detoxification of electrophilic compounds by glutathione S-transferase catalysis: Determinants of individual response to chemical carcinogenesis and chemotherapeutic drugs? *Biofactors* **17,** 115–130.

Coles, B. F., Anderson, K. E., Doerge, D. R., Churchwell, M. I., Lang, N. P., and Kadlubar, F. F. (2000). Quantitative analysis of interindividual variation of glutathione S-transferase expression in human pancreas and the ambiguity of correlating genotype with phenotype. *Cancer Res.* **60,** 573–579.

Coles, B. F., Morel, F., Rauch, C., Huber, W. W., Yang, M., Teitel, C. H., Green, B., Lang, N. P., and Kadlubar, F. F. (2001a). Effect of polymorphism in the human glutathione S-transferase A1 (*hGSTA1*) promoter on hepatic GSTA1 and GSTA2 expression. *Pharmacogenetics* **11,** 663–669.

Coles, B., Nowell, S. A., MacLeod, S. L., Sweeney, C., Lang, N. P., and Kadlubar, F. F. (2001b). The role of human glutathione S-transferases (hGSTs) in the detoxification of the food-derived carcinogen metabolite N-acetoxy-PhIP, and the effect of a polymorphism in *hGSTA1* on colorectal cancer risk. *Mutat. Res.* **482,** 3–10.

Coles, B. F., Chen, G., Kadlubar, F. F., and Radominska-Pandya, A. (2002). Interindividual variation and organ-specific patterns of glutathione S-transferase alpha mu and pi expression in gastrointestinal tract mucosa of normal individuals. *Arch. Biochem. Biophys.* **403,** 270–276.

Czerwinski, M., Gibbs, J. P., and Slattery, J. T. (1996). Busulfan conjugation by glutathione S-transferases α, μ, and π. *Drug Metab. Disp.* **24,** 1015–1019.

Desmots, F., Rauch, C., Henry, C., Guillouzo, A., and Morel, F. (1998). Genomic organization, 5'-flanking region and chromosomal location of the human glutathione transferase A4 gene. *Biochem. J.* **336,** 437–442.

Desmots, F., Rissel, M., Loyer, P., Turlin, B., and Guillouzo, A. (2001). Immunohistochemical analysis of glutathione transferase A4 distribution in several human tissues using a specific polyclonal antibody. *J. Histochem. Cytochem.* **49,** 1573–1579.

Dirven, H. A. A. M., van Ommen, B., and van Bladeren, P. J. (1996). Glutathione conjugation of alkylating cytostatic drugs with a nitrogen mustard group and the role of glutathione S-transferases. *Chem. Res. Toxicol.* **9,** 351–360.

Fryer, A. A., Bianco, A., Hepple, M., Jones, P. W., Strange, R. C., and Spiteri, M. A. (2000). Polymorphism at the glutathione S-transferase GSTP1 locus. A new marker for bronchial hyperresponsiveness and asthma. *Am. J. Respir. Crit. Care. Med.* **161,** 1437–1442.

Gardner, J. L., and Gallagher, E. P. (2001). Development of a peptide antibody specific to human glutathione S-transferase alpha4-4 (hGSTA4-4) reveals preferential localization in human liver mitochondria. *Arch. Biochem. Biophys.* **390,** 19–27.

Germain, I., Tetu, B., Brisson, J., Mondor, M., and Cherian, M. G. (1996). Markers of chemoresistance in ovarian carcinomas: An immunohistochemical study of 86 cases. *Int. J. Gynecol. Pathol.* **15,** 54–62.

Gilliland, F. D., Li, Y.-F., Dubeau, L., Berhane, K, Avol, E., McConnell, R., Gauderman, W. J., and Peters, J. M. (2002a). Effects of glutathione S-transferase M1, maternal smoking during pregnancy, and environmental tobacco smoke on asthma and wheezing in children. *Am. J. Respir. Crit. Care. Med.* **166,** 457–463.

Gilliland, F. D., Gauderman, W. J., Vora, H., Rappaport, E., and Dubeau, L. (2002b). Effects of glutathione S-transferase M1, T1, and P1 on childhood lung function growth. *Am. J. Respir. Crit. Care. Med.* **166,** 710–716.

Gilliland, F. D., Li, Y.-F., Saxon, A., and Diaz-Sanchez, D. (2004). Effect of glutathione S-transferase M1 and P1 genotypes on xenobiotic enhancement of allergic responses: Randomized, placebo-controlled crossover study. *Lancet* **363,** 119–125.

Gilot, D., Loyer, P., Corlu, A., Glaise, D., Lagadic-Gossmann, D., Atfi, A., Morel, F., Ichijo, H., and Guguen-Guillouzo, C. (2002). Liver protection from apoptosis requires both blockage of initiator caspase activities and inhibition of ASK1/JNK pathway via glutathione S-transferase regulation. *J. Biol. Chem.* **277,** 49220–49229.

Guy, C. A., Hoogendoorn, B., Smith, S. K., Coleman, S., O'Donovan, M. C., and Buckland, P. R. (2004). Promoter polymorphisms in glutathione S-transferase genes affect transcription. *Pharmacogenetics* **14,** 45–51.

Habdous, M., Siest, G., Herbeth, B., Vincent-Viry, M., and Visvikis, S. (2004). Polymorphismes des glutathion S-transférases et pathologies humanies: Bilan des études épidémiologiques. *Ann. Biol. Clin.* **62,** 15–24.

Hayes, J. D., and Pulford, D. J. (1995). The glutathione S-transferase supergene family: Regulation of GST and the contribution of the isoenzymes to cancer chemoprotection and drug resistance. *Crit. Rev. Biochem. Molec. Biol.* **30,** 445–600.

Hayes, J. D., and Strange, R. C. (2000). Glutathione S-transferase polymorphisms and their biological consequences. *Pharmacology* **61,** 154–166.

Hayes, J. D., Flanagan, J. U., and Jowsey, I. R. (2005). Glutathione transferases. *Ann. Rev. Pharmacol. Toxicol.* **45,** 51–88.

Hengstler, J. G., Böttger, T., Tanner, B., Dietrich, B., Henrich, M., Knapstein, P. G., Junginger, T., and Oesch, F. (1998). Resistance factors in colon cancer tissue and the adjacent normal colon tissue: Glutathione S-transferases α and π, glutathione and aldehyde dehydrogenase. *Cancer Lett.* **128,** 105–112.

Hoensch, H., Morgenstern, I., Petereit, G., Siepmann, M., Peters, W. H. M., Roelofs, H. M. J., and Kirch, W. (2002). Influence of clinical factors, diet, and drugs on the human upper gastrointestinal glutathione system. *Gut* **50,** 235–240.

Howie, A. F., Forrester, L. M., Glancey, M. J., Schlager, J. J., Powis, G., Beckett, G. J., Hayes, J. D., and Wolf, C. R. (1990). Glutathione S-transferase and glutathione peroxidase expression in normal and tumour human tissues. *Carcinogenesis* **11,** 451–458.

Hubatsch, I., Ridderström, M., and Mannervik, B. (1998). Human glutathione transferase A4-4: An alpha class enzyme with high catalytic efficiency in the conjugation of 4-hydroxynonenal and other genotoxic products of lipid peroxidation. *Biochem. J.* **330,** 175–179.

Ivaschenko, T. E., Sideleva, O. G., and Baranov, V. S. (2002). Glutathione S-transferase μ and theta gene polymorphisms as new risk factors for atopic bronchial asthma. *J. Mol. Med.* **80,** 39–43.

Johansson, A.-S., and Mannervik, B. (2001). Human glutathione transferase A3-3, a highly efficient catalyst of double bond isomerization in the biosynthetic pathway of steroid hormones. *J. Biol. Chem.* **276**, 33061–33065.

Kabesch, M., Hoefler, C., Carr, D., Leupold, W., Weiland, S. K., and von Mutius, E. (2004). Glutathione *S* transferase deficiency and passive smoking increase childhood asthma. *Thorax* **59**, 569–573.

Kodera, Y., Isobe, K., Yamauchi, M., Kondo, K., Akiyama, S., Ito, K., Nakashima, I., and Takagi, H. (1994). Expression of glutathione transferase alpha and pi in gastric cancer: A correlation with cisplatin resistance. *Cancer. Chemother. Pharmacol.* **34**, 203–208.

Kolm, R. H., Danielson, U. H., Zhang, Y., Talalay, P., and Mannervik, B. (1995). Isothiocyanates as substrates for human glutathione *S*-transferases: Structure-activity studies. *Biochem. J.* **311**, 453–459.

Lampe, J. W., Chen, C., Li, S., Prunty, J., Grate, M. T., Meehan, D. E., Barale, K. V., Dightman, D. A., Feng, Z., and Potter, J. D. (2000). Modulation of human glutathione *S*-transferases by botanically defined vegetable diets. *Cancer Epidem. Biomarkers Prevention* **9**, 787–793.

Langouët, S., Corcos, L., Abdel-Razzak, Z., Loyer, P., Ketterer, B., and Guillouzo, A. (1995). Up-regulation of glutathione *S*-transferases alpha by interleukin 4 in human hepatocytes in primary culture. *Biochem. Biophys. Res. Commun.* **216**, 793–800.

Landi, S. (2000). Mammalian class theta GST and differential susceptibility to carcinogens: A review. *Mutat. Res.* **463**, 247–283.

Li, R., Boerwinkle, E., Olshan, A. F., Chambless, L. E., Pankow, J. S., Tyroler, H. A., Bray, M., Pittman, G. S., Bell, D. A., and Heiss, G. (2000). Glutathione *S*-transferase genotype as a susceptibility factor in smoking-related coronary heart disease. *Atherosclerosis* **149**, 451–462.

Liu, S., Stoesz, S. P., and Pickett, C. B. (1998). Identification of a novel human glutathione *S*-transferase using bioinformatics. *Arch. Biochem. Biophys.* **352**, 306–313.

Lörper, M., Schulz, W. A., Morel, F., Warskulat, U., and Sies, H. (1996). Positive and negative regulatory regions in promoters of human glutathione transferase alpha genes. *Biol. Chem. Hoppe-Seyler* **377**, 39–46.

Magagnotti, C., Pastorelli, R., Pozzi, S., Andreoni, B., Fanelli, R., and Airoldi, L. (2003). Genetic polymorphisms and modulation of 2-amino-1-methyl-6-phenylimidazo[4,5-*b*] pyridine adducts in human lymphocytes. *Int. J. Cancer* **107**, 878–884.

Mapp, C. E., Fryer, A. A., De Marzo, N., Pozzato, V., Padoan, M., Boschetto, P., Strange, R. C., Hemmingsen, A., and Spiteri, M. A. (2002). Glutathione *S*-transferase GSTP1 is a susceptibility gene for occupational asthma induced by isocyanates. *J. Allergy Clin. Immunol.* **109**, 867–872.

Mark, R. J., Lovell, M. A., Markesbery, W. R., Uchida, K., and Mattson, M. P. (1997). A role for 4-hydroxynonenal, an aldehydic product of lipid peroxidation, in disruption of ion homeostasis and neuronal death induced by amyloid *β*-peptide. *J. Neurochem.* **68**, 255–264.

Masetti, S., Botto, N., Manfredi, S., Colombo, M. G., Rizza, A., Vassalle, C., Clerico, A., Biagini, A., and Andreassi, M. G. (2003). Interactive effect of the glutathione *S*-transferase genes and cigarette smoking on occurrence and severity of coronary artery risk. *J. Mol. Med.* **81**, 488–494.

Matsuno, K., Kubota, T., Matsukura, Y., Ishikawa, H., and Iga, T. (2004). Genetic analysis of glutathione *S*-transferase A1 and T1 polymorphisms in a Japanese population. *Clin. Chem. Lab. Med.* **42**, 560–562.

McGlynn, K. A., Hunter, K., Le Voyer, T., Roush, J., Wise, P., Michielli, R. A., Shen, F.-M., Evans, A. A., London, W. T., and Buetow, K. H. (2003). Susceptibility to aflatoxin B1-related primary hepatocellular carcinoma in mice and humans. *Cancer Res.* **63**, 4594–4601.

Meyer, D. J., and Ketterer, B. (1995). Purification of soluble human glutathione S-transferases. *Methods Enzymol.* **252**, 53–65.

Morel, F., Fardel, O., Meyer, D. J., Langouët, S., Gilmore, K. S., Meunier, B., Tu, C-P.D, Kensler, T. W., Ketterer, B., and Guillouzo, A. (1993). Preferential increase of glutathione S-transferase class α transcripts in cultured human hepatocytes by phenobarbital, 3-methylcholanthrene, and dithiolethiones. *Cancer Res.* **53**, 231–234.

Morel, F., Schulz, W. A., and Sies, H. (1994). Gene structure and regulation of expression of human glutathione S-transferases alpha. *Biol. Chem. Hoppe-Seyler* **375**, 641–649.

Morel, F., Rauch, C., Coles, B., Le Ferrec, E., and Guillouzo, A. (2002). The human glutathione transferase alpha locus: Genomic organization of the gene cluster and functional characterization of the genetic polymorphism in the *hGSTA1* promoter. *Pharmacogenetics* **12**, 277–286.

Mulder, T. P. J., Court, D. A., and Peters, W. H. M. (1999). Variability of glutathione S-transferase α in human liver and plasma. *Clin. Chem.* **45**, 355–359.

Nijhoff, W. A., Mulder, T. P. J., Verhagen, H., van Poppel, G., and Peters, W. H. M. (1995). Effects of consumption of Brussels sprouts on plasma and urinary glutathione S-transferase class-α and -π in humans. *Carcinogenesis* **16**, 955–957.

Ning, B., Wang, C., Morel, F., Nowell, S., Ratnasinghe, D. L., Carter, W., Kadlubar, F. F., and Coles, B. (2004). Human glutathione S-transferase A2 polymorphisms: Variant expression, distribution in prostate cancer cases/controls and a novel form. *Pharmacogenetics* **14**, 35–44.

Olshan, A. F., Li, R., Pankow, J. S., Bray, M., Tyroler, H. A., Chambless, L. E., Boerwinkle, E., Pittman., G. S., and Bell, D. A. (2003). Risk of atherosclerosis: Interaction of smoking and glutathione S-transferase genes. *Epidemiology* **14**, 321–327.

Palinski, W., Rosenfeld, M. E., Ylä-Herttuala, S., Gurtner, G. C., Socher, S. S., Butler, S. W., Parthasarathy, S., Carew, T. E., Steinberg, D., and Witztum, J. L. (1989). Low density lipoprotein undergoes oxidative modification *in vivo*. *Proc. Natl. Acad. Sci. USA* **86**, 1372–1376.

Peters, W. H. M., Roelofs, H. M. J., Hectors, M. P. C., Nagengast, F. M., and Jansen, J. B. M. J. (1993). Glutathione and glutathione S-transferases in Barrett's epithelium. *Br. J. Cancer* **67**, 1413–1417.

Piirilä, P., Wikman, H., Luukkonen, R., Kääriä, K., Rosenberg, C., Nordman, H., Norppa, H., Vainio, H., and Hirvonen, A. (2001). Glutathione S-transferase genotypes and allergic responses to diisocyanate exposure. *Pharmacogenetics* **11**, 437–445.

Poonkuzhali, B., Chandy, M., Srivastava, A., Dennison, D., and Krishnamoorthy, R. (2001). Glutathione S-transferase activity influences busulfan pharmacokinetics in patients with beta thalassemia major undergoing bone marrow transplantation. *Drug Metab. Disp.* **29**, 264–267.

Rebbeck, T. R. (1997). Molecular epidemiology of the human glutathione S-transferase genotypes GSTM1 and GSTT1 in cancer susceptibility. *Cancer Epidemiol. Biomarkers Prev.* **6**, 733–743.

Ritter, C. A., Bohnenstengel, F., Hofmann, U., Kroemer, H. K., and Sperker, B. (1999). Determination of tetrahydrothiophene formation as a probe of *in vitro* busulfan metabolism by human glutathione S-transferase A1-1: Use of a highly sensitive gas chromatographic-mass spectrometric method. *J. Chromatogr. B Biomed. Sci. Appl.* **730**, 25–31.

Romieu, I., Sienra-Monge, J. J., Ramirez-Aguilar, M., Moreno-Macias, H., Reyes-Ruiz, N. I., Estela del Rio-Navarro, B., Hernández-Avila, M., and London, S. J. (2004). Genetic polymorphism of GSTM1 and antioxidant supplementation influence lung function in relation to ozone exposure in asthmatic children in Mexico City. *Thorax* **59**, 8–10.

Rowe, J. D., Nieves, E., and Listowsky, I. (1997). Subunit diversity and tissue distribution of human glutathione S-transferases: Interpretations based on electrospray ionization-MS and peptide-sequence-specific antisera. *Biochem. J.* **325**, 481–486.

Ryoo, K., Huh, S.-H., Lee, Y. H., Yoon, K. W., Cho, S.-G., and Choi, E.-J. (2004). Negative regulation of MEKK1-induced signaling by glutathione S-transferase Mu. *J. Biol. Chem.* **279**, 43589–43594.

Sargent, J. M., Williamson, C., Hall, A. G., Elgie, A. W., and Taylor, C. G. (1999). Evidence for the involvement of the glutathione pathway in drug resistance in AML. *Adv. Exp. Med. Biol.* **457**, 205–209.

Sreerama, L., Hedge, M. W., and Sladek, N. E. (1995). Identification of a class 3 aldehyde dehydrogenase in human saliva and increased levels of this enzyme, glutathione S-transferases, and DT-diaphorase in the saliva of subjects who continually ingest large quantities of coffee or broccoli. *Clin. Cancer Res.* **1**, 1153–1163.

Suzuki, T., Johnston, P. N., and Board, P. G. (1993). Structure and organization of the human alpha class glutathione S-transferase genes and related pseudogenes. *Genomics* **18**, 680–686.

Suzuki, T., Smith, S., and Board, P. G. (1994). Structure and function of the 5′-flanking sequences of the human alpha class glutathione S-transferase genes. *Biochem. Biophys. Res. Commun.* **200**, 1665–1671.

Sweeney, C., Coles, B. F., Nowell, S., Lang, N. P., and Kadlubar, F. F. (2002). Novel markers of susceptibility to carcinogens in diet: Associations with colorectal cancer. *Toxicology* **181–182**, 83–87.

Sweeney, C., Ambrosone, C. B., Joseph, L., Stone, A., Hutchins, L. F., Kadlubar, F. F., and Coles, B. F. (2003). Association between a glutathione S-transferase A1 promoter polymorphism and survival after breast cancer treatment. *Int. J. Cancer* **103**, 810–814.

Tamer, L., Ercan, B., Camsari, A., Yildirim, H., Çiçek, D., Sucu, N., Ates, N. A., and Atik, U. (2004). Glutathione S-transferase gene polymorphism as a susceptibility factor in smoking-related coronary artery disease. *Basic. Res. Cardiol.* **99**, 223–229.

Tetlow, N., and Board, P. G. (2004). Functional polymorphism of human glutathione transferase A2. *Pharmacogenetics* **14**, 111–116.

Tetlow, N., Liu, D., and Board, P. (2001). Polymorphism of human alpha class glutathione transferases. *Pharmacogenetics* **11**, 609–617.

Tetlow, N., Coggan, M., Casarotto, M. G., and Board, P. G. (2004). Functional polymorphism of human glutathione transferase A3: Effects on xenobiotic metabolism and steroid biosynthesis. *Pharmacogenetics* **14**, 657–663.

Tew, K. D. (1994). Glutathione-associated enzymes in anticancer drug resistance. *Cancer Res.* **54**, 4313–4320.

Townsend, D. M., and Tew, K. D. (2003). The role of glutathione-S-transferase in anti-cancer drug resistance. *Oncogene* **22**, 7369–7375.

van der Logt, E. M. J., Bergevoet, S. M., Roelofs, H. M. J., van Hooijdonk, Z., te Morsche, R. H. M., Wobbes, T., de Kok, J. B., Nagengast, F. M., and Peters, H. M. (2004). Genetic polymorphisms in UDP-glucuronosyltransferases and glutathione S-transferases and colorectal cancer risk. *Carcinogenesis* **25**, 2407–2415.

Van Schooten, F. J., Hirvonen, A, Maas, L. M., De Mol, B. A., Kleinjans, J. C. S., Bell, D. A., and Durrer, J. D. (1998). Putative susceptibility markers of coronary artery disease: Association between VDR genotype, smoking, and aromatic DNA adduct levels in human right atrial tissue. *FASEB J.* **12**, 1409–1417.

Verhulst, M. L., Van Oijen, A. H. A. M., Roelofs, H. M. J., Peters, W. H. M., and Jansen, J. B. M. J. (2000). Antral glutathione concentration and glutathione S-transferases activity in patients with and without *Helicobacter pylori.* *Dig. Dis. Sci.* **45**, 629–632.

Wahab, P. J., Peters, W. H. M., Roelofs, H. M. J., and Jansen, J. B. M. J. (2001). Glutathione S-transferases in small intestinal mucosa of patients with coeliac disease. *Jpn. J. Cancer Res.* **92,** 279–284.

Wang, X. L., Greco, M., Sim, A. S., Duarte, N., Wang, J., and Wilcken, D. E. (2002). Glutathione S-transferase mu 1 deficiency, cigarette smoking and coronary artery disease. *J. Cardiovasc. Risk* **9,** 25–31.

Wheatley, J. B., Montali, J. A., and Schmidt, D. E., Jr. (1994). Coupled affinity-reversed-phase high-performance liquid chromatography systems for the measurement of glutathione S-transferases in human tissues. *J. Chromatogr. A* **676,** 65–79.

Wilson, M. H., Grant, P. J., Kain, K., Warner, D. P., and Wild, C. P. (2003). Association between the risk of coronary artery disease in South Asians and a deletion polymorphism in glutathione S-transferase M1. *Biomarkers* **8,** 43–50.

Yoritaka, A., Hattori, N., Uchida, K., Tanaka, M., Stadtman, E. R., and Mizuno, Y. (1996). Immunohistochemical detection of 4-hydroxynonenal protein adducts in Parkinson disease. *Proc. Natl. Acad. Sci. USA* **93,** 2696–2701.

[3] Regulation of GST-P Gene Expression During Hepatocarcinogenesis

By Masaharu Sakai and Masami Muramatsu

Abstract

Placental glutathione S-transferase (GST-P), a member of glutathione S-transferase, is known for its specific expression during rat hepatocarcinogenesis and has been used as a reliable tumor marker for experimental rat hepatocarcinogenesis. To explain the molecular mechanism underlying its specific expression concomitant with the malignant transformation, we have analyzed the regulatory element of the GST-P gene and the transcription factor that binds to this element. From the extensive analyses by the establishment of the transgenic rat lines having various regions of GST-P gene, we could identify the GPE1 as an essential enhancer element for specific GST-P expression. Next, we examined the transcription factor that binds and activates the GPE1, specifically in the early stage of hepatocarcinogenesis and in the hepatoma. Electrophoresis gel mobility shift assay, reporter transfection analysis, and the chromatin immunoprecipitation analysis indicate that the Nrf2/MafK heterodimer binds and activates GPE1 element in preneoplastic lesions and hepatomas but not in the normal liver cells.

In this chapter, we describe details of the transgenic rat analyses and the identification of a factor responsible for the specific expression of the GST-P gene and discuss a possible molecular scenario for malignant transformation and tumor marker gene expression.

METHODS IN ENZYMOLOGY, VOL. 401
Copyright 2005, Elsevier Inc. All rights reserved.
0076-6879/05 $35.00
DOI: 10.1016/S0076-6879(05)01003-7

Introduction

Glutathione S-transferase (GST) is a family of phase II detoxification enzymes that catalyze GSH conjugation of a wide variety of exogenous and endogenous compounds (Pickett and Lu, 1989; Tsuchida and Sato, 1992). On the other hand, alteration of phase II detoxification enzymes, such as epoxide hydrolase, UDP-glucuronyltransferase, and GST is a major biochemical characteristic of preneoplastic transformation of hepatocytes (Levin *et al.*, 1978; Lindahl and Feinstein, 1976). Among them, rat pi-class GST (GST-P) is markedly and specifically induced in preneoplastic foci, irrespective of spontaneous or experimental hepatocarcinogenesis. Therefore, GST-P has been used as a reliable tumor marker in experimental rat hepatocarcinogenesis (Sato *et al.*, 1984; Satoh *et al.*, 1985; Sugioka *et al.*, 1985a) and the bioassay screening system for carcinogens (Ito *et al.*, 1996, 2003). Although tumor markers are valuable for early diagnosis of various tumors and for posttherapeutic management, the molecular mechanisms underlying their expression during neoplastic transformation are unknown. Explanation of the mechanisms by which those genes are activated concomitantly with the initial step of malignant transformation may shed some light on the molecular mechanisms of carcinogenesis. GST-P in rat is highly induced in precancerous foci and nodules, even being detected in a single cell in the rat liver as early as 2–3 days after administration of chemical carcinogen (Moore *et al.*, 1987; Satoh *et al.*, 1989). This protein is highly expressed in the neoplastic foci induced by almost all of the chemical carcinogens, with the rare exception of nongenotoxic carcinogen such as clofibrate (Rao *et al.*, 1986; Sakai *et al.*, 1995). GST-P is not expressed in fetal liver and is not induced during liver regeneration and, therefore, is not deemed a so-called oncofetal protein nor is its expression related to cellular proliferation (Sato, 1989; Sugioka *et al.*, 1985a). These findings indicate that the molecular mechanism of this gene regulation may be closely related to the fundamental process of early hepatocarcinogenesis in the rat.

To explain the regulation mechanism of GST-P gene expression during hepatocarcinogenesis, we cloned this gene and analyzed the transcription regulatory regions (Okuda *et al.*, 1987, 1989, 1990; Sakai *et al.*, 1988; Sugioka *et al.*, 1985b). Figure 1A schematically represents the regulatory regions of the rat GST-P gene. The strong enhancer element, GST-P enhancer1 (GPE1) is present at -2.5 kb from the cap site. GPE1 consists of two phorbol 12-o–tetradecanoate 13-acetate-responsive elements (TPA responsive element, TRE) with palindromic orientation (5′-TCAG-*TCAGTCACTATGATTCAG*CAA-3′, TRE-like sequences are underlined). TRE is a well-characterized enhancer element that binds AP-1 family transcription factors, such as Jun and Fos family proteins (Angel

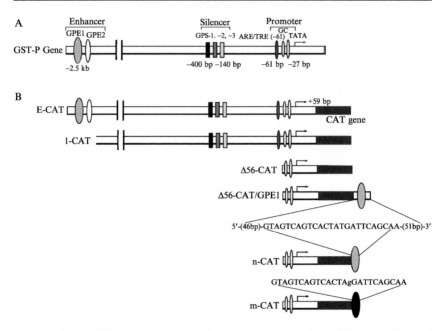

FIG. 1. Rat GST-P gene and transgenes for generating transgenic rat. (A) The position s of *cis*-acting elements of rat GST-P gene is schematically shown. (B) Six transgenes for the production of transgenic rats are shown. Mutated nucleotide of GPE1 in m-CAT is indicated by small letter.

et al., 1987; Okuda *et al.*, 1990; Oridate *et al.*, 1994; Sakai *et al.*, 1992). However, GST-P expression is not always correlated with AP-1 expression (e.g., it is active in F9 embryonal carcinoma cells that are considered to lack AP-1 activity), suggesting that it may be activated by some transcription activator other than AP-1 (Okuda *et al.*, 1990).

In contrast to other genes of GST isozymes, the GST-P gene is almost completely suppressed in normal rat liver as mentioned previously. The mechanisms of down-regulation might also be important for regulation of this gene. In addition to the positive control element, multiple negative control regions (GST-P silencers, GPSs) were found in between −140 and −400 bp region (Fig. 1A) (Sakai *et al.*, 1988), and at least three factors bind to this region (Silencer factor A, -B, and -C, SF-A, -B, and -C) (Imagawa *et al.*, 1991a). Osada, Imagawa, and their colleagues identified the nuclear factor-1 (NF-1) family as SF-A, the CAAT enhancer binding protein (C/EBP) family as SF-B, and zinc finger proteins (BTEB2, LKLF, TIEG1, MZFP, and TFIIIA) as SF-C (Imagawa *et al.*, 1991b; Osada *et al.*, 1995, 1997; Tanabe *et al.*, 2002). These factors may contribute to the silencing of

GST-P gene as the repressors in the normal liver in a cooperative manner, but details of physiological mechanisms are not yet known.

Among these enhancer and silencer elements, GPE1 is a major control element that regulates GST-P expression in an early stage of hepatocarcinogenesis, which has clearly been demonstrated by transgenic rat analyses (Morimura *et al.*, 1993; Suzuki *et al.*, 1995). Then how is the GST-P gene expressed concomitantly with malignant transformation? If it can be assumed that some of the changes in the transcription regulation system lead to malignant transformation, theoretically two kinds of mechanisms may be considered for the simultaneous events, malignant transformation and GST-P gene activation. One is *cis*-activation, in which two closely located genes are coactivated by local activation of genes. The other possible mechanism is *trans*activation, in which one common regulator transactivates both the putative gene that led to the malignant transformation and the GST-P gene expression. To determine whether the GST-P gene and putative transforming gene(s) are activated by a *cis*- or *trans*- mechanism during hepatocarcinogenesis, we established transgenic rats harboring the upstream regulatory region of the GST-P gene. The transgenic approach will also clarify the precise regulatory region required for the specific expression of the GST-P gene during the preneoplastic transformation *in vivo*. Although transgenic animal technologies are commonly used in the mouse system, the pi-class GST genes (GST-P and GST-P1 for rat and mouse, respectively) are expressed rather in a species-specific manner. Mouse GST-P1 is expressed in normal liver, predominantly in male mice, and it is even androgen dependent (Hatayama *et al.*, 1986; Ikeda *et al.*, 2002). Although mouse GST-P1 is also induced in the preneoplastic foci of the mouse liver, the increase in expression is not so marked as in rat liver (Hatayama *et al.*, 1993). Although the main reason for this discrepancy is probably due to the absence of a strong enhancer element GPE1 in mouse pi-class GST gene (Bammler *et al.*, 1994; Xu and Stambrook, 1994), some other mechanisms may be present for this phenotype. Therefore, we have established the transgenic rat rather than mouse.

In this chapter, we first describe the transgenic rat experiments that revealed that the GST-P gene is activated by a *trans*activation mechanism, but not by a coactivation mechanism, during hepatocarcinogenesis (Morimura *et al.*, 1993). This experiment also shows that the GPE1, an enhancer element for GST-P gene, is essential and nearly sufficient for the expression of this gene in the early stage of hepatocarcinogenesis (Suzuki *et al.*, 1995). Next, we describe that a heterodimer consisting of the Nrf2 (NF-E2 related factor 2) and the MafK is a specific *trans*activator that binds to the GPE1, in fact activating the GST-P gene during preneoplastic transformation and in hepatomas (Ikeda *et al.*, 2004).

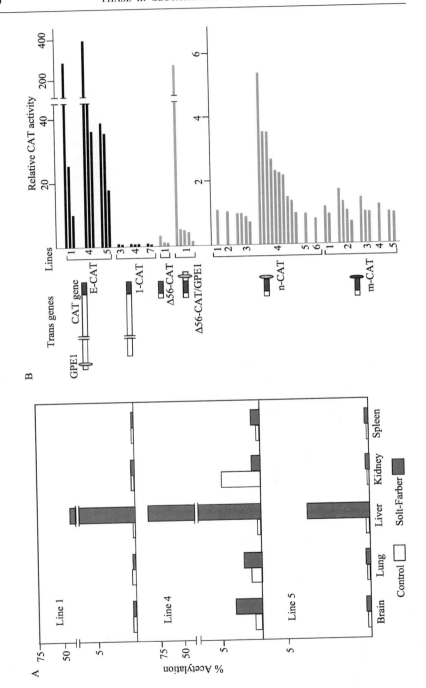

Methods and Results

Establishment of Transgenic Rats Harboring the Regulatory Region of GST-P Gene

Transactivation of GST-P Gene during Chemical Hepatocarcinogenesis. A 2.95-kb fragment of the GST-P gene from −2.9 kb to +59 bp relative to the cap site were joined to the promoterless chloramphenicol acetyltransferase gene plasmid (pSV0CAT, Nippon gene, Toyama Japan) and designated as E-CAT (Sakai *et al.*, 1988). Plasmid DNA was linearized and injected into the male pronuclei of fertilized eggs from Wister rats. Injected eggs were transferred to the oviducts of pseudopregnant rats. The wild and transgenic rats were subjected to the Solt-Farber protocol for efficient hepatocarcinogenesis (Solt and Farber, 1976). Experiments were initiated by injecting 200 mg/kg of diethylnitrosamine into 6-week-old Wister male rats. After feeding the rats the basal diet for 2 weeks, the diet was changed to a basal diet containing 0.01% 2-acetylaminofluorene. Partial hepatectomy was performed at the beginning of the third week, and most of the rats were killed at the end of the eighth week, and the chloramphenicol acetyltransferase (CAT) activity was measured. The livers were homogenized in 0.25 M Tris-HCl (pH7.5) and centrifuged at 10,000g. The supernatants were heated at 65° for 10 min to inactivate endogenous deacetylase activity, and 100 μg proteins were subjected to CAT assay as described (Gorman *et al.*, 1982). The CAT activity was quantified with the Fuji-BAS 200 image analyzer.

The data in Fig. 2A clearly indicate that the Solt-Farber treated livers from all three transgenic lines containing 2.9-kb 5′-flanking sequence of GST-P gene (E-CAT) exhibited high CAT activities, whereas those from the control rat were completely negative. Interestingly, the degree of the inducibility was somehow dependent on the transgenic line, although little variation was noted among individual animals within a given line. The GST-P expression of the Solt-Farber–treated transgenic rat liver was determined by Western blotting (data not shown) simultaneously with CAT assay. The CAT assay and Western blotting analysis indicated that the

---

FIG. 2. Expression of transgenes in normal and Solt-Farber protocol–treated transgenic rats. (A) CAT activities from E-CAT transgenic rats with (closed box) or without (open box) Solt-Farber protocol treatment. Tissues were taken 8 weeks after the protocol started. (B) Expression of the transgenes in transgenic rats. CAT activities of the livers at 8 weeks of the protocol (E-CAT, 1-CAT, black bars) or excised foci from indicated transgenic rat lines at 16 weeks of the protocol (Δ56-CAT, Δ56-CAT/GPE1, n-CAT and m-CAT, gray bars) were analyzed. Each bar represents the value obtained from individual animal.

expression of CAT in the transgenic line during the Solt-Farber protocol was almost exactly parallel to that of GST-P expression (i.e., CAT activity appeared at 3 weeks when the number of GST-P positive cells became significant and increased gradually to a high activity at 6 weeks when the number and size of the foci increase [data not shown]). The integration of transgenes was analyzed by Southern blotting, which showed the integration of the transgenes as tandem arrays at different positions for each line. The copy numbers of transgenes were 13, 30, and 25 for lines 1, 4, and 5, respectively (data not shown). These data clearly indicated that the E-CAT gene was activated during the course of hepatocarcinogenesis by a *trans*mechanism and also indicated that the 2.9-kb upstream region of GST-P gene contains almost all the elements required for specific expression in preneoplastic lesions of the liver and those for repression in normal liver.

  Identification of Control Element for Tumor-Specific Expression of GST-P Gene. To identify the crucial regulatory elements controlling the specific expression of the GST-P gene, we established five other independent lines bearing various regions of the GST-P gene as shown in Fig. 1B (Suzuki *et al.*, 1995). A 5'-terminal 0.7–kb fragment containing GPE1 of the E-CAT was deleted and designated as 1-CAT. A 122–bp fragment containing GPE1 was inserted into Δ56-CAT containing the minimum promoter region ($-56/+59$) of GST-P gene (Δ56-CAT/GPE1). Synthetic oligonucleotide of GPE1 core sequence and its mutated sequence, shown in Fig. 1B, were inserted into Δ56-CAT to produce n-CAT and m-CAT, respectively. Using these constructs, transgenic rats were generated and treated by the Solt-Farber protocol as mentioned previously. Because CAT activity in transgenic rat livers having short transgenes was considerably lower than that in E-CAT, we used excised liver foci for Δ56-CAT, Δ56-CAT/GPE1, n-CAT, and m-CAT transgenic rat. All of the normal livers (without the Solt-Farber protocol) from transgenic lines did not show any significant CAT activity (data not shown). The 1-CAT lines, in which the 0.7-kb region (-2.9 to -2.2 kb) containing GPE1 enhancer was deleted from E-CAT had no significant CAT activities in all of three lines. In contrast to the 1-CAT, the construct of the 122-bp region containing GPE1 joined to the minimal promoter (containing GC- and TATA-box) and the CAT gene (termed Δ56-CAT/GPE1) showed significant CAT activities when rats were treated with the Solt-Farber protocol (Fig. 2B). The control construct, Δ56-CAT, having only minimal promoter, had no CAT activity at all. These results have indicated that the 122-bp GPE1 region is essential for tumor-specific expression of GST-P gene. To further narrow down the essential region of the GPE1, we established transgenic rats bearing synthetic 17-bp GPE1 core sequence (n-CAT) and its point mutant (m-CAT) joined to the minimal GST-P gene promoter. The m-CAT contains one point mutation at the first T of the 3'-side TRE-like sequence as shown in Fig. 1B. This mutation completely abolished enhancer activity in the reporter transfection analysis as reported previously (Okuda *et al.*, 1989, 1990). One of the six lines of n-CAT rats (line 4) showed a significant increase of CAT activity during hepatocarcinogenesis. In this line, CAT activity at 16 weeks increased three to five times greater than in the control liver in 3 of 10 independent experiments. This increase is apparently significant and establishes that the GPE1 core sequence is effective as an enhancer *in vivo* under favorable conditions. No significant CAT activity was seen in any of the foci of the remaining lines. As expected from reporter transfection analysis, none of the livers from the five m-CAT lines showed any increase in CAT activity.

Taken together with the previous data obtained with reporter transfection analyses, we interpret these data to mean that the 17-bp core sequence that is essential for the activation of the GST-P gene during hepatocarcinogenesis is effective but not always sufficient to exert stable enhancing activity in various genomic contexts. The GPE1 core sequence might not be able to work when inserted in repressed regions such as heterochromatin, albeit the GPE1 enhancer with longer flanking sequences could be active in such an environment. The flanking sequences that are present in Δ56-CAT/GPE1 but deleted in n-CAT may help the GPE1 core to exert the full enhancing activity, because Δ56-CAT/GPE1 showed a similar enhancing activity to E-CAT under favorable conditions.

In any event, these transgenic rat analyses unequivocally establish that the GPE1, a 122-bp stretch of DNA containing 17-bp palindromic GPE1 core sequence, is the major regulator of the GST-P gene that is activated at the early phase of hepatocarcinogenesis of the rat. Although there are other control elements such as GPE2 and silencers that were identified between the GPE1 and the promoter of the GST-P gene by DNA transfection experiments, the transgenic experiments demonstrate that they are rather minor in effect *in vivo*.

Nrf2/MafK Is a Responsive Factor for GPE1 Enhancer Activity

Nrf2/MafK Binds to the GPE1 Element of GST-P Gene. The next obvious step is to identify the factors that interact with GPE1 in hepatocarcinogenesis. GPE1 consists of two TRE-like sequences with palindromic orientation, and this sequence is also similar to ARE (antioxidant responsive element, -GTGACTTGGCA-) (Favreau and Pickett, 1991; Jaiswal, 1994) and the MARE (Maf recognition element, -TGCTGACTCAGCT-) (Kataoka *et al.*, 1994). These similarities suggest those transcription factors,

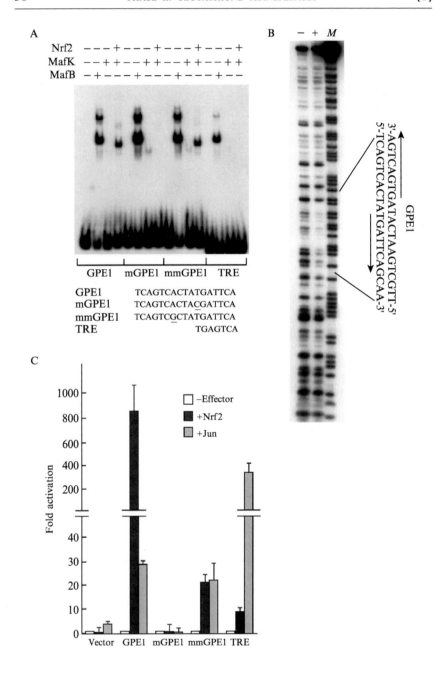

A

Nrf2	–	–	–	+	–	–	–	+	–	–	–	+	–	–	–	+
MafK	–	–	+	+	–	–	+	+	–	–	+	+	–	–	+	+
MafB	–	+	–	–	+	–	–	–	+	–	–	–	+	–	–	–

GPE1   mGPE1   mmGPE1   TRE

GPE1	TCAGTCACTATGATTCA
mGPE1	TCAGTCACTA<u>C</u>GATTCA
mmGPE1	TCAGTC<u>G</u>CTATGATTCA
TRE	TGAGTCA

B   – + *M*

5'-TCAGTCACTATGATTCAGCAA-3'
3'-AGTCAGTGATACTAAGTCGTT-5'

GPE1

C

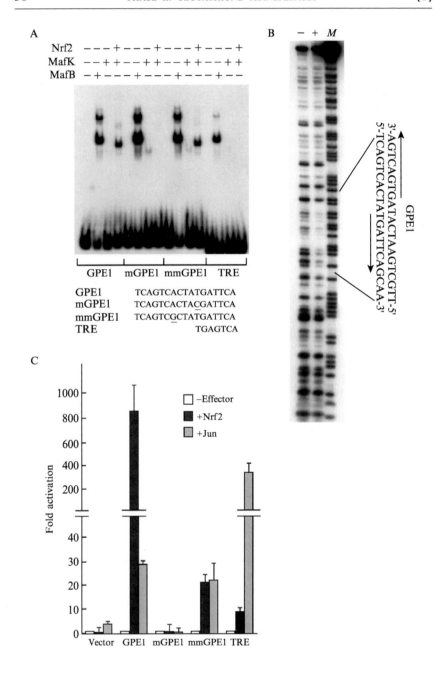

such as Jun and Fos, Nrf2, Maf, and their family members, may bind to GPE1. Among these transcription factors, we focused on Nrf2 as the candidate factor responsible for GPE1 activation, because the other factors did not correlate with the expression of GST-P.

Nrf2 (NF-E2 related factor 2) is a member of the CNC (cap 'n' collar) family of transcription factors (Moi *et al.*, 1994). It dimerizes with small Maf proteins such as MafK, MadG, and MafF (Andrews *et al.*, 1993; Igarashi *et al.*, 1994). Small Maf proteins lack the transcription activation domain present in the large Maf proteins, including c-Maf, MafB, L-Maf/MafA, and Nrl. The Nrf2/small Maf heterodimer binds to ARE and MARE and regulates the genes having ARE or MARE, including those of phase II detoxifying enzymes, such as NQO1 and GST-Ya gene (Itoh *et al.*, 1997; Jaiswal, 2000).

To demonstrate the heterodimer of Nrf2/small Maf proteins bound to the GPE1, we performed electrophoretic mobility shift assay (EMSA) using recombinant Nrf2 and MafK proteins (Ikeda *et al.*, 2004). The cDNA fragment encoding the DNA binding domain of three transcription factors (amino acids 318–598 for Nrf2, 1–158 for MafK, and 82–323 for MafB) were fused to the *E. coli* maltose binding protein (MBP) of a pMALc2 vector (New England Biolabs, Beverly MA). Recombinant proteins were synthesized in *E. coli* and purified according to the supplier's protocol (New England Biolabs, Beverly MA). The synthetic double-stranded oligonucleotides were labeled by γ-^{32}P-ATP and polynucleotide kinase for EMSA probes (Fig. 3A). The probe for footprinting was labeled by filling in the overhanging end of the restriction fragment with Klenow DNA polymerase and α-^{32}P dCTP. Approximately 50 ng of the recombinant protein (Nrf2, MafK, or MafB) was preincubated with 0.5 μg of poly (dI/dC) (Pharmacia Biochemicals, Stockholm) for 10 min in 10 μl of binding buffer (20 mM HEPES, pH 7.9, 20 mM KCl, 1 mM EDTA, 5 mM DTT, 4 mM MgCl$_2$, 15% glycerol, and 100 μg/ml bovine serum albumin). ^{32}P-labeled DNA probe (1 × 10^4 cpm) was added, and the mixture was

---

FIG. 3. Nrf2/MafK binds and activates GPE1, a strong enhancer element of GST-P gene. (A) EMSA was performed with Nrf2/MafK and MafB proteins and GPE1, mGPE1, and mmGPE1. The nucleotide sequences of the probes are shown. Mutated positions are indicated by underlines. (B) DNase I footprinting analysis of Nrf2/MafK with GPE1 probe. (+) and (-) indicate the probe with or without Nrf2/MafK proteins. Guanine and adenine residues of the same probe were cleaved by Maxam and Gilbert methods (M). Arrows indicate TRE-like sequences. (C) Reporter transfection analysis of wild- and mutated-GPE1 in F9 cells. Indicated reporter plasmid or promoter-less luciferase plasmid (Vector, pGVB2) was cotransfected with expression plasmid of Nrf2 (black), c-Jun (gray), or without expression vector (open).

incubated another 10 min. Protein–DNA complexes were analyzed by electrophoresis on 4% (w/v) polyacrylamide gel in a buffer of 25 mM Tris-borate and 0.5 mM EDTA, pH 8.2. For the DNase I footprinting analysis, the probe (2×10^5 cpm) was incubated with or without Nrf2- and MafK-fusion proteins (500 ng each) in 50 μl of binding buffer. Fifty microliters of DNase buffer (5 mM $MgCl_2$, 5 mM $CaCl_2$, 200 mM KCl, and 4 μg/ml sheared salmon sperm DNA) and 20 ng of DNase I (Takara, Kyoto) were added to the binding mixture and incubated for 3 min at room temperature. The reaction was terminated by the addition of 100 μl of stop solution (20 mM EDTA, 1% SDS, and 0.2 M NaCl). After proteinase K treatment (100 μg/ml for 15 min), DNA was extracted and analyzed on 6% (w/v) polyacrylamide gel containing 8 M urea. The guanine and adenine bases of the same probes were modified, digested by the Maxam-Gilbert method (Sambrook, 1989), and loaded for a marker ladder.

EMSA and DNase I footprinting analyses show that Nrf2/MafK heterodimer bound specifically to the GPE1 (Fig. 3A, B). Homodimers of MafK and MafB bound to the GPE1 with weak and strong affinities, respectively. Nrf2/MafK protected the entire GPE1 core sequence from DNase I digestion. To analyze further the DNA-binding specificity of Nrf2/MafK, we used point-mutated GPE1 sequences. Our previous studies had indicated that each 5′-end nucleotide of the two inverted TRE-like sequences 5′-TCAGTCActaTGATTCA-3′ (TRE-like sequences in capital letters and 5′-ends of TRE-like sequences underlined) was essential for the strong enhancer activity of GPE1 (Okuda et al., 1989, 1990). We prepared two GPE1 mutants mutated at these nucleotides and designated to mGPE1 and mmGPE1 as shown in Fig. 3A. EMSA showed that the binding of Nrf2/MafK to the mGPE1 probe was completely abolished, but the binding to mmGPE1 was almost the same as the wild-type GPE1 (Fig. 3A). Both mGPE1 and mmGPE1 bound MafB homodimer to a similar extent to wild-type GPE1. TRE was found to interact with MafB but not with Nrf2/MafK.

Nrf2 Strongly Activates GPE1 Enhancer Element. To determine the transcriptional activity of Nrf2 on the GPE1 and mutated GPE1s, we carried out the reporter transfection assay. Mouse F9 embryonal carcinoma cell lines were used for these experiments instead of the hepatoma cell line, because a reporter gene containing GPE1 was highly expressed in the hepatoma cells without an effector plasmid because of the presence of the two TRE-like sequences in GPE1. In F9 cells, which are devoid of AP-1 activity, basal expression of GPE1 reporter gene is so low that the effects of an exogenous factor may clearly be demonstrated.

For the reporter transfection assay, a total of 4 μg DNA, containing 1 μg reporter plasmid, 0.5 μg β-galactosidase expression plasmid (pSVβgal, Promega, Madison WI) and pUC18 DNA with or without 1 μg Nrf2 expression plasmid was cotransfected according to the method of Chen and Okayama (Chen and Okayama, 1987). At 45 h after transfection, cells were harvested and assayed for luciferase activity using a luciferase assay kit (Nippon Gene, Toyama) and for β-galactosidase activity (Sambrook, 1989).

The gene containing GPE1 was dramatically stimulated ($\sim$1,000 times) by cotransfection of Nrf2 expression plasmid (Fig. 3C). Although Nrf2 usually functioned as a heterodimer with a small-Maf protein (e.g., MafK), cotransfection of MafK-expression vector resulted in repression of GPE1-mediated luciferase gene expression in a dose-dependent manner (data not shown). As previously reported, we assume that small Maf proteins are ubiquitously expressed and that the presence of excess small Maf protein leads to the formation of homodimers that do not contain a transactivation domain, resulting in the inhibition of transcription of Nrf2-dependent genes by competing for the binding site (Motohashi et al., 2000). Therefore, we omitted the MafK expression vector in this experiment. As expected from EMSA, Nrf2 did not activate the mGPE1 reporter gene. Nrf2-dependent activation of mmGPE1 was also remarkably lower than expected, despite the strong binding of Nrf2/MafK to mmGPE1 as shown with EMSA (Fig. 3A). EMSA and reporter transfection analysis with mutated GPE1s thus revealed that the Nrf2/MafK heterodimer binds to the 3′-side of the TRE-like sequences (5′-TCAGTCActaT̲-G̲A̲T̲T̲C̲A̲-3′) of the GPE1 core element. The 5′-side of the TRE-like sequences (5′-T̲C̲A̲G̲T̲C̲A̲ctaTGATTCA-3′) is also important for full enhancer activity of the GPE1. Finding that the 5′-end of both TRE-like sequences is important for binding and activation by Nrf2 is consistent with the previous findings (Okuda et al., 1990). Furthermore, these results indicate that Nrf2 is a factor that is responsible for the enhancer activity of GPE1.

Nrf2 and MafK Bind to GPE1 in Preneoplastic Liver Lesions and Hepatoma Cells In Vivo. To examine the binding of Nrf2 and MafK to the GPE1 of GST-P gene *in vivo*, chromatin immunoprecipitation analyses (ChIP) were performed (Crane-Robinson et al., 1999; O'Neill and Turner, 1996). We analyzed chromatins from cells of normal liver, of livers bearing hyperplastic nodules induced by the Solt-Farber protocol, and those of rat hepatoma cell lines (H4IIE and dRLh84). Cells were fixed in 1% formaldehyde at 37° for 10 min. After washing twice with ice-cold phosphate-buffered saline (PBS) containing protease inhibitors (1 mM phenyl-methylsulfonyl fluoride, 1 μg/ml aprotinin, and 1 μg/ml pepstatin), cells were harvested and sonicated by an ultrasonic generator (Tomy Seiko, Tokyo, Japan) in SDS lysis buffer (1% SDS, 10 mM EDTA, and 50 mM Tris-HCl, pH 8.1). After centrifugation at 15,000 rpm for 10 min, the supernatant chromatin fraction was diluted 10-fold with IP buffer (0.01% SDS, 1.1% Triton X-100, 1.2 mM EDTA, 16.7 mM Tris-HCl, pH 8.1, and 167 mM NaCl) and then precleared by mixing with 30 μl of Protein A-Sepharose (50% Protein A slurry containing 20 μg/ml sheared salmon sperm DNA and 1 mg/ml BSA, Pharmacia Biochemicals, Stockholm), gently rotated at 4° for 30 min, and Protein A beads were removed by centrifugation at 13,000 rpm for 30 sec. Precleared chromatin fraction was incubated with anti-Nrf2 (Ikeda *et al.*, 2004), anti-MafK (kindly provided by Dr. K. Igarashi, Hiroshima University), or preimmune serum for 16 h at 4° with gentle rotation. Immune complexes were mixed with 50 μl of Protein A-Sepharose slurry and washed sequentially with wash buffer 1 (0.1% SDS, 1% Triton X-100, 2 mM EDTA, 20 mM Tris-HCl, pH8.1) containing 150 mM NaCl, wash buffer 1 containing 500 mM NaCl, wash buffer 2 (0.25 M LiCl, 1% NP40, 1% sodium deoxycholate, 1 mM EDTA, 10 mM Tris-HCl, pH8.1) and finally twice with TE (10 mM Tris-HCl, pH8.1, and 1 mM EDTA). Complexes were eluted from Protein A-Sepharose in 0.1 M NaHCO$_3$ containing 1% SDS, and cross-linking was reversed by heating to 65° for 4 h. DNAs were purified from total sonicated nuclei, Protein A bound chromatin fraction without antibody (precleared step), and final Protein A bound fraction with antibodies or preimmune serum. The GPE1 region (−2.5 kb) and promoter region containing ARE/TRE (−61 bp) of the GST-P gene were amplified by PCR. In addition, the ARE sequence of NQO1 (NAD(P)H:quinone oxidoreductase1) gene, a well-characterized Nrf2 target gene, was amplified as a control. Amplified DNAs were analyzed on 3% agarose gel electrophoresis.

Figure 4A clearly shows that anti-Nrf2 antibody precipitated the GST-P GPE1 from the chromatin of cells from liver with hyperplastic nodules and of hepatoma cells, but not from the chromatin of normal liver cells. Detection of GPE1 by ChIP analysis correlated well with GST-P expression; the Nrf2 was bound to the GPE1 in the cells expressing GST-P but not in the cells not expressing this gene. The anti-Nrf2 antibody did not precipitate the proximal ARE/TRE-like sequence of the GST-P gene.

The anti-Nrf2 antibody precipitated the *NQO1* ARE from the chromatin of all samples examined, consistent with the notion that the NQO1, a phase II detoxification enzyme, was expressed in the normal liver and hyperplastic nodule–bearing liver as well as in the hepatoma cell lines.

ChIP analysis was also carried out using the anti-MafK antibody (Fig. 4B). The results were completely the same as those seen with anti-Nrf2 antibody. GPE1 was precipitated by anti-MafK antibody from the cells of hyperplastic nodules and of hepatoma cells but not from the normal liver cells. Anti-MafK antibody, as well as anti-Nrf2 antibody, did not precipitate the proximal ARE/TRE-like element of the GST-P gene, which

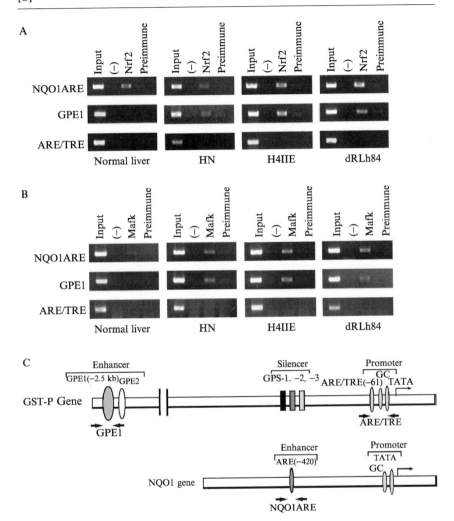

FIG. 4. Nrf2 and MafK bind to GPE1 in hyperplastic nodules and hepatoma cells *in vivo*. ChIP analyses were performed with anti-Nrf2 (A) and anti-MafK (B) antibodies on normal rat liver (Normal Liver), livers bearing hyperplastic nodules (HN), H4IIE cells, and dRLh84 cells as described in the text. DNAs extracted from total sonicated nuclei (Input), those from Protein A–bound chromatin without antibody (−), with the indicated antibodies (Nrf2 or MafK) and with pre-immune serum (Preimmune) were analyzed, respectively. Specific enhancer and promoter regions were amplified by PCR (33 cycles) with specific primers of the ARE region of the NQO1 gene (NQO1 ARE, 5'-AGACCCAAGCGTGTACACCC-3' and 5'-GTCCTTGGTCAGATGTGGGA-3'), GPE1 region of GST-P gene (GPE1, 5'-TGATTCTGCCATCTTTCTGC-3' and 5'-CCAGCTTCTCTGGACAAACC-3'), and proximal ARE/TRE region of GST-P gene (ARE/TRE, 5'-CAGACTCCGGTCCAGCTGCT-3' and 5'-CGCGAACTTACTAGCTGCTG-3'), respectively. The amplified regions of GST-P and NQO1 genes were schematically indicated in (C).

suggests that the Nrf2/MafK heterodimer did not bind and activate the proximal ARE/TRE-like element. Anti-MafK antibody precipitated *NQO1*ARE from the chromatin fraction of all samples. These data suggest that of the small-Maf protein family members (MafK, G and F), MafK is the predominant partner molecule of Nrf2 in liver and hepatoma cells. We performed ChIP analysis using anti-cJun antibody, but the antibody did not precipitate either GPE1 or the proximal ARE/TRE-like element from any sample (data not shown).

The data from EMSA, reporter transfection, and ChIP analyses strongly suggest that Nrf2/MafK is the activator responsible for GST-P expression during hepatocarcinogenesis.

Conclusion and Discussion

The placental form of the GST of the rat, GST-P, has been known as the most reliable tumor marker of rat hepatocarcinogenesis. This high correlation between hepatocarcinogenesis and GST-P expression might suggest that the mechanism of malignant transformation is closely related to GST-P expression. We have long been analyzing the transcription mechanism of GST-P expression in the early stage of rat hepatocarcinogenesis. As a result, we obtained firm lines of evidence that the GPE1 is the major enhancer element, and Nrf2/MafK heterodimer is the major factor binding to this enhancer element, activating this gene during neoplastic transformation of rat liver. Transgenic rats having a 2.9-kb 5′-upstream region of the GST-P gene joined to the reporter gene (CAT) completely coexpressed together with GST-P gene when rats were treated with carcinogen. Different lines of transgenic rats have different copies (10–30 copies) of transgenes at different positions on the chromosome, indicating that this 2.9-kb fragment is essential and sufficient for the specific and strong expression of GST-P gene during hepatocarcinogenesis. Further analyses of transgenic rats having different upstream regions of the GST-P gene have indicated that a 17-bp palindromic enhancer element, GPE1, is the major control element of this gene. The factor binding to this control element was identified to be Nrf2/MafK heterodimer. In addition to the large body of evidence from rather "artificial" analyses, such as EMSA and reporter transfection analyses, chromatin immunoprecipitation analysis (ChIP), an *in vivo*-based study, has unequivocally indicated that these factors really exist on the GPE1 of the carcinogen-treated liver but not of the normal liver. However, a number of questions remain to be solved. First, for the binding of Nrf2/MafK to the GPE1, only the 3′-half of GPE1 is required, but the 5′-half of this element is still required for full enhancer activity. The expression of GST-P in the preneoplastic foci is extremely high, even higher than in the hepatoma cells (Ikeda *et al.*, 2004). These results suggest that another factor(s) may interact with the 5'-half of the element, working for the strong activation of this gene in the preneoplastic lesions. Second, because considerable amounts of the Nrf2 are expressed in the normal liver (Ikeda *et al.*, 2004), the gene-silencing mechanism must be functioning in the normal liver. Although the silencer elements were identified in the upstream of GST-P gene (−140 to −400 bp) and several factors were reported to bind to these elements, transgenic analyses suggest that the function of these silencers is rather minor. Because all of the transgenes containing GPE1 coexpressed with the endogenous GST-P gene in carcinogen-treated liver but repressed in the normal liver, the GPE1 may also have a silencer activity in normal liver, in addition to the enhancer activity in hyperplastic foci. This idea is further supported by the fact that the mouse homolog of rat GST-P gene, lacking GPE1 element, is expressed significantly in the normal liver. Recently, we found that the GPE1 sequence is overlapped with the consensus-binding sequence of C/EBP family proteins. Both C/EBPα and Nrf2/MafK bound to GPE1 in a mutually exclusive manner. On the other hand, C/EBPα strongly inhibited GPE1 in the hepatoma cell but not in F9 embryonal carcinoma cell in transient transfection analysis. Chromatin immunoprecipitation analysis using anti-C/EBPα showed that the C/EBPα was bound to the GPE1 in the normal liver but was not bound in the preneoplastic hepatocytes *in vivo* (Ikeda, H., Sakai, M. manuscript in preparation). These preliminary results suggest that the C/EBPα may bind to the GPE1 and interfere with the strong positive factor Nrf2/MafK in the normal liver, and, therefore, the C/EBPα is the most promising candidate for the repressor in normal liver. However, because C/EBP is primarily a positive transcription factor, C/EBP–GPE1 interaction alone cannot account for the complete repression in normal liver. Involvement of another factor may not be excluded. Although silencing factors interacting with silencers of GST-P gene may contribute to the repression, physiological details of repression mechanism in normal liver remain to be explained.

    As described previously, the molecular mechanism of the GST-P expression has now been gradually clarified; however, our final goal is to understand the molecular mechanism of malignant transformation through the study of GST-P expression. Recently, Higashi and his colleagues reported that GST-P-positive foci are closely related to tumor susceptibility, but not directly associated with final malignant transformation, by extensive genetic linkage analyses of a carcinogen-resistant rat strain DRH (Denda *et al.*, 1999; Higashi *et al.*, 2004; Yan *et al.*, 2002; Zeng *et al.*, 2000). How the alterations of the gene expression system in the early stage of malignant transformation, involving the GPE1, Nrf2/MafK, and C/EBP expression, correlate with the direct transforming mechanism is the crucial problem to be solved in the future.

Acknowledgments

We thank our colleagues from Tokyo University, School of Medicine, Saitama Medical School, and Hokkaido University School of Medicine for the valuable discussion and help for this work. This work was supported by a grant from Saitama Medical School Research Center for Genomic Medicine and from grants-in-aid from Ministry of Education, Science, Sports and Culture of Japan.

References

Andrews, N. C., Kotkow, K. J., Ney, P. A., Erdjument-Bromage, H., Tempst, P., and Orkin, S. H. (1993). The ubiquitous subunit of erythroid transcription factor NF-E2 is a small basic-leucine zipper protein related to the v-maf oncogene. *Proc. Natl. Acad. Sci. USA* **90**, 11488–11492.

Angel, P., Imagawa, M., Chiu, R., Stein, B., Imbra, R. J., Rahmsdorf, H. J., Jonat, C., Herrlich, P., and Karin, M. (1987). Phorbol ester-inducible genes contain a common cis element recognized by a TPA-modulated trans-acting factor. *Cell* **49**, 729–739.

Bammler, T. K., Smith, C. A., and Wolf, C. R. (1994). Isolation and characterization of two mouse Pi-class glutathione S-transferase genes. *Biochem. J.* **298**(Pt. 2), 385–390.

Chen, C., and Okayama, H. (1987). High-efficiency transformation of mammalian cells by plasmid DNA. *Mol. Cell Biol.* **7**, 2745–2752.

Crane-Robinson, C., Myers, F. A., Hebbes, T. R., Clayton, A. L., and Thorne, A. W. (1999). Chromatin immunoprecipitation assays in acetylation mapping of higher eukaryotes. *Methods Enzymol.* **304**, 533–547.

Denda, A., Kitayama, W., Konishi, Y., Yan, Y., Fukamachi, Y., Miura, M., Gotoh, S., Ikemura, K., Abe, T., Higashi, T., and Higashi, K. (1999). Genetic properties for the suppression of development of putative preneoplastic glutathione S-transferase placental form-positive foci in the liver of carcinogen-resistant DRH strain rats. *Cancer Lett.* **140**, 59–67.

Favreau, L. V., and Pickett, C. B. (1991). Transcriptional regulation of the rat NAD(P)H: Quinone reductase gene. Identification of regulatory elements controlling basal level expression and inducible expression by planar aromatic compounds and phenolic antioxidants. *J. Biol. Chem.* **266**, 4556–4561.

Gorman, C. M., Moffat, L. F., and Howard, B. H. (1982). Recombinant genomes which express chloramphenicol acetyltransferase in mammalian cells. *Mol. Cell Biol.* **2**, 1044–1051.

Hatayama, I., Nishimura, S., Narita, T., and Sato, K. (1993). Sex-dependent expression of class pi glutathione S-transferase during chemical hepatocarcinogenesis in B6C3F1 mice. *Carcinogenesis* **14**, 537–538.

Hatayama, I., Satoh, K., and Sato, K. (1986). Developmental and hormonal regulation of the major form of hepatic glutathione S-transferase in male mice. *Biochem. Biophys. Res. Commun.* **140**, 581–588.

Higashi, K., Hiai, H., Higashi, T., and Muramatsu, M. (2004). Regulatory mechanism of glutathione S-transferase P-form during chemical hepatocarcinogenesis: Old wine in a new bottle. *Cancer Lett.* **209**, 155–163.

Igarashi, K., Kataoka, K., Itoh, K., Hayashi, N., Nishizawa, M., and Yamamoto, M. (1994). Regulation of transcription by dimerization of erythroid factor NF-E2 p45 with small Maf proteins. *Nature* **367,** 568–572.

Ikeda, H., Nishi, S., and Sakai, M. (2004). Transcription factor Nrf2/MafK regulates rat placental glutathione S-transferase gene during hepatocarcinogenesis. *Biochem. J.* **380,** 515–521.

Ikeda, H., Serria, M. S., Kakizaki, I., Hatayama, I., Satoh, K., Tsuchida, S., Muramatsu, M., Nishi, S., and Sakai, M. (2002). Activation of mouse Pi-class glutathione S-transferase gene by Nrf2(NF-E2-related factor 2) and androgen. *Biochem. J.* **364,** 563–570.

Imagawa, M., Osada, S., Koyama, Y., Suzuki, T., Hirom, P. C., Diccianni, M. B., Morimura, S., and Muramatsu, M. (1991a). SF-B that binds to a negative element in glutathione transferase P gene is similar or identical to trans-activator LAP/IL6-DBP. *Biochem. Biophys. Res. Commun.* **179,** 293–300.

Imagawa, M., Osada, S., Okuda, A., and Muramatsu, M. (1991b). Silencer binding proteins function on multiple cis-elements in the glutathione transferase P gene. *Nucleic Acids Res.* **19,** 5–10.

Ito, N., Hasegawa, R., Imaida, K., Hirose, M., and Shirai, T. (1996). Medium-term liver and multi-organ carcinogenesis bioassays for carcinogens and chemopreventive agents. *Exp. Toxicol. Pathol.* **48,** 113–119.

Ito, N., Tamano, S., and Shirai, T. (2003). A medium-term rat liver bioassay for rapid *in vivo* detection of carcinogenic potential of chemicals. *Cancer Sci.* **94,** 3–8.

Itoh, K., Chiba, T., Takahashi, S., Ishii, T., Igarashi, K., Katoh, Y., Oyake, T., Hayashi, N., Satoh, K., Hatayama, I., Yamamoto, M., and Nabeshima, Y. (1997). An Nrf2/small Maf heterodimer mediates the induction of phase II detoxifying enzyme genes through antioxidant response elements. *Biochem. Biophys. Res. Commun.* **236,** 313–322.

Jaiswal, A. K. (1994). Antioxidant response element. *Biochem. Pharmacol.* **48,** 439–444.

Jaiswal, A. K. (2000). Regulation of genes encoding NAD(P)H: Quinone oxidoreductases. *Free Radic. Biol. Med.* **29,** 254–262.

Kataoka, K., Noda, M., and Nishizawa, M. (1994). Maf nuclear oncoprotein recognizes sequences related to an AP-1 site and forms heterodimers with both Fos and Jun. *Mol. Cell Biol.* **14,** 700–712.

Levin, W., Lu, A. Y., Thomas, P. E., Ryan, D., Kizer, D. E., and Griffin, M. J. (1978). Identification of epoxide hydrase as the preneoplastic antigen in rat liver hyperplastic nodules. *Proc. Natl. Acad. Sci. USA* **75,** 3240–3243.

Lindahl, R., and Feinstein, R. N. (1976). Purification and immunochemical characterization of aldehyde dehydrogenase from 2-acetylaminofluorene-induced rat hepatomas. *Biochim. Biophys. Acta* **452,** 345–355.

Moi, P., Chan, K., Asunis, I., Cao, A., and Kan, Y. W. (1994). Isolation of NF-E2-related factor 2 (Nrf2), a NF-E2-like basic leucine zipper transcriptional activator that binds to the tandem NF-E2/AP1 repeat of the beta-globin locus control region. *Proc. Natl. Acad. Sci. USA* **91,** 9926–9930.

Moore, M. A., Nakagawa, K., Satoh, K., Ishikawa, T., and Sato, K. (1987). Single GST-P positive liver cells–putative initiated hepatocytes. *Carcinogenesis* **8,** 483–486.

Morimura, S., Suzuki, T., Hochi, S., Yuki, A., Nomura, K., Kitagawa, T., Nagatsu, I., Imagawa, M., and Muramatsu, M. (1993). Trans-activation of glutathione transferase P gene during chemical hepatocarcinogenesis of the rat. *Proc. Natl. Acad. Sci. USA* **90,** 2065–2068.

Motohashi, H., Katsuoka, F., Shavit, J. A., Engel, J. D., and Yamamoto, M. (2000). Positive or negative MARE-dependent transcriptional regulation is determined by the abundance of small Maf proteins. *Cell* **103,** 865–875.

O'Neill, L. P., and Turner, B. M. (1996). Immunoprecipitation of chromatin. *Methods Enzymol.* **274**, 189–197.

Okuda, A., Imagawa, M., Maeda, Y., Sakai, M., and Muramatsu, M. (1989). Structural and functional analysis of an enhancer GPEI having a phorbol 12-O-tetradecanoate 13-acetate responsive element-like sequence found in the rat glutathione transferase P gene. *J. Biol. Chem.* **264**, 16919–16926.

Okuda, A., Imagawa, M., Sakai, M., and Muramatsu, M. (1990). Functional cooperativity between two TPA responsive elements in undifferentiated F9 embryonic stem cells. *EMBO J.* **9**, 1131–1135.

Okuda, A., Sakai, M., and Muramatsu, M. (1987). The structure of the rat glutathione S-transferase P gene and related pseudogenes. *J. Biol. Chem.* **262**, 3858–3863.

Oridate, N., Nishi, S., Inuyama, Y., and Sakai, M. (1994). Jun and Fos related gene products bind to and modulate the GPE I, a strong enhancer element of the rat glutathione transferase P gene. *Biochim. Biophys. Acta* **1219**, 499–504.

Osada, S., Daimon, S., Ikeda, T., Nishihara, T., Yano, K., Yamasaki, M., and Imagawa, M. (1997). Nuclear factor 1 family proteins bind to the silencer element in the rat glutathione transferase P gene. *J. Biochem. (Tokyo)* **121**, 355–363.

Osada, S., Takano, K., Nishihara, T., Suzuki, T., Muramatsu, M., and Imagawa, M. (1995). CCAAT/enhancer-binding proteins alpha and beta interact with the silencer element in the promoter of glutathione S-transferase P gene during hepatocarcinogenesis. *J. Biol. Chem.* **270**, 31288–31293.

Pickett, C. B., and Lu, A. Y. (1989). Glutathione S-transferases: Gene structure, regulation, and biological function. *Annu. Rev. Biochem.* **58**, 743–764.

Rao, M. S., Tatematsu, M., Subbarao, V., Ito, N., and Reddy, J. K. (1986). Analysis of peroxisome proliferator-induced preneoplastic and neoplastic lesions of rat liver for placental form of glutathione S-transferase and gamma-glutamyltranspeptidase. *Cancer Res.* **46**, 5287–5290.

Sakai, M., Matsushima-Hibiya, Y., Nishizawa, M., and Nishi, S. (1995). Suppression of rat glutathione transferase P expression by peroxisome proliferators: Interaction between Jun and peroxisome proliferator-activated receptor alpha. *Cancer Res.* **55**, 5370–5376.

Sakai, M., Muramatsu, M., and Nishi, S. (1992). Suppression of glutathione transferase P expression by glucocorticoid. *Biochem. Biophys. Res. Commun.* **187**, 976–983.

Sakai, M., Okuda, A., and Muramatsu, M. (1988). Multiple regulatory elements and phorbol 12-O-tetradecanoate 13-acetate responsiveness of the rat placental glutathione transferase gene. *Proc. Natl. Acad. Sci. USA* **85**, 9456–9459.

Sambrook, J., Fritsch, E. F., and Maniatis, Y. (1989). Assay for b-galactosidase in extracts of mammalian cells. "Molecular Cloning: A Laboratory Manual."

Sato, K. (1989). Glutathione transferases as markers of preneoplasia and neoplasia. *Adv. Cancer Res.* **52**, 205–255.

Sato, K., Kitahara, A., Satoh, K., Ishikawa, T., Tatematsu, M., and Ito, N. (1984). The placental form of glutathione S-transferase as a new marker protein for preneoplasia in rat chemical hepatocarcinogenesis. *Gann* **75**, 199–202.

Satoh, K., Hatayama, I., Tateoka, N., Tamai, K., Shimizu, T., Tatematsu, M., Ito, N., and Sato, K. (1989). Transient induction of single GST-P positive hepatocytes by DEN. *Carcinogenesis* **10**, 2107–2111.

Satoh, K., Kitahara, A., Soma, Y., Inaba, Y., Hatayama, I., and Sato, K. (1985). Purification, induction, and distribution of placental glutathione transferase: A new marker enzyme for preneoplastic cells in the rat chemical hepatocarcinogenesis. *Proc. Natl. Acad. Sci. USA* **82**, 3964–3968.

Solt, D., and Farber, E. (1976). A new principle for the analysis of chemical carcinogenesis. *Nature* **263**, 701–703.

Sugioka, Y., Fujii-Kuriyama, Y., Kitagawa, T., and Muramatsu, M. (1985a). Changes in polypeptide pattern of rat liver cells during chemical hepatocarcinogenesis. *Cancer Res.* **45**, 365–378.

Sugioka, Y., Kano, T., Okuda, A., Sakai, M., Kitagawa, T., and Muramatsu, M. (1985b). Cloning and the nucleotide sequence of rat glutathione S-transferase P cDNA. *Nucleic Acids Res.* **13**, 6049–6057.

Suzuki, T., Imagawa, M., Hirabayashi, M., Yuki, A., Hisatake, K., Nomura, K., Kitagawa, T., and Muramatsu, M. (1995). Identification of an enhancer responsible for tumor marker gene expression by means of transgenic rats. *Cancer Res.* **55**, 2651–2655.

Tanabe, A., Kurita, M., Oshima, K., Osada, S., Nishihara, T., and Imagawa, M. (2002). Functional analysis of zinc finger proteins that bind to the silencer element in the glutathione transferase P gene. *Biol. Pharm. Bull.* **25**, 970–974.

Tsuchida, S., and Sato, K. (1992). Glutathione transferases and cancer. *Crit. Rev. Biochem. Mol. Biol.* **27**, 337–384.

Xu, X., and Stambrook, P. J. (1994). Two murine GSTpi genes are arranged in tandem and are differentially expressed. *J. Biol. Chem.* **269**, 30268–30273.

Yan, Y., Zeng, Z. Z., Higashi, S., Denda, A., Konishi, Y., Onishi, S., Ueno, H., Higashi, K., and Hiai, H. (2002). Resistance of DRH strain rats to chemical carcinogenesis of liver: Genetic analysis of later progression stage. *Carcinogenesis* **23**, 189–196.

Zeng, Z. Z., Higashi, S., Kitayama, W., Denda, A., Yan, Y., Matsuo, K., Konishi, Y., Hiai, H., and Higashi, K. (2000). Genetic resistance to chemical carcinogen-induced preneoplastic hepatic lesions in DRH strain rats. *Cancer Res.* **60**, 2876–2881.

[4] Human Glutathione Transferase Zeta

By PHILIP G. BOARD and M. W. ANDERS

Abstract

Zeta-class glutathione transferases (GSTZs) were recently discovered by a bioinformatics approach and the availability of human expressed sequence tag databases. Although GSTZ showed little activity with conventional GST substrates (1-chloro-2,4-dinitrobenzene; organic hydroperoxides), GSTZ was found to catalyze the oxygenation of dichloroacetic acid (DCA) to glyoxylic acid and the *cis-trans* isomerization of maleylacetoacetate to fumarylacetoacetate. Hence, GSTZ plays a critical role in the tyrosine degradation pathway and in α-haloacid metabolism. The GSTZ-catalyzed biotransformation of DCA is of particular interest, because DCA is used in the human clinical management of congenital lactic acidosis and because DCA is a common drinking water contaminant. Substrate selectivity studies showed that GSTZ catalyzes the glutathione-dependent biotransformation of a range of dihaloacetic acids along with fluoroacetic acid,

METHODS IN ENZYMOLOGY, VOL. 401
Copyright 2005, Elsevier Inc. All rights reserved.
0076-6879/05 $35.00
DOI: 10.1016/S0076-6879(05)01004-9

2-halopropanoic acids, and 2,2-dichloropropanoic acid. Human clinical studies showed that the elimination half-life of DCA increases with repeated doses of DCA; also, rats given DCA show low GSTZ activity with DCA as the substrate. DCA was found to be a mechanism-based inactivator of GSTZ, and proteomic studies showed that Cys-16 of human GSTZ1-1 is covalently modified by a reactive intermediate that contains glutathione and the carbon skeleton of DCA. Bioinformatics studies also showed the presence of at least four polymorphic variants of human GSTZ; these variants differ considerably in the rates of catalysis and in their susceptibility to inactivation by DCA. Finally, $Gstz1^{-/-}$ mouse strains have been developed; these mice fail to biotransform DCA or maleylacetone. Although the mice have no obvious phenotype, a high incidence of lethality is observed in young mice given phenylalanine in their drinking water. $Gstz1^{-/-}$ mice should prove useful in expanding the role of GSTZ in α-haloacid metabolism and in the tyrosine degradation pathway.

Introduction

The glutathione transferases (GSTs; EC 2.5.1.18) are members of at least three gene families: the cytosolic GSTs, the mitochondrial GST, and the membrane-associated proteins involved in eicosanoid and glutathione metabolism (MAPEG). Although the GSTs play important roles in the detoxication and bioactivation of xenobiotics, some GSTs are involved in eicosanoid synthesis, steroid hormone metabolism, and amino acid degradation. Several reviews about the glutathione transferases have appeared (Armstrong, 1997; Hayes and Pulford, 1995; Hayes et al., 2005; Salinas and Wong, 1999; Sherratt and Hayes, 2002; Sheehan et al., 2001).

The soluble or cytosolic GSTs constitute the largest family of GSTs. The mammalian cytosolic GSTs are members of a superfamily of homodimeric and heterodimeric proteins whose subunits molecular masses range from 23–28 kDa. Seven classes of mammalian, cytosolic GSTs are known: alpha (GSTA), mu (GSTM), pi (GSTP), sigma (GSTS), theta (GSTT), omega (GSTO), and zeta (GSTZ). The cytosolic GSTs within a class share >40% identity, whereas there is only approximately 25% identity between classes. Although the cytosolic GSTs catalyze the reaction of glutathione with a range of electrophilic compounds, some cytosolic GSTs also have nonenzymatic functions. For example, GSTP1-1 regulates the activity of Jun N-terminal kinase (Adler et al., 1999), GSTM regulates apoptosis stimulating kinase (ASK1) (Cho et al., 2001), GSTO1-1 regulates the activity of the ryanodine receptor Ca^{2+} ion channel (Dulhunty et al., 2001), and CLIC1, which is structurally homologous to the GSTs but lacks

GST catalytic activity, is a member of the intracellular chloride ion channel proteins (Harrop *et al.*, 2001).

GST kappa (GSTK) is presently the only known mitochondrial GST. Although GSTK is not found in cytosol, it is present in peroxisomes. GSTK catalyzes the reaction of glutathione with 1-chloro-2,4-dinitrobenzene, a prototypical GST substrate, and with some organic hydroperoxides, it seems to be a distinct transferase closely related in structure to 2-hydroxychromene-2-carboxylase isomerase (Ladner *et al.*, 2004; Robinson *et al.*, 2004).

The MAPEG proteins represent a unique family of GSTs that share no sequence identity with the cytosolic or mitochondrial GSTs (Jakobsson *et al.*, 1999). Four MAPEG subgroups (I–IV) are known, and the proteins within a subgroup share >20% sequence identity. The catalytic activities of MGST-1, -2, and -3 are similar to that of the cytosolic GSTs in that they play a major role in detoxication of endobiotics and xenobiotics. Other MAPEG proteins (LTC$_4$S, PGES1, FLAP) are involved in the synthesis of eicosanoids.

Discovery of GSTZ

The zeta-class GSTs were discovered by exploiting advances in bioinformatics and the availability of human expressed sequence tag (EST) databases (Board *et al.*, 1997, 2001). Most GSTs were discovered by conventional protein chemistry approaches and the ability of GSTs to catalyze the reaction of glutathione with 1-chloro-2,4-dinitrobenzene. The use of immobilized glutathione agarose columns also facilitated the discovery and purification of new GSTs. After the discovery of theta-class GSTs (GSTT), it became apparent that not all GSTs bind to glutathione agarose columns or conjugate 1-chloro-2,4-dinitrobenzene (Meyer *et al.*, 1991; Pemble *et al.*, 1994).

The human EST database was searched with the BLAST program with sequence alignment strategies to identify sequences similar to GST-like sequences (Altschul *et al.*, 1997). This strategy resulted in the discovery of GSTZ (Board *et al.*, 1997) and, later, GST omega (GSTO) (Board *et al.*, 2000). The EST database, which now contains more than 5 million entries, can be applied to identify all members of gene families and to detect polymorphisms in gene families of interest.

Zeta-class GSTs are widely distributed in eukaryotic species and have been found in plants, fungi, and mammals (Board *et al.*, 1997; Fernández-Cañón and Peñalva, 1998; Wagner *et al.*, 2002). The widespread conservation of zeta-class GSTs indicated that the enzyme might catalyze an important homeostatic reaction. Indeed, the work of Fernández-Cañón and Penalva showed that GSTZ and maleylacetoacetate isomerase (MAAI) are identical. MAAI catalyzes the penultimate step in the tyrosine degradation pathway, a pathway that is associated with a range of metabolic disorders (Mitchell et al., 1995).

Enzymology of GSTZ

Substrate Selectivity of GSTZ

Although the protein identified by a BLAST search of the human EST database clearly indicated that GSTZ was a member of the GST superfamily, the enzyme lacked significant activity with 1-chloro-2,4-dinitrobenzene and, like most cytosolic GSTs, showed some activity with organic hydroperoxides (Board et al., 1997). Shortly thereafter, an enzyme was purified from rat liver that catalyzed the biotransformation of dichloroacetic acid (DCA) to glyoxylic acid (Tong et al., 1998). A search of the protein database with an amino acid sequence from the isolated protein revealed a high degree of similarity with hGSTZ, indicating that this protein was the rat ortholog of hGSTZ. Indeed, rabbit antibodies raised against hGSTZ reacted with the rat liver enzyme, confirming the identity of the protein as rat GSTZ. Fernández-Cañón and Penalva (1998) demonstrated that GSTZ also possessed MAAI activity, and subsequent studies showed that hGSTZ also catalyzed the isomerization of maleylacetoacetate to fumarylacetoacetate (Blackburn et al., 2000). Although GSTZ catalyzes the oxygenation of DCA to glyoxylic acid and the isomerization of maleylacetoacetate to fumarylacetoacetate, the efficiency of the isomerization reaction is much greater than that of the oxygenation reaction: the k_{cat}/K_m for the isomerization reaction is approximately 1000-fold greater than the reaction with α-haloacids (Board et al., 2003; Lantum et al., 2002).

Maleylacetoacetate Isomerase Activity

Maleylacetoacetate, which is the endobiotic substrate for GSTZ/MAAI, is formed as a product of tyrosine degradation, and GSTZ/MAAI catalyzes the glutathione-dependent cis-trans isomerization of maleylacetoacetate to give fumarylacetoacetate as a product (Fig. 1). Maleylacetoacetate has apparently not been obtained by synthesis but can be formed by the action of homogentistate dioxygenase on homogentistate (Edwards and Knox, 1956). Hence, studies on the mechanism of the reaction have relied on model substrates, particularly maleylacetone and cis-β-acetylacrylate (Fowler and Seltzer, 1970; Lee and Seltzer, 1989). With maleylacetone as the substrate, the mechanism of the reaction involves an enzyme-catalyzed

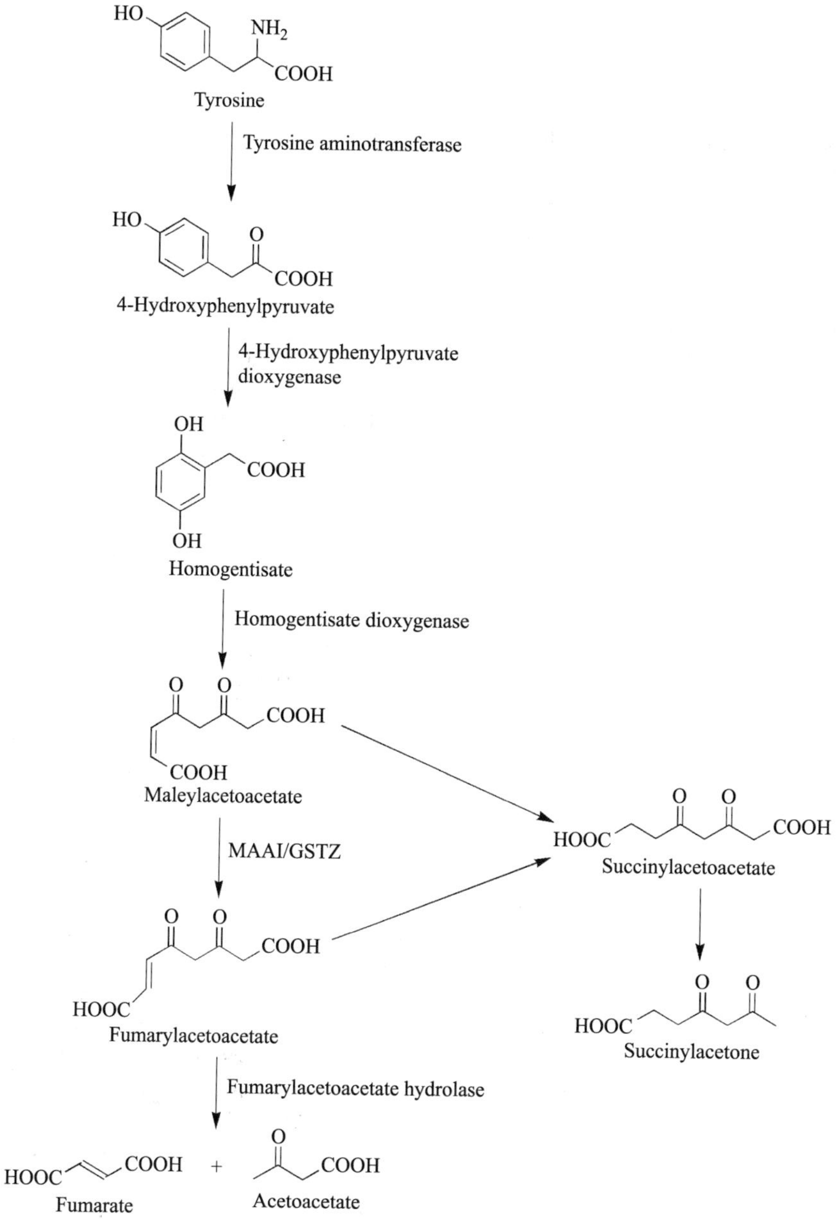

FIG. 1. Tyrosine degradation pathway.

nucleophilic attack of glutathione to give 2-glutathion-S-yl-4,6-heptanoic acid as an intermediate (Seltzer and Lin, 1979); this intermediate undergoes bond rotation, elimination of glutathione, and protonation to give fumarylacetone.

In addition to its role in the degradation of tyrosine, MAAI may catalyze the cis-trans isomerization of the benzene metabolite cis,cis-muconic acid to give trans, trans-muconic acid, which is excreted in the urine of mice given benzene (Seltzer and Hane, 1988).

Xenobiotic α-Haloacid Substrates

Early studies showed that the biotransformation of DCA to glyoxylic acid is catalyzed by a cytosolic enzyme that required glutathione for activity (James et al., 1997; Lipscomb et al., 1995). Subsequent studies showed that the biotransformation of DCA to glyoxylic acid is catalyzed by a rat liver enzyme that was identified as GSTZ (Tong et al., 1998). This observation clarified the role of GSTZ in the biotransformation of xenobiotic α-haloacids and paved the way for studies to investigate the substrate selectivity of GSTZ.

GSTZ catalyzes the glutathione-dependent biotransformation of a range of α-haloacetic acids, including DCA, bromochloroacetic acid, chlorofluoroacetic acid, and dibromoacetic acid, but not difluoroacetic acid, to glyoxylic acid. In addition, GSTZ catalyzes the addition of glutathione to (R)-, (S)-, and (R,S)-2-chloropropanoic acid to give S-(α-methylcarboxymethyl)glutathione; (R,S)-2-bromopropanoic acid, but not (R,S)-2-fluoropropanoic acid, was also a substrate. Fluoroacetic acid is biotransformed to S-(carboxymethyl)glutathione, thereby clarifying the role of GSTZ in the previously observed glutathione-dependent defluorination of fluoroacetic acid (Kostyniak et al., 1978). 2,2-Dichloropropanoic acid is also a substrate for GSTZ and is converted to pyruvic acid. 3,3-Dichloropropanoic acid, fluoroacetamide, and ethyl fluoroacetate are not substrates, indicating that the enzyme catalyzes an attack of glutathione on the carbon alpha to the carboxylic acid group and that an unblocked carboxylic acid group is required for activity. In addition, with DCA as the substrate, glutathione was neither consumed nor converted to glutathione disulfide. The rate of S-(α-methylcarboxymethyl)glutathione conjugate formation was greater with (S)-2-chloropropanoic acid as the substrate than with (R)-2-chloropropanoic acid as the substrate, indicating a significant degree of enantioselectivity. Finally, significant species differences were observed in the biotransformation of DCA to glyoxylic acid: the V_{max}/K_m for human, rat, and mouse liver cytosol were $(8.25 \pm 1.37) \times 10^3$, $(32.4 \pm 4.87) \times 10^3$, and $(52.9 \pm 2.46) \times 10^3$, respectively.

Other studies showed that with [2-^{2}H]DCA and [2-^{2}H]chlorofluoroacetic acid as substrates, the deuterium present in the [2-^{2}H]dihaloacetic acid is retained in the [2-^{2}H]glyoxylic acid formed, indicating that the enol of the dihaloacetic acid is not a substrate for the enzyme (Wempe et al., 1999).

The stereochemistry of the GSTZ-catalyzed biotransformation of (+)- and (−)-bromochloroacetic acid has been investigated (Schultz and Sylvester, 2001). These studies showed that the rate of the GSTZ-dependent biotransformation of (−)-bromochloroacetic acid was much greater than that of (+)-bromochloroacetic acid. Apparently, the absolute configuration of the stereoisomers of bromofluoroacetic acid has not been reported, which prevents correlations with the biotransformation of (R)- and (S)-2-chloropropanoic acid.

Mechanism-Based Inactivation of GSTZ by α-Haloacids

The plasma half-life of DCA in humans is increased with successive doses of DCA (Curry et al., 1991). Similarly, the biotransformation of DCA in rat liver cytosol is reduced in rats previously given DCA (James et al., 1997). The explanation for these observations was provided by studies that showed that GSTZ is inactivated in rats given fluorine-lacking α-haloacetic acids, including DCA (Anderson et al., 1999). Also, chlorofluoroacetic acid, which does not inactivate GSTZ, has a higher specific activity with GSTZ than does fluorine-lacking α-haloacetic acids. The role of fluorine substituents in governing the inactivation of GSTZ is likely to be complicated but may rely on the nucleofugicity of the leaving group (Wempe et al., 1999).

The mechanism of the dichloroacetic-acid–induced inactivation of GSTZ has been investigated. Tzeng et al. (2000) confirmed that rat liver cytosolic GSTZ is inactivated by DCA in the presence of glutathione and that recombinant hGSTZ1-1 is also inactivated by DCA. Species differences were observed for the DCA-dependent inactivation of GSTZ: the half-lives for the DCA-induced inactivation of GSTZ in rat, mouse, and human liver cytosol were 5.44, 6.61, and 22 min, respectively, but the half-maximal inhibitory concentration (K_{inact}) of DCA did not differ among the species studied. Differences were also observed in the rate constants for polymorphic variants of hGSTZ1-1 (see "GSTZ Polymorphic Variants" following). Tzeng et al. also showed that a metabolite of DCA covalently modifies GSTZ, thereby inactivating the enzyme. These studies also showed that both glutathione and DCA are required for inactivation: when both [2-^{14}C]DCA and glutathione or when both DCA and [^{35}S]glutathione were incubated with GSTZ, covalent modification was observed, but no covalent modification was observed when [2-^{14}C]DCA or [^{35}S]glutathione was incubated with GSTZ in the absence of glutathione or DCA, respectively.

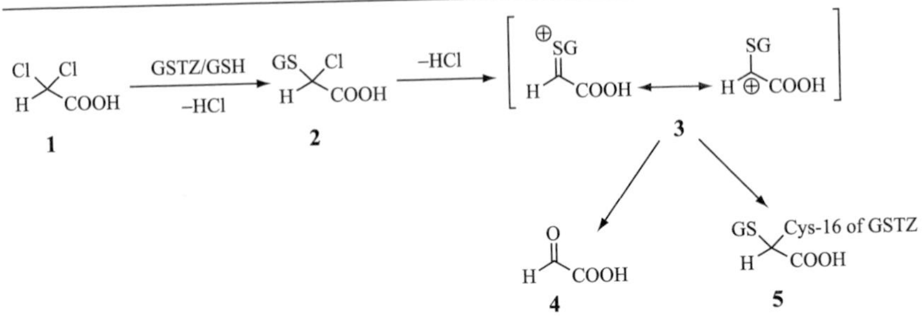

FIG. 2. GSTZ-catalyzed biotransformation of dichloroacetic acid and mechanism of the dichloroacetic acid–induced inactivation of GSTZ.

The studies by Tzeng *et al.* (2000) afforded an explanation for the observed DCA-induced inactivation of GSTZ. The GSTZ-catalyzed attack of glutathione on the alpha carbon of dichloroacetic acid 1 would displace chloride to give rise to S-(α-chlorocarboxymethyl)glutathione 2, which may lose chloride to give a carbonium/sulfonium intermediate 3 that may be hydrolyzed to give glyoxylic acid 4 and glutathione or may react directly with a nucleophilic site in GSTZ 5 to give a covalently modified and inactivated enzyme (Fig. 2).

Proteomic studies revealed the site and mechanism of the DCA-induced modification of GSTZ (Anderson *et al.*, 2002). The partition ratio for the DCA-induced inactivation of hGSTZ1c-1c was $(5.7 \pm 0.5) \times 10^2$, and the k_{cat} for the biotransformation of DCA was 39 min^{-1}. The stoichiometry of modification of hGSTZ1c-1c by DCA was ~0.5 mol of DCA/mol of enzyme monomer. The use of both matrix-assisted laser-desorption-ionization time-of-flight and electrospray-ionization quadrupole ion-trap mass spectrometry showed the presence of a single DCA-derived adduct, which was assigned to cysteine-16. The observed DCA-derived adduct contained both glutathione and the carbon skeleton of DCA, presumably as a dithioacetal (5, Fig. 2). Interestingly, cysteine-16 is not required for activity: the C16A mutant of GSTZ showed excellent catalytic activity and is not inactivated by DCA (Board *et al.*, 2003).

As noted previously, GSTZ catalyzes the *cis-trans* isomerization of maleylacetone to fumarylacetone. Incubation of GSTZ with either maleylacetone or fumarylacetone in the absence of glutathione led to inactivation of the enzyme (Lantum *et al.*, 2002). Electrospray ionization-tandem mass spectrometry and SALSA (Scoring Algorithm for Spectral Analysis) analyses of tryptic digests of GSTZ showed that the active site (SSCSWR) cysteine residue was covalently modified by both maleylacetone and fumarylacetone. The inactivation of GSTZ by both compounds was blocked by glutathione. C-terminal (LLVLEAFQVSHPCR) cysteine residues of GSTZ were also covalently modified by maleylacetone and fumarylacetone.

GSTZ Polymorphic Variants

Polymorphic variants are common among the xenobiotic-metabolizing enzymes and may give rise to significant functional differences that affect drug and chemical action and toxicity. The initial identification of GSTZ from the EST database indicated the presence of polymorphisms in the human *GSTZ* gene (Board *et al.*, 1997). Accordingly, a bioinformatics approach was developed to detect genetic polymorphisms (Board *et al.*, 1998; Tetlow *et al.*, 2001). The EST database is compiled from cDNA sequences derived from a large number of cDNA libraries, each of which represents the genome of a different individual; hence, searching the EST database is equivalent to sequencing the genes in a small, multiracial sample.

The use of the BLAST alignment program and the "Flat query-anchored with identities" output format (http://www.ncbi.nlm.nih.gov/) allows the alignment of the human GSTZ sequence with a range of ESTs. The use of this approach has allowed the identification of several polymorphisms in the human *GSTZ* (Blackburn *et al.*, 2000, 2001). In addition, the single nucleotide polymorphism (SNP) finder program (http://gai.nci. nih.gov/) can be also used to search the EST database and allows confirmation of observed sequence variations. Furthermore, compilations of SNPs can also be searched (for a list of databases, see Board *et al.*, 2004).

It has become apparent, however, that no single program or database will suffice to identify all the variants present in the EST database; hence, false positives must be eliminated by genotyping population samples. Furthermore, genes that are poorly expressed in some tissues or are conditionally expressed are underrepresented in the EST database. Also, because most ESTs are derived from the 5'- and 3'-ends, SNPs in the center of large cDNAs may be poorly represented in the database. Finally, the bioinformatics approach described previously does not identify SNPs found in promoter regions, which may be important sources of variation.

Four polymorphic variants of human GSTZ have been identified by bioinformatics analysis (Table I) (Blackburn *et al.*, 2000, 2001). The frequency of hGSTZ1c is the highest and has been taken to represent the wild-type enzyme. The polymorphic variants exhibit different catalytic activities with a range of substrates and also differ in their rates of inactivation by DCA. The GSTZ1a-1a variant shows the highest specific activity with (±)-2-bromo-3-(4-nitrophenyl)propanoic acid as the substrate, whereas the specific activities for the other variants are similar. With chlorofluoroacetic acid

TABLE I
FREQUENCY OF GSTZ1 HAPLOTYPES IN A NORMAL EUROPEAN POPULATION ($N = 128$)

	Amino acid residue			
Haplotype	32	42	82	Frequency
Z1A	Lys	Arg	Thr	.086
Z1B	Lys	Gly	Thr	.285
Z1C	Glu	Gly	Thr	.473
Z1D	Glu	Gly	Met	.156

Data are from Blackburn *et al.* (2001).

as the substrate, only small differences in specific activities were observed among the four GSTZ variants. With enantiomers of 2-chloropropanoic acid as the substrate, GSTZ1a-1a shows a high specific activity with the (R)-enantiomer and a much lower specific activity with the (S)-enantiomer. Finally, GSTZ1a-1a undergoes DCA-induced inactivation at a rate that is much slower than the other variants (Table II).

GSTZ1a-1a clearly has distinct functional properties compared with the other allelic variants. The GSTZ1a subunit differs from the other allelic variants by the substitution of arginine for glycine at position 42 (Blackburn *et al.*, 2000). This substitution is some distance from the active site and is on the side of the N-terminal domain that is exposed to solvent (Polekhina *et al.*, 2001). This substitution occurs in a region that connects β-strand 2 with α-helix 2 and contains some irregular structures consisting of a loop with β-turns and a turn of 3_{10} helix. Although a Gly to Arg substitution is not conservative, it is not clear how this remote substitution modifies the active site to cause the observed differences in substrate selectivity and inactivation kinetics. Glycine is highly conserved in this position and occurs in 13 of 18 zeta sequences aligned by Thom *et al.* (2001). Although the structure of human GSTZ1c-1c containing Gly-42 is not yet available, a comparison of the human GSTZ1a-1a structure containing Arg-42 with a zeta-class GST from *Arabidopsis thaliana* that contains glycine in the equivalent position is of interest. There are three very highly conserved residues (hGSTZ1:$S_{14}S_{15}C_{16}$; *At*GSTZ1: $S_{17}S_{18}C_{19}$) in the active site of the zeta-class GSTs, and in multiple alignments Ser-14 aligns with the active site Ser in the theta and delta classes (Board *et al.*, 1997). In the *At*GSTZ1 structure (Thom *et al.*, 2001), Ser-17 (equivalent to Ser-14 in human GSTZ1-1) is modeled at the N-terminus of α-helix 1 and is well positioned to stabilize the glutathione thiolate ion required for effective catalysis. Mutation of this residue largely inactivates *At*GSTZ1-1 (Thom *et al.*, 2001). In contrast, the structure of human GSTZ1a-1a indicates that Ser-14 is poorly oriented and Ser-15 seems to be the most likely residue to

TABLE II
SPECIFIC ACTIVITIES AND RATES OF INACTIVATION OF GSTZ POLYMORPHIC VARIANTS[a]

Enzyme	BNPP[b] (μmol/min/mg)	CFA[c] (μmol/min/mg)	(R)-2-Chloropropanoic acid[d] (μmol/min/mg)	(S)-2-Chloropropanoic acid[d] (μmol/min/mg)	MA[c] (μmol/min/mg)	Inactivation half-life[d] (min)
1a-1a	2.3 ± 0.15	1.35 ± 0.05	1.11 ± 0.04	0.07 ± 0.002	318 ± 91	23 ± 1
1b-1b	1.5 ± 0.03	1.34 ± 0.03	0.28 ± 0.009	0.21 ± 0.004	1010 ± 217	9.6 ± 0.3
1c-1c	1.6 ± 0.04	1.29 ± 0.05	0.29 ± 0.009	0.22 ± 0.005	1856 ± 716	10.1 ± 0.5
1d-1d	1.2 ± 0.08	1.27 ± 0.025	0.26 ± 0.002	0.25 ± 0.004	464 ± 215	9.5 ± 0.3

[a] Table modified from Board et al. (2004).
[b] BNPP, (±)-2-bromo-3-(4-nitrophenyl)propanoic acid; data from Board et al. (2003).
[c] CFA, chlorofluoroacetic acid; MA, maleylacetone; data from Blackburn et al. (2001).
[d] Data from Tzeng et al. (2000).

stabilize the glutathione thiolate ion required for catalysis. Thus, the available data indicate that the Gly42Arg substitution in GSTZ1a-1a may result in changes in the active site that repositions Ser-15 relative to glutathione and the second substrate. As a consequence, Cys-16 will also be realigned to a position that diminishes its potential for reaction with the carbonium/sulfonium ion of the S-(carboxymethyl)glutathione intermediate and subsequent inactivation of the enzyme. The prediction that the Gly42Arg substitution between GSTZ1c and GSTZ1a results in a switch in the active-site residue from Ser-14 in GSTZ1c to Ser-15 in GSTZ1a is also supported by the observation that mutation of Ser-14 to Ala results in the inactivation of GSTZ1c-1c (Board et al., 2003).

Crystallographic Analysis of GSTZ

To date, the crystal structures for human GSTZ1a-1a and *Arabidopsis thaliana* GSTZ1-1 have been determined (Polekhina et al., 2001; Thom et al., 2001) and have shed considerable light on the catalytic functions of the zeta-class GSTs. Both structures confirm that the zeta-class GSTs are dimers and adopt the canonical glutathione transferase fold. The active site of the human enzyme is located in a deep crevice between the N- and C-terminal domains. The depth of this crevice may explain the failure of zeta-class GSTs to bind to immobilized glutathione affinity chromatography media. Glutathione binds in the zeta-class GSTs in a similar position to the one it occupies in the other GST classes. Several features of the structures have provided insight into the enzyme's catalytic function. Reactions catalyzed by GSTs normally promote catalysis by lowering the pK_a of the glutathione thiol. In other GSTs, the glutathione thiolate ion is typically stabilized by a hydrogen bond from the hydroxyl of a tyrosine or serine residue (Armstrong, 1997). In the beta and omega class GSTs, a cysteine residue is thought to fulfill this role (Board et al., 2000; Rossjohn et al., 1998). Both GSTZ structures contain a SSC motif in the active-site region and, as discussed in the preceding section, there may be differences in the role of these residues in the two enzymes. In GSTZ1a-1a, the second Ser residue and the Cys residue seem to be suitably placed to play a catalytic role. In contrast, the first Ser residue is more favorably oriented in the *At*GSTZ1-1 structure. The crystal structure has demonstrated the presence of Cys-16 in the active site region of GSTZ1a-1a. This is consistent with the modification of this residue in the suicide inactivation that occurs with DCA as the substrate (Anderson et al., 2002; Board et al., 2003; Tzeng et al., 2000).

The position of a bound sulfate ion in the human GST1a-1a structure indicated that Arg-175 forms a salt bridge with the carboxylate of α-haloacid substrates and plays a crucial role in their orientation for attack by glutathione on the α-carbon and the displacement of one of the halide atoms (Polekhina *et al.*, 2001). Mutation of Arg-175 to Ala significantly lowers the k_{cat}, which is consistent with the proposed role of this residue (Board *et al.*, 2003).

GSTZ$^{-/-}$ Mouse Strains

Deficiencies of all the enzymes in the pathway responsible for the catabolism of phenylalanine and tyrosine have been well documented in humans, except for GST zeta (MAAI). The disease phenotype associated with these deficiencies varies considerably with the most severe resulting from tyrosinemia type 1 (fumarylacetoacetate hydrolase deficiency, FAH; Fig. 1) (Grompe, 2001; Tanguay *et al.*, 1996). The absence of clearly characterized cases of MAAI deficiency indicates that the disorder may be either very severe, or alternately, so benign that the deficiency is not detected. To examine this question and to provide a model to study the effects of α-haloacids, we and others have developed strains of *Gstz1* knockout mice (Fernández-Cañón *et al.*, 2002; Lim *et al.*, 2004).

Under normal dietary conditions, *Gstz1*$^{-/-}$ mice appear normal and breed successfully. Anatomical investigations show, however, enlarged liver and kidneys as well as splenic atrophy. Light and electron microscopic examination revealed multifocal hepatitis and ultrastructural changes in the kidney (Lim *et al.*, 2004). The administration of tyrosine, phenylalanine, or homogentisate clearly overloads the tyrosine catabolic pathway and is lethal in young *Gstz1*$^{-/-}$ mice (Fernández-Cañón *et al.*, 2002; Lim *et al.*, 2004). In older mice (>28 days), addition of 3% phenylalanine to the drinking water results in severe liver and kidney damage as well as a striking loss of circulating leukocytes (Lim *et al.*, 2004). Liver extracts from *Gstz1*$^{-/-}$ mice were devoid of MAAI activity, indicating that other GSTs or other enzymes do not catalyze this reaction (Lim *et al.*, 2004). The mild phenotype and survival of these mice under standard dietary conditions indicated that the nonenzymatic isomerization of MAA to FAA might be sufficient to bypass this deficiency and to cope with low levels of tyrosine catabolism (Fernández-Cañón *et al.*, 2002).

Although other GSTs do not catalyze the isomerization of maleylacetone, GSTZ1-1–deficient mice showed constitutive induction of alpha-, mu-, and pi class GSTs, as well as NAD(P)H:quinone oxidoreductase 1 (Lim *et al.*, 2004). Expression of these enzymes is regulated by the Nrf2-Keap1 pathway (Nguyen *et al.*, 2003), and their constitutive expression is consistent with the constant production of an electrophilic metabolite (Hayes *et al.*, 2005). An accumulation of succinylacetone was detected in the serum of GSTZ1-1–deficient mice, but the possibility that maleyl-acetoacetate and maleylacetone may also accumulate and activate this pathway cannot be excluded (Lim *et al.*, 2004). The Nrf2-Keap1 pathway regulates transcription of genes with antioxidant response elements (ARE), and the observation that this pathway is constitutively active in *Gstz1*[-] mice indicates that GSTZ1-1 negatively regulates the transcriptional activation of ARE-containing genes by endogenous metabolic products (Hayes *et al.*, 2005).

The repeated administration of DCA alters its pharmacokinetics as a result of its irreversible inactivation of GSTZ1-1 (Tzeng *et al.*, 2000). The pharmacokinetics of DCA has recently been studied in the absence of GSTZ (Ammini *et al.*, 2003). When DCA is given to *Gstz1*[-/-] mice, plasma and urine DCA concentrations remain high, and the mice fail to biotransform DCA to glyoxylic acid to a significant extent. These results indicate that GSTZ is the only significant enzyme that catalyzes the biotransformation of DCA.

References

Adler, V., Yin, Z., Fuchs, S. Y., Benezra, M., Rosario, L., Tew, K. D., Pincus, M. R., Sardana, M., Henderson, C. J., Wolf, C. R., Davis, R. J., and Ronai, Z. (1999). Regulation of JNK signaling by GSTp. *EMBO J.* **18**, 1321–1334.

Altschul, S. F., Madden, T. L., Schaffer, A. A., Zhang, J., Zhang, Z., Miller, W., and Lipman, D. J. (1997). Gapped BLAST and PSI-BLAST: A new generation of protein database search programs. *Nucleic Acids Res.* **25**, 3389–3402.

Ammini, C. V., Fernandez-Canon, J., Shroads, A. L., Cornett, R., Cheung, J., James, M. O., Henderson, G. N., Grompe, M., and Stacpoole, P. W. (2003). Pharmacologic or genetic ablation of maleylacetoacetate isomerase increases levels of toxic tyrosine catabolites in rodents. *Biochem. Pharmacol.* **66**, 2029–2038.

Anderson, W. B., Board, P. G., Gargano, B., and Anders, M. W. (1999). Inactivation of glutathione transferase zeta by dichloroacetic acid and other fluorine-lacking a-haloalkanoic acids. *Chem. Res. Toxicol.* **12**, 1144–1149.

Anderson, W. B., Liebler, D. C., Board, P. G., and Anders, M. W. (2002). Mass spectral characterization of dichloroacetic acid-modified human glutathione transferase zeta. *Chem. Res. Toxicol.* **15**, 1387–1397.

Armstrong, R. N. (1997). Structure, catalytic mechanism, and evolution of the glutathione transferases. *Chem. Res. Toxicol.* **10**, 2–18.

Blackburn, A. C., Coggan, M., Tzeng, H. F., Lantum, H., Polekhina, G., Parker, M. W., Anders, M. W., and Board, P. G. (2001). GSTZ1d: A new allele of glutathione transferase zeta and maleylacetoacetate isomerase. *Pharmacogenetics* **11**, 671–678.

Blackburn, A. C., Tzeng, H.-F., Anders, M. W., and Board, P. G. (2000). Discovery of a functional polymorphism in human glutathione transferase zeta by expressed sequence tag database analysis. *Pharmacogenetics* **10**, 49–57.

Board, P., Blackburn, A., Jermiin, L. S., and Chelvanayagam, G. (1998). Polymorphism of phase II enzymes: Identification of new enzymes and polymorphic variants by database analysis. *Toxicol. Lett.* **102–103**, 149–154.

Board, P. G., Anders, M. W., and Blackburn, A. C. (2004). Catalytic function and expression of glutathione transferase zeta. *In* "Methods in Pharmacology and Toxicology, Drug Metabolism and Transport: Molecular Methods and Mechanisms" (L. H. Lash, ed.), pp. 85–107. Humana Press, Totowa, NJ.

Board, P. G., Baker, R. T., Chelvanayagam, G., and Jermiin, L. S. (1997). Zeta, a novel class of glutathione transferases in a range of species from plants to humans. *Biochem. J.* **328**, 929–935.

Board, P. G., Chelvanayagam, G., Jermiin, L. S., Tetlow, N., Tzeng, H. F., Anders, M. W., and Blackburn, A. C. (2001). Identification of novel glutathione transferases and polymorphic variants by expressed sequence tag database analysis. *Drug Metab. Dispos.* **29**, 544–547.

Board, P. G., Coggan, M., Chelvanayagam, G., Easteal, S., Jermiin, L. S., Schulte, G. K., Danley, D. E., Hoth, L. R., Griffor, C. M., Kamath, A. V., Rosner, M. H., Chrunyk, B. A., Perregaux, D. E., Gabel, C. A., Geoghegan, K. F., and Pandit, J. (2000). Identification, characterization and crystal structure of the omega class glutathione transferases. *J. Biol. Chem.* **275**, 24798–24806.

Board, P. G., Taylor, M. C., Coggan, M., Parker, M. W., Lantum, H. B., and Anders, M. W. (2003). Clarification of the role of key active site residues of glutathione transferase zeta/maleylacetoacetate isomerase by a new spectrophotometric technique. *Biochem. J.* **374**, 731–737.

Cho, S. G., Lee, Y. H., Park, H. S., Ryoo, K., Kang, K. W., Park, J., Eom, S. J., Kim, M. J., Chang, T. S., Choi, S. Y., Shim, J., Kim, Y., Dong, M. S., Lee, M. J., Kim, S. G., Ichijo, H., and Choi, E. J. (2001). Glutathione *S*-transferase mu modulates the stress-activated signals by suppressing apoptosis signal-regulating kinase 1. *J. Biol. Chem.* **276**, 12749–12755.

Curry, S. H., Lorenz, A., Chu, P.-I., Limacher, M., and Stacpoole, P. W. (1991). Disposition and pharmacodynamics of dichloroacetate (DCA) and oxalate following oral DCA doses. *Biopharm. Drug Dispos.* **12**, 375–390.

Dulhunty, A., Gage, P., Curtis, S., Chelvanayagam, G., and Board, P. (2001). The glutathione transferase structural family includes a nuclear chloride channel and a ryanodine receptor calcium release channel modulator. *J. Biol. Chem.* **276**, 3319–3323.

Edwards, S. W., and Knox, W. E. (1956). Homogentisate metabolism: The isomerization of maleylacetoacetate by an enzyme which requires glutathione. *J. Biol. Chem.* **220**, 79–91.

Fernández-Cañón, J. M., Baetscher, M. W., Finegold, M., Burlingame, T., Gibson, K. M., and Grompe, M. (2002). Maleylacetoacetate isomerase (MAAI/GSTZ)-deficient mice reveal a glutathione-dependent nonenzymatic bypass in tyrosine catabolism. *Mol. Cell. Biol.* **22**, 4943–4951.

Fernández-Cañón, J. M., and Peñalva, M. A. (1998). Characterization of a fungal maleylacetoacetate isomerase gene and identification of its human homologue. *J. Biol. Chem.* **273**, 329–337.

Fowler, J., and Seltzer, S. (1970). The synthesis of model compounds for maleylacetoacetic acid. Maleylacetone. *J. Org. Chem.* **35**, 3529–3532.

Grompe, M. (2001). The pathophysiology and treatment of hereditary tyrosinemia type 1. *Semin. Liver Dis.* **21**, 563–571.

Harrop, S. J., De Maere, M. Z., Fairlie, W. D., Reztsova, T., Valenzuela, S. M., Mazzanti, M., Tonini, R., Qiu, M. R., Jankova, L., Warton, K., Bauskin, A. R., Wu, W. M., Pankhurst, S., Campbell, T. J., Breit, S. N., and Curmi, P. M. (2001). Crystal structure of a soluble form of the intracellular chloride ion channel CLIC1 (NCC27) at 1.4-Å resolution. *J. Biol. Chem.* **276**, 44993–45000.

Hayes, J. D., Flanagan, J. U., and Jowsey, I. R. (2005). Glutathione transferases. *Annu. Rev. Pharmacol. Toxicol.* **45**, 51–88.

Hayes, J. D., and Pulford, D. J. (1995). The glutathione S-transferase supergene family: Regulation of GST and the contribution of the isoenzymes to cancer chemoprotection and drug resistance. *Crit. Rev. Biochem. Mol. Biol.* **30**, 445–600.

Jakobsson, P. J., Morgenstern, R., Mancini, J., Ford-Hutchinson, A., and Persson, B. (1999). Common structural features of MAPEG—a widespread superfamily of membrane associated proteins with highly divergent functions in eicosanoid and glutathione metabolism. *Protein Sci.* **8**, 689–692.

James, M. O., Cornett, R., Yan, Z., Henderson, G. N., and Stacpoole, P. W. (1997). Glutathione-dependent conversion to glyoxylate, a major metabolite of dichloroacetate biotransformation in hepatic cytosol from humans and rats, is reduced in dichloroacetate-treated rats. *Drug Metab. Dispos.* **25**, 1223–1227.

Kostyniak, P. J., Bosmann, H. B., and Smith, F. A. (1978). Defluorination of fluoroacetate *in vitro* by rat liver subcellular fractions. *Toxicol. Appl. Pharmacol.* **44**, 89–97.

Ladner, J. E., Parsons, J. F., Rife, C. L., Gilliland, G. L., and Armstrong, R. N. (2004). Parallel evolutionary pathways for glutathione transferases: Structure and mechanism of the mitochondrial class kappa enzyme rGSTK1-1. *Biochemistry* **43**, 352–361.

Lee, H. E., and Seltzer, S. (1989). *cis*-b-Acetylacrylate is a substrate for maleylacetoacetate *cis-trans* isomerase. Mechanistic implications. *Biochem. Int.* **18**, 91–97.

Lim, C. E., Matthaei, K. I., Blackburn, A. C., Davis, R. P., Dahlstrom, J. E., Koina, M. E., Anders, M. W., and Board, P. G. (2004). Mice deficient in glutathione transferase zeta/maleylacetoacetate isomerase exhibit a range of pathological changes and elevated expression of alpha, mu, and pi class glutathione transferases. *Am. J. Pathol.* **165**, 679–693.

Lipscomb, J. C., Mahle, D. A., Brashear, W. T., and Barton, H. A. (1995). Dichloroacetic acid: Metabolism in cytosol. *Drug Metab. Dispos.* **23**, 1202–1205.

Meyer, D. J., Coles, B., Pemble, S. E., Gilmore, K. S., Fraser, G. M., and Ketterer, B. (1991). Theta, a new class of glutathione transferases purified from rat and man. *Biochem. J.* **274**, 409–414.

Mitchell, G. A., Lambert, M., and Tanguay, R. M. (1995). Hypertyrosinemia. *In* "The Metabolic and Molecular Basis of Inherited Disease" (C. R. Scriver, A. L. Beaudet, W. Sly, and D. Vallee, eds.), pp. 1077–1106. McGraw-Hill, New York.

Nguyen, T., Sherratt, P. J., and Pickett, C. B. (2003). Regulatory mechanisms controlling gene expression mediated by the antioxidant response element. *Annu. Rev. Pharmacol. Toxicol.* **43**, 233–260.

Pemble, S., Schroeder, K. R., Spencer, S. R., Meyer, D. J., Hallier, E., Bolt, H. M., Ketterer, B., and Taylor, J. B. (1994). Human glutathione S-transferase theta (GSTT1): CDNA cloning and the characterization of a genetic polymorphism. *Biochem. J.* **300**, 271–276.

Polekhina, G., Board, P. G., Blackburn, A. C., and Parker, M. W. (2001). Crystal structure of maleylacetoacetate isomerase/glutathione transferase zeta reveals the molecular basis for its remarkable catalytic promiscuity. *Biochemistry* **40**, 1567–1576.

Robinson, A., Huttley, G. A., Booth, H. S., and Board, P. G. (2004). Modelling and bioinformatics studies of the human kappa class glutathione transferase predict a novel third glutathione transferase family with homology to prokaryotic 2-hydroxychromene-2-carboxylate (HCCA) isomerases. *Biochem. J.* **379**, 541–552.

Rossjohn, J., Polekhina, G., Feil, S. C., Allocati, N., Masulli, M., De Illio, C., and Parker, M. W. (1998). A mixed disulfide bond in bacterial glutathione transferase: Functional and evolutionary implications. *Structure* **6**, 721–734.

Salinas, A. E., and Wong, M. G. (1999). Glutathione S-transferases—A review. *Curr. Med. Chem.* **6**, 279–309.

Schultz, I. R., and Sylvester, S. R. (2001). Stereospecific toxicokinetics of bromochloro- and chlorofluoroacetate: Effect of GST-z depletion. *Toxicol. Appl. Pharmacol.* **175**, 104–113.

Seltzer, S., and Hane, J. (1988). Maleylacetoacetate *cis-trans* isomerase: One-step double *cis-trans* isomerization of monomethyl muconate and the enzyme's probable role in benzene metabolism. *Bioorg. Chem.* **16**, 394–407.

Seltzer, S., and Lin, M. (1979). Maleylacetone *cis-trans*-isomerase. Mechanism of the interaction of coenzyme glutathione and substrate maleylacetone in the presence and absence of enzyme. *J. Am. Chem. Soc.* **101**, 3091–3097.

Sheehan, D., Meade, G., Foley, V. M., and Dowd, C. A. (2001). Structure, function and evolution of glutathione transferases: Implications for classification of non-mammalian members of an ancient enzyme superfamily. *Biochem. J.* **360**, 1–16.

Sherratt, P. J., and Hayes, J. D. (2002). Glutathione *S*-transferases. *In* "Enzyme Systems that Metabolise Drugs and Other Xenobiotics" (C. Ioannides, ed.). Wiley, New York.

Tanguay, R. M., Jorquera, R., Poudrier, J., and St-Louis, M. (1996). Tyrosine and its catabolites: From disease to cancer. *Acta Biochim. Pol.* **43**, 209–216.

Tetlow, N., Liu, D., and Board, P. (2001). Polymorphism of human alpha class glutathione transferases. *Pharmacogenetics* **11**, 609–617.

Thom, R., Dixon, D. P., Edwards, R., Cole, D. J., and Lapthorn, A. J. (2001). The structure of a zeta class glutathione *S*-transferase from *Arabidopsis thaliana*: Characterisation of a GST with novel active-site architecture and a putative role in tyrosine catabolism. *J. Mol. Biol.* **308**, 949–962.

Tong, Z., Board, P. G., and Anders, M. W. (1998). Glutathione transferase zeta catalyzes the oxygenation of the carcinogen dichloroacetic acid to glyoxylic acid. *Biochem. J.* **331**, 371–374.

Tzeng, H.-F., Blackburn, A. C., Board, P. G., and Anders, M. W. (2000). Polymorphism- and species-dependent inactivation of glutathione transferase zeta by dichloroacetate. *Chem. Res. Toxicol.* **13**, 231–236.

Wagner, U., Edwards, R., Dixon, D. P., and Mauch, F. (2002). Probing the diversity of the *Arabidopsis* glutathione *S*-transferase gene family. *Plant Mol. Biol.* **49**, 515–532.

Wempe, M. F., Anderson, W. B., Tzeng, H.-F., Board, P. G., and Anders, M. W. (1999). Glutathione transferase zeta-catalyzed biotransformation of deuterated dihaloacetic acids. *Biochem. Biophys. Res. Commun.* **261**, 779–783.

Further Reading

Lantum, H. B., Board, P. G., and Anders, M. W. (2002). Kinetics of the biotransformation of maleylacetone and chlorofluoroacetic acid by polymorphic variants of human glutathione transferase zeta (hGSTZ1-1). *Chem. Res. Toxicol.* **15**, 957–963.

Lantum, H. B., Liebler, D. C., Board, P. G., and Anders, M. W. (2002). Alkylation and inactivation of human glutathione transferase zeta (hGSTZ1-1) by maleylacetone and fumarylacetone. *Chem. Res. Toxicol.* **15**, 707–716.

[5] Characterization of the Omega Class of Glutathione Transferases

By Astrid K. Whitbread, Amir Masoumi, Natasha Tetlow, Erica Schmuck[†], Marjorie Coggan, and Philip G. Board

Abstract

The Omega class of cytosolic glutathione transferases was initially recognized by bioinformatic analysis of human sequence databases, and orthologous sequences were subsequently discovered in mouse, rat, pig, *Caenorhabditis elegans*, *Schistosoma mansoni*, and *Drosophila melanogaster*. In humans and mice, two *GSTO* genes have been recognized and their genetic structures and expression patterns identified. In both species, GSTO1 mRNA is expressed in liver and heart as well as a range of other tissues. GSTO2 is expressed predominantly in the testis, although moderate levels of expression are seen in other tissues. Extensive immunohistochemistry of rat and human tissue sections has demonstrated cellular and subcellular specificity in the expression of GSTO1-1. The crystal structure of recombinant human GSTO1-1 has been determined, and it adopts the canonical GST fold. A cysteine residue in place of the catalytic tyrosine or serine residues found in other GSTs was shown to form a mixed disulfide with glutathione. Omega class GSTs have dehydroascorbate reductase and thioltransferase activities and also catalyze the reduction of monomethylarsonate, an intermediate in the pathway of arsenic biotransformation. Other diverse actions of human GSTO1-1 include modulation of ryanodine receptors and interaction with cytokine release inhibitory drugs. In addition, *GSTO1* has been linked to the age at onset of both Alzheimer's and Parkinson's diseases. Several polymorphisms have been identified in the coding regions of the human *GSTO1* and *GSTO2* genes. Our laboratory has expressed recombinant human GSTO1-1 and GSTO2-2 proteins, as well as a number of polymorphic variants. The expression and purification of these proteins and determination of their enzymatic activity is described.

Introduction

The cytosolic glutathione transferases (GSTs) are a superfamily of phase II enzymes that use glutathione in reactions contributing to the biotransformation and disposition of a wide range of exogenous and

[†] Deceased.

METHODS IN ENZYMOLOGY, VOL. 401   0076-6879/05 $35.00
Copyright 2005, Elsevier Inc. All rights reserved.   DOI: 10.1016/S0076-6879(05)01005-0

endogenous compounds (Hayes *et al.*, 2005). These include chemical carcinogens, therapeutic drugs, the products of oxidative stress (Hayes *et al.*, 2004), steroid hormones such as Δ^5-androstenedione (Johansson and Mannervik, 2001) and metabolic intermediates in the tyrosine degradation pathway (Lim *et al.*, 2004). The glutathione transferase (GST) superfamily can be subdivided into a number of phylogenetic classes on the basis of their amino acid sequence (Mannervik and Danielson, 1988). Within mammals, previous studies have defined the Alpha, Mu, Pi, Sigma, Theta, Zeta, and Omega classes (Board *et al.*, 2000). In addition, a subfamily of *Ch*loride *I*ntracellular *C*hannel proteins (CLIC) has been shown to be members of the cytosolic GST structural family but have no known enzymatic activity (Dulhunty *et al.*, 2001). Several other soluble GST classes have been reported in insects: Delta, Epsilon (Chelvanayagam *et al.*, 1997); plants: Phi, Tau, Lambda, DHAR (Dixon *et al.*, 2002); and bacteria: Beta (Allocati *et al.*, 2000). Another soluble enzyme with glutathione conjugating activity has been detected in mammalian mitochondria and peroxisomes and is known as GST Kappa (Morel *et al.*, 2004; Pemble *et al.*, 1996). Recent studies have shown that GST Kappa is the product of a separate evolutionary pathway and has significant structural differences in comparison with the other soluble GSTs (Ladner *et al.*, 2004; Robinson *et al.*, 2004).

The Omega class GSTs have recently been the subject of considerable attention after the discovery of their role in the metabolism of arsenic (Zakharyan *et al.*, 2001) and the linkage of the Omega GST genes to the age at onset of both Alzheimer's and Parkinson's diseases (Kolsch *et al.*, 2004; Li *et al.*, 2003).

Discovery of the Omega Class GSTs

The Omega class of GSTs is one of the most recently described GST classes and was recognized by bioinformatic analysis of the human *E*xpressed *S*equence *T*ag (EST) database (Board *et al.*, 2000). Related sequences from mice and rats had previously been thought to be a member of the Theta class (Kodym *et al.*, 1999) or a dehydroascorbate reductase (Ishikawa *et al.*, 1998). Omega class GSTs seem to be widespread in the animal kingdom and have recently been identified in the pig (Rouimi *et al.*, 2001), *Caenorhabditis elegans* (Wilson *et al.*, 1994), *Schistosoma mansoni* (Girardini *et al.*, 2002), and *Drosophila melanogaster* (GenBank accession AAF50405). Although the Omega class GSTs only share approximately 20% amino acid sequence identity with members of the other classes, structural studies confirmed that the Omega class GSTs are members of the GST structural family (Board *et al.*, 2000).

Organization of the Omega Class GST Genes

GSTO *Genes in Humans*

In humans, there are two actively transcribed genes termed *GSTO1* and *GSTO2* located on chromosome 10q 24.3, and a reverse transcribed pseudogene (*hGSTO3p*) has been identified on chromosome 3 by bioinformatic analysis (Whitbread *et al.*, 2003). The *GSTO1* gene contains 6 exons and spans 12.5 kb. The *GSTO2* gene lies approximately 7.5 kb downstream of *GSTO1* and is also composed of six exons and spans 24.5 kb (Fig. 1). So far the genes encoding each of the different GST classes have been located in clusters on different chromosomes (Table I). The CLIC genes seem to be an exception and are distributed on a range of chromosomes.

Gsto *Genes in Mice*

Analysis of the EST database and the publicly available mouse genome sequence has revealed two transcriptionally active Omega class GST genes. A mouse GST originally considered to be a member of the Theta class and termed p28 (accession No U80819) (Kodym *et al.*, 1999) has 73% amino acid sequence identity and seems to be orthologous to human *GSTO1*. This cDNA was overexpressed in cells resistant to radiation-induced apoptosis. We suggest that the gene encoding p28 be termed *Gsto1* and the encoded protein be renamed mGSTO1-1 in agreement with the established nomenclature for GSTs (Mannervik *et al.*, 1992). The *Gsto1* gene has been located on a sequence contig (accession No NM_039692) from chromosome 19 in a region that is syntenic with the region containing the human Omega class genes on chromosome 10. Based on the p28 (U80819) cDNA sequence, *Gsto1* is composed of six exons and five introns and spans 9.8 kb (Fig. 1). All the splice sites conform to the GT/AG splicing rule and occur in the same position in the coding sequence as occurs in the human *GSTO1* gene (Whitbread *et al.*, 2003). A second GST gene homologous to human *GSTO2* is located downstream of *Gsto1* on the same contig and has been termed *Gsto2*. Examination of the cDNA sequences in the mouse EST database suggests that a range of alternate splicing occurs among three to five 5' noncoding exons that precede six coding region exons that are located in the same positions as the exons in *GSTO1*. The gene shown in Fig. 1 reflects the most common transcript (accession NM_026619) and starts approximately 1 kb downstream of *Gsto1* and spans approximately 20.5 kb. GSTO2 cDNAs are not common in the database, and most are derived from testis libraries reflecting the abundant expression of *GSTO2* transcripts detected in the human (Whitbread *et al.*, 2003) and mouse testis (Fig. 2). The different size and abundance of *GSTO2* transcripts observed

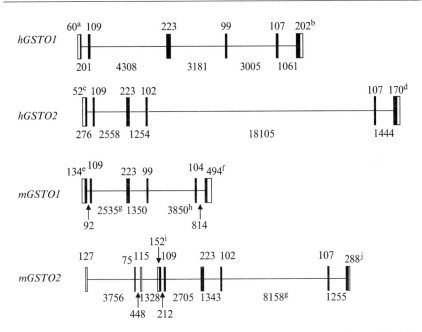

FIG. 1. Structure of Omega class genes. Solid black segments represent the protein-coding regions of the genes. Unshaded segments represent noncoding regions. Exon and intron sizes (bp) are shown above each exon and below each intron. *hGSTO1* consists of 6 exons spanning 12.5 kb. The structure shown is based on the hGSTO1 cDNA AF212303. (a) *hGSTO1* exon 1 consists of 26 bp of 5'-untranslated sequence and 34 bp of coding sequence. (b) *hGSTO1* exon 6 consists of 48 bp of 3'-untranslated sequence and 154 bp of coding sequence. *hGSTO2* consists of 6 exons spanning 24.5 kb. The structure shown is based on the hGSTO2 cDNA XM_058395. (c) *hGSTO2* exon 1 consists of 18 bp of 5'-untranslated sequence and 34 bp of coding sequence. (d) *hGSTO2* exon 6 consists of 13 bp of 3'-untranslated sequence and 157 bp of coding sequence. Mouse genomic structures were determined using the mouse genomic sequence contig NT_039692, except where noted below. *mGSTO1* consists of 6 exons spanning 9.8 kb. The structure shown is based on the mGSTO1 cDNA U80819. (e) *mGSTO1* exon 1 consists of 100 bp of 5'-untranslated sequence plus 34 bp of coding sequence. (f) *mGSTO1* exon 6 consists of 154 bp of coding sequence plus 340 bp of 3'-untranslated sequence. (g) the sizes of *mGSTO1* intron 2 and *mGSTO2* intron 7 were calculated from the mouse genomic sequence AC126679 because of discontinuity in the NT_039392 sequence. (h) the size of *mGSTO1* intron 4 is 3850 bp in NT_039692, but only 3848 bp in the mouse genomic sequence AC126679. *mGSTO2* consists of three 5'-noncoding exons and six coding exons spanning 20.5 kb. The structure shown is based on the mGSTO2 cDNA NM_026619. (i) *mGSTO2* coding exon 1 consists of 118 bp of 5'-untranslated sequence plus 34 bp of coding sequence. (j) *mGSTO2* coding exon 6 consists of 172 bp of coding sequence and 116 bp of 3'-untranslated sequence.

TABLE I
LOCATION OF HUMAN GST GENE CLUSTERS

GST class	Chromosome	Reference
Alpha	6p12	Board and Webb (1987)
Mu	1p13	Ross et al. (1993)
Pi	11q13	Board et al. (1989)
Theta	22q11	Webb et al. (1996)
Sigma	4q21–22	Kanaoka et al. (2000)
Zeta	14q 24–3	Blackburn et al. (1998)
Omega	10q 24–3	Whitbread et al. (2003)

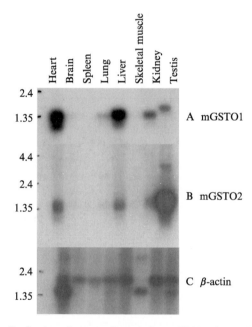

FIG. 2. Tissue distribution of mouse Omega class mRNAs. A Northern blot containing 2 μg mRNA from a range of Balb/c mouse tissues was hybridized with three probes: (A) mGSTO1, (B) mGSTO2, and (C) β-actin. The mGSTO1 probe hybridized with a band of 1.35 kb in heart, lung, liver, and kidney, and a weaker 1.1-kb band in heart and liver. The mGSTO1 band in the testis is 1.6 kb in size. The mGSTO2 probe hybridized with a 1.4-kb band in heart and liver and a 1.5-kb band in kidney and testis.

in Northern blots of RNA from testis and other tissues in mice and in humans are consistent with alternate splicing and suggest that the regulation of GSTO2 expression is complex. With one exception, the splicing sites in mouse *Gsto2* transcripts are consistent with the GT/AG rule. The

GC/AG splice site between noncoding exon 3 and coding exon 1 does not conform strictly to the GT/AG rule. This may reflect flexibility in the splicing mechanism or an inaccuracy in the sequence data. Not all splice junctions conform to the GT/AG consensus, and GC/AG is the most common nonconsensus site, representing 0.5–1.0% of all mammalian splice sites (Burset *et al.*, 2000; Mount, 2000; Thanaraj and Clark, 2001).

The existence of tandemly repeated *GSTO* genes with identical coding region exon/intron boundaries in humans and mice suggests that the duplication of the Omega class GST genes occurred before the divergence of the human and mouse lineages.

Characteristics of Omega Class GSTs

Omega Class GST Sequences

The amino acid sequences of the known Omega class GSTs are aligned in Fig. 3. Although two distinct Omega class GSTs have been identified in humans and mice, only one has been identified in other species. The human GSTO1 cDNA encodes a protein of 241 amino acids with 64% identity to the 243 residues of GSTO2 (Board *et al.*, 2000; Whitbread *et al.*, 2003). Although the pig GSTO1 monomer migrates on sodium dodecyl sulfate-polyacrylamide gel electrophoresis (SDS-PAGE) at approximately 31 kDa, mass spectroscopy determined its mass to be 27328 Da (Rouimi *et al.*, 2001), comparable to the predicted molecular mass of 27419 Da (Rouimi *et al.*, 2001). Similarly, although the predicted size of the human GSTO1 monomer is 27.6 kDa, it migrates at approximately 31 kDa (Board *et al.*, 2000). The Omega class GSTs in mice are similar to those in humans; the *Gsto1* gene encodes a protein of 240 amino acids and the *Gsto2* gene encodes a protein of 248 residues. Size exclusion chromatography and sedimentation equilibrium studies indicate that the human and pig GSTO1 subunits form homodimers (hGSTO1-1, pGSTO1-1) (Board *et al.*, 2000; Rouimi *et al.*, 2001). This is consistent with the quaternary structure of most cytosolic GSTs, the exception being the CLIC proteins that seem to be monomeric (Board *et al.*, 2004; Harrop *et al.*, 2001). The quaternary structure of GSTO2 enzymes has not been studied.

A notable feature of the GSTO2 subunits is their high cysteine content. There are 11 cysteine residues in hGSTO2 (4.5%) and 15 in mGSTO2 (6.1%) compared with 5 (2.1%) and 4 (1.6%) in hGSTO1 and mGSTO1, respectively. The high cysteine content is reminiscent of proteins such as the keratins, where their structural properties are partly due to their high proportion of cysteine residues. In comparison soft/epidermal keratins contain 2.9% cysteine and hard/hair keratins contain 7.6% (Yu *et al.*,

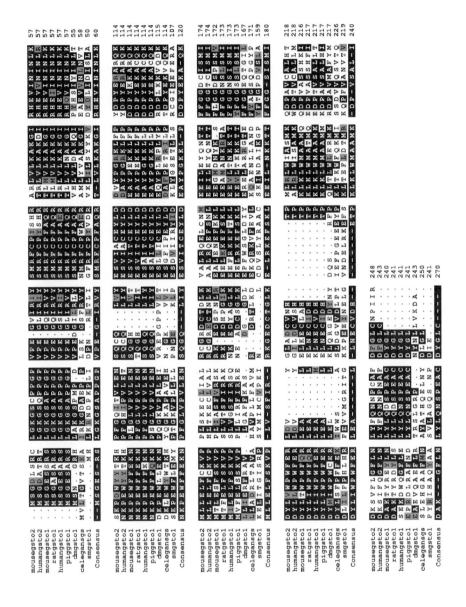

1993). Although a structural role for GSTO2 subunits cannot be assumed on the basis of their cysteine content alone, it is interesting to note that other members of the GST structural family act as crystallins and play a structural role in the lens of cephalopods (Tomarev *et al.*, 1993).

The Crystal Structure of the Omega Class GSTs

The crystal structure of recombinant human GSTO1-1 has been determined at 2 Å resolution (Fig. 4), and the co-ordinates are available in the Protein Databank under the code 1eem (Board *et al.*, 2000). GSTO1-1 adopts the canonical GST fold and is composed of an N-terminal thioredoxin-like domain and a C-terminal domain that is composed entirely of α-helices. GSTO1 has several distinguishing features including an N-terminal extension of approximately 19 residues. This extension contains a proline-rich segment that combines with the C-terminus to create a structural element not seen in other members of the GST structural family (Board *et al.*, 2000).

The active site cysteine is another notable feature of GSTO1, and it is conserved in GSTO2. In multiple alignments of GST sequences, Cys-32 aligns with tyrosine and serine residues that play significant catalytic roles in other GSTs. Typically, a hydrogen bond from the hydroxyl of active site tyrosine and serine residues stabilizes bound glutathione as a thiolate (Armstrong, 1997). In GSTO1-1, Cys-32 formed a disulfide bond with glutathione bound in the active site (Board *et al.*, 2000). Because the enzyme is sensitive to alkylating agents (Board *et al.*, 2000), and because mutation of Cys-32 to Ala eliminates its thioltransferase activity (Table II), it is evident that Cys-32 plays a novel catalytic role in the Omega class and that the formation of the disulfide is part of the catalytic mechanism in the thioltransferase reaction. GSTO1-1 has very weak glutathione conjugating activity with the classical GST substrate 1-chloro-2,4-dinitrobenzene (Board *et al.*, 2000). Surprisingly, mutation of Cys-32 to Ala strongly elevates activity with 1-chloro-2,4-dinitrobenzene (Table II). Presumably, the inability to form a disulfide allows the bound GSH to form a stable thiolate that readily participates in glutathione conjugation reactions. It is also likely in this case that the formation of a thiolate is promoted by the positioning of

FIG. 3. Alignment of Omega class GSTs from a range of species. The alignment was carried out with Clustal W using the following sequences: mouse GSTO2 (NP_080895), human GSTO2 (NP_899062), mouse GSTO1 (AAB70110), rat GSTO1 (Q9Z339), human GSTO1 (AAF73376), pig GSTO1 (AAF71994), *Drosophila melanogaster* GSTO1 (AAF50405), *Caenorhabditis elegans* GSTO1 (P34345), and *Schistosoma mansoni* GSTO1 (AAO49385). Identical residues are shaded in black and similar residues are shaded in gray.

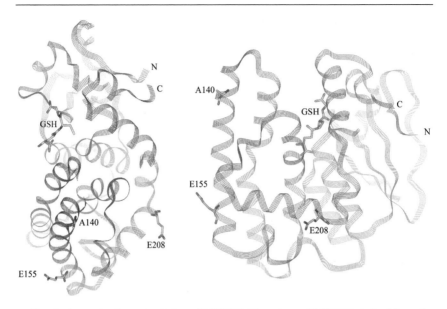

FIG. 4. A structural representation of hGSTO1. Two views of hGSTO1 derived from the protein database coordinates 1eem (Board *et al.*, 2000). The position of genetically variable residues and GSH are indicated. (See color insert.)

TABLE II
THE EFFECT OF A CYS 32 ALA MUTATION IN GST01-1

			Thioltransferase
GST conjugation	CDNB	Ethacrynic acid	2-hydroxyethyldisulfide
GST01-Cys 32[a]	0.18 ± 0.006	ND	2.9 ± 0.12
GST01-Ala 32	2.8 ± 0.27	0.14 ± 0.001	ND

[a] Data from Board *et al.* (2000). All values expressed as μM/min/mg. ND, not detectable.

the GSH thiol at the N-terminal of α-helix 1. The positioning of a Cys side chain at the positive end of a helix dipole substantially lowers the pK_a of the thiol (Kortemme and Creighton, 1995).

The GSTs typically possess an "H" site where hydrophobic substrates can bind in proximity to GSH bound in the "G" site (Mannervik and Danielson, 1988). GSTO1-1 possesses a relatively large well-defined "H" site pocket that is notably lined with some polar residues, making it less hydrophobic than the "H" site in other GSTs (Board *et al.*, 2000). This feature, together with a large open cleft between the subunits, suggests that the natural substrate or binding partner of GSTO1-1 may not be particularly hydrophobic and could be another protein.

Tissue and Cellular Distribution of Omega Class GSTs

Northern blots of human tissues suggest that *GSTO1* is transcribed at a basal rate in a wide range of tissues (Board *et al.*, 2000). Relatively high mRNA levels were observed in liver, heart, and skeletal muscle. Similarly, in mice, high levels of Gsto1 mRNA were detected in liver and heart compared with other tissues (Fig. 2). Northern blots of RNA from human tissues probed with a GSTO2 specific probe suggest that this isoenzyme is also widely expressed with relatively high levels of expression in liver, kidney, and skeletal muscle (Whitbread *et al.*, 2003). The highest levels of GSTO2 expression were clearly detected in the testis, and there was also hybridization to a larger transcript than occurs in other tissues. Similar results were obtained with RNA derived from mouse tissues (Fig. 2). Northern blotting gives an overview of gene expression in a particular tissue but may obscure significant variations in expression in particular cells or cell regions. An immunohistochemical study of GSTO1-1 in human tissues has revealed considerable cellular specificity within different tissues (Yin *et al.*, 2001). For example, in the brain, neurons did not stain, but there was strong cross-reaction with the nucleus in glial cells (Yin *et al.*, 2001). Localization of human GSTO1 to the nucleus and nuclear membrane has been demonstrated by immunohistochemistry in a number of cell types (Yin *et al.*, 2001), and translocation of mouse GSTO1-1 to the nucleus after heat stress has been reported in cell culture (Kodym *et al.*, 1999). This nuclear localization may indicate additional roles for Omega class GSTs unconnected with xenobiotic metabolism.

On the basis of its sequence, rat glutathione–dependent dehydroascorbate reductase (DHAR) seems to be orthologous to human GSTO1-1 (Fig. 3). Northern blotting demonstrated its expression in liver, kidney, testis, and brain (Ishikawa *et al.*, 1998). The expression of rGSTO1-1 (DHAR) has been extensively investigated in rat tissues by immunohisto-chemical techniques where it is found in the testis, liver, and kidney (Paolicchi *et al.*, 1996) and is abundant in the cerebellum, striatum, and hippocampus (Fornai *et al.*, 1999, 2001b). Detailed evaluation indicated that GSTO1 is expressed in several parts of the substantia nigra, including the pars compacta, which is affected in Parkinson's disease (Fornai *et al.*, 2001a). GSTO1-1 expression in nigral neurons demonstrates perinuclear and nuclear membrane localization. It has been suggested that rat GSTO1-1 may account for up to 65% of the glutathione-dependent dehydroascorbate reductase activity in the brain (Fornai *et al.*, 1999). Because ascorbate is considered an important antioxidant in the brain (Rice, 2000) and oxidative stress has been implicated in the pathogenesis of both Alzheimer's and Parkinson's diseases (Fahn and Cohen, 1992; Mattson, 2004; Simonian and Coyle, 1996), a variation in the expression of Omega class GSTs in critical cells in the brain could be a common factor that has led to the linkage of Omega class GST genes to the age at onset of these important neurological disorders.

A glutathione-dependent dehydroascorbate reductase has also been purified from human erythrocytes, suggesting that GSTO1-1 and or GSTO2-2 may be expressed in those cells and may prove to be a ready source of the nonrecombinant human enzyme (Xu *et al.*, 1996).

Purification of Omega Class GSTs

Purification of GSTO1-1

Some of the Omega class GSTs that have been studied have been recombinant enzymes expressed in *Escherichia coli* by fusion to a poly-histidine tag in the pQE 30 vector (Qiagen Hilden Germany) (Board *et al.*, 2000; Girardini *et al.*, 2002; Whitbread *et al.*, 2003). The expressed proteins were purified by Ni-agarose affinity chromatography. Although this approach is straightforward and provides good yields of highly purified protein, it has the disadvantage of leaving a residual poly-histidine tag at the N-terminus of the purified protein. However, because the N-terminus of GSTO1 is external and some distance from the active site, the additional residues do not seem to affect the function of the final product. Fusion of GSTO1 to a schistosome GST in a pGEX vector, followed by glutathione agarose chromatography and cleavage of the fused protein by digestion with thrombin has also been used as an effective purification strategy (Board *et al.*, 2000). However, this latter procedure has the disadvantage that traces of the schistosome GST can potentially contaminate the final product. Many GSTs belonging to the Alpha, Mu, and Pi classes may be purified by affinity chromatography on glutathione agarose (Simons and Vander Jagt, 1977). However, in our hands, human GSTO1-1 was not retained by glutathione agarose where the GSH was coupled to the agarose support by means of its sulfhydryl group (Board *et al.*, 2000). Pig liver GSTO1-1 does not bind to glutathione affinity columns but was purified along with other GSTs by affinity chromatography on S-hexylglutathione immobilized on sepharose (Rouimi *et al.*, 2001). The Omega class GST from *Schistosoma mansoni* has also been shown to bind to S-hexylglu-tathione immobilized on sepharose (Girardini *et al.*, 2002). Thus, it seems that immobilized glutathione and its derivatives have varying capacities to bind effectively to Omega class GSTs.

There are two studies in the literature reporting the purification of glutathione-dependent dehydroascorbate reductase from rat liver and human erythrocytes (Maellaro *et al.*, 1994; Xu *et al.*, 1996). It is likely that these enzymes were Omega class GSTs, and they required a range of precipitation and chromatographic steps to achieve a significant level of purification. Although affinity chromatography on S-hexylglutathione sepharose has a number of advantages over conventional purification procedures, it is evident that if Omega GSTs are purified from cytosolic extracts by S-hexylglutathione sepharose affinity chromatography, they still need to be separated from other GSTs and glutathione-binding proteins. This problem was effectively overcome in the purification of pig liver GSTO1-1 by first passing the cytosol through glutathione agarose to remove other GSTs and glutathione-binding proteins. GSTO1-1 was then selectively retained on S-hexylglutathione sepharose and eluted in a high yield with only limited contamination by other proteins that could be removed by gel filtration (Rouimi *et al.*, 2001).

Expression and Purification of GSTO2-2

The purification of recombinant human and mouse GSTO2-2 has been difficult to achieve because of the enzyme's insolubility when expressed in *Escherichia coli* (Whitbread *et al.*, 2003). This problem has recently been solved by the use of a low-temperature bacterial expression protocol and the development of a cyclic refolding strategy (Schmuck *et al.*, 2005). A full-length cDNA fragment encoding human GSTO2-2 (Whitbread *et al.*, 2003) was cloned into the expression vector pQE30 (Qiagen Hilden Germany) and transfected into the M15-rep4 strain of *Escherichia coli*. Bacterial cultures were grown at 37° overnight in Luria broth medium supplemented with 100 μg/ml ampicillin and 30 μg/ml kanamycin. A 1:100 dilution of overnight culture was used to inoculate 2-L subcultures that were grown at 37° for 2 h and then chilled to 8°. Expression of recombinant protein was induced with isopropyl-β-D-thiogalactopyranoside (IPTG) at a final concentration of 0.1 mM. Incubation was carried out with shaking at 8° for 24 h. Cells were pelleted by centrifugation at 4° and stored at $-20°$.

For the purification and refolding of GSTO2-2, cell pellets were resuspended in 25 ml of ice-cold buffer A (8 M urea; 1 mM GSH; 50 mM NaH$_2$PO$_4$; 300 mM NaCl; pH 7.5) and lysed by sonication. Cell debris was removed by centrifugation at 10,000g for 15 min at 4°, and the cleared lysates were incubated with 3 ml of Ni-sepharose beads equilibrated in buffer A for at least 1 h with constant gentle mixing on a rotating wheel to

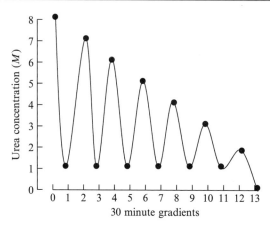

FIG. 5. Refolding of hGSTO2-2. A schematic diagram showing the time course of the cyclic changes in urea used to refold hGSTO2-2 immobilized on Ni-agarose by a poly-His tag.

allow the binding of the recombinant His-tagged GSTO2-2. Subsequent steps were performed at room temperature. The Ni-sepharose beads were washed with buffer A and collected by brief centrifugation twice to remove nonspecific unbound protein. The beads were then transferred onto a sintered glass funnel and washed with 500 ml of buffer A. The Ni-sepharose beads were then packed into a column, and the refolding of bound GSTO2-2 and the removal of urea was carried out on a Pharmacia FPLC apparatus at a buffer flow rate of 0.5 ml/min. A program was used to generate 13×30 minute gradients to cyclically lower, and partially raise, the urea concentration, leading to the eventual removal of all urea from the equilibrating buffer. This process is illustrated graphically in Fig. 5. Refolded enzyme was eluted from the column using a 0–500 mM imidazole gradient. Fractions of 1 ml were collected, and those containing expressed refolded GSTO2-2 were identified by gel electrophoresis and concentrated in a Centricon YM-10 centrifugal filter (Millipore Bedford, MA). Glycerol and buffer A minus urea were added to the stocks to a final glycerol concentration of 30% and a final protein concentration of 0.5 mg/ml. Protein stocks were stored at $-20°$. Under these conditions, human GSTO2-2 does not precipitate and remains active over several months.

Determination of Enzymatic Activity

Because of the presence of Cys-32 in the active site, the Omega class GSTs exhibit a range of catalytic activities that are distinct from those of other mammalian GSTs. These activities are all glutathione-dependent thiol exchange or reduction reactions. Human GSTO2-2 and GSTO1-1 from human, pig, and *Schistosoma mansoni* have been shown to catalyze thioltransferase and dehydroascorbate reductase reactions (Board *et al.*, 2000; Girardini *et al.*, 2002; Rouimi *et al.*, 2001; Schmuck *et al.*, 2005). These activities are not associated with other GSTs but are typical of the glutaredoxins, enzymes with structural similarity to the N-terminal domain of cytosolic GSTs. Reduction of dehydroascorbate by rat GSTO1 has also been demonstrated (Maellaro *et al.*, 1994). The reduction of monomethylarsonic acid, an intermediate in the arsenic biotransformation pathway, by human GSTO1-1 and GSTO2-2 has also been demonstrated (Schmuck *et al.*, 2005; Zakharyan *et al.*, 2001). Each of these activities can be measured by spectrophotometric assays.

Thioltransferase

Both GSTO1-1 and GSTO2-2 catalyze thioltransferase reactions (Board *et al.*, 2000; Schmuck *et al.*, 2005). Because the active site of GSTO1-1 is open and not particularly hydrophobic, it has been suggested that it may accommodate protein substrates and play a role in glutathionylation reactions. However, no protein substrates have been identified so far. A number of thioltransferase assays have been described, and in our laboratory 2-hydroxyethyldisulfide has proven to be a convenient and readily available substrate. The assay measures the rate of formation of oxidized glutathione (GSSG) through a reaction coupled by glutathione reductase to the oxidation of NADPH. The method we use is based on a previously published technique (Holmgren and Aslund, 1995).

The reactions are carried out in a final volume of 1 ml containing 0.3 mM NADPH, 1 mM GSH, 4 units glutathione reductase from *Saccharomyces cerevisiae* (Sigma-Aldrich), 0.1 mg/ml bovine serum albumin, 1.5 mM EDTA, 0.75 mM 2-hydroxyethyldisulfide, and 0.1 M Tris HCl, pH8.0. The reaction mix is preincubated at 30° for 3 min to allow the formation of a HEDS-GSH mixed disulfide and the reduction of any GSSG present in the GSH. The reaction is started by the addition of approximately 25 μg of enzyme sample, and the reaction is recorded at 340 nm. Blank reactions contain sample buffer instead of the enzyme. Specific activities are calculated using an extinction coefficient of 6.22 m$M^{-1}\cdot$cm^{-1}

Dehydroascorbate Reductase

Although the initial studies of Board *et al.* (2000) demonstrated that GSTO1-1 exhibited dehydroascorbate reductase activity, our recent studies have shown that GSTO2-2 has 70–100 times greater activity (Schmuck *et al.*, 2005). Dehydroascorbate is relatively unstable and is prepared immediately before use by the oxidation of L-ascorbic acid with bromine and should be kept out of the light and used within 3 h (Wells *et al.*, 1995).

Reactions are carried out in a final volume of 1 ml containing 200 mM sodium phosphate, pH 6.85, 3 mM GSH, 1.5 mM dehydroascorbate, and 1–25 μg of enzyme. Reactions are initiated by addition of dehydroascorbate, and the rate of change in absorbance at 265 nm was measured at 30° for 2 min. Blank reactions contain sample buffer instead of the enzyme. Because the activity of GSTO2-2 is approximately 100-fold that of GSTO1-1, care is required in adjusting the enzyme concentration to a level where the reaction rate is linear and the substrate does not become limiting.

Specific activities were calculated using an extinction coefficient of 14.7 m$M^{-1} \cdot$cm^{-1}.

Monomethylarsonate Reductase

Caution: Inorganic arsenic is classified as a human carcinogen and the following chemicals should be handled with caution.

Both GSTO1-1 and GSTO2-2 catalyze the reduction of pentavalent methylated arsenic species (Fig. 6) (Schmuck et al., 2005; Zakharyan et al., 2001). In previous studies, monomethylarsonate (MMA^V) reduction was measured by the use of [^{14}C]MMA^V and the separation of MMA^V from MMA^{III} by an organic extraction procedure (Zakharyan et al., 1999). This method has the major disadvantage that [^{14}C]MMA^V is not commercially available. Another previously reported method used ICP-MS to quantify the MMA^{III} produced (Tanaka-Kagawa et al., 2003).

We have used a spectrophotometric assay that uses commercially available substrates and a standard UV/VIS spectrophotometer. The enzymatic reduction of pentavalent monomethylarsonate (disodium methylarsonate; Chem Service, West Chester, PA) or dimethylarsinic acid (DMA^V, sodium

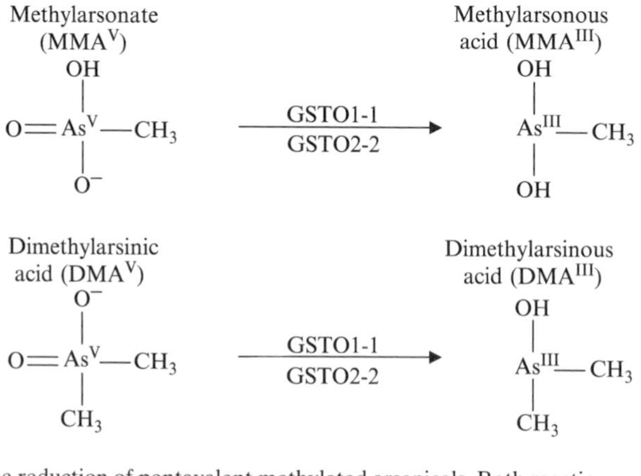

Methylarsonate (MMA^V)

Methylarsonous acid (MMA^{III})

$$O = As^V - CH_3 \xrightarrow[\text{GSTO2-2}]{\text{GSTO1-1}} As^{III} - CH_3$$

Dimethylarsinic acid (DMA^V)

Dimethylarsinous acid (DMA^{III})

$$O = As^V - CH_3 \xrightarrow[\text{GSTO2-2}]{\text{GSTO1-1}} As^{III} - CH_3$$

FIG. 6. The reduction of pentavalent methylated arsenicals. Both reactions are glutathione dependent and are catalyzed by hGSTO1-1 and GSTO2-2.

cacodylate; BDH Chemical, Poole England) results in the oxidation of reduced glutathione (GSH) to its disulfide GSSG. The production of GSSG is
followed spectrophotometrically at 340 nm by linking its reduction by glutathione reductase to the oxidation of NADPH. The assay is based on
a previously described method (Denton *et al.*, 2004) and contained
the following reagents in a final volume of 1 ml: Bis-Tris, 0.1 M, pH5.5;
GSH, 10 mM; NADPH, 0.3mM; EDTA, 5 mM, glutathione reductase, 4 units;
and either MMA^V or DMA^V at a final concentration of 10 mM. The reaction
mix is preincubated for 10 min at 30° before the addition of the enzyme
sample. The reactions are recorded for 3 min, and the initial linear reaction
rate over the first minute is used for the calculation of activity. Specific
activities are calculated using an extinction coefficient of 6.22 mM^{-1}·cm^{-1}.

Genetic Polymorphism in Omega Class GSTs

Arsenic is a highly toxic and carcinogenic environmental contaminant
in many areas of the world (Zakharyan and Aposhian, 1999). In addition,
arsenic trioxide (ATO) is used therapeutically in the treatment of acute
promyelocytic leukemia (Westervelt *et al.*, 2001). Individual differences in
response to arsenic, including fatal adverse reactions during ATO therapy,
have been reported (Loffredo *et al.*, 2003; Westervelt *et al.*, 2001). The basis
for the observed differences in individual response has not been established; however, the critical role played by Omega class GSTs in the
methylation pathway of arsenic biotransformation suggests that genetic
polymorphism in the Omega class enzymes may be an important factor.

Recent genetic linkage studies have implicated polymorphism in the
Omega class GSTs as a contributing factor influencing the age at onset of
both Alzheimer's and Parkinson's diseases (Li *et al.*, 2003; Kolsch *et al.*,
2004). Polymorphisms that cause amino acid substitutions or deletions in
the Omega class GSTs have recently been investigated in a number of
laboratories (Marnell *et al.*, 2003; Whitbread *et al.*, 2003; Yu *et al.*, 2003).
We have expressed several variant isoforms of GSTO1-1 and GSTO2-2 in
E. coli, and a summary of their properties is shown in Table III.

The 3-bp genomic deletion that causes the deletion of a glutamic
residue at position 155 in GSTO1 occurs at the junction of exon 4 and
intron 4 (Whitbread *et al.*, 2003). The deletion occurs in such a way that the
splice site is potentially reformed, giving rise to a spliced transcript that is
missing Glu155. The presence of ESTs with this deletion suggests that the
reformed splice site is functional, and it is likely that the Glu155 deleted
protein is expressed in individuals with this variant allele. Although
the ΔGlu155 variant has elevated activity, its heat stability is impaired
(Whitbread *et al.*, 2003). Further studies are required to determine whether
this deletion causes instability and deficiency *in vivo*.

TABLE III
CHARACTERISTICS OF POLYMORPHIC VARIANTS OF OMEGA CLASS GSTs

GSTO1-1	Thioltransferase activity	MMA^V reductase activity	Heat stability
Ala 140 Asp	NC	NC	NC
Del Glu 155	×3	×2	↓
Glu 208 Lys	NC	NC	NC
GSTO2-2			
Asn 142 Asp	NC	NC	NC

NC, No change.
The information in this table was compiled from Whitbread et al. (2003) and Schmuck et al. (2005).

A single nucleotide polymorphism that encodes a Thr-217Asn substitution in GSTO1-1 has been reported in dbSNP, and the recombinant protein has decreased activity (Tanaka-Kagawa et al., 2003). However, this variant has not been detected in any population survey and must be very rare or the result of a sequencing error (Kolsch et al., 2004; Marnell et al., 2003; Whitbread et al., 2003; Yu et al., 2003).

Other Functions of Omega Class GSTs

Modulation of Calcium Channels

GSTO1-1 has been shown to modulate ryanodine receptors, which are calcium channels in the endoplasmic reticulum of cells (Dulhunty et al., 2001). The mechanism of this effect is unclear, but it is interesting to note that mutation of the active site Cys-32 to alanine abolishes the effect. Although the physiological role of this effect is unclear, it has been speculated that because mGSTO1-1 is overexpressed in a radiation-resistant lymphoma cell line (Kodym et al., 1999), it may reduce apoptosis normally resulting from Ca^{2+} mobilization through ryanodine receptor–sensitive stores (Pan et al., 2000). Similarly, GSTO1-1 may also protect cancer cells from apoptosis caused by Ca^{2+} mobilization through ryanodine receptors (Mariot et al., 2000). In addition, GSTO1-1 could modulate immune responses that depend on a sustained increase in $[Ca^{2+}]$ through activity of ryanodine receptors in T and B lymphocytes (Sei et al., 1999; Xu et al., 1998).

Interaction with Cytokine Release Inhibitory Drugs

Interleukin-1(IL-1) is a proinflammatory mediator produced in activated monocytes and macrophages (Dinarello, 1998). IL-1 is not constitutively released and requires posttranslational processing before it is released

(Hogquist *et al.*, 1991; Perregaux *et al.*, 1992). Several diarylsulfonylureas act as cytokine release inhibitory drugs (CRIDS) (Perregaux *et al.*, 1992). These CRIDs inhibit the posttranslational processing of interleukin-1β in activated human monocytes. Recent studies have shown that CRIDs bind GSTO1-1 in monocytes, and this interaction may underlie the mechanism by which CRIDs arrest interleukin-1β processing (Laliberte *et al.*, 2003). The interaction between CRIDs and GSTO1-1 requires Cys-32. Much higher concentrations of CRIDs bind CLIC1, which has a cysteine residue that aligns with Cys-32 of GSTO1 (Laliberte *et al.*, 2003). The precise role that GSTO1-1 plays in this pathway is unclear, but it has been speculated that it could mediate its effect by modulating or creating ion channels, because interleukin-1β processing is associated with significant changes in ionic homeostasis (Laliberte *et al.*, 2003). Although GSTO1-1 has not yet been shown to form ion channels, the CLIC proteins are structurally related to GSTO1-1 (Dulhunty *et al.*, 2001; Harrop *et al.*, 2001). In addition, GSTO1-1 can modulate ryanodine receptor calcium release channels (Dulhunty *et al.*, 2001). Alternately, GSTO1-1 could mediate an effect on interleukin-1β processing by means of its thioltransferase activity, because activation of monocytes can alter redox balance and alter the potential glutathionylation of a range of proteins. Glutathionylation can potentially modulate protein function by blocking functionally important thiols (Laliberte *et al.*, 2003).

Conclusion

The Omega class GSTs have a unique range of enzymatic activities compared with other GSTs, which can probably be attributed to the presence of a catalytic cysteine residue. The capacity of Omega class GSTs to reduce pentavalent methylated arsenicals indicates that they play a key part in the biotransformation of arsenic. Genetic polymorphism in the *GSTO1* and *GSTO2* genes may underlie variability in response to arsenic exposure. The role of GSTO enzymes in the reduction of dehydroascorbate in the brain may be the basis of their genetic linkage to age at onset of Alzheimer's disease and Parkinson's disease. The crystal structure of human GSTO1-1 indicated the potential for protein substrates for this enzyme. Although none have been identified to date, the thioltransferase activity of GSTO1-1 and GSTO2-2 indicates a potential role in the regulation of protein function through glutathionylation. The nuclear localization of GSTO1 in several cell types may indicate additional endogenous roles that remain to be explained.

Acknowledgments

Studies of the Omega class GSTs were supported by Grant 179818 to PB from the Australian National Health and Medical Research Council.

References

Allocati, N., Casalone, E., Masulli, M., Polekhina, G., Rossjohn, J., Parker, M. W., and Di Ilio, C. (2000). Evaluation of the role of two conserved active-site residues in beta class glutathione S-transferases. *Biochem. J.* **351,** 341–346.

Armstrong, R. N. (1997). Structure, catalytic mechanism, and evolution of the glutathione transferases. *Chem. Res. Toxicol.* **10,** 2–18.

Blackburn, A. C., Woollatt, E., Sutherland, G. R., and Board, P. G. (1998). Characterization and chromosome location of the gene GSTZ1 encoding the human Zeta class glutathione transferase and maleylacetoacetate isomerase. *Cytogenet. Cell Genet.* **83,** 109–114.

Board, P. G., Coggan, M., Chelvanayagam, G., Easteal, S., Jermiin, L. S., Schulte, G. K., Danley, D. E., Hoth, L. R., Griffor, M. C., Kamath, A. V., Rosner, M. H., Chrunyk, B. A., Perregaux, D. E., Gabel, C. A., Geoghegan, K. F., and Pandit, J. (2000). Identification, characterization and crystal structure of the Omega class glutathione transferases. *J. Biol. Chem.* **275,** 24798–24806.

Board, P. G., Coggan, M., Watson, S., Gage, P. W., and Dulhunty, A. F. (2004). CLIC-2 modulates cardiac ryanodine receptor Ca2+ release channels. *Int. J. Biochem. Cell Biol.* **36,** 1599–1612.

Board, P. G., and Webb, G. C. (1987). Isolation of a cDNA clone and localization of human glutathione S- transferase 2 genes to chromosome band 6p12. *Proc. Natl. Acad. Sci. USA* **84,** 2377–2381.

Board, P. G., Webb, G. C., and Coggan, M. (1989). Isolation of a cDNA clone and localization of the human glutathione S-transferase 3 genes to chromosome bands 11q13 and 12q13–14. *Ann. Hum. Genet.* **53,** 205–213.

Burset, M., Seledtsov, I. A., and Solovyev, V. V. (2000). Analysis of canonical and non-canonical splice sites in mammalian genomes. *Nucl. Acids Res.* **28,** 4364–4375.

Chelvanayagam, G., Parker, M. W., and Board, P. G. (1997). Fly fishing for GSTs: A unified nomenclature for mammalian and insect glutathione transferases. *Chemico-Biological Interactions* **133,** 256–260.

Dinarello, C. A. (1998). Interleukin-1, interleukin-1 receptors and interleukin-1 receptor antagonist. *Int. Rev. Immunol.* **16,** 457–499.

Dixon, D. P., Davis, B. G., and Edwards, R. (2002). Functional divergence in the glutathione transferase superfamily in plants. Identification of two classes with putative functions in redox homeostasis in *Arabidopsis thaliana. J. Biol. Chem.* **277,** 30859–30869.

Dulhunty, A., Gage, P., Curtis, S., Chelvanayagam, G., and Board, P. (2001). The glutathione transferase structural family includes a nuclear chloride channel and a ryanodine receptor calcium release channel modulator. *J. Biol. Chem.* **276,** 3319–3323.

Fahn, S., and Cohen, G. (1992). The oxidant stress hypothesis in Parkinson's disease: Evidence supporting it. *Ann. Neurol.* **32,** 804–812.

Fornai, F., Gesi, M., Saviozzi, M., Lenzi, P., Piaggi, S., Ferrucci, M., and Casini, A. (2001a). Immunohistochemical evidence and ultrastructural compartmentalization of a new antioxidant enzyme in the rat substantia nigra. *J. Neurocytol.* **30,** 97–105.

Fornai, F., Piaggi, S., Gesi, M., Saviozzi, M., Lenzi, P., Paparelli, A., and Casini, A. F. (2001b). Subcellular localization of a glutathione-dependent dehydroascorbate reductase within specific rat brain regions. *Neuroscience* **104,** 15–31.

Fornai, F., Saviozzi, M., Piaggi, S., Gesi, M., Corsini, G. U., Malvaldi, G., and Casini, A. F. (1999). Localization of a glutathione-dependent dehydroascorbate reductase within the central nervous system of the rat. *Neuroscience* **94,** 937–948.

Girardini, J., Amirante, A., Zemzoumi, K., and Serra, E. (2002). Characterization of an omega-class glutathione S-transferase from Schistosoma mansoni with glutaredoxin-like dehydroascorbate reductase and thiol transferase activities. *Eur. J. Biochem.* **269,** 5512–5521.

Harrop, S. J., De Maere, M. Z., Fairlie, W. D., Reztsova, T., Valenzuela, S. M., Mazzanti, M., Tonini, R., Qiu, M. R., Jankova, L., Warton, K., Bauskin, A. R., Wu, W. M., Pankhurst, S., Campbell, T. J., Breit, S. N., and Curmi, P. M. (2001). Crystal structure of a soluble form of the intracellular chloride ion channel CLIC1 (NCC27) at 1.4-A resolution. *J. Biol. Chem.* **276**, 44993–45000.

Hogquist, K. A., Nett, M. A., Unanue, E. R., and Chaplin, D. D. (1991). Interleukin 1 is processed and released during apoptosis. *Proc. Natl. Acad. Sci. USA* **88**, 8485–8489.

Holmgren, A., and Aslund, F. (1995). Glutaredoxin. *Methods Enzymol.* **252**, 283–292.

Ishikawa, T., Casini, A. F., and Nishikimi, M. (1998). Molecular cloning and functional expression of rat liver glutathione-dependent dehydroascorbate reductase. *J. Biol. Chem.* **273**, 28708–28712.

Johansson, A. S., and Mannervik, B. (2001). Human glutathione transferase A3-3, a highly efficient catalyst of double-bond isomerization in the biosynthetic pathway of steroid hormones. *J. Biol. Chem.* **276**, 33061–33065.

Kanaoka, Y., Fujimori, K., Kikuno, R., Sakaguchi, Y., Urade, Y., and Hayaishi, O. (2000). Structure and chromosomal localization of human and mouse genes for hematopoietic prostaglandin D synthase. Conservation of the ancestral genomic structure of sigma-class glutathione S-transferase. *Eur. J. Biochem.* **267**, 3315–3322.

Kodym, R., Calkins, P., and Story, M. (1999). The cloning and characterization of a new stress response protein. A mammalian member of a family of theta class glutathione s-transferase-like proteins. *J. Biol. Chem.* **274**, 5131–5137.

Kolsch, H., Linnebank, M., Lutjohann, D., Jessen, F., Wullner, U., Harbrecht, U., Thelen, K. M., Kreis, M., Hentschel, F., Schulz, A., von Bergmann, K., Maier, W., and Heun, R. (2004). Polymorphisms in glutathione S-transferase omega-1 and AD, vascular dementia, and stroke. *Neurology* **63**, 2255–2260.

Kortemme, T., and Creighton, T. E. (1995). Ionisation of cysteine residues at the termini of model alpha-helical peptides. Relevance to unusual thiol pKa values in proteins of the thioredoxin family. *J. Mol. Biol.* **253**, 799–812.

Ladner, J. E., Parsons, J. F., Rife, C. L., Gilliland, G. L., and Armstrong, R. N. (2004). Parallel evolutionary pathways for glutathione transferases: Structure and mechanism of the mitochondrial class kappa enzyme rGSTK1-1. *Biochemistry* **43**, 352–361.

Laliberte, R. E., Perregaux, D. G., Hoth, L. R., Rosner, P. J., Jordan, C. K., Peese, K. M., Eggler, J. F., Dombroski, M. A., Geoghegan, K. F., and Gabel, C. A. (2003). Glutathione s-transferase omega 1-1 is a target of cytokine release inhibitory drugs and may be responsible for their effect on interleukin-1beta posttranslational processing. *J. Biol. Chem.* **278**, 16567–16578.

Li, Y. J., Oliveira, S. A., Xu, P., Martin, E. R., Stenger, J. E., Scherzer, C. R., Hauser, M. A., Scott, W. K., Small, G. W., Nance, M. A., Watts, R. L., Hubble, J. P., Koller, W. C., Pahwa, R., Stern, M. B., Hiner, B. C., Jankovic, J., Goetz, C. G., Mastaglia, F., Middleton, L. T., Roses, A. D., Saunders, A. M., Schmechel, D. E., Gullans, S. R., Haines, J. L., Gilbert, J. R., Vance, J. M., and Pericak-Vance, M. A. (2003). Glutathione S-transferase omega-1 modifies age-at-onset of Alzheimer disease and Parkinson disease. *Hum. Mol. Genet.* **12**, 3259–3267.

Lim, C. E. L., Matthaei, K. I., Blackburn, A. C., Davis, R. P., Dahlstrom, J. E., Koina, M. E., Anders, M. W., and Board, P. G. (2004). Mice deficient in glutathione transferase zeta/maleylacetoacetate isomerase exhibit a range of pathological changes and elevated expression of glutathione transferases regulated by antioxidant response elements. *Am. J. Pathol.* **165**, 679–693.

Maellaro, E., Del Bello, B., Sugherini, L., Santucci, A., Comporti, M., and Casini, A. F. (1994). Purification and characterization of glutathione-dependent dehydroascorbate reductase from rat liver. *Biochem. J.* **301**, 471–476.

Mannervik, B., Awasthi, Y. C., Board, P. G., Hayes, J. D., Di Ilio, C., Ketterer, B., Listowsky, I., Morgenstern, R., Muramatsu, M., Pearson, W. R., Pickett, C. B., Sato, K., Widersten,

M., and Wolf, C. R. (1992). Nomenclature for human glutathione transferases [letter]. *Biochem. J.* **282**, 305–306.

Mannervik, B., and Danielson, U. H. (1988). Glutathione transferases—structure and catalytic activity. *CRC Crit. Rev. Biochem.* **23**, 283–337.

Mariot, P., Prevarskaya, N., Roudbaraki, M. M., Le Bourhis, X., Van Coppenolle, F., Vanoverberghe, K., and Skryma, R. (2000). Evidence of functional ryanodine receptor involved in apoptosis of prostate cancer (LNCaP) cells. *Prostate* **43**, 205–214.

Marnell, L. L., Garcia-Vargas, G. G., Chowdhury, U. K., Zakharyan, R. A., Walsh, B., Avram, M. D., Kopplin, M. J., Cebrian, M. E., Silbergeld, E. K., and Aposhian, H. V. (2003). Polymorphisms in the human monomethylarsonic acid (MMA V) reductase/hGSTO1 gene and changes in urinary arsenic profiles. *Chem. Res. Toxicol.* **16**, 1507–1513.

Mattson, M. P. (2004). Pathways towards and away from Alzheimer's disease. *Nature* **430**, 631–639.

Morel, F., Rauch, C., Petit, E., Piton, A., Theret, N., Coles, B., and Guillouzo, A. (2004). Gene and protein characterization of the human glutathione S-transferase kappa and evidence for a peroxisomal localization. *J. Biol. Chem.* **279**, 16246–16253.

Mount, S. M. (2000). Genomic sequence, splicing, and gene annotation. *Am. J. Hum. Genet.* **67**, 788–792.

Pan, Z., Damron, D., Nieminen, A. L., Bhat, M. B., and Ma, J. (2000). Depletion of intracellular Ca^{2+} by caffeine and ryanodine induces apoptosis of Chinese hamster ovary cells transfected with ryanodine receptor. *J. Biol. Chem.* **275**, 19978–19984.

Pemble, S. E., Wardle, A. F., and Taylor, J. B. (1996). Glutathione S-transferase class Kappa: Characterization by the cloning of rat mitochondrial GST and identification of a human homologue. *Biochem. J.* **319**, 749–754.

Perregaux, D., Barberia, J., Lanzetti, A. J., Geoghegan, K. F., Carty, T. J., and Gabel, C. A. (1992). IL-1 beta maturation: Evidence that mature cytokine formation can be induced specifically by nigericin. *J. Immunol.* **149**, 1294–1303.

Rice, M. E. (2000). Ascorbate regulation and its neuroprotective role in the brain. *Trends Neurosci.* **23**, 209–216.

Robinson, A., Huttley, G. A., Booth, H. S., and Board, P. G. (2004). Modelling and bioinformatics studies of the human Kappa class Glutathione Transferase predict a novel third Glutathione Transferase family with homology to prokaryotic 2-hydroxychromene-2-carboxylate (HCCA) Isomerases. *Biochem. J.* **379**, 541–552.

Ross, V. L., Board, P. G., and Webb, G. C. (1993). Chromosomal mapping of the human Mu class glutathione S-transferases to 1p13. *Genomics* **18**, 87–91.

Rouimi, P., Anglade, P., Benzekri, A., Costet, P., Debrauwer, L., Pineau, T., and Tulliez, J. (2001). Purification and characterization of a glutathione S-transferase Omega in pig: Evidence for two distinct organ-specific transcripts. *Biochem. J.* **358**, 257–262.

Schmuck, E. M., Board, P. G., Whitbread, A. K., Tetlow, N., Blackburn, A. C., and Masoumi, A. (2005). Characterization of the monomethylarsonate reductase and dehydroascorbate reductase activities of Omega class glutathione transferase variants. Implications for arsenic metabolism and age-at-onset of Alzheimer's and Parkinson's diseases. *Pharmacogenet. Genomics* **15**, 493–501.

Sei, Y., Gallagher, K. L., and Basile, A. S. (1999). Skeletal muscle type ryanodine receptor is involved in calcium signaling in human B lymphocytes. *J. Biol. Chem.* **274**, 5995–6002.

Simonian, N. A., and Coyle, J. T. (1996). Oxidative stress in neurodegenerative diseases. *Annu. Rev. Pharmacol. Toxicol.* **36**, 83–106.

Simons, P. C., and Vander Jagt, D. L. (1977). Purification of glutathione S-transferases from human liver by glutathione-affinity chromatography. *Anal. Biochem.* **82**, 334–341.

Tanaka-Kagawa, T., Jinno, H., Hasegawa, T., Makino, Y., Seko, Y., Hanioka, N., and Ando, M. (2003). Functional characterization of two variant human GSTO 1-1s (Ala140Asp and Thr217Asn). *Biochem. Biophys. Res. Commun.* **301**, 516–520.

Thanaraj, T. A., and Clark, F. (2001). Human GC-AG alternative intron isoforms with weak donor sites show enhanced consensus at acceptor exon positions. *Nucl. Acids Res.* **29**, 2581–2593.

Tomarev, S. I., Zinovieva, R. D., Guo, K., and Piatigorsky, J. (1993). Squid glutathione S-transferase. Relationships with other glutathione S-transferases and S-crystallins of cephalopods. *J. Biol. Chem.* **268**, 4534–4542.

Webb, G., Vaska, V., Coggan, M., and Board, P. (1996). Chromosomal localization of the gene for the human theta class glutathione transferase (GSTT1). *Genomics* **33**, 121–123.

Wells, W. W., Xu, D. P., and Washburn, M. P. (1995). Glutathione: Dehydroascorbate oxidoreductases. *Methods Enzymol.* **252**, 30–38.

Whitbread, A. K., Tetlow, N., Eyre, H. J., Sutherland, G. R., and Board, P. G. (2003). Characterization of the human Omega class glutathione transferase genes and associated polymorphisms. *Pharmacogenetics* **13**, 131–144.

Wilson, R., Ainscough, R., Anderson, K., Baynes, C., Berks, M., Bonfield, J., Burton, J., Connell, M., Copsey, T., Cooper, J., Coulson, A., Craxton, M., Dear, S., Du, Z., Durbin, R., Favello, A., Fraser, A., Fulton, L., Gardner, A., Green, P., Hawkins, T., Hillier, L., Jier, M., Johnston, L., Jones, M., Kershaw, J., Kirsten, J., Laisster, N., Latreille, P., Lightning, J., Lloyd, C., Mortimore, B., O'Callaghan, M., Parsons, J., Percy, C., Rifken, L., Roopra, A., Saunders, D., Shownkeen, R., Sims, M., Smaldon, N., Smith, A., Smith, M., Sonnhammer, E., Staden, R., Sulston, J., Thierry-Mieg, J., Thomas, K., Vaudin, M., Vaughan, K., Waterston, R., Watson, A., Weinstock, L., Wilkinson-Sproat, J., and Wohldman, P. (1994). 2.2 Mb of contiguous nucleotide sequence from chromosome III of *C. elegans*. *Nature* **368**, 32–38.

Xu, D. P., Washburn, M. P., Sun, G. P., and Wells, W. W. (1996). Purification and characterization of a glutathione dependent dehydroascorbate reductase from human erythrocytes. *Biochem. Biophys. Res. Commun.* **221**, 117–121.

Yin, Z. L., Dahlstrom, J. E., Le Couteur, D. G., and Board, P. G. (2001). Immunohistochemistry of omega class glutathione S-transferase in human tissues. *J. Histochem. Cytochem.* **49**, 983–987.

Yu, J., Yu, D. W., Checkla, D. M., Freedberg, I. M., and Bertolino, A. P. (1993). Human hair keratins. *J. Invest. Dermatol.* **101**, 56S–59S.

Yu, L., Kalla, K., Guthrie, E., Vidrine, A., and Klimecki, W. T. (2003). Genetic variation in genes associated with arsenic metabolism: Glutathione S-transferase omega 1-1 and purine nucleoside phosphorylase polymorphisms in European and indigenous Americans. *Environ. Health Perspect.* **111**, 1421–1447.

Zakharyan, R. A., and Aposhian, H. V. (1999). Enzymatic reduction of arsenic compounds in mammalian systems: The rate-limiting enzyme of rabbit liver arsenic biotransformation is MMA(V) reductase. *Chem. Res. Toxicol.* **12**, 1278–1283.

Zakharyan, R. A., Sampayo-Reyes, A., Healy, S. M., Tsaprailis, G., Board, P. G., Liebler, D. C., and Aposhian, H. V. (2001). Human monomethylarsonic acid (MMA(V)) reductase is a member of the glutathione-S-transferase superfamily. *Chem. Res. Toxicol.* **14**, 1051–1107.

Further Reading

Hayes, J. D., Flanagan, J. U., and Jowsey, I. R. (2005). Glutathione transferases. *Annu. Rev. Pharmacol. Toxicol.* **45**, 51–88.

[6] Alternative Splicing of Glutathione S-Transferases

By JANTANA WONGSANTICHON and ALBERT J. KETTERMAN

Abstract

This chapter discusses the alternative splicing of glutathione S-transferase proteins, including current investigations of enzymatic, nonenzymatic functions, as well as structural differences between the alternatively spliced products. The data demonstrate that the different GST splice forms possess different properties, both in their catalytic function and in the effects of their protein–protein interactions.

Introduction

Alternative splicing is a potent regulatory mechanism in higher eukaryotes to generate different transcript isoforms from a single gene by differential incorporation of exons into mature mRNAs. This mechanism is often regulated in temporal patterns depending on either developmental or tissue-specific determinants. In humans, recent genomic and bioinformatic analysis indicates that approximately 35–65% of human genes are alternatively spliced (Graveley, 2001; Mironov *et al.*, 1999; Modrek and Lee, 2002; Sorek *et al.*, 2004). Generally, alternative splice events that take place in the protein-coding region will generate different primary sequences, and, therefore, the resulting proteins exhibit functional diversities. For example, in a study of the *Dscam* gene from *Drosophila melanogaster* that codes for a cell surface protein involved in neuronal connectivity, as many as 38,016 different mRNA isoforms can be generated from the single gene through alternative splicing mechanisms (Neves *et al.*, 2004; Wojtowicz *et al.*, 2004). The unique profile of *Dscam* isoforms in each individual cell type seems to specify cell identity with possible roles in the nervous system.

Glutathione S-transferases (GSTs) are ubiquitous in nature, found in most aerobic eukaryotes and prokaryotes. Cytosolic GSTs have been classified into at least 13 classes from mammals, plants, insects, parasites, fungus, as well as bacteria (Chelvanayagam *et al.*, 2001; Ketterer, 2001; Sheehan *et al.*, 2001). A vast diversity of GSTs allows the enzyme superfamily to perform various enzymatic and nonenzymatic functions. As an example, GSTs comprise a number of various isoforms that recognize at least 100 different xenobiotic chemicals (Hayes and Pulford, 1995). Alternative splicing is, therefore, only one mechanism used to generate functional heterogeneity of GSTs.

METHODS IN ENZYMOLOGY, VOL. 401
Copyright 2005, Elsevier Inc. All rights reserved.
0076-6879/05 $35.00
DOI: 10.1016/S0076-6879(05)01006-2

Alternative Splicing of GSTs

Currently identified spliced transcripts of GSTs seem to share the same N-terminus that is involved in the binding and activation of glutathione. The diverse C-terminal region of the full-length GSTs is responsible for different specificities toward electrophilic hydrophobic compounds. The alternative splicing mechanism generally gives rise to functional diversities of proteins and generates the complex proteome in metazoan organisms. Consequently, the alternatively spliced GSTs may also contribute to different important physiological functions. The alternative splicing mechanism seems to take place at a comparable frequency in mammals, flies, and worms (Brett *et al.*, 2002). Therefore, it is probable that the alternative splicing mechanism for GST genes is also more prevalent than previously thought.

Mu Class

Alternative splicing was first hypothetically suggested in Mu class GSTs based on two partial transcripts from a human testis cDNA library (Ross and Board, 1993), but no further data have been obtained. However, gene structure predictions from the human genome by Ensembl automated annotation pipelines (Curwen *et al.*, 2004; Potter *et al.*, 2004), together with corresponding verifications from external identifiers, suggest alternative splicing events occur in GSTM1 and GSTM4 as illustrated in Fig. 1. In hGSTM1, there is exon skipping at exon 7, which generates two different transcript isoforms with the size of 218 and 181 amino acids. The two isoforms have protein similarity matches in the Genbank database with accession numbers of AAA59203 (Seidegård *et al.*, 1988) and AAH24005 (Strausberg *et al.*, 2002), respectively. In general, the longer transcript from most of the alternatively spliced genes is considered an ancestral form that is produced from constitutively spliced exons, whereas the shorter form arises from an exon-skipping event evolved in a later evolutionary period (Ast, 2004; Kondrashov and Koonin, 2003).

In another splicing mechanism, an alternate in-frame exon in the 3' coding region (exon 8 or 9) each with a different 3' UTR are brought into play in hGSTM4, which generates two different transcripts of 218 and 195 amino acids (Fig. 1B). The matching protein sequences from Genbank database are AAA57346 (Comstock *et al.*, 1993) and AAA58623 (Ross *et al.*, 1993), respectively. In fact, predicted splice variants from genomic and EST data could represent either real functional forms or perhaps noise. The noise or aberrant splicing from the prediction possibly enables the evolution of new functional forms (Lareau *et al.*, 2004). Nevertheless, when the predicted splice isoforms have corresponding sequences in public databases, this suggests an alternative splicing in these genes does occur.

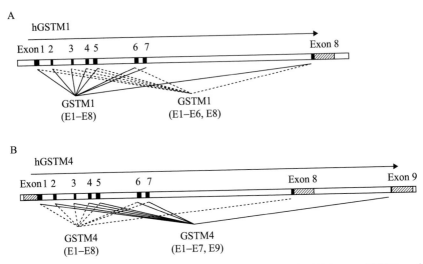

FIG. 1. Schematic diagram of predicted alternative splicing in (A) human GSTM1 and (B) GSTM4. The shaded boxes indicate untranslated regions (UTR), whereas solid boxes are coding regions.

Omega Class

A recent study of the genomic organization of Ov-GST3, classified as an omega class GST, from the human parasitic nematode *Onchocerca volvulus* revealed three different alternatively spliced transcripts (Kampkötter *et al.*, 2003). The differential splicing of exons 4–6 (Fig. 2) generates splicing transcripts of OvGST3-1, OvGST3-2, and OvGST3-3 with sizes of 145, 239, and 198 amino acids, respectively. An investigation of the genomic DNA organization and reverse transcriptase-polymerase chain reaction (RT-PCR) was performed to verify the existence of all three isoforms as functional transcripts. Among the three *O. volvulus* GSTs, OvGST1–OvGST3, OvGST3 was shown to be the stress-responsive transcript that was significantly up-regulated under toxic oxidant stimulation (Liebau *et al.*, 2000). Variable alternatively spliced transcripts of these particular filarial GSTs might confer advantages for the parasitic defense against different environmental stress, as well as serving as host immune effector molecules.

Sigma Class

In an annotation and phylogeny study of the *Anopheles gambiae* genome, an alternatively spliced gene from this Dipteran species was identified (Ding *et al.*, 2003). The resulting transcripts were classified as sigma class GSTs by a comparative analysis of the *Drosophila melanogaster* and *Anopheles gambiae* genomes. The genomic organization in Fig. 3 shows the two splicing products of GSTS1-1 and GSTS1-2 with the sizes of 200 and 195 amino acids, respectively. However, functional studies of these two isoforms still need to be performed.

Delta Class

Perhaps the best-characterized alternatively spliced GSTs are from an orthologous delta class gene in the Anopheline mosquitoes, *Anopheles gambiae* or *Anopheles dirus* (Pongjaroenkit *et al.*, 2001; Ranson *et al.*, 1998). Investigations of the genomic organization (Fig. 4), as well as the putative splice sites (Table I), demonstrate the splicing mechanism in the *An. gambiae aggstlα* gene and the *An. dirus adgstlAS1* gene is highly conserved. However, an investigation of the *Drosophila* genome reveals no such mechanism in any of the cytosolic GSTs, even though the two Diptera, *Anopheles gambiae* and *Drosophila melanogaster*, diverged only approximately 250 million years ago with considerable similarities in their proteomes (Zdobnov *et al.*, 2002). It is thought that *Anopheles gambiae* and

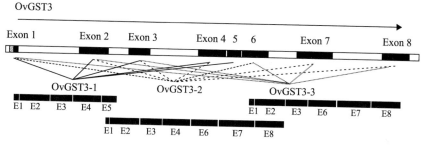

FIG. 2. Schematic diagram of alternative splicing in an Omega class GST from *O. volvulus*. The shaded box indicates a UTR region, whereas solid boxes are coding regions.

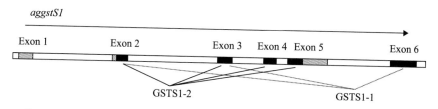

FIG. 3. Schematic diagram of the alternative splicing in a Sigma class GST from *An. gambiae*. The shaded boxes indicate UTR regions whereas solid boxes are coding regions.

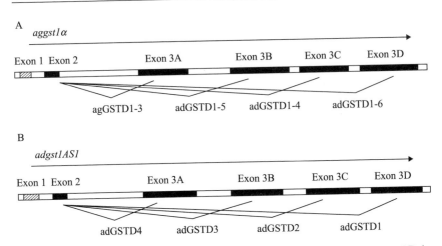

FIG. 4. Schematic diagram of orthologous alternative splicing in a gene member of Delta class GSTs from (A) *An. gambiae* and (B) *An. dirus*. The shaded box indicates 5' UTR region, whereas solid boxes are coding regions.

TABLE I

THE EXON COMPOSITION AND SPLICE SITES OF ALTERNATIVE TRANSCRIPTS
COMPARED BETWEEN THE *AN. GAMBIAE AGGST1A* GENE AND THE *AN. DIRUS ADGST1AS1*
GENE (PONGJAROENKIT *ET AL.*, 2001; RANSON *ET AL.*, 1998)

Gene	Exon composition consensus	5' Splice site (exon/intron) AG/GTRAGT	3' Splice site (intron/exon) Y_nNYAG/NN
aggst1α	exon1....exon2	TG/GTGAGT	CCCCAG/AA
	exon2....exon3A	AG/GTAGGT	AAAG/CT
	exon2....exon3B		CCTAG/AT
	exon2....exon3C		TTTGTAG/CT
	exon2....exon3D		TTTTCTAG/CT
adgst1AS1	exon1....exon2	CG/GTGAGT	CTCGCAG/AA
	exon2....exon3A	AG/GTAAGT	TTTAAAG/CT
	exon2....exon3B		CCCTCAG/AT
	exon2....exon3C		TCCGCAG/CT
	exon2....exon3D		ATTACAG/CT

(R = A or G, Y = C or T, N = A, C, G, or T).

Anopheles dirus have diverged in the last 2 or 3 million years. This suggests that the alternative splicing has recently occurred and is being conserved in the very close evolutionarily related species for physiologically important reasons. Each alternative transcript shares the same exon 2 at the

N-terminus, which is about one fifth of the entire protein, demonstrating a significant role of this domain for glutathione binding. In addition to genomic investigation, each alternatively spliced product was confirmed by RT-PCR to determine the coding sequences and predicted spliced sites. A comparison of the *aggst1α* and *adgst1AS1* genes reveals an identical exon/intron arrangement (Fig. 4). Comparing coding regions between these orthologous genes shows high nucleotide identities ranging between 80–90%, although the size and sequence of the introns vary (Pongjaroenkit *et al.*, 2001).

Structural Impact of Spliced Variants

The spliced variants of the Anopheline GSTs are mainly used for discussion in this and the following section because of the advantages of available structural and functional information over alternatively spliced GSTs from other classes.

Primary Sequences

The primary amino acid sequence alignment of *An. gambiae* and *An. dirus* alternatively spliced GSTs (Fig. 5A) suggests these two species are highly evolutionarily related, although *An. gambiae* is found in Africa and *An. dirus* in South East Asia. Amino acid comparisons of GST splicing products from the two mosquito species show greater than 60% identity (Fig. 5B). Comparisons between species orthologous isoforms showed higher identity than between splicing isoforms within a species, which suggests that the conservation of each splicing isoform might possess a distinctive function or response to different kinds of xenobiotic compounds.

Tertiary Structures

Currently, there are three available crystal structures of the Anopheline GST spliced products: adgstD3, adgstD4, and recently aggstD1-6 (Chen *et al.*, 2003; Oakley *et al.*, 2001). These three GST isoforms represent three of the four splice products from the orthologous genes and possess overall sequence identities greater than 60%. The tertiary structures are incredibly similar (Fig. 6), although the active site conformations are different, as shown in Fig. 7. Active site topologies of adgstD3 and aggstD1-6 are nearly identical, because not only do their amino acid sequences share 80% identity and 90% similarity but also the amino acid residues that form the active site pocket share 95% identity. This means that only 2 of the 44 amino acid residues that make up the pocket are different. In addition, the two different residues are functionally conserved, changing Thr158 and

A

```
                 *        20         *        40         *        60         *
adgstD1   : MDFYYLPGSAPCRAVQMTAAALGVELNLKLTDLMAGEHMKPEFLKLNPQHCIPTLDD-NGFSLWESRAIQIYLVE : 74
aggstD1-6 : MDFYYLPGSAPCRAVQMTAAAVGVELNLKLTDLMAGEHMKPEFLKLNPQHCIPTLVD-NGFALWESRAIQIYLAE : 74
adgstD2   : MDFYYLPGSAPCRAVQMTAAAVGVELNLKLTDLMAGEHMKPEFLKLNPQHCVPTLVD-DGFALCESRAIMCYLVE : 74
aggstD1-4 : MDFYYLPGSAPCRAVQMTAAAVGVELNLKLTDLMAGEHMKPEFLKLNPQHCVPTLVD-SGFALWESRAIMCYLVE : 74
adgstD3   : MDFYYLPGSAPCRAVQMTAAAVGVELNLKLTDLMAGEHMKPEFLKINPQHCIPTLVD-NGFALWESRAICTYLAE : 74
aggstD1-5 : MDFYYLPGSAPCRAVQMTAAAVGVELNLKLTDLMAGEHMKPEFLKINPQHCIPTLVD-NGFALWESRAICTYLAE : 74
adgstD4   : MDFYYLPGSAPCRAVQMTAAAVGVELNLKLTDLMAGEHMKPEFLKLNPQHCIPTLVDEDGFVLWESRAIQIYLVE : 75
aggstD1-3 : MDFYYLPGSAPCRAVQMTAAAVGVELNLKLTDLMKGEHMKPEFLKLNPQHCIPTLVDEDGFVLWESRAIQIYLVE : 75

              80          *        100         *        120         *        140         *
adgstD1   : KYGKDD-----KLYPKDPQKRAVVNQRLFFDMGTLYQAFGDYLYPQIBAKQP--ANAENEKRMKEAVGFLNTFLE : 142
aggstD1-6 : KYGKDD-----KLYPKDPQKRAVVNQRLYFDMGTLYQRFADYHYPQIBAKQP--ANPEDEKRMKDAVGLNTFLE : 142
adgstD2   : KYGKPIE--ADRLLPSDPQRRAIVNQRLYFDMGTLYQRFGDYYYPQIBEGAA--ASEALYARIGEALTFLDTFLE : 145
aggstD1-4 : KYGKPCN--NDSLYPTDPQKRAIVNQRLYFDMGTLYQRFGDYYYPQIBEGAP--ANEAIFARIGEALAFLDTFLE : 145
adgstD3   : KYGKDD-----KLYPKDPQKRAVVNQRLYFDMGTLYQRFADYYYPQIBAKQP--ANAENEKRMKDAVDFLNTFLD : 142
aggstD1-5 : KYGKDD-----KLYPKDPQKRAVVNQRLYFDMGTLYQRFADYYYPQIBAKQP--ANPEDEQRMKDAVGFLNSFLD : 142
adgstD4   : KYCAHDADLAERLYPSDPRRRAVVHQRLFFDVAVLYQRFAEYYYPQIBGQKVPVGDPGRLRSTEQALEFLNTFLE : 150
aggstD1-3 : KYCAHDPALAERLYPGDPRRRAVVHQRLFFDVAILYQRFAEYYYPQIBGKKV-AGDPDRLRSTEQALEFLNTFLE : 149

             160          *        180         *        200         *        220
adgstD1   : GQ-EYAAG-SDLTIADLSLAASNPTYEVAGFDFAPYPNVAAWLARCKANAFCYALNQAGADEDKAKFMS------ : 209
aggstD1-6 : GQ-EYAAG-NDLTIADLSLAATIATYEVAGFDFAPYPNVAAWFARCKANAPCYALNQAGADEDKAKELS------ : 209
adgstD2   : GDAKFVAGGDSFSLADISVYATLTTFEVAGHDFSAIGIVLRWIKSMAGTIEGADMDRSWAEAARPFFDRIKH--- : 217
aggstD1-4 : GE-RFVAGGNGYSLADISLYATLTTFEVAGYDFSAYNVLRWYKSMPELIEASDTDRSWAEAARPFFDKVKH---- : 216
adgstD3   : GH-KYVAG-DSLTIADLTVLATVSTYDVAGFELAKMPHVAAWYERTRKEAPGAAIMEACIEEDRKYBEK------ : 209
aggstD1-5 : GH-KYVAG-DSLTIADLSILATISTYDVAGFDLAKMQHVAAWYENIRKEAPGAAIMQACIEEDKKYBEK------ : 209
adgstD4   : GE-QYVAGGDDPTIADLSILATIATYEVAGYDLRRYENVQRWVERTSAIVEGADKDVEGAKVEGRYBTQK----- : 219
aggstD1-3 : GE-RFVAGGDDPTIADFSILASIAFDAAGYDLRRYENIHRWYEQTGNIARAADKDLAGAKIEGLYBRQK----- : 218
```

B

	aggstD1-6	adgstD2	aggstD1-4	adgstD3	aggstD1-5	adgstD4	aggstD1-3
adgstD1	**91%** **95%**	61% 73%	62% 74%	77% 87%	76% 86%	63% 73%	60% 72%
aggstD1-6		61% 75%	65% 75%	80% 90%	82% 91%	65% 76%	63% 76%
adgstD2			**85%** **91%**	63% 78%	61% 79%	62% 74%	59% 71%
aggstD1-4				62% 78%	63% 80%	59% 74%	60% 74%
adgstD3					**92%** **97%**	64% 78%	61% 77%
aggstD1-5						64% 77%	63% 77%
adgstD4							**86%** **92%**

FIG. 5. An amino acid sequence comparison of alternatively spliced delta class GSTs from *An. dirus* and *An. gambiae*. (A) Sequence alignment. (B) Matrix table of percent identities (top line) and percent similarities (bottom line). Percent identities of orthologous splicing transcripts are shown in bold. (Genbank accession numbers are adgstD1 AF273041; adgstD2 AF273038; adgstD3 AF273039; adgstD4 AF273040; aggstD1-3 AAC79992; aggstD1-4 AAC79994; aggstD1-5 AAC79993; and aggstD1-6 AAC79995.)

FIG. 6. Stereo view of a structural superimposition of splicing products of *Anopheles dirus* adgstD3 (black), *Anopheles dirus* adgstD4 (grey), and *Anopheles gambiae* aggstD1-6 (white). *An. dirus* adgstD3 and adgstD4 as well as *An. gambiae* aggstD1-6 have Protein Data Bank accession numbers 1JLV, 1JLW, and 1PN9, respectively.

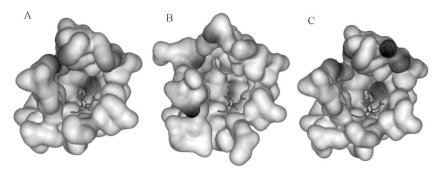

FIG. 7. Active site topology of the splicing products from (A) *Anopheles dirus* adgstD3, (B) *Anopheles dirus* adgstD4, and (C) *Anopheles gambiae* aggstD1-6. The three tertiary structures were superimposed to illustrate active sites in an identical view. Ball and stick representations show the glutathione (or S-hexyl glutathione for aggstD1-6) in the active site pocket.

Tyr206 in adgstD3 to Ser158 and Phe207 in adgstD1-6, respectively. The active site pocket of adgstD4 enzyme is apparently more unique because of the diverse amino acid residues used to form the active site. The distinct active site topology provides diverse enzyme specificity toward xenobiotics and insecticides. Anopheline delta class GSTs, as well as epsilon class, have been implicated in detoxication particularly in insecticide resistance (Ortelli *et al.*, 2003; Prapanthadara *et al.*, 2000; Ranson *et al.*, 1997).

Intersubunit Interaction

Structural investigations of available crystal structures of Anopheline alternatively spliced GSTs demonstrate several dimeric interactions along the subunit interfaces. Amino acid residues involved in the intersubunit interactions in *An. gambiae* aggstD1-6 structure are similar to *An. dirus* adgstD3, although they are not orthologous enzymes; therefore, no representation is shown here. The intersubunit interactions of the observed structures involve two main locations: the center of the twofold axis and both ends of the twofold axis.

Intersubunit Interaction in the Center of the Twofold Axis

TOP REGION. There is an ionic interaction across the dimer interface at the top of the twofold axis of adgstD4 with E116 from one subunit and R134 of the other subunit but absent in the equivalent residues in adgstD3 (Fig. 8A).

MIDDLE REGION. Hydrophobic interactions are prevalent in the middle of the dimeric structure primarily by a counterpart aromatic residue (Y98 in adgstD3, F104 in adgstD4) from both subunits that hook around each other with an offset pi–pi interaction in a "clasp-like" arrangement (Fig. 8B). There are also several hydrophobic residues (A67, L97, and M101 in adgstD3; A68, L103, and V107 in adgstD4) involved generating a lock-and-key motif by surrounding the aromatic "clasp" residues (which form the "key"). However, this motif is particularly different from that demonstrated in alpha/mu/pi GST classes (Hornby *et al.*, 2002; Sayed *et al.*, 2000; Stenberg *et al.*, 2000), in which the lock-and-key motif is formed at either end of the twofold axis. In addition, a unique feature of the delta GST is that the "key" residue itself also acts as the "lock" for the other subunit. The lock-and-key "clasp" interaction is highly conserved in other nonalternatively spliced delta class GSTs such as adgstD5, adgstD6, and LcGST from *Lucilia cuprina*.

BOTTOM REGION. An ionic interaction is not only present in the top region, but also the bottom region in the center of the twofold axis with E75 in adgstD4 interacting with R96 from the other subunit. This interaction is also present in adgstD3 with E74 interacting with R90 from the other subunit (Fig. 8C).

Intersubunit Interaction at Either End of the Twofold Axis. There is an additional ionic interaction for adgstD4 at either end of the twofold axis with R94 from one subunit interacting with D57 and D59 from the other subunit (Fig. 8D). This interaction is absent for the equivalent residues of adgstD3.

The alternatively spliced adgstD3 and adgstD4 seem to possess differences in the dimeric interface such that there are two additional ionic interactions that occur at the dimer interface of adgstD4 but do not occur

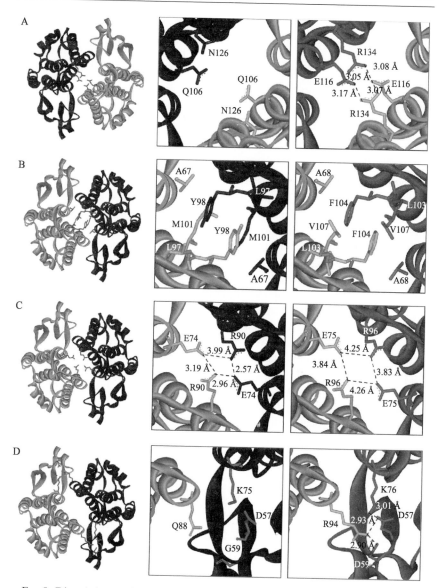

FIG. 8. Dimeric interactions along the subunit interface compared between adgstD3 (left boxes) and adgstD4 (right boxes). The structural representations in B, C, and D were horizontally rotated by 180° for a better view.

in the equivalent positions in adgstD3. The role of ionic interactions in proteins is generally to stabilize the tertiary and quaternary structures. This is supported by previous thermal stability experiments that showed the half-life at 45° to be 3.8 and 12.9 min for adgstD3 and adgstD4, respectively (Wongsantichon *et al.*, 2003; Wongtrakul *et al.*, 2003).

Functional Impact of Splicing

Substrate Specificity

All Anopheline GST spliced isoforms are able to catalyze the common CDNB substrate with different catalytic efficiencies. Table II shows kinetic properties of all four spliced isoforms from the *An. dirus* adgst1AS1 gene and two of the spliced isoforms from the *An. gambiae* aggst1α gene. Kinetic characteristics of each spliced variant are diverse in either affinity toward substrates (K_m) or turn over number (k_{cat}). Available kinetic information for aggstD1-6 and aggstD1-5 are also shown to compare their catalytic properties with the orthologous *An. dirus* enzymes, adgstD1 and adgstD3, respectively. Although amino acid identities of orthologous GSTs from both Anopheline species are greater than 90%, the enzymes display catalytic differences. Moreover, crystal structure shows the active site residues in adgstD1-6 and adgstD3 to be nearly identical. As discussed in the previous section, the 44 active site residues differ by only two functionally conserved amino acids. However, the kinetic data show the enzymes possess very different kinetic properties, illustrating a continuing evolution of the GSTs after species divergence.

Because of a major role of GST enzymes in detoxication processes, many electrophilic hydrophobic compounds have been used to characterize

TABLE II
KINETIC PARAMETERS OF ALTERNATIVELY SPLICED DELTA CLASS GSTS FROM *AN. DIRUS* AND
AN. GAMBIAE. (JIRAJAROENRAT *ET AL.*, 2001; KETTERMAN *ET AL.*, 2001; RANSON *ET AL.*, 1997)

Enzyme	V_{max} (μmol/min/mg)	k_{cat} (s^{-1})	K_m (mM)		k_{cat}/K_m (mM^{-1} s^{-1})	
			GSH	CDNB	GSH	CDNB
adgstD1	12.9 ± 0.63	5.0	0.86 ± 0.18	0.10 ± 0.03	6	48
adgstD2	63.9 ± 3.50	25.9	1.30 ± 0.15	0.21 ± 0.03	20	121
adgstD3	67.5 ± 1.97	26.9	0.40 ± 0.05	0.10 ± 0.01	67	269
adgstD4	40.3 ± 1.89	16.9	0.83 ± 0.08	0.52 ± 0.07	20	32
aggstD1-5	83.51	40.47	0.822	0.099	49	410
aggstD1-6	348.35	136.2	0.807	0.123	120	792

TABLE III
SUBSTRATE SPECIFICITY OF ALTERNATIVELY SPLICED DELTA CLASS GSTs FROM *AN. DIRUS* AND *AN. GAMBIAE*. (RANSON *ET AL.*, 1997; UDOMSINPRASERT *ET AL.*, 2004)

Enzyme	Specific activity (μmol/min/mg of protein)			DDTase[a] activity
	CDNB (1 mM)	DCNB (1 mM)	PNPB (0.1 mM)	
adgstD1	6.54 ± 0.53	0.070 ± 0.002	nd	0.95
adgstD2	45.1 ± 3.41	0.177 ± 0.006	0.047 ± 0.010	1.87 ± 0.82
adgstD3	67.5 ± 1.97	0.312 ± 0.023	nd	2.66 ± 0.29
adgstD4	41.8 ± 1.40	0.042 ± 0.011	0.023 ± 0.002	7.50 ± 1.68
aggstD1-5	56.44 ± 8.71	0.326 ± 0.035	<0.15	4.80 ± 0.09
aggstD1-6	195.14 ± 11.95	0.636 ± 0.026	<0.15	7.71 ± 0.72

[a] The DDTase activity is expressed as nmol DDE formation/mg of enzyme protein. CDNB, 1-chloro 2,4-dinitrobenzene; DCNB, 1,2-chloro 4-nitrobenzene; PNPB, *p*-nitrophenethyl bromide; PNBC, *p*-nitrobenzyl chloride. Nd, no detectable activity.

catalytic specificities of the enzymes. Table III demonstrates that all four alternatively spliced GSTs from *Anopheles dirus*, as well as aggstD1-5 and adgstD1-6 from *Anopheles gambiae*, possess various specificities toward the substrates tested. There is almost negligible activity toward PNPB, a specific substrate for theta class GSTs; hence, these alternatively spliced GSTs do not evidently belong to the theta class as initially classified before the complete *Anopheles gambiae* and *Drosophila melanogaster* genomes became available.

DDTase activity was also investigated because of its potential role in insecticide resistance and was found to be elevated in resistant insect strains (Ortelli *et al.*, 2003; Prapanthadara *et al.*, 1995). Most of the alternatively spliced Anopheline GSTs are able to metabolize DDT to different extents. Noticeably, whereas aggstD1-6 possesses the greatest DDTase activity, the orthologous *An. dirus* enzyme adgstD1 possesses the lowest. Therefore, a high primary sequence identity does not correlate with catalytic properties of the enzymes. Moreover, the nearly identical active site pockets of adgstD3 and aggstD1-6 do not correlate with substrate specificities.

Permethrin inhibition of CDNB activity shows that adgstD2, adgstD3, and adgstD4 possess similar affinities (K_i) in the range of 9–53 μM, and each alternatively spliced isoform displays different types of inhibition kinetics (Jirajaroenrat *et al.*, 2001). The types of inhibition in adgstD2, adgstD3, and adgstD4 are uncompetitive, noncompetitive, and competitive, respectively. Dissimilar inhibition patterns imply that permethrin interacts differently with the individual isoforms and possibly with different binding sites on each isoform.

Regulation of the JNK Pathway

Studies of mammalian GST Pi and Mu revealed that several GSTs were involved in the regulation of signaling pathways by protein–protein interactions. For example, mammalian GST Pi was first demonstrated to inhibit JNK (c-Jun N-terminal kinase, a member of the mitogen activated stress kinase family-MAPK) activity in a dose-dependent manner (Adler *et al.*, 1999), and the JNK inhibition was suggested to occur by a direct interaction of GST Pi to the C-terminal of JNK (Wang *et al.*, 2001). Mammalian GST Mu was shown to inhibit ASK1 (Apoptosis Signal-regulating Kinase 1, a member of the mitogen-activated stress kinase family-MAPKKK) activity by binding of the C-terminus of GST Mu to the N-terminus of ASK1 (Cho *et al.*, 2001). This suggests that different GST classes can possibly interact with different stress kinase proteins in the MAP kinase pathway.

A study of the interactions of the four alternatively spliced delta class GSTs from *Anopheles dirus* to components of the Diptera JNK pathway revealed that individual isoforms differentially interact with *Drosophila* HEP (Hemipterous, upstream activator of JNK and a member of the mitogen-activated stress kinase family-MAPKK) and *Drosophila* JNK (Udomsinprasert *et al.*, 2004). HEP and JNK seemed to have effects on GST activity by inhibiting activities of adgstD2, adgstD3, and adgstD4 to various extents. The activity of adgstD1 could only be inhibited by HEP, but not JNK. Furthermore, all four alternatively spliced GSTs also had effects on the protein kinase activities. The adgstD2, adgstD3, and adgstD4 increased the ability of JNK to phosphorylate Jun (a substrate for JNK), whereas adgstD1 inhibited JNK activity by 50% in the presence of HEP, JNK, and Jun. Without Jun in the reaction, effects of individual GST isoforms on JNK and HEP phosphorylation also were observed, because JNK is a substrate for HEP and HEP is a substrate for JNK. Results showed that adgstD2 and adgstD3 can increase phosphorylation of both JNK and HEP, whereas adgstD1 and adgstD4 inhibit the phosphorylation of HEP but not JNK. This suggests that the different alternatively spliced isoforms of GST can act as positive and negative regulators of the JNK signaling pathway. The different protein–protein interactions between the alternatively spliced GSTs and components of the JNK pathway might be significant in regulation of the stress kinase proteins in response to stress and under normal conditions for maintenance of basal activity.

Conclusion

Alternative splicing is one mechanism used in the GST superfamily to increase the diversity of isoforms that can metabolize a broad spectrum of compounds and perform nonenzymatic functions such as a regulatory role in a signal transduction pathway. The data from the splice forms show that not only are the amino acid residues in the active site pocket significant to the catalytic function of the enzymes, but residues outside the active site pocket also are crucial for catalytic efficiencies and determining substrate specificities. These data again highlight the functional diversity of the GST splice forms. Other than the enzymatic and nonenzymatic properties, discussed herein, alternative splicing mechanisms of GSTs that have taken place during the course of evolution might have significant roles in additional functions that still remain to be explained.

Acknowledgments

This work was supported by the Thailand Research Fund (TRF). JW was supported by a Royal Golden Jubilee (RGJ) scholarship.

References

Adler, V., Yin, Z., Fuchs, S. Y., Benezra, M., Rosario, L., Tew, K. D., Pincus, M. R., Sardana, M., Henderson, C. J., Wolf, C. R., Davis, R. J., and Ronai, Z. (1999). Regulation of JNK signaling by GSTp. *EMBO J.* **18,** 1321–1334.
Ast, G. (2004). How did alternative splicing evolve? *Nat. Rev. Genet.* **5,** 773–782.
Brett, D., Pospisil, H., Valcárcel, J., Reich, J., and Bork, P. (2002). Alternative splicing and genome complexity. *Nat. Genet.* **30,** 29–30.
Chelvanayagam, G., Parker, M. W., and Board, P. G. (2001). Fly fishing for GSTs: A unified nomenclature for mammalian and insect glutathione transferases. *Chem. Biol. Interact.* **133,** 256–260.
Chen, L., Hall, P. R., Zhou, X. E., Ranson, H., Hemingway, J., and Meehan, E. J. (2003). Structure of an insect d-class glutathione S-transferase from a DDT-resistant strain of the malaria vector *Anopheles gambiae. Acta Cryst. D.* **59,** 2211–2217.
Cho, S.-G., Lee, Y. H., Park, H.-S., Ryoo, K., Kang, K. W., Park, J., Eom, S.-J., Kim, M. J., Chang, T.-S., Choi, S.-Y., Shim, J., Kim, Y., Dong, M.-S., Lee, M.-J., Kim, S. G., Ichijo, H., and Choi, E.-J. (2001). Glutathione *S*-transferase Mu modulates the stress-activated signals by suppressing apoptosis signal-regulating kinase 1. *J. Biol. Chem.* **276,** 12749–12755.
Comstock, K. E., Johnson, K. J., Rifenbery, D., and Henner, W. D. (1993). Isolation and analysis of the gene and cDNA for a human Mu class glutathione S-transferase, GSTM4. *J. Biol. Chem.* **268,** 16958–16965.
Curwen, V., Eyras, E., Andrews, T. D., Clarke, L., Mongin, E., Searle, S. M. J., and Clamp, M. (2004). The Ensembl automatic gene annotation system. *Genome Res.* **14,** 950.
Ding, Y., Ortelli, F., Rossiter, L. C., Hemingway, J., and Ranson, H. (2003). The *Anopheles gambiae* glutathione transferase supergene family: Annotation, phylogeny and expression profiles. *BMC Genomics* **4,** 35–50.
Graveley, B. R. (2001). Alternative splicing: Increasing diversity in the proteomic world. *Trends Genet.* **17,** 100–107.
Hayes, J. D., and Pulford, D. J. (1995). The glutathione S-transferase supergene family: Regulation of GST and the contribution of the isoenzymes to cancer chemoprotection and drug resistance. *CRC Crit. Rev. Biochem. Molec. Biol.* **30,** 445–600.

Hornby, J. A. T., Codreanu, S. G., Armstrong, R. N., and Dirr, H. W. (2002). Molecular recognition at the dimer interface of a class Mu glutathione transferase: Role of a hydrophobic interaction motif in dimer stability and protein function. *Biochemistry* **41**, 14238–14247.

Jirajaroenrat, K., Pongjaroenkit, S., Krittanai, C., Prapanthadara, L., and Ketterman, A. J. (2001). Heterologous expression and characterization of alternatively spliced glutathione S-transferases from a single *Anopheles* gene. *Insect Biochem. Mol. Biol.* **31**, 867–875.

Kampkötter, A., Volkmann, T. E., de Castro, S. H., Leiers, B., Klotz, L.-O., Johnson, T. E., Link, C. D., and Henkle-Dührsen, K. (2003). Functional analysis of the glutathione S-transferase 3 from *Onchocerca volvulus* (Ov-GST-3): A parasite GST confers increased resistance to oxidative stress in *Caenorhabditis elegans*. *J. Mol. Biol.* **325**, 25–37.

Ketterer, B. (2001). A bird's eye view of the glutathione transferase field. *Chem. Biol. Interact.* **138**, 27–42.

Ketterman, A. J., Prommeenate, P., Boonchauy, C., Chanama, U., Leetachewa, S., Promtet, N., and Prapanthadara, L. (2001). Single amino acid changes outside the active site significantly affect activity of glutathione S-transferases. *Insect Biochem. Mol. Biol.* **31**, 65–74.

Kondrashov, F. A., and Koonin, E. V. (2003). Evolution of alternative splicing: Deletions, insertions and origin of functional parts of proteins from intron sequences. *Trends Genet.* **19**, 115–119.

Lareau, L. F., Green, R. E., Bhatnagar, R. S., and Brenner, S. E. (2004). The evolving roles of alternative splicing. *Curr. Opin. Struct. Biol.* **14**, 273–282.

Liebau, E., Eschbach, M.-L., Tawe, W., Sommer, A., Fischer, P., Walter, R. D., and Henkle-Dührsen, K. (2000). Identification of a stress-responsive *Onchocerca volvulus* glutathione S-transferase (*Ov*-GST-3) by RT-PCR differential display. *Mol. Biochem. Parasitol.* **109**, 101–110.

Mironov, A. A., Fickett, J. W., and Gelfand, M. S. (1999). Frequent alternative splicing of human genes. *Genome Res.* **9**, 1288–1293.

Modrek, B., and Lee, C. (2002). A genomic view of alternative splicing. *Nat. Genet.* **30**, 13–19.

Neves, G., Zucker, J., Daly, M., and Chess, A. (2004). Stochastic yet biased expression of multiple *Dscam* splice variants by individual cells. *Nat. Genet.* **36**, 240–246.

Oakley, A. J., Harnnoi, T., Udomsinprasert, R., Jirajaroenrat, K., Ketterman, A. J., and Wilce, M. C. J. (2001). The crystal structures of glutathione S-transferases isozymes 1–3 and 1–4 from *Anopheles dirus* species B. *Protein Science* **10**, 2176–2185.

Ortelli, F., Rossiter, L. C., Vontas, J., Ranson, H., and Hemingway, J. (2003). Heterologous expression of four glutathione transferase genes genetically linked to a major insecticide-resistance locus from the malaria vector *Anopheles gambiae*. *Biochem. J.* **373**, 957–963.

Pongjaroenkit, S., Jirajaroenrat, K., Boonchauy, C., Chanama, U., Leetachewa, S., Prapanthadara, L., and Ketterman, A. J. (2001). Genomic organization and putative promotors of highly conserved glutathione S-transferases originating by alternative splicing in *Anopheles dirus*. *Insect Biochem. Mol. Biol.* **31**, 75–85.

Potter, S. C., Clarke, L., Curwen, V., Keenan, S., Mongin, E., Searle, S. M. J., Stabenau, A., Storey, R., and Clamp, M. (2004). The Ensembl analysis pipeline. *Genome Res.* **14**, 934–941.

Prapanthadara, L., Ketterman, A. J., and Hemingway, J. (1995). DDT-resistance in *Anopheles gambiae* giles from Zanzibar Tanzania based on increased DDT-dehydrochlorinase activity of glutathione S-transferases. *Bull. Entomol. Res.* **85**, 267–274.

Prapanthadara, L., Promtet, N., Koottathep, S., Somboon, P., and Ketterman, A. J. (2000). Isoenzymes of glutathione S-transferase from the mosquito Anopheles dirus species B: The purification, partial characterization and interaction with various insecticides. *Insect Biochem. Mol. Biol.* **30**, 395–403.

Ranson, H., Collins, F., and Hemingway, J. (1998). The role of alternative mRNA splicing in generating heterogeneity within the *Anopheles gambiae* class I glutathione S-transferase family. *Proc. Natl. Acad. Sci. USA* **95**, 14284–14289.

Ranson, H., Prapanthadara, L., and Hemingway, J. (1997). Cloning and characterization of two glutathione S-transferases from a DDT-resistant strain of *Anopheles gambiae*. *Biochem. J.* **324**, 97–102.

Ross, V. L., and Board, P. G. (1993). Molecular cloning and heterologous expression of an alternatively spliced human Mu class glutathione S-transferase transcript. *Biochem. J.* **294**, 373–380.

Sayed, Y., Wallace, L. A., and Dirr, H. W. (2000). The hydrophobic lock-and-key intersubunit motif of glutathione transferase A1-1: Implications for catalysis, ligandin function and stability. *FEBS Lett.* **465**, 169–172.

Seidegård, J., Vorachek, W. R., Pero, R. W., and Pearson, W. R. (1988). Hereditary differences in the expression of the human glutathione transferase active on *trans*-stilbene oxide are due to a gene deletion. *Proc. Natl. Acad. Sci. USA* **85**, 7293–7297.

Sheehan, D., Meade, G., Foley, V. M., and Dowd, C. A. (2001). Structure, function and evolution of glutathione transferases: Implications for classification of non-mammalian members of an ancient enzyme superfamily. *Biochem. J.* **360**, 1–16.

Sorek, R., Shamir, R., and Ast, G. (2004). How prevalent is functional alternative splicing in the human genome. *Trends Genet.* **20**, 68–71.

Stenberg, G., Abdalla, A.-M., and Mannervik, B. (2000). Tyrosine 50 at the subunit interface of dimeric human glutathione transferase P1-1 is a structural key residue for modulating protein stability and catalytic function. *Biochem. Biophys. Res. Comm.* **271**, 59–63.

Strausberg, R. L., Feingold, E. A., Grouse, L. H., Derge, J. G., Klausner, R. D., Collins, F. S., Wagner, L., Shenmen, C. M., Schuler, G. D., Altschul, S. F., Zeeberg, B., Buetow, K. H., Schaefer, C. F., Bhat, N. K., Hopkins, R. F., Jordan, H., Moore, T., Max, S. I., Wang, J., Hsieh, F., Diatchenko, L., Marusina, K., Farmer, A. A., Rubin, G. M., Hong, L., Stapleton, M., Soares, M. B., Bonaldo, M. F., Casavant, T. L., Scheetz, T. E., Brownstein, M. J., Usdin, T. B., Toshiyuki, S., Carninci, P., Prange, C., Raha, S. S., Loquellano, N. A., Peters, G. J., Abramson, R. D., Mullahy, S. J., Bosak, S. A., Mcewan, P. J., McKernan, K. J., Malek, J. A., Gunaratne, P. H., Richards, S., Worley, K. C., Hale, S., Garcia, A. M., Gay, L. J., Hulyk, S. W., Villalon, D. K., Muzny, D. M., Sodergren, E. J., Lu, X. H., Gibbs, R. A., Fahey, J., Helton, E., Ketteman, M., Madan, A., Rodrigues, S., Sanchez, A., Whiting, M., Madan, A., Young, A. C., Shevchenko, Y., Bouffard, G. G., Blakesley, R. W., Touchman, J. W., Green, E. D., Dickson, M. C., Rodriguez, A. C., Grimwood, J., Schmutz, J., Myers, R. M., Butterfield, Y. S. N., Kryzywinski, M. I., Skalska, U., Smailus, D. E., Schnerch, A., Schein, J. E., Jones, S. J. M., and Marra, M. A. (2002). Generation and initial analysis of more than 15,000 full-length human and mouse cDNA sequences. *Proc. Natl. Acad. Sci. USA* **99**, 16899–16903.

Udomsinprasert, R., Bogoyevitch, M. A., and Ketterman, A. J. (2004). Reciprocal regulation of glutathione *S*-transferase spliceforms and the *Drosophila* c-Jun N-terminal Kinase pathway components. *Biochem. J.* **383**, 483–490.

Wang, T., Arifoglu, P., Ronai, Z., and Tew, K. D. (2001). Glutathione *S*-transferase P1-1 (GSTP1-1) inhibits c-Jun N-terminal kinase (JNK1) signaling through interaction with the C terminus. *J. Biol. Chem.* **276**, 20999–21003.

Wojtowicz, W. M., Flanagan, J. J., Millard, S. S., Zipursky, S. L., and Clemens, J. C. (2004). Alternative splicing of *Drosophila* Dscam generates axon guidance receptors that exhibit isoform-specific homophilic binding. *Cell* **118**, 619–633.

Wongsantichon, J., Harnnoi, T., and Ketterman, A. J. (2003). A sensitive core region in the structure of glutathione S-transferases. *Biochem. J.* **373**, 759–765.

Wongtrakul, J., Sramala, I., and Ketterman, A. (2003). A non-active site residue, cysteine 69, of glutathione S-transferase adGSTD3-3 has a role in stability and catalytic function. *Protein Peptide Lett.* **10,** 375–385.

Zdobnov, E. M., von Mering, C., Letunic, I., Torrents, D., Suyama, M., Copley, R. R., Christophides, G. K., Thomasova, D., Holt, R. A., Subramanian, G. M., Mueller, H. M., Dimopoulos, G., Law, J. H., Wells, M. A., Birney, E., Charlab, R., Halpern, A. L., Kokoza, E., Kraft, C. L., Lai, Z. W., Lewis, S., Louis, C., Barillas-Mury, C., Nusskern, D., Rubin, G. M., Salzberg, S. L., Sutton, G. G., Topalis, P., Wides, R., Wincker, P., Yandell, M., Collins, F. H., Ribeiro, J., Gelbart, W. M., Kafatos, F. C., and Bork, P. (2002). Comparative genome and proteome analysis of *Anopheles gambiae* and *Drosophila melanogaster. Science* **298,** 149–159.

Further Reading

Prapanthadara, L., Koottathep, S., Promtet, N., Hemingway, J., and Ketterman, A. J. (1996). Purification and characterization of a major glutathione S-transferase from the mosquito *Anopheles dirus* (species B). *Insect Biochem. Mol. Biol.* **26,** 277–285.

[7] Disruption of the Glutathione Transferase Pi Class Genes

By COLIN J. HENDERSON and C. ROLAND WOLF

Abstract

Glutathione transferases are a multi-gene family of enzymes responsible for the metabolism of a wide range of both endogenous and exogenous substrates. These polymorphic enzymes, which form part of an adaptive response to chemical and oxidative stress, are widely distributed and ubiquitously expressed and are subject to regulation by a number of structurally unrelated chemicals. One of these enzymes, GST P, has been the focus of much research in recent years in relation to its involvement in the etiology of disease, particularly cancer. As part of our research efforts into GST P, we have developed a mouse line that lacks this enzyme and have used this model to investigate the consequences of the absence of GST P on tumorigenesis, drug metabolism, and toxicity.

Introduction

Glutathione Transferases

Glutathione transferases, historically referred to as glutathione S-transferases, hence the common abbreviation GST, are a family of dimeric enzymes involved in phase II detoxification reactions (EC 2.5.1.18).

METHODS IN ENZYMOLOGY, VOL. 401
Copyright 2005, Elsevier Inc. All rights reserved.
0076-6879/05 $35.00
DOI: 10.1016/S0076-6879(05)01007-4

Cytosolic GSTs have been grouped according to their primary amino acid sequence into seven classes—alpha, mu, pi, sigma, theta, zeta, and omega—although a further, mitochondrial, class exists in mammalian cells (kappa), and other groups have also been described in nonmammalian systems (For a review, see Hayes *et al.* [2005]). The nomenclature system for GSTs has recently been updated (Mannervik *et al.*, 2005). Typically, GSTs in the same class share >40% sequence identity, and between classes this is <25%; GST display a broad range of overlapping substrate specificities, making identification of individual forms by catalytic activity problematic. Although GSTs are ubiquitously expressed, tissue distribution can vary widely between species; there is also a difference between fetal expression and that found in the adult. Furthermore, GST may be induced by a wide range of structurally unrelated compounds, some of which may be substrates, giving credence to the view that GST are part of an adaptive response to cellular stress (Hayes and McLellan, 1999; Hayes *et al.*, 2005; Townsend and Tew, 2003; Wolf, 2002). Overall, expression and regulation of GST is complex and highly species-dependent; additional complexities arise from the findings that many of the GST isoforms are polymorphic in man (see later).

GSTs catalyze the conjugation of the reduced tripeptide glutathione (GSH), the most common intracellular non-protein thiol, to a wide range of electrophilic compounds, both endogenous and exogenous. Together, GST and GSH form a vital defensive system against chemical agents and oxidative stress, in addition to maintaining the tertiary structure of proteins through thiol–disulfide exchange by interacting with protein disulfide isomerases and glutaredoxin (Eaton and Bammler, 1999; Hayes *et al.*, 2005; Salinas and Wong, 1999).

Although many chemical reactions have been ascribed to GST (Hayes *et al.*, 2005), much remains to be learned, particularly about their endogenous role(s). For many years, our laboratory has been interested in the relationship between a particular GST isoform, GST P, and carcinogenesis and drug resistance (Hayes and McLellan, 1999; Lewis *et al.*, 1988a,b; McLellan and Wolf, 1999). GST P was first described as a placental isoform, and later it was found, in rats, to be elevated in hepatic preneoplastic foci induced by chemical treatment (Satoh *et al.*, 1989). Since then, GST P has been the focus, first in animal models and subsequently in human tumors, of studies that have shown that this protein seems to play a pivotal role in the carcinogenic process and the development of drug resistance. This chapter will review the work done with a *GSTP* null mouse line, illustrating the power of this mouse model in characterizing GST P functions.

GST P

GST P was first identified as a placental isoform and subsequently shown by immunoblotting and immunohistochemistry to be present at variable levels in most, if not all, tissues; it is the most ubiquitous and highly expressed GST in mammals and, unlike other GST, is expressed in cell lines in culture. In human and rat, GST P is not expressed constitutively in liver, although it is found in mouse hepatocytes (Eaton and Bammler, 1999; Sherratt and Hayes, 2002). It is, however, inducible in human and rat hepatocytes by exogenous agents and in disease states. Sato and co-workers demonstrated nearly 20 years ago that GST P, which is found at very low levels in rat liver, was present at elevated levels in hepatic preneoplastic foci in rats treated with chemical agents (Satoh *et al.*, 1989) and could be used as an immunohistochemical marker for colon carcinoma (Kodate *et al.*, 1986) and possibly a wide range of other tumors (Tsuchida *et al.*, 1989). Howie *et al.* (1990), using a specific radioimmunoassay, showed that GST P was the predominant GST expressed in range of human tumors, including breast, stomach, lung, colon, kidney, and liver). Concomitant with these studies, GST P was found to be elevated in a wide range of cell lines made resistant to a variety of chemicals (Black and Wolf, 1991; Wareing *et al.*, 1993; Wolf *et al.*, 1990). Conversely, overexpression of GSTP in cell lines, or yeast, rendered them more resistant to the toxic effects of chemical agents such as Adriamycin or chlorambucil than the parental, wild-type, line (Black *et al.*, 1990; Whelan *et al.*, 1992). These studies were followed throughout the 1990s by a number of reports in humans, both *in vivo* and *in vitro*, which demonstrated that elevated GST P expression was found in a range of tumor types and that such elevated expression was invariably associated with a more aggressive tumor type, with a poor patient prognosis, and with resistance to chemotherapeutic treatments (Ali-Osman *et al.*, 1997; Arai *et al.*, 2000; Berendsen *et al.*, 1997; Gajewska and Szczypka, 1992; Hanada *et al.*, 1991; Mulder *et al.*, 1995; Schipper *et al.*, 1997; Zhang *et al.*, 1994) (Table I).

It should, however, be pointed out that not all reports have demonstrated such a relationship between GST P and tumorigenesis or drug resistance; indeed, a number of other studies failed to find elevated GST P expression in certain human tumors, and in one particular cancer, prostate, GST P expression was completely absent in malignant disease in contrast to normal or benign hyperplastic tissue because of hypermethylation of the GST P promotor (Lee *et al.*, 1994, 1997). Treatment of human prostate cancer cells with procainamide, a non-nucleoside inhibitor of DNA methyltransferases, reversed GSTP CpG island hypermethylation and restored GSTP expression (Lin *et al.*, 2001).

TABLE I
GST EXPRESSION IN HUMAN TUMORS

Disease	Significance	Reference
Prostate cancer	GST P1 may have diagnostic and prognostic utility in prostate cancer.	(Woodson *et al.*, 2004)
	Absence of GST P1 expression in (malignant) prostate cancer.	(Cookson *et al.*, 1997)
Head and neck squamous cell carcinoma (HNSCC)	GST P1 may be a clinically important predictor of survival.	(Shiga *et al.*, 1999)
	GST P1 up-regulated to protect HNSCC from cytotoxic effects of NO• in tumor.	(Bentz *et al.*, 2000)
Acute nonlymphoblastic leukemia	Significant correlation between GST P1 expression at diagnosis and outcome.	(Tidefelt *et al.*, 1992)
Nasopharyngeal carcinoma	GST P1 expression associated with metastasis.	(Jayasurya *et al.*, 2002)
Breast cancer	GST P1 positive tumors more aggressive with poorer prognosis.	(Huang *et al.*, 2003)
	GST P1 elevated in breast cancer cells resistant to anticancer agents.	(Molina *et al.*, 1993)
	GST P1 positivity = poor prognosis.	(Su *et al.*, 2003)
Barrett's esophagus and esophageal carcinoma	Loss of GST P1 expression may have an important role in the development and progression of this disease.	(Brabender *et al.*, 2002)
Colorectal cancer	Increased levels of GST P1 = poor prognosis	(Mulder *et al.*, 1995)
Skin cancer	GST P1 involved in the process of skin carcinogenesis.	(Shimizu *et al.*, 1995b)
Non-small cell lung cancer	GST P1 expression related to chemotherapy response.	(Arai *et al.*, 2000)
	Increased activity levels of GST P1 in tumors.	(Di Ilio *et al.*, 1988)
Ovarian cancer	GST P1 levels may be useful for monitoring patients with ovarian cancer during the course of treatment.	(Ghalia *et al.*, 2000)
Gastric carcinoma	GST P1 levels in normal and carcinomatous gastric tissue have prognostic impact on patient survival.	(Schipper *et al.*, 1997)
Malignant mesothelioma	Expression of GST P1 correlates positively with increased survival.	(Segers *et al.*, 1996)
Testicular cancer	GST P1 may contribute to drug resistance in testicular cancer.	(Katagiri *et al.*, 1993)
Renal cell carcinoma	GST P1 as a prognostic indicator in renal cell carcinoma.	(Grignon *et al.*, 1994)
Malignant melanoma	GST P1 expressed in malignant melanoma.	(Moral *et al.*, 1997)
Bladder cancer	Over-expression of GST P1 suggested tumor acquires enhanced detoxification properties.	(Berendsen *et al.*, 1997)
Glioma	Positive correlation between level of GST P1 expression and tumor grade.	(Ali-Osman *et al.*, 1997)
Endometrial carcinoma	GST P1 expression reflects metastatic potential of endometrial carcinomas.	(Yokoyama *et al.*, 1998)

GST P has also been associated with other diseases; for instance, the Fanconi anemia complementation group C (FANCC) protein has been shown to increase GST P activity after apoptosis by preventing disulfide bond formation and thus protein oxidation (Brodeur *et al.*, 2004; Cumming *et al.*, 2001). Table I lists a selection of cancers with which GST P has been found to be associated, either positively or negatively.

The finding that *GSTP* is a single gene, highly conserved in evolution, that the promotor region contains a CpG island, that it is invariably expressed in cell lines in culture, and that it is expressed in most somatic cells *in vivo* indicates that this gene has important physiological functions.

GST P is present in most species as a single gene, *GSTP1*; however, in mice, there are two *GSTP* genes, *mGSTP1* and *mGSTP2*, each containing seven exons and approximately 3 kb in length, which lie adjacent to one another on chromosome 19 separated by less than 3 kb (Bammler *et al.*, 1994). Interestingly, the presence of a third, pseudo, *GSTP* gene in the mouse has been suspected (T. Bammler and C. R. Wolf, 1993, unpublished), and recent bioinformatic analysis seems to support this with the finding of a *GSTP*-like gene approximately 14 kb upstream of *mGSTP1* and *mGSTP2* on chromosome 19 (M. Mitchell and C.J.H., 2005, unpublished). Close examination of the syntenic chromosome region in human, chromosome 11, reveals the possibility of a similar arrangement of *GSTP* genes, although this remains unproven as yet (M. Mitchell and C.J.H., 2005, unpublished).

In the mouse, GST P1 is much more catalytically active than GST P2, the six amino acid changes between GST P1 and GST P2 rendering the latter essentially catalytically inactive (Bammler *et al.*, 1995). In addition, GST P2 is expressed at significantly lower levels than GST P1, and GST P1 is expressed in a sexually differentiated manner, at least in the liver, with female mice exhibiting at least tenfold lower GST P1 than males (McLellan and Hayes, 1987). GST P is not known to be sexually differentiated in other species.

In humans, *GSTP1* is located on chromosome 11q13 and is polymorphic, with four major alleles reported—*GSTP1*A* (105I, 114A), *GSTP1*B* (105V, 114A), *GSTP1*C* (105V, 114V), and *GSTP1*D* (105I, 114V). The most common allele, Val^{105}, has higher catalytic activity toward polycyclic aromatic hydrocarbon diol epoxides by sevenfold but lowers the reaction with 1-chloro-2,4-dinitrobenzene by a factor of three (Hu *et al.*, 1997; Sundberg *et al.*, 1998). A large number of epidemiological studies have been carried out to link *GSTP1* allelic variants with disease. Fryer *et al.* (2000), investigating the chromosome 11q13 region as a "hot-spot" for asthma-related genes, found the frequency of *GSTP1* Val^{105}/Val^{105} was significantly lower in asthmatic subjects than in controls, a finding subsequently confirmed in another ethnic grouping (Aynacioglu et al., 2004). Henrion-Caude and co-workers (2002) found a link between liver disease associated with cystic fibrosis and homozygosity for Ile at position 105; patients with such a genotype were at eightfold higher risk of liver disease developing than other GSTP1 genotypes (Henrion-Caude et al., 2002). Similarly, in a study of chronic obstructive pulmonary disease (COPD), a significantly higher number of patients were found to be homozygous for GSTP1 Ile^{105} than control subjects (Ishii et al., 1999). Harries et al. (1997) found an increased number of individuals homozygous for the GSTP1*B allele in patients with testicular or bladder cancer and a decreased frequency of GSTP1*A homozygotes in prostate cancer samples (Harries et al., 1997). The same study also found an increased frequency of GSTP1*B homozygosity in lung cancer and COPD, although the changes were not statistically significant; however, a subsequent investigation found a significantly higher proportion of patients with lung cancer were homozygous for GSTP1*B (Ryberg et al., 1997).

Deletion of GST P

Targeting Strategy

To further characterize the potential role(s) of GST P in chemical toxicity, drug resistance, and carcinogenesis, we have created a mouse line in which this enzyme was deleted. The targeting construct was devised in such a manner as to delete either one or both GSTP genes in the mouse (Henderson et al., 1998; Fig. 1) and incorporated an IRES-βGEO element. The Internal Ribosome Binding Site (IRES) element (Mountford and Smith, 1995) is used in a promotorless construct and allows better expression of the βGEO unit, which expresses a fusion protein containing both a selectable marker (neomycin resistance) to allow tracking of homologous recombination events after gene targeting and a reporter (*lacZ*) that permits investigation, in the resultant mouse line, of GST P expression and regulation.

GSTP null mice were made where either one (GSTP1 null) or both GSTP genes (GSTP1/2 null mice) were inactivated; the latter model has been used in subsequent studies and investigations. The original mouse line was on a mixed (129xMF1) background, and it is carefully maintained by random breeding. Both GSTP null mouse lines have been crossed onto a defined genetic background (C57BL/6J) by back-crossing for greater than six generations; although we have worked on this line, nothing has yet been published from it.

GSTP locus on chromosome 19

Targeted locus in *GSTP1/2* null mice

FIG. 1. Genomic layout of *GSTP* genes in the mouse (top) and targeted deletion of *GSTP1* and *GSTP2* (bottom). En2a, splice acceptor site; IRES, internal ribosome entry site; *lacZ*, β-galactosidase reporter; SvpA, polyadenylation site.

Initial characterization, and subsequent detailed observations suggest that there are no obvious phenotypic changes in the *GSTP1/2* null mouse, relative to wild-type (Henderson *et al.*, 1998 and unpublished). There is no apparent difference in lifespan in the *GSTP1/2* null mouse, and it does not have an increased incidence of any particular physiological condition or tumor (C.J.H., 2000, unpublished). Histologically, those tissues that have been examined appear normal relative to wild-type mice, and although we have noted an increased lung/body weight in the *GSTP1/2* null mice, we have been unable to establish a reason for this (C.J.H., unpublished).

All tissues tested in *GSTP1/2* null mice exhibited a complete deletion of GstP, including skin (Henderson *et al.*, 1998; Fig. 2). Immunoblotting (Henderson *et al.*, 1998) looked at a number of other hepatic GST to see whether there were compensatory changes and failed to find any, a finding that was subsequently confirmed by proteomic investigation (Kitteringham *et al.*, 2003). However, recently, we have carried out Affymetrix microarray analysis and found that, at least in terms of RNA expression, other hepatic GST are expressed at higher levels in *GSTP1/2* null mice (C.J.H. and M.P. Chamberlain, 2004, unpublished); given the absence of increased GST protein, however, it is unlikely that these findings are of any great significance. There was no significant change in hepatic GSH levels or in any of the enzymes involved in GSH biosynthesis, including glutathione synthetase and γ-glutamylcysteine ligase, between GSTP1/P2 null mice and wild-types (Henderson *et al.*, 1998).

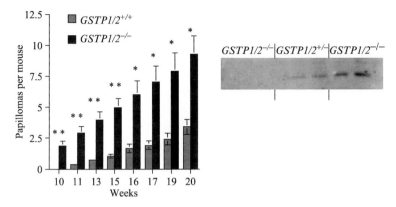

FIG. 2. Increased papilloma formation in *GSTP1/2* null mice in a two-stage skin tumorigenesis bioassay. 7,12-Dimethylbenz[a]anthracene (DMBA) (25 μg in 200 ml of acetone). The tumor-promoting agent 12-*O*-tetradodecanoyl-13-acetate (TPA) was then applied twice weekly (5×10^{-5} *M* in 200 ml acetone; 10 nmol) beginning 1 week after initiation and continuing for an additional 20 weeks. The papilloma incidence was monitored weekly. $n = 20$; ** $p < 0.0001$; * $p < 0.001$. Immunoblot shows skin cytosols from $GSTP1/2^{-/-}$, $GSTP1/2^{+/-}$, and $GSTP1/2^{-/-}$ mice probed with antiserum to rat GST P1.

GST P and Tumorigenesis

GST P is the predominant GST in human, rat, and mouse skin (Konohana *et al.*, 1990; Raza *et al.*, 1991; Shimizu *et al.*, 1995a). Given the suspected role of GST P in the carcinogenic process, we studied the effect of the absence of this protein in a two-stage skin tumorigenesis assay (Brown and Balmain, 1995) using 7,12-dimethylbenz[a]anthracene (DMBA) as initiator (single treatment of 25 μg) and 12-*O*-tetradecanoyl-phorbol-13-acetate (TPA) as a promotor (10 nmol weekly for 20 weeks). By 20 weeks after starting treatment, *GSTP1/2* null mice had a significantly higher number (3.5-fold) of skin papillomas than wild-type mice (Fig. 2; Henderson *et al.*, 1998). By 40 weeks, all the *GSTP1/2* null mice in the study had been sacrificed because of tumor burden, whereas approximately half of the wild-type mice were still alive, further emphasizing the difference between the two groups and illustrating the influence of GST P on chemically induced skin tumorigenesis (C.J.H., unpublished). The mechanism behind this phenomenon remains unclear; although the primary hypothesis would be that GST P is playing a detoxification role, this remains to be formally shown. We are currently using "genetically initiated" mice, such as the mutant *Ras* TgAC line (Leder *et al.*, 1990) to define which part of the tumorigenic process the absence of GSTP affects. Interestingly, subsequent analysis of DNA adducts in the skin of mice treated as described previously failed to demonstrate any significant difference between GSTP1/2 null and wild-type mice (C.J.H. and David Phillips, 1999, unpublished).

Niitsu and co-workers have shown a difference in chemically induced colon carcinogenesis between the *GSTP* null animals and their wild-type counterparts (Niitsu *et al.*, 2001). After treatment of mice with azoxymethane, a marked reduction in tumorigenesis was observed: *GSTP1/2* null ryptic foci (ACF) and no polyp formation in the colon compared with a significant incidence of ACF and polyps in the colon of wild-type mice (Niitsu *et al.*, 2001). A subsequent study showed that the presence of GST P1 seemed to protect ACF from apoptosis induced by deoxycholic acid, thus prolonging their survival and allowing the accumulation of mutations (Nobuoka *et al.*, 2004). These data indicate that GST P may influence carcinogenesis in different ways, depending on the site and the carcinogen. This is consistent with findings in prostate cancer where, unlike in most other tumors, a complete absence of GST P expression in malignant tissue is observed. The data do, however, demonstrate that GST P is involved in tumorigenesis, possibly as a consequence of the multiple functions of this protein.

We have recently investigated the role of GST P in lung tumorigenesis. GST P is the major form of GST in the human lung, being found predominantly in the Clara cells of the bronchiolar epithelium (Fryer *et al.*, 1986), and is also present in a similar localization in mouse pulmonary tissues (Forkert *et al.*, 1999). Interestingly, total pulmonary GST activity, measured by activity toward the broad GST substrate 1-chloro-2,4-dintrobenzene, was significantly reduced in the lungs of *GSTP1/2* null mice, indicating that GST P1 is also the major GST isoform in the mouse lung (Henderson *et al.*, 1998). GST P1 is known to be involved in the detoxification of the classes of chemicals found in tobacco smoke (i.e., polycyclic aromatic hydrocarbons) (Hu *et al.*, 1997; Sundberg *et al.*, 1998), and overexpression of GST P1 in human lung fibroblasts led to inhibition of the cytotoxic effects of cigarette smoke (Ishii *et al.*, 2001). Furthermore, epidemiological studies have shown linked polymorphisms in the *GSTP1* gene to certain subtypes of lung cancer (Harries *et al.*, 1997; Ryberg *et al.*, 1997), whereas GST P1 was found to have a protective effect against camptothecin-induced necrosis in a subset of human lung adenocarcinoma cells (Ishii *et al.*, 2004). *GSTP1/2* null mice were treated with benzo[a]pyrene, 3-methylcholanthrene (3MC) and urethane (vinyl carbamate), and the incidence of pulmonary adenomas assessed in these mice, relative to a group of wild-type mice treated in the same way. With all three chemicals tested, the mice lacking *GSTP1/2* were found to have significantly higher numbers of adenomas in the lungs (Henderson *et al.*, manuscript in preparation). Again, evidence was found that GST P may be functioning through more than one mechanism, because pulmonary DNA adducts were found to be higher in *GSTP1/2* null mice after administration of benzo[a]pyrene but not after treatment with 3MC. Interestingly, Ishii *et al.* (2003), using *GSTP1* anti-sense expression in human lung fibroblasts, found that depletion of *GSTP1* led to increased levels of apoptosis and necrosis (Ishii *et al.*, 2003).

GST P, Drug Metabolism, and Toxicity

In light of reports indicating that GST P is involved in the detoxification of chemical electrophiles, we have challenged mice with drugs or chemicals to determine whether their sensitivity to such agents was altered in the *GSTP1/2* null animals.

Acetaminophen (paracetamol) is a commonly used analgesic that has been thoroughly investigated over several decades (Anker and Smilkstein, 1994; Prescott, 1996). In addition to extensive sulfation and glucuronidation, a small proportion is metabolized, by the cytochrome P450 system (Patten *et al.*, 1993; Raucy *et al.*, 1989), to an active intermediate, N-acetylbenzoquinonimine, which is detoxified by conjugation to glutathione (Bergman *et al.*, 1996; Dahlin *et al.*, 1984). Under conditions of acetaminophen overdose, the hepatic cellular pool of reduced glutathione is exhausted, leading to an accumulation of N-acetylbenzoquinonimine. This reactive metabolite can bind irreversibly to protein and DNA, leading ultimately to cellular necrosis and death. The detoxification of N-acetylbenzoquinonimine, by conjugation with GSH, is very rapid; however, it was suggested that this reaction was catalyzed by GST P (Coles *et al.*, 1988). Treatment of the *GSTP1/2* null mice with acetaminophen would therefore be predicted to result in increased sensitivity to this compound; however, when treated, *GSTP1/2* null mice exhibited a profound resistance to a hepatotoxic dose of this drug (Henderson *et al.*, 2000). The mechanism of this effect is still unclear (see later) but may be due to a change in the ability of the *GSTP1/2* null mice to regulate glutathione homeostasis.

In certain tumor cell lines made resistant to platinum drugs, GST P1 is elevated. However, the role of GST P in the detoxification of such compounds is still unclear. Initial investigations of cisplatin toxicity in the *GSTP1/2* null mouse model have shown that treatment resulted in greater weight loss in mice lacking GSTP, with a concomitant elevation of serum markers for liver and kidney damage (alanine aminotransferase and creatinine, respectively) (Fig. 3). More recently, we have also shown similar differences with a newer platinum drug, oxaliplatin (Ritchie *et al.*, manuscript in preparation).

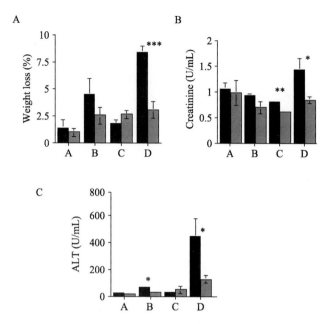

FIG. 3. Changes in body weight and serum toxicity markers in *GSTP1/2* null mice treated with cisplatin. Adult male *GSTP1/2* null mice (black bars) and wild-type mice (light bars) were treated intraperitoneally with cisplatin once (on day 1) at a dose of 5 mg (A) or 10 mg (B) or daily for 3 days at a dose of 5 mg (C) or 10 mg (D). Mice were sacrificed on day 4, body weight loss measured (A) and serum toxicity markers for kidney—creatinine—or liver—alanine aminotransferase (ALT)—were determined (B and C, respectively). $n = 5$; *** $p < 0.001$; ** $p < 0.005$; * $p < 0.05$.

GST P, Oxidative Stress, and Cellular Signaling

GST P has the capacity to make cells either more sensitive or resistant to chemicals, and the reason for this dichotomy remains unclear, However, in 1998 Ronai *et al.* published the results of a study in which they had isolated a small protein bound to Jun N-terminal kinase (JNK) that seemed to negatively regulate this enzyme (Adler *et al.*, 1999). Further analysis identified this protein as GST P1 in its monomeric form. GST P1 has subsequently been shown to bind to the C-terminus of JNK1 (Wang *et al.*, 2001). In a series of subsequent experiments, Ronai and co-workers demonstrated that oxidative stress could cause the dissociation of GST P1 from JNK, resulting in oligomerization of GST P1 and concomitant reversal of JNK inhibition with consequent phosphorylation of c-jun. In previous studies, we have shown that expression of the *GSTP1* gene is controlled by the binding of c-Jun to an AP-1 site in the promotor (Moffat *et al.*, 1994, 1996). Therefore, the

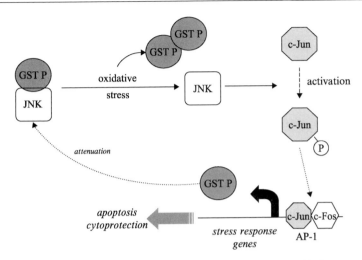

FIG. 4. Potential role for GST P in regulation of JNK function. GST P binds as a monomer to JNK, inhibiting function. Under conditions of oxidative stress, GST P is released from JNK and oligomerizes, allowing JNK to phosphorylate (and activate) c-Jun, which in turn activates a battery of genes, including GST P, involved in apoptosis and cytoprotection. Increased levels of (monomeric) GST P lead to attenuation of the response by rebinding to JNK and inhibiting function.

activation of c-jun on oxidative stress would, in theory, increase GST P1 expression with concomitant attenuation of the c-Jun signal (Fig. 4). This hypothesis, however, remains to be proven. Bernardini *et al.* demonstrated the appearance of a dimeric form of GST P1 after the induction of apoptosis by hydrogen peroxide or etoposide in Jurkat cells or a neuroblastoma cell line (Bernardini *et al.*, 2000, 2002), whereas Ishisaki *et al.* (2001) showed that suppression of *GSTP1* in PC12 cells by antisense expression resulted in augmentation of both dopamine-induced JNK activation and apoptosis (Ishisaki *et al.*, 2001). It has also been reported in mouse embryonic fibroblasts derived from JNK1 or JNK2 null mice that GST P1 expression is significantly suppressed (Chen *et al.*, 2002). As mentioned previously, on depletion of GST P1 in human lung fibroblasts, Ishii *et al.* (2003) found increased cell death, although there was no evidence of altered JNK activation). However, none of these studies have addressed the key issue of how the cell maintains a pool of monomeric GST P1, an enzyme previously shown to require dimerization for catalytic (i.e., glutathionylation) activity (Hayes *et al.*, 2005). Posttranslational modification, in the form of phosphorylation of a serine residue in the C-terminal region of GST P1, has been suggested as a possible mechanism whereby monomers of the enzyme could be formed (Ranganathan *et al.*, 2005). Interestingly, another GST, GST

M1, has been shown to bind to apoptosis signal-regulating kinase 1 (ASK1) and suppress its ability to activate the JNK and p38 signaling pathways (Cho et al., 2001). It is, therefore, likely that GST P1, and other GST, have additional, regulatory functions, which have yet to be fully characterized. In this context, it is of interest that GST P1 has recently been shown to be present in the nucleus under certain conditions (Kamada et al., 2004; Kelly et al., 2000) as has GST T1 (Sherratt et al., 1997, 1998).

Further support for a role for GST P in regulating the function of JNK came from Elsby et al. (2003), who found a significant increase in JNK activity in the liver and lung of GSTP1/2 null mice). Concomitant with this increase, the authors found significantly higher levels of AP-1 binding activity in the liver of GSTP1/2 null mice, as demonstrated by electrophoretic mobility shift analysis, and higher expression of mRNA for heme oxygenase-1, a known AP-1 target gene and a gene central to the oxidative stress response.

Novel Antitumor Agents Exploiting GST P Overexpression

The increased expression of GST P1 in many human tumors has been exploited to develop novel pro-drugs that interact with this enzyme. TLK286, also known as Telcyta, is a compound that, on metabolism by GST P1, releases a phosphorodiamidate moiety that is toxic to tumor cells and a glutathione analogue (glutathione vinyl sulfone) that remains bound to enzyme, thus preventing it from further activity toward other anticancer drugs (Morgan et al., 1998). Indeed, TLK286 has been shown to act synergistically with a variety of chemotherapeutic agents, including platinum drugs, taxanes, and anthracyclines. TLK286 apparently acts through induction of the pro-apoptotic stress response pathway—MEK4, JNK, and caspase 3—culminating in apoptosis. Using mouse embryo fibroblasts from the GSTP1/2 null mice, Rosario et al. demonstrated a twofold increase in resistance of tumor cells to TLK286 compared with wild-type cells, and a human cell line made resistant to TLK286 by chronic exposure to the compound had significantly reduced levels of GST P1 (Rosario et al., 2000). In *nude* mice carrying a human breast tumor xenograft, TLK286 inhibited tumor growth by 80% after daily treatment with the drug for 5 days (Kauvar et al., 1998). Recently, TLK286 has been evaluated in phase 1 clinical trials in ovarian, non-small cell lung, breast, and colorectal cancer with promising results (Rosen et al., 2003, 2004), and phase 2 trials are now in progress.

A peptidomimetic inhibitor of GST P1, TLK199 (γ-glutamyl-S-(benzyl) cysteinyl-R-phenyl glycine diethyl ester), has been shown to increase susceptibility to the chemotherapeutic effects of melphalan, both in cell lines *in vitro* and in tumor xenografts in *scid* mice (Morgan *et al.*, 1996). This further illustrates the potential for manipulation of GST P in augmenting the action of anticancer drugs.

Conclusions

Although much has been learned about GST in general, and GST P in particular, over the past two decades, much remains to be discovered about these enzymes and their functions, particularly in relation to polymorphisms, the regulation of expression, and their functions. It is clear that GST P is a key part of a cellular adaptive response to chemical or oxidative stress. Work to date has shown that GST P is potentially linked to a number of disease states, including cancer, although the mechanisms underlying the interaction with GST P remain to be resolved. The *GSTP1/2* null mouse provides a unique and powerful model system with which to further investigate the endogenous activities of this enzyme. Of particular interest is a potential role for GST P, and other GST, in the regulation of the cellular stress response and cell signaling pathways. Such effects would provide a link between GST P and fundamental cellular processes such as apoptosis.

Acknowledgments

The Molecular Pharmacology Unit is funded by Cancer Research UK. We thank Dr. Mark Chamberlain for Table I. Mike Mitchell of the Cancer Research UK Computational Genome Analysis Laboratory is thanked for bioinformatic analysis.

References

Adler, V., Yin, Z., Fuchs, S. Y., Benezra, M., Rosario, L., Tew, K. D., Pincus, M. R., Sardana, M., Henderson, C. J., Wolf, C. R., Davis, R. J., and Ronai, Z. (1999). Regulation of JNK signaling by GSTp. *EMBO J.* **18,** 1321–1334.

Ali-Osman, F., Brunner, J. M., Kutluk, T. M., and Hess, K. (1997). Prognostic significance of glutathione S-transferase pi expression and subcellular localization in human gliomas. *Clin. Cancer Res.* **3,** 2253–2261.

Anker, A. L., and Smilkstein, M. J. (1994). Acetaminophen—Concepts and controversies. *Concepts Controversies Toxicol.* **12,** 335–349.

Arai, T., Yasuda, Y., Takaya, T., Hayakawa, K., Toshima, S., Shibuya, C., Kashiki, Y., Yoshimi, N., and Shibayama, M. (2000). Immunohistochemical expression of glutathione transferase-pi in untreated primary non-small-cell lung cancer. *Cancer Detect. Prev.* **24,** 252–257.

Aynacioglu, A. S., Nacak, M., Filiz, A., Ekinci, E., and Roots, I. (2004). Protective role of glutathione s-transferase P1(ABTP1) Val105Val genotype in patients with bronchial asthma. *Br. J. Clin. Pharmacol.* **57,** 213–217.

Bammler, T. K., Driessen, H., Finnstrom, N., and Wolf, C. R. (1995). Amino acid differences at positions 10, 11, and 104 explain the profound catalytic differences between two murine pi-class glutathione S-transferases. *Biochemistry* **34,** 9000–9008.

Bammler, T. K., Smith, C. A., and Wolf, C. R. (1994). Isolation and characterization of two mouse Pi-class glutathione S-transferase genes. *Biochem. J.* **298**(Pt 2), 385–390.

Bentz, B. G., Haines, G. K., 3rd., and Radosevich, J. A. (2000). Glutathione S-transferase pi in squamous cell carcinoma of the head and neck. *Laryngoscope* **110**, 1642–1647.

Berendsen, C. L., Peters, W. H., Scheffer, P. G., Bouman, A. A., Boven, E., and Newling, D. W. (1997). Glutathione S-transferase activity and subunit composition in transitional cell cancer and mucosa of the human bladder. *Urology* **49**, 644–651.

Bergman, K., Muller, L., and Teigen, S. W. (1996). Series: Current issues in mutagenesis and carcinogenesis, No. 65. The genotoxicity and carcinogenicity of paracetamol: A regulatory (re)view. *Mutat. Res.* **349**, 263–288.

Bernardini, S., Bellincampi, L., Ballerini, S., Ranalli, M., Pastore, A., Cortese, C., and Federici, G. (2002). Role of GST P1-1 in mediating the effect of etoposide on human neuroblastoma cell line Sh-Sy5y. *J. Cell Biochem.* **86**, 340–347.

Bernardini, S., Bernassola, F., Cortese, C., Ballerini, S., Melino, G., Motti, C., Bellincampi, L., Iori, R., and Federici, G. (2000). Modulation of GST P1-1 activity by polymerization during apoptosis. *J. Cell Biochem.* **77**, 645–653.

Black, S. M., Beggs, J. D., Hayes, J. D., Bartoszek, A., Muramatsu, M., Sakai, M., and Wolf, C. R. (1990). Expression of human glutathione S-transferases in *Saccharomyces cerevisiae* confers resistance to the anticancer drugs Adriamycin and chlorambucil. *Biochem. J.* **268**, 309–315.

Black, S. M., and Wolf, C. R. (1991). The role of glutathione-dependent enzymes in drug resistance. *Pharmacol. Ther.* **51**, 139–154.

Brabender, J., Lord, R. V., Wickramasinghe, K., Metzger, R., Schneider, P. M., Park, J. M., Holscher, A. H., De Meester, T. R., Danenberg, K. D., and Danenberg, P. V. (2002). Glutathione S-transferase-pi expression is downregulated in patients with Barrett's esophagus and esophageal adenocarcinoma. *J. Gastrointest. Surg.* **6**, 359–367.

Brodeur, I., Goulet, I., Tremblay, C. S., Charbonneau, C., Delisle, M. C., Godin, C., Huard, C., Khandjian, E. W., Buchwald, M., Levesque, G., and Carreau, M. (2004). Regulation of the Fanconi anemia group C protein through proteolytic modification. *J. Biol. Chem.* **279**, 4713–4720.

Brown, K., and Balmain, A. (1995). Transgenic mice and squamous multistage skin carcinogenesis. *Cancer Metastasis Rev.* **14**, 113–124.

Chen, N., She, Q. B., Bode, A. M., and Dong, Z. (2002). Differential gene expression profiles of Jnk1- and Jnk2-deficient murine fibroblast cells. *Cancer Res.* **62**, 1300–1304.

Cho, S. G., Lee, Y. H., Park, H. S., Ryoo, K., Kang, K. W., Park, J., Eom, S. J., Kim, M. J., Chang, T. S., Choi, S. Y., Shim, J., Kim, Y., Dong, M. S., Lee, M. J., Kim, S. G., Ichijo, H., and Choi, E. J. (2001). Glutathione S-transferase mu modulates the stress-activated signals by suppressing apoptosis signal-regulating kinase 1. *J. Biol. Chem.* **276**, 12749–12755.

Coles, B., Wilson, I., Wardman, P., Hinson, J. A., Nelson, S. D., and Ketterer, B. (1988). The spontaneous and enzymatic reaction of N-acetyl-p-benzoquinonimine with glutathione: a stopped-flow kinetic study. *Arch. Biochem. Biophys.* **264**, 253–260.

Cookson, M. S., Reuter, V. E., Linkov, I., and Fair, W. R. (1997). Glutathione S-transferase PI (GST-pi) class expression by immunohistochemistry in benign and malignant prostate tissue. *J. Urol.* **157**, 673–676.

Cumming, R. C., Lightfoot, J., Beard, K., Youssoufian, H., O'Brien, P. J., and Buchwald, M. (2001). Fanconi anemia group C protein prevents apoptosis in hematopoietic cells through redox regulation of GSTP1. *Nat. Med.* **7**, 814–820.

Dahlin, D. C., Miwa, G. T., Lu, A. Y. H., and Nelson, S. D. (1984). N-acetyl-p-benzoquinone imine: A cytochrome P450-mediated oxidation product of acetaminophen. *Proc. Natl. Acad. Sci. USA* **81**, 1327–1331.

Di Ilio, C., Del Boccio, G., Aceto, A., Casaccia, R., Mucilli, F., and Federici, G. (1988). Elevation of glutathione transferase activity in human lung tumor. *Carcinogenesis* **9**, 335–340.

Eaton, D. L., and Bammler, T. K. (1999). Concise review of the glutathione S-transferases and their significance to toxicology. *Toxicol. Sci.* **49**, 156–164.

Elsby, R., Kitteringham, N. R., Goldring, C. E., Lovatt, C. A., Chamberlain, M., Henderson, C. J., Wolf, C. R., and Park, B. K. (2003). Increased constitutive c-Jun N-terminal kinase signaling in mice lacking glutathione S-transferase Pi. *J. Biol. Chem.* **278**, 22243–22249.

Forkert, P. G., D'Costa, D., and El-Mestrah, M. (1999). Expression and inducibility of alpha, pi, and Mu glutathione S-transferase protein and mRNA in murine lung. *Am. J. Respir. Cell Mol. Biol.* **20**, 143–152.

Fryer, A. A., Bianco, A., Hepple, M., Jones, P. W., Strange, R. C., and Spiteri, M. A. (2000). Polymorphism at the glutathione S-transferase GSTP1 locus. A new marker for bronchial hyperresponsiveness and asthma. *Am. J. Respir. Crit. Care Med.* **161**, 1437–1442.

Fryer, A. A., Hume, R., and Strange, R. C. (1986). The development of glutathione S-transferase and glutathione peroxidase activities in human lung. *Biochim. Biophys. Acta* **883**, 448–453.

Gajewska, J., and Szczypka, M. (1992). Role of pi form of glutathione S-transferase (GST-pi) in cancer: A minireview. *Mater Med. Pol.* **24**, 45–49.

Ghalia, A. A., Rabboh, N. A., el Shalakani, A., Seada, L., and Khalifa, A. (2000). Estimation of glutathione S-transferase and its Pi isoenzyme in tumor tissues and sera of patients with ovarian cancer. *Anticancer Res.* **20**, 1229–1235.

Grignon, D. J., Abdel-Malak, M., Mertens, W. C., Sakr, W. A., and Shepherd, R. R. (1994). Glutathione S-transferase expression in renal cell carcinoma: A new marker of differentiation. *Mod. Pathol.* **7**, 186–189.

Hanada, K., Ishikawa, H., Tamai, K., Hashimoto, I., and Sato, K. (1991). Expression of glutathione S-transferase-pi in malignant skin tumors. *J. Dermatol. Sci.* **2**, 18–23.

Harries, L. W., Stubbins, M. J., Forman, D., Howard, G. C., and Wolf, C. R. (1997). Identification of genetic polymorphisms at the glutathione S-transferase Pi locus and association with susceptibility to bladder, testicular and prostate cancer. *Carcinogenesis* **18**, 641–644.

Hayes, J. D., Flanagan, J. U., and Jowsey, I. R. (2005). Glutathione transferases. *Annu. Rev. Pharmacol. Toxicol.* **45**, 51–88.

Hayes, J. D., and McLellan, L. I. (1999). Glutathione and glutathione-dependent enzymes represent a co-ordinately regulated defence against oxidative stress. *Free Radic. Res.* **31**, 273–300.

Henderson, C. J., Smith, A. G., Ure, J., Brown, K., Bacon, E. J., and Wolf, C. R. (1998). Increased skin tumorigenesis in mice lacking pi class glutathione S-transferases. *Proc. Natl. Acad. Sci. USA* **95**, 5275–5280.

Henderson, C. J., Wolf, C. R., Kitteringham, N., Powell, H., Otto, D., and Park, B. K. (2000). Increased resistance to acetaminophen hepatotoxicity in mice lacking glutathione S-transferase Pi. *Proc. Natl. Acad. Sci. USA* **97**, 12741–12745.

Henrion-Caude, A., Flamant, C., Roussey, M., Housset, C., Flahault, A., Fryer, A. A., Chadelat, K., Strange, R. C., and Clement, A. (2002). Liver disease in pediatric patients with cystic fibrosis is associated with glutathione S-transferase P1 polymorphism. *Hepatology* **36**, 913–917.

Howie, A. F., Forrester, L. M., Glancey, M. J., Schlager, J. J., Powis, G., Beckett, G. J., Hayes, J. D., and Wolf, C. R. (1990). Glutathione S-transferase and glutathione peroxidase expression in normal and tumour human tissues. *Carcinogenesis* **11**, 451–458.

Hu, X., Xia, H., Srivastava, S. K., Herzog, C., Awasthi, Y. C., Ji, X., Zimniak, P., and Singh, S. V. (1997). Activity of four allelic forms of glutathione S-transferase hGSTP1–1 for diol epoxides of polycyclic aromatic hydrocarbons. *Biochem. Biophys. Res. Commun.* **238**, 397–402.

Huang, J., Tan, P. H., Thiyagarajan, J., and Bay, B. H. (2003). Prognostic significance of glutathione S-transferase-pi in invasive breast cancer. *Mod. Pathol.* **16**, 558–565.

Ishii, T., Fujishiro, M., Masuda, M., Nakajima, J., Teramoto, S., Ouchi, Y., and Matsuse, T. (2003). Depletion of glutathione S-transferase P1 induces apoptosis in human lung fibroblasts. *Exp. Lung Res.* **29**, 523–536.

Ishii, T., Matsuse, T., Igarashi, H., Masuda, M., Teramoto, S., and Ouchi, Y. (2001). Tobacco smoke reduces viability in human lung fibroblasts: Protective effect of glutathione S-transferase P1. *Am. J. Physiol. Lung Cell Mol. Physiol.* **280**, L1189–L1195.

Ishii, T., Matsuse, T., Teramoto, S., Matsui, H., Miyao, M., Hosoi, T., Takahashi, H., Fukuchi, Y., and Ouchi, Y. (1999). Glutathione S-transferase P1 (GSTP1) polymorphism in patients with chronic obstructive pulmonary disease. *Thorax* **54**, 693–696.

Ishii, T., Teramoto, S., and Matsuse, T. (2004). GSTP1 affects chemoresistance against camptothecin in human lung adenocarcinoma cells. *Cancer Lett.* **216**, 89–102.

Ishisaki, A., Hayashi, H., Suzuki, S., Ozawa, K., Mizukoshi, E., Miyakawa, K., Suzuki, M., and Imamura, T. (2001). Glutathione S-transferase Pi is a dopamine-inducible suppressor of dopamine-induced apoptosis in PC12 cells. *J. Neurochem.* **77**, 1362–1371.

Jayasurya, A., Yap, W. M., Tan, N. G., Tan, B. K., and Bay, B. H. (2002). Glutathione S-transferase pi expression in nasopharyngeal cancer. *Arch. Otolaryngol. Head Neck Surg.* **128**, 1396–1399.

Kamada, K., Goto, S., Okunaga, T., Ihara, Y., Tsuji, K., Kawai, Y., Uchida, K., Osawa, T., Matsuo, T., Nagata, I., and Kondo, T. (2004). Nuclear glutathione S-transferase pi prevents apoptosis by reducing the oxidative stress-induced formation of exocyclic DNA products. *Free Radic. Biol. Med.* **37**, 1875–1884.

Katagiri, A., Tomita, Y., Nishiyama, T., Kimura, M., and Sato, S. (1993). Immunohistochemical detection of P-glycoprotein and GSTP1-1 in testis cancer. *Br. J. Cancer* **68**, 125–129.

Kauvar, L. M., Morgan, A. S., Sanderson, P. E., and Henner, W. D. (1998). Glutathione based approaches to improving cancer treatment. *Chem. Biol. Interact.* **111–112**, 225–238.

Kelly, V. P., Ellis, E. M., Manson, M. M., Chanas, S. A., Moffat, G. J., McLeod, R., Judah, D. J., Neal, G. E., and Hayes, J. D. (2000). Chemoprevention of aflatoxin B1 hepatocarcinogenesis by coumarin, a natural benzopyrone that is a potent inducer of aflatoxin B1-aldehyde reductase, the glutathione S-transferase A5 and P1 subunits, and NAD(P)H:quinone oxidoreductase in rat liver. *Cancer Res.* **60**, 957–969.

Kitteringham, N. R., Powell, H., Jenkins, R. E., Hamlett, J., Lovatt, C., Elsby, R., Henderson, C. J., Wolf, C. R., Pennington, S. R., and Park, B. K. (2003). Protein expression profiling of glutathione S-transferase pi null mice as a strategy to identify potential markers of resistance to paracetamol-induced toxicity in the liver. *Proteomics* **3**, 191–207.

Kodate, C., Fukushi, A., Narita, T., Kudo, H., Soma, Y., and Sato, K. (1986). Human placental form of glutathione S-transferase (GST-pi) as a new immunohistochemical marker for human colonic carcinoma. *Jpn. J. Cancer Res.* **77**, 226–229.

Konohana, A., Konohana, I., Schroeder, W. T., O'Brien, W. R., Amagai, M., Greer, J., Shimizu, N., Gammon, W. R., Siciliano, M. J., and Duvic, M. (1990). Placental glutathione-S-transferase-pi mRNA is abundantly expressed in human skin. *J. Invest. Dermatol.* **95**, 119–126.

Leder, A., Kuo, A., Cardiff, R. D., Sinn, E., and Leder, P. (1990). v-Ha-ras transgene abrogates the initiation step in mouse skin tumorigenesis: Effects of phorbol esters and retinoic acid. *Proc. Natl. Acad. Sci. USA* **87**, 9178–9182.

Lee, W. H., Isaacs, W. B., Bova, G. S., and Nelson, W. G. (1997). CG island methylation changes near the GSTP1 gene in prostatic carcinoma cells detected using the polymerase chain reaction: A new prostate cancer biomarker. *Cancer Epidemiol. Biomarkers Prev.* **6**, 443–450.

Lee, W. H., Morton, R. A., Epstein, J. I., Brooks, J. D., Campbell, P. A., Bova, G. S., Hsieh, W. S., Isaacs, W. B., and Nelson, W. G. (1994). Cytidine methylation of regulatory sequences near the pi-class glutathione S-transferase gene accompanies human prostatic carcinogenesis. *Proc. Natl. Acad. Sci. USA* **91**, 11733–11737.

Lewis, A. D., Hayes, J. D., and Wolf, C. R. (1988a). Glutathione and glutathione-dependent enzymes in ovarian adenocarcinoma cell lines derived from a patient before and after the onset of drug resistance: Intrinsic differences and cell cycle effects. *Carcinogenesis* **9**, 1283–1287.

Lewis, A. D., Hickson, I. D., Robson, C. N., Harris, A. L., Hayes, J. D., Griffiths, S. A., Manson, M. M., Hall, A. E., Moss, J. E., and Wolf, C. R. (1988b). Amplification and increased expression of alpha class glutathione S-transferase-encoding genes associated with resistance to nitrogen mustards. *Proc. Natl. Acad. Sci. USA* **85**, 8511–8515.

Lin, X., Asgari, K., Putzi, M. J., Gage, W. R., Yu, X., Cornblatt, B. S., Kumar, A., Piantadosi, S., De Weese, T. L., De Marzo, A. M., and Nelson, W. G. (2001). Reversal of GSTP1 CpG island hypermethylation and reactivation of pi-class glutathione S-transferase (GSTP1) expression in human prostate cancer cells by treatment with procainamide. *Cancer Res.* **61**, 8611–8616.

Mannervik, B., Board, P. G., Hayes, J. D., Listowsky, I., and Pearson, W. R. (2005). "Nomenclatures for mammalian soluble glutathione transferases" *Meth. Enzymol.* **401**, 1–8.

McLellan, L. I., and Hayes, J. D. (1987). Sex-specific constitutive expression of the pre-neoplastic marker glutathione S-transferase, YfYf, in mouse liver. *Biochem. J.* **245**, 399–406.

McLellan, L. I., and Wolf, C. R. (1999). Glutathione and glutathione-dependent enzymes in cancer drug resistance. *Drug Resist. Updat.* **2**, 153–164.

Moffat, G. J., McLaren, A. W., and Wolf, C. R. (1994). Involvement of Jun and Fos proteins in regulating transcriptional activation of the human pi class glutathione S-transferase gene in multidrug-resistant MCF7 breast cancer cells. *J. Biol. Chem.* **269**, 16397–16402.

Moffat, G. J., McLaren, A. W., and Wolf, C. R. (1996). Sp1-mediated transcriptional activation of the human Pi class glutathione S-transferase promoter. *J. Biol. Chem.* **271**, 1054–1060.

Molina, R., Oesterreich, S., Zhou, J. L., Tandon, A. K., Clark, G. M., Allred, D. C., Townsend, A. J., Moscow, J. A., Cowan, K. H., McGuire, W. L., *et al.* (1993). Glutathione transferase GST pi in breast tumors evaluated by three techniques. *Dis. Markers* **11**, 71–82.

Moral, A., Palou, J., Lafuente, A., Molina, R., Piulachs, J., Castel, T., and Trias, M. (1997). Immunohistochemical study of alpha, mu and pi class glutathione S transferase expression in malignant melanoma. MMM Group. Multidisciplinary Malignant Melanoma Group. *Br. J. Dermatol.* **136**, 345–350.

Morgan, A. S., Ciaccio, P. J., Tew, K. D., and Kauvar, L. M. (1996). Isozyme-specific glutathione S-transferase inhibitors potentiate drug sensitivity in cultured human tumor cell lines. *Cancer Chemother. Pharmacol.* **37**, 363–370.

Morgan, A. S., Sanderson, P. E., Borch, R. F., Tew, K. D., Niitsu, Y., Takayama, T., Von Hoff, D. D., Izbicka, E., Mangold, G., Paul, C., Broberg, U., Mannervik, B., Henner, W. D., and Kauvar, L. M. (1998). Tumor efficacy and bone marrow-sparing properties of TER286, a cytotoxin activated by glutathione S-transferase. *Cancer Res.* **58**, 2568–2575.

Mountford, P. S., and Smith, A. G. (1995). Internal ribosome entry sites and dicistronic RNAs in mammalian transgenesis. *Trends Genet.* **11**, 179–184.

Mulder, T. P., Verspaget, H. W., Sier, C. F., Roelofs, H. M., Ganesh, S., Griffioen, G., and Peters, W. H. (1995). Glutathione S-transferase pi in colorectal tumors is predictive for overall survival. *Cancer Res.* **55**, 2696–2702.

Niitsu, Y., Takayama, T., Miyanishi, K., Nobuoka, A., Hayashi, T., Nakajima, T., Miyake, S., Henderson, C. J., and Wolf, C. R. (2001). Implication of GST-pi expression in colon carcinogenesis. *Chem. Biol. Interact.* **133**, 287–290.

Nobuoka, A., Takayama, T., Miyanishi, K., Sato, T., Takanashi, K., Hayashi, T., Kukitsu, T., Sato, Y., Takahashi, M., Okamoto, T., Matsunaga, T., Kato, J., Oda, M., Azuma, T., and Niitsu, Y. (2004). Glutathione-S-transferase P1–1 protects aberrant crypt foci from apoptosis induced by deoxycholic acid. *Gastroenterology* **127**, 428–443.

Patten, C. J., Thomas, P. E., Guy, R. L., Lee, M., Gonzalez, F. J., Guengerich, F. P., and Yang, C. S. (1993). Cytochrome P450 enzymes involved in acetaminophen activation by rat and human liver microsomes and their kinetics. *Chem. Res. Toxicol.* **6**, 511–518.

Prescott, L. F. (1996). "Paracetamol (Acetaminophen)—A Critical Bibliographic Review." Taylor & Francis, London.

Ranganathan, P. N., Whalen, R., and Boyer, T. D. (2005). Characterization of the molecular forms of glutathione S-transferase P1 in human gastric cancer cells (Kato III) and in normal human erythrocytes. *Biochem. J.* **386**, 525–533.

Raucy, J. L., Lasker, J. M., Lieber, C. S., and Black, M. (1989). Acetaminophen activation by human liver cytochromes P450IIE1 and P4501A2. *Arch. Biochem. Biophys.* **271**, 270–283.

Raza, H., Awasthi, Y. C., Zaim, M. T., Eckert, R. L., and Mukhtar, H. (1991). Glutathione S-transferases in human and rodent skin: Multiple forms and species-specific expression. *J. Invest. Dermatol.* **96**, 463–467.

Rosario, L. A., O'Brien, M. L., Henderson, C. J., Wolf, C. R., and Tew, K. D. (2000). Cellular response to a glutathione S-transferase P1-1 activated prodrug. *Mol. Pharmacol.* **58**, 167–174.

Rosen, L. S., Brown, J., Laxa, B., Boulos, L., Reiswig, L., Henner, W. D., Lum, R. T., Schow, S. R., Maack, C. A., Keck, J. G., Mascavage, J. C., Dombroski, J. A., Gomez, R. F., and Brown, G. L. (2003). Phase I study of TLK286 (glutathione S-transferase P1-1 activated glutathione analogue) in advanced refractory solid malignancies. *Clin. Cancer Res.* **9**, 1628–1638.

Rosen, L. S., Laxa, B., Boulos, L., Wiggins, L., Keck, J. G., Jameson, A. J., Parra, R., Patel, K., and Brown, G. L. (2004). Phase 1 study of TLK286 (Telcyta) administered weekly in advanced malignancies. *Clin. Cancer Res.* **10**, 3689–3698.

Ryberg, D., Skaug, V., Hewer, A., Phillips, D. H., Harries, L. W., Wolf, C. R., Ogreid, D., Ulvik, A., Vu, P., and Haugen, A. (1997). Genotypes of glutathione transferase m1 and p1 and their significance for lung DNA adduct levels and cancer risk. *Carcinogenesis* **18**, 1285–1289.

Salinas, A. E., and Wong, M. G. (1999). Glutathione S-transferases—a review. *Curr. Med. Chem.* **6**, 279–309.

Satoh, K., Hatayama, I., Tateoka, N., Tamai, K., Shimizu, T., Tatematsu, M., Ito, N., and Sato, K. (1989). Transient induction of single GST-P positive hepatocytes by DEN. *Carcinogenesis* **10**, 2107–2111.

Schipper, D. L., Wagenmans, M. J. H., Wagener, D. J. T., and Peters, W. H. M. (1997). Gluathine S-transferases and cancer (review). *Int. J. Oncol.* **10**, 1261–1264.

Segers, K., Kumar-Singh, S., Weyler, J., Bogers, J., Ramael, M., Van Meerbeeck, J., and Van Marck, E. (1996). Glutathione S-transferase expression in malignant mesothelioma and non-neoplastic mesothelium: An immunohistochemical study. *J. Cancer Res. Clin. Oncol.* **122**, 619–624.

Sherratt, P. J., and Hayes, J. D. (2002). "Glutathione S-Transferases." M Wiley & Sons, New York.

Sherratt, P. J., Manson, M. M., Thomson, A. M., Hissink, E. A., Neal, G. E., van Bladeren, P. J., Green, T., and Hayes, J. D. (1998). Increased bioactivation of dihaloalkanes in rat liver due to induction of class theta glutathione S-transferase T1-1. *Biochem. J.* **335**(Pt 3), 619–630.

Sherratt, P. J., Pulford, D. J., Harrison, D. J., Green, T., and Hayes, J. D. (1997). Evidence that human class theta glutathione S-transferase T1-1 can catalyse the activation of dichloromethane, a liver and lung carcinogen in the mouse. Comparison of the tissue distribution of GST T1-1 with that of classes Alpha, Mu and Pi GST in human. *Biochem. J.* **326**(Pt 3), 837–846.

Shiga, H., Heath, E. I., Rasmussen, A. A., Trock, B., Johnston, P. G., Forastiere, A. A., Langmacher, M., Baylor, A., Lee, M., and Cullen, K. J. (1999). Prognostic value of p53, glutathione S-transferase pi, and thymidylate synthase for neoadjuvant cisplatin-based chemotherapy in head and neck cancer. *Clin. Cancer Res.* **5**, 4097–4104.

Shimizu, K., Toriyama, F., and Yoshida, H. (1995a). The expression of placental-type glutathione S-transferase (GST-pi) in human cutaneous squamous cell carcinoma and normal human skin. *Virchows Arch.* **425**, 589–592.

Shimizu, K., Toriyama, F., Zhang, H. M., and Yoshida, H. (1995b). The expression of placental-type glutathione S-transferase (GST-pi) in human cutaneous carcinoma *in situ,* that is, actinic keratosis and Bowen's disease, compared with normal human skin. *Carcinogenesis* **16**, 2327–2330.

Su, F., Hu, X., Jia, W., Gong, C., Song, E., and Hamar, P. (2003). Glutathione S transferase pi indicates chemotherapy resistance in breast cancer. *J. Surg. Res.* **113**, 102–108.

Sundberg, K., Johansson, A. S., Stenberg, G., Widersten, M., Seidel, A., Mannervik, B., and Jernstrom, B. (1998). Differences in the catalytic efficiencies of allelic variants of glutathione transferase P1-1 towards carcinogenic diol epoxides of polycyclic aromatic hydrocarbons. *Carcinogenesis* **19**, 433–436.

Tidefelt, U., Elmhorn-Rosenborg, A., Paul, C., Hao, X. Y., Mannervik, B., and Eriksson, L. C. (1992). Expression of glutathione transferase pi as a predictor for treatment results at different stages of acute nonlymphoblastic leukemia. *Cancer Res.* **52**, 3281–3285.

Townsend, D. M., and Tew, K. D. (2003). The role of glutathione-S-transferase in anti-cancer drug resistance. *Oncogene* **22**, 7369–7375.

Tsuchida, S., Sekine, Y., Shineha, R., Nishihira, T., and Sato, K. (1989). Elevation of the placental glutathione S-transferase form (GST-pi) in tumor tissues and the levels in sera of patients with cancer. *Cancer Res.* **49**, 5225–5229.

Wang, T., Arifoglu, P., Ronai, Z., and Tew, K. D. (2001). Glutathione S-transferase P1-1 (GSTP1-1) inhibits c-Jun N-terminal kinase (JNK1) signaling through interaction with the C terminus. *J. Biol. Chem.* **276**, 20999–21003.

Wareing, C. J., Black, S. M., Hayes, J. D., and Wolf, C. R. (1993). Increased levels of alpha-class and pi-class glutathione s-transferases in cell-lines resistant to 1-chloro-2,4-dinitrobenzene. *Eur. J. Biochem.* **217**, 671–676.

Whelan, R. D., Waring, C. J., Wolf, C. R., Hayes, J. D., Hosking, L. K., and Hill, B. T. (1992). Over-expression of P-glycoprotein and glutathione S-transferase pi in MCF-7 cells selected for vincristine resistance *in vitro. Int. J. Cancer* **52**, 241–246.

Wolf, C. R. (2002). The Gerhard Zbinden memorial lecture: Application of biochemical and genetic approaches to understanding pathways of chemical toxicity. *Toxicol Lett.* **127**, 3–17.

Wolf, C. R., Wareing, C. J., Black, S. M., and Hayes, J. D. (1990). Glutathione S-transferases in resistance to chemotherapeutic drugs. *In* "Glutathione S-Transferases and Drug Resistance" (J. D. Hayes, C. B. Pickett, and T. J. Mantle, eds.), pp. 297–307. Taylor & Francis, London.

Woodson, K., Hanson, J., and Tangrea, J. (2004). A survey of gene-specific methylation in human prostate cancer among black and white men. *Cancer Lett.* **205**, 181–188.

Yokoyama, Y., Sagara, M., Sato, S., and Saito, Y. (1998). Value of glutathione S-transferase pi and the oncogene products c-Jun, c-Fos, c-H-Ras, and c-Myc as a prognostic indicator in endometrial carcinomas. *Gynecol. Oncol.* **68**, 280–287.

Zhang, L., Xiao, Y., and Priddy, R. (1994). Increase in placental glutathione S-transferase in human oral epithelial dysplastic lesions and squamous cell carcinomas. *J. Oral Pathol. Med.* **23**, 75–79.

[8] Microsomal Glutathione Transferase 1

By RALF MORGENSTERN

Abstract

Microsomal glutathione transferase 1 (MGST1) is an abundant membrane-bound glutathione transferase and peroxidase constituting 3% of the endoplasmic reticulum protein in rat liver (and 5% of the outer mitochondrial membrane). The enzyme is most well studied in mammals and belongs to a large and widely distributed superfamily. Cellular and organelle protection versus oxidative stress has been demonstrated. The enzyme displays activity to a multitude of reactive substrates ranging from products of lipid peroxidation to cytostatic drugs. The methods developed for the study of MGST1 by necessity differs from that of cytosolic glutathione transferases, because detergents or lipids are included. Here, purification, assay, and preparation procedures that maintain the enzyme in its native functional state during isolation and characterization are described.

Microsomal glutathione transferase 1 is activated by sulfhydryl reagents (and proteolysis), and procedures for activation and study of the activated enzyme are described. In new developments, the enzyme is studied by pre-steady state methods, as well as mass spectrometry involving direct observation of the native enzyme.

Introduction

Microsomal glutathione transferase 1 (MGST1) is a member of a superfamily (MAPEG, see Chapter 9) of proteins involved in cellular protection and eicosanoid signaling (Jakobsson et al., 1999). So far, many members have been defined in prokaryotes and eukaryotes but none in archaea (Bresell et al., 2005). MGST1 displays glutathione transferase and glutathione peroxidase activity (Morgenstern and DePierre, 1983) and protects membranes (Mosialou et al., 1993b) and cells from oxidative stress (Maeda et al., 2005). Human MGST1 is most closely related to PGE synthase and less similar to the other four human MAPEG proteins. As one traces back in evolution, a clearly distinct MGST1 can be found in insects (*Drosophila melanogaster*).

The simple fact that MGST1 is a membrane protein sets it apart from cytosolic GSTs, but functionally it has the same broad and overlapping substrate specificity (Andersson et al., 1994). Uniquely, though, MGST1 is activated by sulfhydryl reagents and proteolysis (Aniya et al., 2001;

METHODS IN ENZYMOLOGY, VOL. 401
Copyright 2005, Elsevier Inc. All rights reserved.
0076-6879/05 $35.00
DOI: 10.1016/S0076-6879(05)01008-6

Morgenstern *et al.*, 1979, 1989). This characteristic has been observed in mammalian MGST1 but not in enzymes from lower organisms that, in contrast, often display much higher basal activity (Sun *et al.*, 1998; and unpublished observations).

The fact that MGST1 is a membrane protein governs the method used in its study, where detergents or lipids are used to keep the enzyme in its native functional state during purification and characterization. It has been observed that the detergent Triton X-100 substitutes for lipids in a way that does not change the overall catalytic and dynamic behavior of MGST1 (Busenlehner *et al.*, 2004; Morgenstern and DePierre, 1983; Morgenstern *et al.*, 1983). All procedures described in the following apply to rat liver MGST1 but in general can be applied to most MGSTs.

Reagents

Reagents are of highest purity from common commercial sources. The following is a list of selected reagents and suppliers: Triton X-100, GSH, GSH reductase, N-ethyl maleimide (NEM), trypsin inhibitor, Sigma; lauryl diaminoxide, Calbiochem; 1-chloro-2,4-dinitrobenzene (CDNB), Merck; 4-chloro-3-nitrobenzamide, Alfred Baeder Library of Rare Chemicals, Division of Aldrich; 1,3,5.trinitrobezene was a kind gift from Nobel Chemicals, Karlskoga, Sweden; trypsin, Boehringer; hydroxy apatite Bio-Gel HTP, Bio-Rad; CM-Sepharose CL-6B and G-25 Fine, Amersham Pharmacia Biotech; Centricon tubes, Amicon Inc.; Econo-Pac 10DG Columns; BioRad, Hercules, CA; Dialysis membranes, SpectraPor 4 from SpectraPor Medical Industries Inc., Houston, TX, USA.

Methodology

Preparation of Enzyme

Microsomal glutathione transferase 1 was first purified from rat liver (Morgenstern *et al.*, 1982) and subsequently with similar methodology from mouse and human liver (Andersson *et al.*, 1988; McLellan *et al.*, 1989; Mosialou *et al.*, 1993a).

Liver Microsome Preparation. Livers from male Sprague-Dawley rats (180–240 g, starved overnight) are homogenized in 0.25 M sucrose by four up and down strokes in a Potter Elvehjem glass Teflon homogenizer to yield a 20% homogenate. All procedures are at 4°. The homogenate is spun 15 min at 10,000g and the supernatant a further 60 min at 100,000g. The microsomes are resuspended and washed twice in 0.15 M Tris-HCl, pH 8, by centrifugation at 100,000g for 30 min. The resulting pellet is resuspended in 0.25 M sucrose at 10 mg/ml and activity determined on fresh material. If the microsomes are used for purification, they are resuspended in 10 mM potassium phosphate, pH 7, 20% glycerol, 1 mM GSH, 0.1 mM EDTA, and 1% Triton X-100 (buffer A). The rat liver microsome preparation is essentially that described by Ernster *et al.* (1962) with the addition of an extra wash to diminish cytosolic GST contamination.

Enzyme Purification. Material from between 10 and 35 rats is used in routine purifications, and the amounts given in the following apply to 15 rat livers.

The resuspended microsomes (60 ml of buffer A) are solubilized by addition of buffer A containing 8% Triton X-100 (100 ml)) for 15 min. Next, a purification step involving hydroxyapatite is performed in two alternate ways: (1) the solubilized material is loaded on a hydroxyapatite column 5 × 40 cm pre-equilibrated with buffer A and eluted with a linear gradient of 10 mM to 0.4 M potassium phosphate, pH 7, in 2 l at 1 ml/min. Fractions of 5 ml are collected, assayed for GST activity, and pooled (the enzyme elutes at approximately 0.2 M potassium phosphate); (2) the solubilized microsomes are added directly to hydroxyapatite pre-equilibrated with buffer A (30 g hydroxyapatite slurry + an additional 140 ml buffer A directly in 1-l centrifuge flasks) in batch and mixed for 20 min by occasional gentle swirling. The hydroxyapatite is pelleted by low-speed centrifugation (a pulse of 3000 rpm is used throughout) and the supernatant discarded. The hydroxyapatite is first washed twice with 200 ml of buffer A, then with 200 ml buffer A containing 50 mM potassium phosphate, pH 7, and subsequently with 2–3 100 ml washes of buffer A containing 200 mM potassium phosphate, pH 7, in which the enzyme activity elutes.

The eluates from procedure 1 or 2 are combined and subjected to gel filtration on a G-25 (fine) column (10 × 80 cm) equilibrated with buffer A to reduce potassium phosphate concentration. The protein-containing fractions usually are colored yellow and can be collected by visual inspection; alternately, fractions are collected and assayed for enzyme activity.

The G-25 pool is added to a CM-Sepharose column (2.5 × 15 cm) equilibrated with buffer A at 1 ml/min. The column is eluted at the same rate with a gradient of 0–0.2 M KCl in 200 ml buffer A where the enzyme elutes at approximately 0.1 M KCl; 3-ml fractions are collected, assayed for activity, and pooled. If additional purification is needed (routinely when the hydroxyapatite batch procedure is used), this step is repeated but with all buffers and the enzyme pool adjusted to pH 8. We usually pool top fractions and side fractions separately to yield one highly concentrated (1–3 mg/ml) and one less concentrated pool. Protein is determined by the method of Peterson (Peterson, 1977) including the precipitation step (because GSH responds in the assay). This protein assay has been calibrated with amino acid analysis and found to yield the true protein content with bovine serum albumin as a standard. Purity is determined by SDS-PAGE according to Laemmli (1970) but including the double concentration of Tris-HCl in the separation gel. More SDS (6%) is sometimes needed in the sample buffer to balance Triton X-100. The pools are frozen in portions of 1–5 ml under nitrogen and stored in the freezer. Enzyme pools kept in this way are stable for years.

Detergent Exchange, Achieving High Protein Concentration, and Ligand Removal

The last purification step can be used to exchange Triton X-100 for an equivalent concentration of other detergents (that the enzyme tolerates) in the elution buffer, such as lauryl diaminoxide, which is UV transparent. An absorbance of 0.55 at 280 nm corresponds to 1 mg/ml of MGST1. We also lower the Triton X-100 concentration to 0.2% in the last purification step for applications such as stop-flow spectrophotometric observation of thiolate anion stabilization by MGST1 (Morgenstern et al., 2001). This amount of Triton X-100 does not compromise measurements at 240 nm where the thiolate absorbs.

The enzyme can be concentrated up to 10 mg/ml using, for instance, Centricon-10 tubes. One has to keep in mind that the detergent does not pass through most filters, because the micellar size corresponds to a Mr of 300,000 and is therefore concentrated together with the protein.

For some applications, it is desirable to remove GSH from the purified MGST1. Gel filtration is a convenient way to accomplish this, and we routinely use pre-packed Econo-Pac 10DG Columns according to the manufacturer's instructions. Because of the slow off rate of GSH, it is important to determine that all GSH is removed by measuring Meisenheimer complex formation (see later). We routinely use two gel filtration steps in tandem for complete GSH removal.

Preparation of MGST1 for Electrospray Mass Spectrometry

Electrospray mass spectrometry requires low ionic strength and minimal detergent in samples for efficient ionization. A solution of 2 mg/ml MGST1 in 10 mM potassium phosphate (pH 8), 1 mM GSH, 0.2% (v/v) Triton X-100, 0.1 mM EDTA, 0.2 M KCl, and 20% glycerol is subjected to size-exclusion chromatography on a 10DG column using 1 mM potassium phosphate (pH 8), 0.1 mM GSH as a running buffer to obtain ≈2 mg/ml MGST1 in the latter buffer and approximately 0.2% Triton X-100 (not removed in the size-exclusion procedure). The pooled sample is then diluted fivefold with the same buffer to give 0.3–0.4 mg/ml MGST1 in

1 mM potassium phosphate (pH 8), 0.1 mM GSH and approximately 0.04% Triton X-100. This solution can be used in electrospray as described (Lengqvist *et al.*, 2004).

Preparation of Proteoliposomes

The study of MGST1 in its native phospholipid environment is accomplished by reconstitution of the enzyme into proteoliposomes (Mosialou *et al.*, 1995). Two milligrams of phospholipid (e.g., phosphatidylcholine or rat liver microsomal phospholipids) is dissolved in 20 μl of 20% sodium cholate and sonicated under nitrogen for 5 min (30-s intervals with 10 s cooling) with a Branson 2.200 Sonifier at room temperature. Thereafter, 0.18 ml of buffer E, 10 mM potassium phosphate, pH 7, 20% glycerol (v/v), 50 mM KCl, and 0.1 mM EDTA and 0.2 mg of purified MGST1, is added. Proteoliposomes are formed on detergent removal by dialysis against buffer E containing 1 mM GSH and 0.05% sodium cholate for 96 h (two changes each 24 h) and an additional 96 h against the same buffer lacking cholate (two changes each 24 h). The removal of Triton X-100 is very slow (approximately 7 days) and can be followed by the decrease in absorbance at 275 nm where the detergent absorbs ($E_{1\%\ \text{solution}} = 21$). During dialysis, the enzyme becomes activated to a certain degree, and as a result, the activation achieved by NEM is diminished. In essence, the production of two-dimensional crystals (described elsewhere in this volume) is a development of this procedure.

Heterologous Expression and Purification from Bacterial Membranes

MGST1 has been expressed in *Escherichia coli* and mammalian cells such as COS cells. Expression in mammalian cells involved different vectors and standard procedures (Maeda *et al.*, 2005; Weinander *et al.*, 1995).

Because the production of mammalian membrane proteins in *E. coli* is less routine, I will focus on this aspect. cDNA for MGST1 is inserted into the bacterial expression vector pSP19T7LT that was developed for successful expression of mammalian cytochrome P450s. The bacteria used for expression is *E. coli* BL 21 (DE 3) harboring the plasmid pLysSL (that produces lysozyme that prevents leaky expression from the genomic copy of the T7 promoter).

Bacterial expression is started by (5–10 μl) bacterial glycerol stock grown in 1.5 ml 2xYT overnight at 37°. The culture is diluted 1:100 in TB (terrific broth) (Tartof and Hobbs, 1987) and grown until OD$_{600}$ of 0.4–1.2. Expression is induced by the addition of 1 mM IPTG, the temperature is switched to 30°, and the culture is allowed to grow for another 2–12 h. All steps are performed in the presence of ampicillin (75 μg/ml) and chloramphenicol

(10 μg/ml) with 240 rpm shaking. Cells are pelleted and resuspended in 15 mM Tris-HCl, pH 8.0, 0.25 M sucrose, 0.1 mM EDTA, and 1 mM GSH. The cells are lysed by sonication using four 30-s pulses from a MSE Soniprep 150 sonifier at 40–60% of maximum power. Magnesium chloride is added to a final concentration of 6 mM, and DNA and RNA are hydrolyzed by incubation with DNaseI (4 μg/ml) and RNase A (4 μg/ml) for 30 min at 4° with gentle stirring. Cell debris is removed by centrifugation at 5000g for 10 min. The supernatant is then centrifuged at 250,000g for 60 min, and the membrane pellets are suspended in 10 mM potassium phosphate, pH 7.0, 20% glycerol, 0.1 mM EDTA, and 1 mM GSH.

Purification of rat liver microsomal glutathione transferase expressed in bacteria is performed as follows: Membranes (isolated from 1 l of culture) are solubilized by addition of an equal volume of 10 mM potassium phosphate, pH 7.0, 20% glycerol, 0.1 mM EDTA, 1 mM glutathione, and 6% Triton X-100 and incubated 15 min on ice. Hydroxyapatite chromatography is performed by a batch procedure, where solubilized membranes are adsorbed to 5 g hydroxyapatite equilibrated with buffer A for 15 min. The hydroxyapatite is pelleted by a low-speed centrifugation pulse as described previously and washed with 2 volumes of buffer A followed by 1 volume of 50 mM potassium phosphate in buffer A. Microsomal glutathione transferase is then eluted with 0.4 M potassium phosphate in buffer A and desalted by dialysis for 24 h against 50 volumes of buffer A. Further purification was performed by ion-exchange chromatography on a CM-Sepharose column equilibrated with buffer A. After washing the column with 5 column volumes buffer A, the enzyme is eluted with 0.2 M KCl in buffer A; 0.2 ml fractions are collected at a flow rate of 0.3 ml/min. It is estimated that approximately 1% of bacterial membrane protein is MGST1 and 1 mg of pure protein can be obtained from 1 l of bacterial culture.

Enzyme Assays

Glutathione Transferase

The classical CDNB assay used for most GSTs is also the most common for MGST1 (Habig *et al.*, 1974). The only alteration is the inclusion of detergent and adjustment of substrate concentrations to the specific characteristics of MGST1 (relatively high K_m for GSH and a very low K_m for CDNB) (Morgenstern and DePierre, 1983). The assay at 30° is performed in 0.1 M potassium phosphate, pH 6.5, and 0.1 % Triton X-100 including 5 mM GSH and 0.5 mM CDNB (added from a 20-mM stock in ethanol; final concentration of ethanol tolerated by MGST1 is 5%) and activity recorded at 340 nm with an extinction coefficient of 9.6 mM^{-1}. When subcellular fractions are measured, detergent can be omitted (although the decrease in turbidity on detergent inclusion does increase sensitivity considerably), and if a new system is examined, it is important to compare rates with and without detergent. For instance, in the case of rat liver microsomes, the rates are comparable, whereas human liver microsomes displayed severalfold higher activity in detergent. An assay similar to the one with CDNB involves the analog 4-chloro-3-nitro-benzamide (Morgenstern *et al.*, 1988). The procedure is exactly as the previous one, but the wavelength used is 370 nm and the extinction coefficient 3.1 mM^{-1}. This assay is particularly useful for MGST1, because it does not respond to the degree of activation of the enzyme.

Glutathione Peroxidase

The glutathione peroxidase activity of MGST1 is measured in a coupled spectrophotometric assay system containing 0.1 M potassium phosphate, pH 6.5, 0.1% Triton X-100, 0.2 mM NADPH, 5 mM GSH, GSH reductase (2.4 U/ml), and 1 mM (final) cumene hydroperoxide (30 mM stock dissolved in ethanol). NADPH oxidation is followed at 340 nm with an extinction coefficient of 6.22 mM^{-1}. In principle, this assay can be used with most hydroperoxides, provided that the activity is reasonable compared with the background oxidation rate of GSH and NADPH.

Fluorescent Assay

In a newly developed fluorescent assay for MGST1 (and several cytosolic GSTs), the enzyme is assayed in 0.1 M potassium phosphate buffer (pH 6.5) containing 0.1% Triton X-100 and 5 mM GSH (on a Shimadzu fluorescence spectromonitor RF-510LC, Analytical Instruments Division, Kyoto, Japan) (Svensson *et al.*, 2002). 6-Chloroacetyl-2-dimethylaminonaphthalene is synthesized as described (Svensson *et al.*, 2002), dissolved in methanol, and added to give a final concentration of 2 μM. The excitation wavelength is 380 nm with the emission at 530 nm (regular phosphate buffer) or 520 nm (phosphate buffer containing Triton X-100). This assay responds to activation and has the potential to score MGST1 activation in cellular systems where a high sensitivity is required. Furthermore, the fluorescence of the product is environmentally sensitive and changes on binding to MGST1 and can, therefore, be used to study enzyme ligand interactions as well.

Meisenheimer Complex Formation

Glutathione transferases stabilize reversible dead-end complexes between 1,3,5-trinitrobenzene and GSH (Andersson *et al.*, 1995; Graminski *et al.*, 1989). The complex is intensely colored and absorbs between

400–600 nm, with a peak around 450 nm with an extinction coefficient of 27 mM^{-1}. Thus, addition of 800 μM (final concentration from a 10-mM stock in EtOH) 1,3,5-trinitrobenzene to a solution of MGST1 containing GSH (5 mM) is a convenient way to determine the concentration of enzyme. It seems that only one of the three subunits stabilizes the dead-end complex, and as a result, the concentration of trimers is obtained. A spectrum without enzyme determines the background that needs to be subtracted.

MGST1 Activation

Activation of MGST1 by N-Ethylmaleimide

Each subunit in MGST1 contains a single cysteine. Once these are modified, the enzyme activity to many substrates is increased substantially (Morgenstern and DePierre, 1983; Mosialou et al., 1995). We have shown that activation increases the rate of GSH thiolate formation on the enzyme (Svensson et al., 2004). Therefore, reactive substrates where thiolate anion formation and not the chemical reaction is rate limiting undergo activation. The enzyme can be activated by many reagents that attack the cysteine-49 sulfhydryl, and we routinely use N-ethylmaleimide (NEM). Enzyme as purified is treated with 5 mM NEM (final concentration from a 0.1-M stock dissolved in water or buffer) on ice, and the activity is recorded in the CDNB assay. After a few minutes, a maximum (15–30-fold activation) is reached, and the reaction is stopped with 5 mM GSH (final concentration added from a 0.2-M stock in water or buffer). All solutions of GSH are acidic as prepared and must be pH adjusted to an appropriate pH; we routinely add an equimolar amount of KOH to diminish the acidity and buffering capacity of GSH stock solutions). The activity of the activated enzyme slowly declines, so it is recommended to use each batch the same day. If the excess NEM that has reacted with GSH is a concern, the gel filtration procedure described previously is used to remove this compound. Activation of MGST1 can also be accomplished with many different sulf-hydryl reactive substances (Aniya and Anders, 1989; Aniya et al., 2001; Ji and Bennett, 2003; Ji et al., 2002; Morgenstern et al., 1979), as well as thiol disulfide interchange (Dafre et al., 1996).

Activation of MGST1 in subcellular fractions involves addition of 1 mM NEM (final) and subsequent assay with CDNB (or other substrate to be tested) (Morgenstern et al., 1979). The activation is very rapid (when GSH is not present). A high activation capacity is observed in freshly prepared liver microsomes (up to sevenfold) but declines on storage and freezing; therefore, the use of freshly prepared fractions is important. Activation capacity is lower in rat liver microsomes (compared with pure enzyme), because cytosolic GSTs are present and account for approximately 70% of the basal activity (Morgenstern et al., 1983). Taking this into account, the enzyme is activated to the same degree in microsomal membranes as in the purified state in Triton X-100.

Activation of MGST1 by Proteolysis

An alternate activation protocol involves treatment of MGST1 with trypsin that cleaves the native enzyme at Lys-4 and Lys-41, where cleavage at the latter site results in activation (Morgenstern et al., 1989). MGST1 in purification buffer is treated with trypsin (≈ 2 mg/mg), at room temperature and CDNB activity is followed until a maximum is reached (≈ 60 min); a twofold excess of trypsin inhibitor is added to stop the reaction. The activated preparation is studied the same day as the degree of activation slowly declines. If rat liver microsomes are studied, the required amount of trypsin is much lower (10 μg/mg microsomal protein), which yields maximal activation in 15 min in 70 mM potassium phosphate, pH 7.6 and 0.25 M sucrose. The absence of GSH probably explains this observation.

Notes on the Study of MGST1 Pre-Steady State Kinetics

To study MGST1 thiolate anion formation, Meisenheimer complex stabilization, bursts of product formation (Morgenstern et al., 2001; Svensson et al., 2004), and single turnovers using a stop-flow apparatus large amounts of enzyme need to be prepared as described previously. The enzyme contains no tryptophan, which makes study of protein fluorescence insensitive. GSH-free enzyme, activated enzyme, and enzyme in various buffer compositions are prepared by the 10DG gel filtration procedure. We have chosen to perform all pre-steady state studies of MGST1 at 4°, because the activated enzyme is not stable in the absence of GSH at higher temperatures normally used for assay (but aggregates out of solution).

Acknowledgments

Studies in the author's laboratory were supported by the Swedish Cancer Society, the Swedish National Board for Laboratory Animals, and Funds from Karolinska Institutet. The important experimental contributions by Gerd Lundqvist and all co-workers involved in research on MGST1 over the years are gratefully acknowledged.

References

Andersson, C., Mosialou, E., Weinander, R., and Morgenstern, R. (1994). Enzymology of microsomal glutathione S-transferase. In "Conjugation-Dependent Carcinogenicity and

Toxicity of Foreign Compounds" (M. W. Anders and W. Dekant, eds.), Vol. 27, pp. 19–35. Academic Press, San Diego.

Andersson, C., Piemonte, F., Mosialou, E., Weinander, R., Sun, T.-H., Lundqvist, G., Adang, A. E. P., and Morgenstern, R. (1995). Kinetic studies on rat liver microsomal glutathione transferase, consequences of activation. *Biochim. Biophys. Acta* **1247**, 277–283.

Andersson, C., Söderström, M., and Mannervik, B. (1988). Activation and inhibition of microsomal glutathione transferase from mouse liver. *Biochem. J.* **249**, 819–823.

Aniya, Y., and Anders, M. W. (1989). Activation of rat liver microsomal glutathione S-transferase by reduced oxygen species. *J. Biol. Chem.* **264**, 1998–2002.

Aniya, Y., Kunii, D., and Yamazaki, K. (2001). Oxidative and proteolytic activation of liver microsomal glutathione S-transferase (GSTm) of rats. *Chem. Biol. Interact.* **133**, 144–147.

Bresell, A., Weinander, R., Lundqvist, G., Raza, H., Shimoji, M., Sun, T.-H., Balk, L., Wiklund, R., Eriksson, J., Jansson, C., Persson, B., Jakobsson, P.-J., and Morgenstern, R. (2005). Bioinformatic and enzymatic characterization of the MAPEG super-family. *FEBS J.* **272**, 1688–1703.

Busenlehner, L. S., Codreanu, S. G., Holm, P. J., Bhakat, P., Hebert, H., Morgenstern, R., and Armstrong, R. N. (2004). Stress sensor triggers conformational response of the integral membrane protein microsomal glutathione transferase 1. *Biochemistry* **43**, 11145–11152.

Dafre, A. L., Sies, H., and Akerboom, T. (1996). Protein S-thiolation and regulation of microsomal glutathione transferase activity by the glutathione redox couple. *Arch. Biochem. Biophys.* **332**, 288–294.

Ernster, L., Siekevitz, P., and Palade, G. E. (1962). Enzyme-structure relationships in the endoplasmic reticulum of rat liver: Morphological and biochemical study. *J. Cell Biol.* **15**, 541–562.

Graminski, G. F., Zhang, P., Sesay, M. A., Ammon, H. L., and Armstrong, R. N. (1989). Formation of the 1-(S-glutathionyl)-2,4,6-trinitrocyclohexadiente anion at the active site of glutathione S-transferase: Evidence for enzymic stabilization of s-complex intermediates in nucleophilic aromatic substitution reactions. *Biochemistry* **28**, 6252–6258.

Habig, W. H., Pabst, M. J., and Jakoby, W. B. (1974). Glutathione S-transferases. The first enzymatic step in mercapturic acid formation. *J. Biol. Chem.* **249**, 7130–7139.

Jakobsson, P. J., Morgenstern, R., Mancini, J., Ford-Hutchinson, A., and Persson, B. (1999). Common structural features of MAPEG—a widespread superfamily of membrane associated proteins with highly divergent functions in eicosanoid and glutathione metabolism. *Protein Sci.* **8**, 689–692.

Ji, Y., and Bennett, B. M. (2003). Activation of microsomal glutathione s-transferase by peroxynitrite. *Mol. Pharmacol.* **63**, 136–146.

Ji, Y., Toader, V., and Bennett, B. M. (2002). Regulation of microsomal and cytosolic glutathione S-transferase activities by S-nitrosylation. *Biochem. Pharmacol.* **63**, 1397–1404.

Laemmli, U. K. (1970). Cleavage of structural proteins during the assembly of the head of bacteriophage T4. *Nature* **227**, 680–685.

Lengqvist, J., Svensson, R., Evergren, E., Morgenstern, R., and Griffiths, W. J. (2004). Observation of an intact noncovalent homotrimer of detergent-solubilized rat microsomal glutathione transferase-1 by electrospray mass spectrometry. *J. Biol. Chem.* **279**, 13311–13316.

Maeda, A., Crabb, J. W., and Palczewski, K. (2005). Microsomal glutathione S-transferase 1 in the retinal pigment epithelium: Protection against oxidative stress and a potential role in aging. *Biochemistry* **44**, 480–489.

McLellan, L. I., Wolf, C. R., and Hayes, J. D. (1989). Human microsomal glutathione S-transferase, Its involvment in the conjugation of hexachloro-1.3-butadiene with glutathione. *Biochem. J.* **258**, 87–93.

Morgenstern, R., and DePierre, J. W. (1983). Microsomal glutathione transferase, Purification in unactivated form and further characterization of the activation process, substrate specificity and amino acid composition. *Eur. J. Biochem.* **134**, 591–597.

Morgenstern, R., DePierre, J. W., and Ernster, L. (1979). Activation of microsomal glutathione transferase activity by sulfhydryl reagents. *Biochem. Biophys. Res. Commun.* **87**, 657–663.

Morgenstern, R., Guthenberg, C., and DePierre, J. W. (1982). Microsomal glutathione transferase. Purification, initial characterization and demonstration that it is not identical to the cytosolic glutathione transferases A, B and C. *Eur. J. Biochem.* **128**, 243–248.

Morgenstern, R., Guthenberg, C., Mannervik, B., and DePierre, J. W. (1983). The amount and nature of glutathione transferases in rat liver microsomes determined by immunochemical methods. *FEBS Lett.* **160**, 264–268.

Morgenstern, R., Lundqvist, G., Hancock, V., and DePierre, J. W. (1988). Studies on the activity and activation of rat liver microsomal glutathione transferase, in particular with a substrate analogue series. *J. Biol. Chem.* **263**, 6671–6675.

Morgenstern, R., Lundqvist, G., Jörnvall, H., and DePierre, J. W. (1989). Activation of rat liver microsomal glutathione transferase by limited proteolysis. *Biochem. J.* **260**, 577–582.

Morgenstern, R., Svensson, R., Bernat, B. A., and Armstrong, R. N. (2001). Kinetic analysis of the slow ionization of glutathione by microsomal glutathione transferase MGST1. *Biochemistry* **40**, 3378–3384.

Mosialou, E., Andersson, C., Lundqvist, G., Andersson, G., Bergman, T., Jornvall, H., and Morgenstern, R. (1993a). Human liver microsomal glutathione transferase - Substrate specificity and important protein sites. *FEBS Lett.* **315**, 77–80.

Mosialou, E., Ekström, G., Adang, A. E. P., and Morgenstern, R. (1993b). Evidence that rat liver microsomal glutathione transferase is responsible for glutathione-dependent protection against lipid peroxidation. *Biochem. Pharmacol.* **45**, 1645–1651.

Mosialou, E., Piemonte, F., Andersson, C., Vos, R., Van Bladeren, P. J., and Morgenstern, R. (1995). Microsomal glutathione transferase—lipid-derived substrates and lipid dependence. *Arch. Biochem. Biophys.* **320**, 210–216.

Peterson, G. L. (1977). A simplification of the protein assay method of Lowry *et al.* which is more generally applicable. *Anal. Biochem.* **83**, 346–356.

Sun, T. H., Ling, X., Persson, B., and Morgenstern, R. (1998). A highly active microsomal glutathione transferase from frog (*Xenopus laevis*) liver that is not activated by N-ethylmaleimide. *Biochem. Biophys. Res. Commun.* **246**, 466–469.

Svensson, R., Alander, J., Armstrong, R. N., and Morgenstern, R. (2004). Kinetic characterization of thiolate anion formation and chemical catalysis of activated microsomal glutathione transferase 1. *Biochemistry* **43**, 8869–8877.

Svensson, R., Greno, C., Johansson, A. S., Mannervik, B., and Morgenstern, R. (2002). Synthesis and characterization of 6-chloroacetyl-2-dimethylaminonaphthalene as a fluorogenic substrate and a mechanistic probe for glutathione transferases. *Anal Biochem.* **311**, 171–178.

Tartof, K. D., and Hobbs, C. A. (1987). Improved media for growing plasmid and cosmid clones. *FOCUS* **9**, 12.

Weinander, R., Mosialou, E., Dejong, J., Tu, C. P. D., Dypbukt, J., Bergman, T., Barnes, H. J., Hoog, J. O., and Morgenstern, R. (1995). Heterologous expression of rat liver microsomal glutathione transferase in simian cos cells and *Escherichia coli. Biochem. J.* **311**, 861–866.

[9] Human Microsomal Prostaglandin E Synthase 1: A Member of the MAPEG Protein Superfamily

By Pär L. Pettersson, Staffan Thorén, and Per-Johan Jakobsson

Abstract

In this chapter, we briefly review the MAPEG superfamily (membrane associated proteins in eicosanoid and glutathione metabolism), a family of proteins in which all human members except one possess glutathione conjugating capacity. Recent findings regarding the biological functions of MAPEG proteins are highlighted. More extensively, the characterization of human microsomal prostaglandin E synthase 1 is presented, including results and applied methodology.

Introduction

After the identification of 5-lipoxygenase activating protein (FLAP) and leukotriene (LT) C_4 synthase as two closely related proteins involved in leukotriene biosynthesis, a series of gene databank searches and subsequent protein characterizations revealed two novel microsomal glutathione transferases (MGST2 and MGST3) (Jakobsson *et al.*, 1996, 1997). A third novel protein was also identified as an orphan homolog to MGST1 with 38% sequence identity on the amino acid level and named MGST1-like 1 (MGST1-L1). At the point at which the identity of these six human proteins were known, the MAPEG superfamily was defined based on enzymatic activities, structural properties, and sequence similarities (Jakobsson *et al.*, 1999a). Subsequently, MGST1-L1 was found to possess high catalytic activity as a prostaglandin (PG) E_2 synthase and renamed more appropriately to microsomal prostaglandin E synthase (MPGES) (Jakobsson *et al.*, 1999b). Since the discovery of MPGES, additional enzymes with PGE synthase activity have been described, one of which constitutes another membrane protein referred to as MPGES2 (Tanikawa *et al.*, 2002) and one that is cytosolic (CPGES) (Tanioka *et al.*, 2000). These latter enzymes lack structural similarities with the MAPEG proteins, but the nomenclature has extended the name of the MAPEG prostaglandin E synthase to MPGES1. Table I recapitulates the human MAPEG proteins known to date.

METHODS IN ENZYMOLOGY, VOL. 401
Copyright 2005, Elsevier Inc. All rights reserved.
0076-6879/05 $35.00
DOI: 10.1016/S0076-6879(05)01009-8

TABLE I

THE MAPEG PROTEINS, YEAR OF IDENTIFICATION, AND CHROMOSOMAL LOCALIZATION

Abbreviation	Full name	Discovery	Chromosome
MGST1	Microsomal Glutathione Transferase 1	1982	12p12
FLAP	5-Lipoxygenase Activating Protein	1990	13q12
LTC4S	Leukotriene C4 Synthase	1993	5q35
MGST2	Microsomal Glutathione Transferase 2	1996	4q28–31
MGST3	Microsomal Glutathione Transferase 3	1997	1q23
MPGES1	Microsomal Prostaglandin E Synthase	1999	9q34

MAPEG and Arachidonic Metabolism

Eicosanoids constitute a group of bioactive lipids derived from polyunsaturated fatty acids, mainly arachidonic acid, acting as hormones with various physiological and pathophysiological effects. The prostaglandin H synthase pathway leads to the formation of PGs and thromboxanes (TX), whereas LTs are formed by means of the 5-lipoxygenase pathway; for recent review see Funk (2001). Figure 1 outlines the described pathways. Prostaglandin H synthase 1 and 2 (PGHS1 and -2) both catalyze the formation of the endoperoxide prostaglandin H_2 (PGH_2), the common substrate for various terminal synthases including MPGES1. 5-Lipoxygenase, with the assistance of FLAP (see section later), catalyzes the intracellular biosynthesis of the allelic epoxide LTA_4 that may become conjugated with glutathione (GSH) by certain glutathione transferases (see later) hence forming LTC_4. Alternately, LTA_4 may be hydrolyzed by LTA_4 hydrolase, producing LTB_4 (Haeggstrom, 2004). In addition, MGST2 and MGST3 are both capable of catalyzing the GSH-dependent reduction of the fatty acid hydroperoxide, 5-hydroperoxyeicosatetraenoic acid (5-HpETE), into the corresponding alcohol, 5-hydroxyeicosatetraenoic acid (5-HETE) (Jakobsson et al., 1997).

MGST1, FLAP, and LTC_4 Synthase

MGST1 was the first microsomal glutathione transferase discovered and isolated (Morgenstern et al., 1982). MGST1 is described separately in this volume. In 1990, Miller et al. purified an 18-kDa protein required for the endogenous LT formation. The isolation of this protein was a result of the finding that MK-886, an inhibitor of 5-lipoxygenase activity, did not have any direct effect on 5-lipoxygenase (Miller et al., 1990). Subsequently, the cDNA was identified, and coexpression with 5-lipoxygenase demonstrated the importance of this protein for LT biosynthesis in intact cells

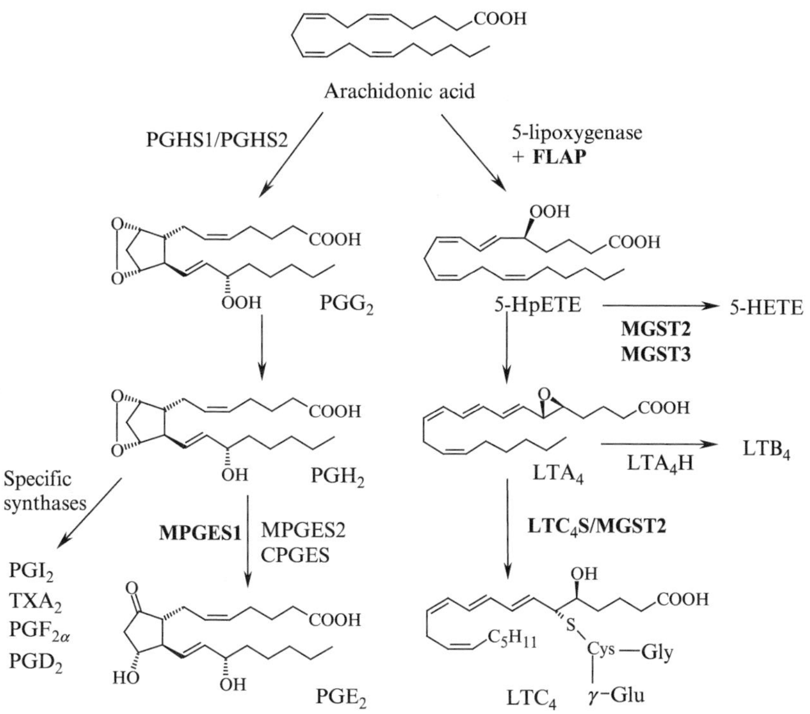

FIG. 1. MAPEG proteins (bold) in arachidonic acid metabolism.

(Dixon et al., 1990). The protein was denominated FLAP, and later research has shown that the mode of action of FLAP is likely to be transportation/presentation of arachidonic acid to 5-lipoxygenase (Abramovitz et al., 1993). As of today, no other enzymatic function has been described for FLAP. Characterizations of various FLAP mutants have demonstrated that the region corresponding to the C-terminal end of the first hydrophilic part binds MK-886 or arachidonic acid (Mancini et al., 1993, 1994). Because MK-886 also inhibits other MAPEG enzymes such as LTC$_4$ synthase, MPGES1, MGST2, and MGST3 (Lam et al., 1994; Quraishi et al., 2002; Schroder et al., 2003), this region of the different MAPEG proteins may represent a hypothetical binding site for fatty acid substrates and may consequently explain and distinguish the substrate specificity of these proteins. Recently, a single-nucleotide polymorphism (SNP) haplotype in the FLAP gene was reported associated with significant increased risk of myocardial infarctions and stroke (Helgadottir et al., 2004), an intriguing finding that together with other recent reports relating to this area of research motivate further investigations on the role of the 5-LO pathway, its enzymes, products, and their receptors for atherogenesis; for recent review see Jala and Haribabu (2004). LTC_4 synthase specifically catalyzes the conjugation of GSH to LTA_4 forming LTC_4, a potent mediator of airway obstruction. LTC_4 synthase was first purified by in 1993 (Nicholson et al., 1993) and was subsequently cloned and characterized independently by two groups (Lam et al., 1994; Welsch et al., 1994). After the cloning of LTC_4 synthase, the homology with FLAP was revealed, justifying further searches for additional proteins as discussed earlier. In white blood cells, the dominant source of this activity originates from LTC_4 synthase (Lam, 2003), but in umbilical vein endothelial cells, this activity seems predominantly derived from MGST2 (Scoggan et al., 1997; Sjöström et al., 2001). In opposition to data reported for MGST2 and MGST3, LTC_4 synthase protein is regulated by various interleukins, and, recently, lipopolysaccharide was reported to increase mRNA and protein expression in rats (Schroder et al., 2005). For further readings on the biology and significance of FLAP and LTC_4 synthase, the reader is referred to one of several reviews of the field (e.g., Lam, 2003; Peters-Golden and Brock, 2003).

The MAPEG Proteins and Glutathione

MGST1, MGST2, and MPGES1 have all been shown to conjugate GSH with 1-chloro-2,4-dinitrobenzene (CDNB), a model substrate for activity measurement of many GSTs. Both MGST2 and MGST3 have also been shown to catalyze the formation of LTC_4 from GSH and LTA_4, although the activity observed for MGST3 was significantly lower compared with MGST2 using the same type of expression system (baculovirus/Sf9 cells) and assay methods (Jakobsson et al., 1997). In addition, MGST1, -2, and -3 all catalyze glutathione-dependent peroxidase reactions (Jakobsson et al., 1997). Like MGST1, both MGST2 and MGST3 are expressed in many various organs and cells and seem not to increase during inflammation. They are also demonstrating wider substrate specificities and catalytic functions. Therefore, apart from the putative specific role of MGST2 for endothelial LTC_4 production, MGST2 and MGST3 seem more related to MGST1 and may possess more general functions (e.g., in the protection from compounds produced during oxidative stress). For example, Ma and coworkers recently showed in a microarray study (Ma et al., 2004) that the MGST2 gene contributes to the resistance of T cells to tert-butylhydroperoxide (TBOOH) adapted to grow in the presence of the hydroperoxide. Their results also suggest a similar function for MGST3, however, not sufficient for TBOOH resistance. On the other hand, one cannot exclude the possibility of specific functions, and clearly further studies are required to better understand the physiology of these enzymes. Another aspect of the MAPEG protein functions are their interactions. On the product level, it has been shown that LTC_4 binds to and constitutes a tight binding inhibitor of MGST1 (Bannenberg *et al.*, 1999; Metters *et al.*, 1994). On the protein level, MGST1 has been found to interact with LTC_4 synthase (Soderstrom *et al.*, 1995; Surapureddi *et al.*, 1996) and to co-localize intracellularly (Surapureddi *et al.*, 2000). Furthermore, FLAP and LTC_4 synthase have been proposed to form heterocomplexes (Mandal *et al.*, 2004). For further reading on the oligomerization of MAPEG proteins, the reader is referred to Chapter 10.

MPGES1

The inducible enzyme MPGES1 is linked to the severalfold increase of PGE_2 formation observed in cells and organs during inflammation. There are several reports on the role of MPGES1 in inflammation coupling with prostaglandin H synthases, expression in the blood–brain barrier, overexpression in cancers, etc; for some recent reviews see Engblom *et al.* (2002), Jakobsson *et al.* (2002), and Murakami and Kudo (2004). In brief, the enzyme was discovered in 1999 (Jakobsson *et al.*, 1999b), and soon thereafter a report was published on the preferential coupling of MPGES1 with PGHS2 (Murakami *et al.*, 2000). The enzyme is induced by proinflammatory cytokines and down-regulated by glucocorticosteroids. In experimental models of inflammation, taking advantage of MPGES1 null mice, the enzyme has been found to be critically involved in the development of collagen-induced arthritis (Kamei *et al.*, 2004; Trebino *et al.*, 2003), endotoxin and immune-dependent induced fever (Engblom *et al.*, 2003; Saha *et al.*, 2005), and both inflammatory (Trebino *et al.*, 2003) and neuropathic pain (Mabuchi *et al.*, 2004). In patients with rheumatoid arthritis, the protein is overexpressed in the synovial tissue, particularly by macrophages and fibroblasts in the synovial lining layer (Westman *et al.*, 2004). Current work in our laboratory focuses on a better understanding of the role of MPGES1 in rheumatoid arthritis, possible shunting of PGH_2 into antiinflammatory pathways in mice lacking this enzyme, and the chemical mechanism of catalysis.

Characterization of MPGES1

Expression and Purification

Cloning of Human MPGES1. The cDNA coding for human microsomal MPGES1 was isolated by polymerase chain reaction (PCR) from an EST-clone (143735) using sequence specific primers (GAGAGACA-

TATGCCTGCCCACAGCCTG and GAGAGAAAGCTTCACAGGTG GCGGGCCGC (Jakobsson et al., 1999b). The cDNA was subcloned into the pSP19T7LT vector, and an N-terminal tag of six histidines was engineered into the nucleotide sequence to enable facilitated purification (Thoren et al., 2003). The resulting pSP19T7LT vector containing the nucleotide sequence for the His-tagged MPGES1 was transformed into Escherichia coli [BL21 (DE3)] also harboring the pLysSL vectors used to aid the expression in E. coli and to lysate the host when harvesting (Studier, 1991).

Expression of Human MPGES1 in E. coli. Expression was started by inoculating a 5-ml culture of LB-media containing 100 μg/ml ampicillin (for the pSP19T7LT vector) and 20 μg/ml chloramphenicol (for the pLysSL vectors) with one single colony from a fresh agar plate containing the same antibiotics (used in all culturing media). The culture was grown for 6–8 h, and then 0.5 ml was transferred to 50-ml LB-media and cultured overnight, in both cases at 37° and shaking. From the overnight culture, 20 ml was used to inoculate 2 l of TB-media in large Erlenmeyer flasks. These cultures were incubated at 37° with shaking (200 rpm). The OD_{600} was monitored and at OD_{600} 0.5–0.6, the expression culture was induced with 3 mM isopropyl-β-D-thiogalactopyranoside (IPTG), and further cultured for 3 h. The E. coli were harvested by centrifugation at approximately 3,000g for 10 min at 4°, washed once with ice-cold phosphate-buffered saline (PBS), and centrifuged again. The pellets were then frozen at −20° until purification.

Solubilization of Recombinant Human MPGES1. A pellet equivalent to 1 l of expression culture was thawed on ice and resuspended in 20 ml 20 mM phosphate buffer, pH 8.0, containing 150 mM NaCl, 10% glycerol, and 1 mM GSH (lysis buffer). The internal T7 lysozyme of the pLysSL will lyse the cells without any additions. The samples were kept on ice at all steps. The cell lysate was sonicated until no longer viscous to break present DNA that will interfere with subsequent steps. One volume of lysis buffer containing 8% of Triton X-100 (Sigma T-9284, final concentration 4%) was added, and the mixture was incubated on ice with gentle stirring for 30 min. After the solubilization step, the mixture was centrifuged at 100,000g for 30 min at 4° to remove any insoluble cell debris. The supernatant containing the solubilized MPGES1 was then ready for the purification process.

Purification of Solubilized Human MPGES1. The purification of MPGES1 in our laboratory follows a two-step procedure with subsequent desalting. The solubilized proteins were mixed with 1.2 g hydroxyapatite (Bio-Rad; Bio-Gel HTP), equilibrated three times with 30 ml of 10 mM phosphate buffer, pH 8.0 (start buffer), containing 150 mM NaCl, 10 mM imidazole (Merck 1.04716), 10% glycerol, 1 mM GSH, and 0.2% reduced Triton X-100 (Sigma X-100-RS or Aldrich 28,210-3). The use of re-duced Triton X-100 enables UV detection of eluted proteins, because its absorbance is approximately 100-fold lower than that of regular Triton X-100 at 280 nm. This mixture was incubated on ice with stirring every 2 min for a total of 10 min. The hydroxyapatite was removed by a short centrifugation at 400g, the supernatant decanted, and centrifuged for an additional 3 min at 3000g before filtration through a 0.45-μm filter. To get the filtrate ready for the next purification step, 10 mM imidazole was added, and the pH was adjusted to 7.8–8.0.

In the second step of purification, immobilized metal affinity chromatography (IMAC) was used. Nickel ions were immobilized onto the Chelating Sepharose Fast Flow gel (Amersham Biosciences, Sweden), prepacked in HiTrap columns according to the manufacturer, and subsequently washed with water before mounting in a ÄKTA purifier 10 system (Amersham Biosciences, Sweden). The column mounted on the ÄKTA system was equilibrated with 5 volumes of start buffer, 5 volumes of elution buffer (same as start buffer but containing 400 mM imidazole), and finally another 10 volumes of start buffer. The filtrate from the hydroxyapatite step was subsequently applied to the column after which 5–10 volumes of start buffer were run through. A washing step of 60 mM imidazole (ÄKTA system: 85% start buffer and 15% elution buffer) was performed, and finally MPGES1 was eluted with 350 mM imidazole (ÄKTA system: 12% start buffer and 88% elution buffer).

Immediately after the elution from the Ni-IMAC, the collected fractions of MPGES1 were transferred into a 20 mM phosphate buffer, pH 7.5, containing 50 mM NaCl, 10% glycerol, 1 mM GSH, and 0.2% reduced Triton X-100 by means of a gel filtration step (HiPrep 26/10 Desalting; Amersham Biosciences, Sweden).

Evaluation of Purity and Protein Amounts. The purity of MPGES1 was evaluated by standard sodium dodecyl sulfate-polyacrylamide gel electrophoresis (SDS-PAGE) and silver staining or Western blot using antibodies raised toward human MPGES1 with subsequent ECL detection (Amersham Biosciences).

Protein amounts were determined using the Lowry based Bio-Rad DC protein assay (Bio-Rad, Hercules, CA) for use with detergents. Also, with the addition of iodoacetamide, it is possible to accurately determine protein concentration in solutions containing thiols such as GSH (Bio-Rad bulletin 1909).

Kinetic Assays

Prostaglandin E_2 Synthase Activity Assay with PGH_2. The standard prostaglandin E synthase activity is performed in 0.1 M phosphate buffer at pH 7.5 containing 0.1% Triton X-100. Specific activity is assayed with 2.5

TABLE II
KINETIC PARAMETERS OF MPGES1 WITH PGH$_2$ AND PGG$_2$

	k_{cat} (s^{-1})	K_m(μM)	k_{cat}/K_m (mM^{-1} s^{-1})
PGH$_2$	50 ± 6	160 ± 4	310 ± 40
PGG$_2$	75 ± 4	160 ± 3	470 ± 30

mM GSH and 10 μM PGH$_2$. When determining kinetic parameters k_{cat}, K_m and k_{cat}/K_M, the concentrations of PGH$_2$ and GSH were varied between 10–400 μM and 0.1–10 mM, respectively. When varying one of the two substrates, the other substrate was kept at a higher concentration (above K_m) to establish pseudo-first-order kinetic conditions. The amount of purified MPGES1 was 20–40 ng/100-μl assay solution. The assays were usually performed at 37°, kinetic parameters in Table II (but also at 4° [on ice] or 30°), and the amount of enzyme used was set to obtain a maximum of 30% conversion of substrate over a 1-min incubation time. After incubation of substrates and enzyme, the reaction is stopped by the addition of 400 μl (to 100 μl reaction assay) of a solution containing citric acid (50 mM), FeCl$_2$ (25 mM), and an internal standard for HPLC assay (e.g., 11-β-PGE$_2$, 2.5 μM). This acidic solution immediately stops the enzymatic activity of MPGES1, and the ferric ions convert all remaining PGH$_2$ into mainly 12-(S)-hydroxyheptadecatrienoic acid (12-HHT) and malondialdehyde. Each reaction was subsequently solid phase extracted (C$_{18}$), and the sample eluted into acetone, evaporated under N$_2$, and redissolved in 100 μl acetonitrile and 200 μl water (mobile phase for the reversed-phase HPLC without trifluoroacetic acid [TFA]). Samples were then applied to a reversed-phase HPLC system, and the chromatograms collected at 195 nm were analyzed with regard to produced PGE$_2$, internal standard, and, if required, other formed products. Analogous to all samples, there was always a nonenzymatic reaction performed to distinguish between enzymatically and nonenzymatically formed products. Also, an additional test was usually performed with an excess of enzyme to confirm the total amount of available substrate.

Prostaglandin E$_2$ Synthase Activity Assay with PGG$_2$. The main activity of MPGES1 is generally recognized to be the catalysis of PGH$_2$ into PGE$_2$. However, MPGES1 also catalyze the formation of 15-hydroperoxy-PGE$_2$ from PGG$_2$, the intermediate prostaglandin of PGHS1 and -2, with a high rate.

The kinetic assay using PGG$_2$ as substrate is analogous to that for PGH$_2$, with the difference that the reaction was terminated by the addition of 6 μl 1 M HCl and 45 μl acetonitrile. The mixture was immediately centrifuged at 14,000g for 1 min before analysis using reversed-phase HPLC with UV detection at 195 nm.

5-HpETE Peroxidase Activity Assay. Forty μM 5-HpETE was incubated in 0.1 mM phosphate buffer, pH 7.5, containing 5 mM GSH and 0.005% (w/v) bovine serum albumin for 1 min at 37°. The amount of MPGES1 was in the order of 2.5–10 $\mu g/100$-μl assay solution. The reaction was terminated by the addition of 200 μl acetonitrile containing 0.2% (v/v) acetic acid; 100 μl water was added, and the sample was centrifuged at 14,000g for 1 min and analyzed with reversed-phase HPLC with UV detection at 236 nm.

Cumene Hydroperoxide Activity Assay. Peroxidase activity of MPGES1 with cumene hydroperoxide was monitored using a coupled assay, where the linear decrease of NADPH, consumed by an excess of glutathione reductase present, is followed. The reaction mix consists of 0.1 M phosphate buffer, pH 6.5, 1 mM GSH, 0.2 mM NADPH, 0.5 mM cumene hydroperoxide, and 50 μg purified MPGES1 per 1-ml assay solution. The decrease of NADPH is proportional to the formation of product and is followed spectrophotometrically at 340 nm and 37°.

Activity Assay for the Conjugation of GSH to CDNB. The classical glutathione transferase conjugation reaction of CDNB and GSH was assayed using 0.1 M phosphate buffer at pH 6.5 containing 5 mM GSH, 2 mM CDNB, and 0.1% Triton X-100. The formation of the conjugate was followed at 340 nm and 30°.

General Enzymatic Activity Assay Procedures. In all activity assays, if formation of nonenzymatic products occurs, these were measured and subtracted from the enzymatic data. Also, in determinations using HPLC, all chromatographic peaks were identified and quantified by comparing with known amounts of synthetic standards.

Hydrodynamic Studies of MPGES1

Hydrodynamic studies were performed on the MPGES1–Triton X-100 complex to determine the molecular mass and oligomerization of MPGES1 using the Svedberg equation (Chang, 1990).

Gradient Centrifugation. Purified MPGES1 (25 μg) and the marker proteins cytochrome c (0.1 mg) and bovine serum albumin (1 mg) with sedimentation coefficients of 1.7 S and 4.6 S, respectively, were dissolved in 10 mM phosphate buffer, pH 7.4, 2 mM GSH, and 1% Triton X-100 in a total volume of 200 μl. The protein solution was carefully added on top of a 10-ml 5–20% linear sucrose gradient in the same buffer as the sample. The samples were subjected to ultracentrifugation at 160,000g for 45 h at 20°. Fractions of 0.4 ml were collected from the bottom of the tube using a syringe and a pump. PGE$_2$

formation was measured to identify MPGES1-containing fractions, and the protein content was measured according to Pande and Murthy (Pande and Murthy, 1994) to identify fractions containing cytochrome c and BSA. After plotting the activities and protein content, the sedimentation coefficient of MPGES1 was calculated by linear regression.

Density Equilibrium Centrifugation. The partial specific volume of the MPGES1–Triton X-100 complex was determined by density equilibrium centrifugation. Purified MPGES1 (15 μg in 200 μl buffer) was added to a 3.8 ml, 20–50% linear sucrose gradient containing the same buffer as previously described. The samples were centrifuged at 246,000g at 20° until equilibrium had been reached (72 h). Fractions of 0.16 ml were collected, and refractive index and PGE$_2$-formation were measured as previously described. The density of sucrose was then plotted against PGE$_2$ formation. The fraction with the highest PGE$_2$ formation corresponded to the density of the MPGES1–Triton X-100 complex, which is inversely proportional to the partial specific volume.

Size Exclusion Chromatography. Stokes radius was determined by the use of size exclusion chromatography. Purified MPGES1 was loaded on a Sephacryl S-300 HR column together with marker enzymes (HMW gel filtration calibration kit, Amersham Biosciences). The samples were eluted, collected, and analyzed by measuring PGE$_2$ formation and absorbance at 280 and 405 nm (for the marker enzymes, according to the manufacturer's instructions). The square root of the $-\log K_{av}$ values were plotted against the known Stokes radii of the marker enzymes and the Stokes radius of the MPGES1–Triton X-100 complex was obtained by linear regression.

Detergent Content. To determine the amount of bound detergent in the complex, protein content and UV absorbance were measured on the eluted fractions from the immobilized metal ion affinity column. Subtracting the absorbance of the Triton X-100 in the buffer and the calculated absorbance of the amount of enzyme in the sample at 280 nm yielded an absorbance value for the corresponding amount of Triton X-100 bound to MPGES1.

Summary of Results

The PGES Activity of MPGES1. The conversion of PGH$_2$ into PGE$_2$ is catalyzed by MPGES1 (Fig. 2) with a high rate. PGH$_2$ is formed by PGHS, which converts arachidonic acid in a two-step mechanism by means of PGG$_2$. MPGES1 also catalyzed the formation 15-hydroperoxy-PGE$_2$ from the PGHS intermediate PGG$_2$ (Fig. 2). The physiological significance of the formation of 15-hydroperoxy-PGE$_2$ has not yet been determined and requires further investigation.

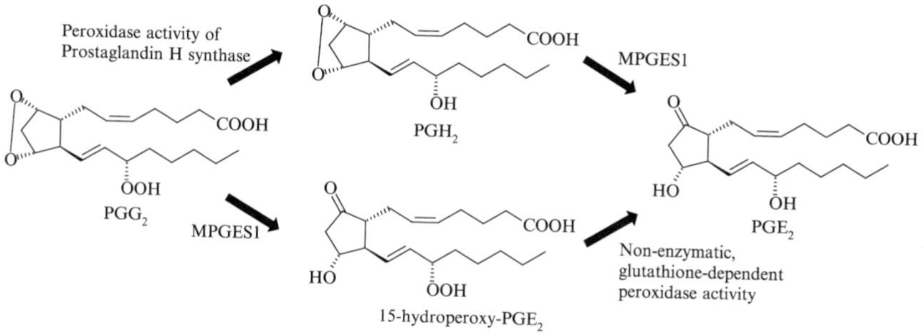

Fig. 2. PGES activity of MPGES1.

TABLE III
PEROXIDASE AND TRANSFERASE ACTIVITY DATA OF MPGES1

	Specific activity (μmol min^{-1} mg^{-1})
5-HpETE	0.043 ± 0.001
Cumene hydroperoxide	0.17 ± 0.02
CDNB	0.81 ± 0.04

The specific activities using PGG$_2$ or PGH$_2$ determined are in the range of 5–10 μmol min^{-1} mg^{-1} when using 10 μM of the prostaglandin substrate and 2.5 mM GSH. Using 400 μM prostaglandin yields activities around 100–200 μmol min^{-1} mg^{-1}. Apparent kinetic parameters k_{cat}, K_m, and k_{cat}/K_m for PGH$_2$ and PGG$_2$ are presented in Table II. K_m^{GSH} is 0.7 ± 0.2 mM. These data were obtained at 37° as reported (Thoren et al., 2003). Another report by Ouellet et al. (2002) presents k_{cat} values for GSH of the same order but with K_m values for PGH$_2$ 10-fold lower. It should be noted that the expression, purification, activity assay, and detergents used were substantially different, and, especially, the selection of detergents may significantly influence kinetic parameters.

Glutathione Peroxidase and Transferase Activity. Table III presents specific activity data derived from Thoren et al. (2003) for 5-HpETE, cumene hydroperoxide, and CDNB.

Quaternary Structure Results. A sedimentation coefficient of 4.1 S, partial specific volume of 0.891 cm^3/g, and a Stokes radius of 5.09 nm were obtained, and the Svedberg equation was used to calculate the molecular mass of the MPGES1–Triton X-100 complex, which was found to be 215,000. The detergent content of the MPGES1–Triton X-100 complex was 2.8 g Triton X-100/g protein, and after subtracting the values for the detergent content, our calculations match with a trimeric quaternary structure. This is in line with studies of the closely related MGST1, which also has been demonstrated to have a trimeric organization; for oligomerization structure, see Chapter 10.

Summary

All members of the MAPEG protein superfamily but FLAP have been characterized as active enzymes, and these enzymes require GSH for their catalytic activity. Several of the MAPEG members catalyze the classical glutathione transferase conjugation of GSH to CDNB as well as a GSH-dependent reduction of fatty acid hydroperoxides (i.e., MGST1, MGST2, and PGES1). The latter reaction is also catalyzed by MGST3. LTC$_4$ synthase and MGST2 both catalyze the conjugation of GSH to LTA$_4$ and thereby produce the proinflammatory compound LTC$_4$. MPGES1 is induced during inflammation and catalyzes the GSH-dependent prostaglandin isomerase reaction of PGG$_2$/PGH$_2$ into 15-hydroperoxy-PGE$_2$/PGE$_2$, respectively. Several of the MAPEG proteins constitute attractive drug targets (MPGES1, LTC$_4$ synthase, and FLAP) for treatment of various diseases such as inflammatory diseases, pain, asthma, and atherosclerosis.

References

Abramovitz, M., Wong, E., Cox, M. E., Richardson, C. D., Li, C., and Vickers, P. J. (1993). 5-Lipoxygenase-activating protein stimulates the utilization of arachidonic acid by 5-lipoxygenase. *Eur. J. Biochem.* **215**, 105–111.

Bannenberg, G., Dahlen, S. E., Luijerink, M., Lundqvist, G., and Morgenstern, R. (1999). Leukotriene C4 is a tight-binding inhibitor of microsomal glutathione transferase-1. Effects of leukotriene pathway modifiers. *J. Biol. Chem.* **274**, 1994–1999.

Chang, R. (1990). "Physical Chemistry with Applications to Biological Systems." Macmillan Publishing Company, New York.

Dixon, R. A., Diehl, R. E., Opas, E., Rands, E., Vickers, P. J., Evans, J. F., Gillard, J. W., and Miller, D. K. (1990). Requirement of a 5-lipoxygenase-activating protein for leukotriene synthesis. *Nature* **343**, 282–284.

Engblom, D., Ek, M., Saha, S., Ericsson-Dahlstrand, A., Jakobsson, P. J., and Blomqvist, A. (2002). Prostaglandins as inflammatory messengers across the blood-brain barrier. *J. Mol. Med.* **80**, 5–15.

Engblom, D., Saha, S., Engström, L., Dahlström, M., Laurent, L. A., Jakobsson, P.-J., and Blomqvist, A. (2003). Microsomal prostaglandin E synthase-1 is the central switch during immune-induced pyresis. *Nat. Neurosci.* **6**, 1137–1138.

Funk, C. D. (2001). Prostaglandins and leukotrienes: Advances in eicosanoid biology. *Science* **294**, 1871–1875.

Haeggstrom, J. Z. (2004). Leukotriene A4 hydrolase/aminopeptidase, the gatekeeper of chemotactic leukotriene B4 biosynthesis. *J. Biol. Chem.* **279**, 50639–50642.

Helgadottir, A., Manolescu, A., Thorleifsson, G., Gretarsdottir, S., Jonsdottir, H., Thorsteinsdottir, U., Samani, N. J., Gudmundsson, G., Grant, S. F., Thorgeirsson, G., Sveinbjornsdottir, S., Valdimarsson, E. M., Matthiasson, S. E., Johannsson, H., Gudmundsdottir, O., Gurney, M. E., Sainz, J., Thorhallsdottir, M., Andresdottir, M., Frigge, M. L., Topol, E. J., Kong, A., Gudnason, V., Hakonarson, H., Gulcher, J. R., and Stefansson, K. (2004). The gene encoding 5-lipoxygenase activating protein confers risk of myocardial infarction and stroke. *Nat. Genet.* **36**, 233–239.

Jakobsson, P. J., Engblom, D., Ericsson-Dahlstrand, A., and Blomqvist, A. (2002). Microsomal prostaglandin E synthase: A key enzyme in PGE2 biosynthesis and inflammation. *Curr. Med. Chem. Anti-Inflammatory Anti-Allergy Agents* **1**, 167–175.

Jakobsson, P.-J., Mancini, J. A., and Ford-Hutchinson, A. W. (1996). Identification and characterization of a novel human microsomal glutathione S-transferase with leukotriene C4 synthase activity and significant sequence identity to 5-lipoxygenase activating protein and leukotriene C4 synthase. *J. Biol. Chem.* **271**, 22203–22210.

Jakobsson, P.-J., Mancini, J. A., Riendeau, D., and Ford-Hutchinson, A. W. (1997). Identification and characterization of a novel microsomal enzyme with glutathione-dependent transferase and peroxidase activities. *J. Biol. Chem.* **272**, 22934.

Jakobsson, P.-J., Morgenstern, R., Mancini, J., Ford-Hutchinson, A., and Persson, B. (1999a). Common structural features of MAPEG—A widespread superfamily of membrane associated proteins with highly divergent functions in eicosanoid and glutathione metabolism. *Protein Sci.* **8**, 689–692.

Jakobsson, P. J., Thoren, S., Morgenstern, R., and Samuelsson, B. (1999b). Identification of human prostaglandin E synthase: A microsomal, glutathione-dependent, inducible enzyme, constituting a potential novel drug target. *Proc. Natl. Acad. Sci. USA* **96**, 7220–7225.

Jala, V. R., and Haribabu, B. (2004). Leukotrienes and atherosclerosis: New roles for old mediators. *Trends Immunol.* **25**, 315–322.

Kamei, D., Yamakawa, K., Takegoshi, Y., Mikami-Nakanishi, M., Nakatani, Y., Oh-Ishi, S., Yasui, H., Azuma, Y., Hirasawa, N., Ohuchi, K., Kawaguchi, H., Ishikawa, Y., Ishii, T., Uematsu, S., Akira, S., Murakami, M., and Kudo, I. (2004). Reduced pain hypersensitivity and inflammation in mice lacking microsomal prostaglandin e synthase-1. *J. Biol. Chem.* **279**, 33684–33695.

Lam, B. K. (2003). Leukotriene C(4) synthase. *Prostaglandins Leukot. Essent. Fatty Acids* **69**, 111–116.

Lam, B. K., Penrose, J. F., Freeman, G. J., and Austen, K. F. (1994). Expression cloning of a cDNA for human leukotriene C_4 synthase, an integral membrane protein conjugating reduced glutathione to leukotriene A_4. *Proc. Natl. Acad. Sci. USA* **91**, 7663–7667.

Ma, W. C., Li, D. Y., Sun, F., Kleiman, N. J., and Spector, A. (2004). The effect of stress withdrawal on gene expression and certain biochemical and cell biological properties of peroxide-conditioned cell lines. *FASEB J.* **18**, 480–488.

Mabuchi, T., Kojima, H., Abe, T., Takagi, K., Sakurai, M., Ohmiya, Y., Uematsu, S., Akira, S., Watanabe, K., and Ito, S. (2004). Membrane-associated prostaglandin E synthase-1 is required for neuropathic pain. *Neuroreport* **15**, 1395–1398.

Mancini, J. A., Abramovitz, M., Cox, M. E., Wong, E., Charleson, S., Perrier, H., Wang, Z., Prasit, P., and Vickers, P. J. (1993). 5-Lipoxygenase-activating protein is an arachidonate binding protein. *FEBS Lett.* **318**, 277–281.

Mancini, J. A., Coppolino, M. G., Klassen, J. H., Charleson, S., and Vickers, P. J. (1994). The binding of leukotriene biosynthesis inhibitors to site-directed mutants of human 5-lipoxygenase-activating protein. *Life Sci.* **54**, PL137–142.

Mandal, A. K., Skoch, J., Bacskai, B. J., Hyman, B. T., Christmas, P., Miller, D., Yamin, T. T., Xu, S., Wisniewski, D., Evans, J. F., and Soberman, R. J. (2004). The membrane organization of leukotriene synthesis. *Proc. Natl. Acad. Sci. USA* **101**, 6587–6592.

Metters, K., Sawyer, N., and Nicholson, D. (1994). Microsomal glutathione S-transferase is the predominant leukotriene C4 binding site in cellular membranes. *J. Biol. Chem.* **269**, 12816–12823.

Miller, D. K., Gillard, J. W., Vickers, P. J., Sadowski, S., Léveillé, C., Mancini, J. A., Charleson, P., Dixon, R. A. F., Ford-Hutchinson, A. W., Fortin, R., Gauthier, J. Y., Rodkey, J., Rosen, R., Rouzer, C., Sigal, I. S., Strader, C. D., and Evans, J. F. (1990). Identification and isolation of a membrane protein necessary for leukotriene production. *Nature* **343**, 278–281.

Morgenstern, R., Guthenberg, C., and DePierre, J. W. (1982). Microsomal glutathione S-transferase. Purification, initial characterization and demonstration that it is not identical to the cytosolic glutathione S-transferases A, B and C. *Eur. J. Biochem.* **128**, 243–248.

Murakami, M., and Kudo, I. (2004). Recent advances in molecular biology and physiology of the prostaglandin E2-biosynthetic pathway. *Prog. Lipid Res.* **43**, 3–35.

Murakami, M., Naraba, H., Tanioka, T., Semmyo, N., Nakatani, Y., Kojima, F., Ikeda, T., Fueki, M., Ueno, A., Oh, S., and Kudo, I. (2000). Regulation of prostaglandin E2 biosynthesis by inducible membrane-associated prostaglandin E2 synthase that acts in concert with cyclooxygenase-2. *J. Biol. Chem.* **275**, 32783–32792.

Nicholson, D. W., Ali, A., Vaillancourt, J. P., Calaycay, J. R., Mumford, R. A., Zamboni, R. J., and Ford-Hutchinson, A. W. (1993). Purification to homogeneity and the N-terminal sequence of human leukotriene C₄ synthase: A homodimeric glutathione S-transferase composed of 18-kDa subunits. *Proc. Natl. Acad. Sci. USA* **90**, 2015–2019.

Ouellet, M., Falgueyret, J.-P., Hien Ear, P., Pen, A., Mancini, J. A., Riendeau, D., and David Percival, M. (2002). Purification and characterization of recombinant microsomal prostaglandin E synthase-1. *Protein Expr. Purif.* **26**, 489–495.

Pande, S. V., and Murthy, M. S. R. (1994). A modified micro-Bradford procedure for elimination of interference from sodium dodecyl sulfate, other detergents, and lipids. *Anal. Biochem.* **220**, 424–426.

Peters-Golden, M., and Brock, T. G. (2003). 5-lipoxygenase and FLAP. *Prostaglandins Leukot. Essent. Fatty Acids* **69**, 99–109.

Quraishi, O., Mancini, J. A., and Riendeau, D. (2002). Inhibition of inducible prostaglandin E (2) synthase by 15-deoxy-Delta(12,14)-prostaglandin J(2) and polyunsaturated fatty acids. *Biochem. Pharmacol.* **63**, 1183–1189.

Saha, S., Engstrom, L., Mackerlova, L., Jakobsson, P. J., and Blomqvist, A. (2005). Impaired febrile responses to immune challenge in mice deficient in microsomal prostaglandin E synthase-1. *Am. J. Physiol. Regul. Integr. Compar. Physiol.* **288**, R1100–R1107.

Schroder, O., Sjostrom, M., Qiu, H., Jakobsson, P. J., and Haeggstrom, J. Z. (2005). Microsomal glutathione S-transferases: Selective up-regulation of leukotriene C(4) synthase during lipopolysaccharide-induced pyresis. *Cell. Mol. Life Sci.* **62**, 87–94.

Schroder, O., Sjostrom, M., Qiu, H., Stein, J., Jakobsson, P. J., and Haeggstrom, J. Z. (2003). Molecular and catalytic properties of three rat leukotriene C(4) synthase homologs. *Biochem. Biophys. Res. Commun.* **312**, 271–276.

Scoggan, K. A., Jakobsson, P. J., and Ford-Hutchinson, A. W. (1997). Production of leukotriene C4 in different human tissues is attributable to distinct membrane bound biosynthetic enzymes. *J. Biol. Chem.* **272**, 10182–10187.

Sjöström, M., Jakobsson, P.-J., Heimburger, M., Palmblad, J., and Haeggstrom, J. (2001). Human umbilical vein endothelial cells generate leukotriene C4 via microsomal glutathione -S-transferase type 2 and express the CysLT1 receptor. *Eur. J. Biochem.* **268**, 2578–2586.

Soderstrom, M., Morgenstern, R., and Hammarstrom, S. (1995). Protein-protein interaction affinity chromatography of leukotriene C4 synthase. *Protein Expr. Purif.* **6**, 352–356.

Studier, F. W. (1991). Use of bacteriophage T7 lysozyme to improve an inducible T7 expression system. *J. Mol. Biol.* **219**, 37–44.

Surapureddi, S., Morgenstern, R., Soderstrom, M., and Hammarstrom, S. (1996). Interaction of human leukotriene c-4 synthase and microsomal glutathione transferase *in vivo*. *Biochem. Biophys. Res. Commun.* **229**, 388–395.

Surapureddi, S., Svartz, J., Magnusson, K. E., Hammarstrom, S., and Soderstrom, M. (2000). Colocalization of leukotriene C synthase and microsomal glutathione S-transferase elucidated by indirect immunofluorescence analysis. *FEBS Lett.* **480**, 239–243.

Tanikawa, N., Ohmiya, Y., Ohkubo, H., Hashimoto, K., Kangawa, K., Kojima, M., Ito, S., and Watanabe, K. (2002). Identification and characterization of a novel type of membrane-associated prostaglandin E synthase. *Biochem. Biophys. Res. Commun.* **291**, 884–889.

Tanioka, T., Nakatani, Y., Semmyo, N., Murakami, M., and Kudo, I. (2000). Molecular identification of cytosolic prostaglandin E2 synthase that is functionally coupled with cyclooxygenase-1 in immediate prostaglandin E2 biosynthesis. *J. Biol. Chem.* **275**, 32775–32782.

Thoren, S., Weinander, R., Saha, S., Jegerschold, C., Pettersson, P. L., Samuelsson, B., Hebert, H., Hamberg, M., Morgenstern, R., and Jakobsson, P. J. (2003). Human microsomal prostaglandin E synthase-1: Purification, functional characterization and projection structure determination. *J. Biol. Chem.* **278**, 22199–22209.

Trebino, C. E., Stock, J. L., Gibbons, C. P., Naiman, B. M., Wachtmann, T. S., Umland, J. P., Pandher, K., Lapointe, J. M., Saha, S., Roach, M. L., Carter, D., Thomas, N. A., Durtschi, B. A., McNeish, J. D., Hambor, J. E., Jakobsson, P. J., Carty, T. J., Perez, J. R., and Audoly, L. P. (2003). Impaired inflammatory and pain responses in mice lacking an inducible prostaglandin E synthase. *Proc. Natl. Acad. Sci. USA* **100**, 9044–9049.

Welsch, D. J., Creely, D. P., Hauser, S. D., Mathis, K. J., Krivi, G. G., and Isakson, P. C. (1994). Molecular cloning and expression of human leukotriene-C4 synthase. *Proc. Natl. Acad. Sci. USA* **91**, 9745–9749.

Westman, M., Korotkova, M., af Klint, E., Stark, A., Audoly, L. P., Klareskog, L., Ulfgren, A.-K., and Jakobsson, P.-J. (2004). Expression of microsomal prostaglandin E synthase in rheumatoid arthritis synovium. *Arthritis Rheum.* **50**, 1774–1780.

[10] Two-Dimensional Crystallization and Electron Crystallography of MAPEG Proteins

By Hans Hebert, Caroline Jegerschöld, Priyaranjan Bhakat, and Peter J. Holm

Abstract

Members of the membrane-associated proteins in the eicosanoid and glutathione metabolism (MAPEG) superfamily have been subjected to two-dimensional crystallization experiments. A common denominator for successful attempts has been the use of a low lipid/protein ratio in the range of 1–9 (mol/mol). Electron crystallography demonstrated either hexagonal or orthorhombic packing of trimeric protein units. Three-dimensional

METHODS IN ENZYMOLOGY, VOL. 401   0076-6879/05 $35.00
Copyright 2005, Elsevier Inc. All rights reserved.
DOI: 10.1016/S0076-6879(05)01010-4

structure analysis of the MAPEG member microsomal glutathione transferase 1 has shown that the monomer for this protein contains a left-handed
bundle of four transmembrane helices. It is likely that this is a common
structural motif for MAPEG proteins, because projection maps of all
structurally characterized members are very similar.

Introduction

Although the information limit of a high-resolution transmission electron microscope (TEM) is in the 1 Å range, it is difficult to achieve atomic
resolution for biological specimens in general. The main reasons for this are
the poor scattering power of low atomic number elements and high sensitivity to beam-induced damage. However, if a specimen can be arranged as a
one-layered ordered array, a two-dimensional (2-D) crystal, the situation
will be improved. Indeed, atomic models have been obtained from such
assemblies. In all but one case (Nogales *et al.*, 1998), the specimens have
been intrinsic membrane proteins (Gonen *et al.*, 2004; Henderson *et al.*,
1990; Kühlbrandt *et al.*, 1994; Miyazawa *et al.*, 2003; Murata *et al.*, 2000).
This is not surprising, because membrane proteins may be difficult to crys-
tallize in 3-D for X-ray crystallography, particularly if they are small and
very hydrophobic. Furthermore, a 2-D crystal of a membrane protein
mimics to some extent the natural environment in a lipid bilayer.

The main reasons for 2-D crystals being ideal objects for structure
analysis at resolution levels that allow building of atomic models are:
(1) specimens are thin, and a well-established theory for image formation
in TEM can be applied, and (2) the dose for data acquisition can be kept
low, and extraction of signals from a very noisy background is possible
through crystallographic methods.

Because most successful applications of electron crystallography in
structural biology have been on small very hydrophobic integral membrane
proteins, members of the MAPEG family were considered potential candidates for obtaining atomic models. As for other structure analysis techniques, production of pure and homogenous protein samples in a
reproducible way is essential. Although a single electron crystallography
experiment requires very little material, a complete project may consume
tens of milligrams along the way from initial crystallization attempts to
collection of a complete data set.

The goal to obtain atomic models requires 2-D crystals of a certain size
and order. Electron diffraction amplitudes can be recorded if crystals are
larger than about 1 μm. Even if crystal disorder can be compensated for to
some extent through lattice unbending (Henderson *et al.*, 1986), large
deviations from a perfect lattice may limit the attainable resolution.

Here, conditions for successful 2-D crystallization of some MAPEG proteins are summarized. Results from electron crystallography analysis in terms of packing arrangement, quaternary structure, and protein fold are also reviewed.

2-D Crystallization

Crystallization experiments have been reported for three members of the MAPEG family: microsomal glutathione transferase 1 (MGST1) (Hebert *et al.*, 1995; Schmidt-Krey *et al.*, 1998) (Fig. 1A), microsomal prostaglandin E synthase 1 (MPGES1) (Thorén *et al.*, 2003) (Fig. 1B), and leukotriene C_4 synthase (LTC$_4$S) (Schmidt-Krey *et al.*, 2004).

The MGST1 protein was purified from liver microsomes of male Sprague-Dawley rats (Morgenstern *et al.*, 1982), whereas the other two were crystallized from heterologously expressed material containing His$_6$-tags at the N- and C-termini, respectively (Lam *et al.*, 1994; Thorén *et al.*, 2003). In all cases, detergent solubilized protein was mixed with phospholipids followed by depletion of detergent through dialysis against detergent-free buffer. Dialysis is the most commonly used method for obtaining 2-D crystals of membrane proteins (for general reviews on the methodology see Hebert [1998], Kühlbrandt [1992], and Mosser [2001]). A large number of parameters can be varied to obtain and improve 2-D crystals. One of the most important is the ratio between lipid and protein content (LPR). For the MAPEG proteins, LPRs have been varied between unusually low values 1–9 (mol/mol). Thus, it is not likely that normal lipid bilayers with inserted protein molecules are formed. Indeed, analysis of the topology of MGST1 molecules related by in-plane symmetry in 2-D crystals shows that the vertical positioning of the molecules is not compatible with what would be expected if the molecules were placed in a continuous lipid bilayer (data

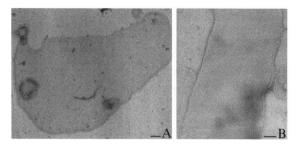

Fig. 1. Electron micrographs of orthorhombic MGST1 (A) and MPGES1 2-D crystals (B). Specimens were stained with 1% uranyl acetate. Scale bars 0.1 μm.

TABLE I
STANDARD CONDITIONS FOR 2-D CRYSTALLIZATION OF MICROSOMAL GLUTATHIONE
TRANSFERASE 1, MICROSOMAL PROSTAGLANDIN E SYNTHASE 1, AND LEUKOTRIENE C₄ SYNTHASE

	MGST1	MPGES1	LTC$_4$S
Protein concentration, mg/ml	0.5–1.75	0.5–1.0	1.0
Buffer	10 mM KPO$_4$	10 mM NaPO$_4$	50 mM HEPES
pH	7.0	7.0	7.6
Salt	50 mM KCl	50 mM NaCl	50 mM KCl
Glutathione, mM	1	1	10
Glycerol, %	20	20	20
Detergent	1% Triton X-100	1% Triton X-100	1% Triton X-100 and 0.5% sodium deoxycholate
Phospholipid	Bovine liver PC	Bovine liver PC	DMPC
Lipid/protein ratio, mol/mol	3	9	1
Crystallization method	Dialysis	Dialysis	Dialysis
Approx. temperature, °C	21	21	21
Time, days	8–10	8–10	10–14

not shown). For MGST1, the LPR also influences the predominance of crystals with hexagonal p6 or orthorhombic p22$_1$2$_1$ symmetry (Schmidt-Krey *et al.*, 1998). The overall conditions for 2-D crystallization of MGST1, MPGES1, and LTC$_4$S are summarized in Table I.

Properties of 2-D Crystals of MAPEG Proteins

With one exception, reported 2-D crystals of MAPEG proteins contain symmetry elements in the plane of the membrane in such a way that up and down orientations are shared equally (Table II). This is a common feature among crystalline membranes obtained from reconstitution of solubilized proteins (Kühlbrandt, 1992). In p6 crystals of MGST1, all directly identified protein molecules are facing the same way, whereas an additional partly disordered density on the sixfold axis may arise from protein units in the other orientation (Schmidt-Krey *et al.*, 1999).

Projection maps clearly demonstrated that protein molecules in both crystal forms of MGST1 are arranged into trimers (Hebert *et al.*, 1995, 1997) (Fig. 2A, B). More recently, the same arrangement has been observed for MPGES1 (Thorén *et al.*, 2003) (Fig. 2C) and LTC$_4$S (Schmidt-Krey *et al.*, 2004), suggesting that the functional unit for at least these MAPEG proteins is a trimer. In the hexagonal crystals, MGST1 p6 and LTC$_4$S p321, the threefold axes are crystallographic, whereas in the

TABLE II
TWO-DIMENSIONAL CRYSTAL PARAMETERS

	MGST1, hexagonal	MGST1, orthorhombic	MPGES1	LTC$_4$S
Two-sided plane group symmetry[a]	p6	p22$_1$2$_1$	p22$_1$2$_1$	p321
Unit cell	a = b = 81.8 Å, $\gamma = 120°$	a = 91.9 Å, b = 90.8 Å, $\gamma = 90°$	a = 97.0 Å, b = 98.0 Å, $\gamma = 90°$	a = b = 73.4 Å, $\gamma = 120°$
Unit cell area, Å^2	5795	8347	9506	4666
Monomers/unit cell	9[b]	12	12	6
Packing density, Da/Å^2	26.4	24.4	22.7	23.1

[a] As defined in Holser (1958).
[b] Two trimers on the threefold axes and one on the sixfold with four possible orientations (Fig. 2A).

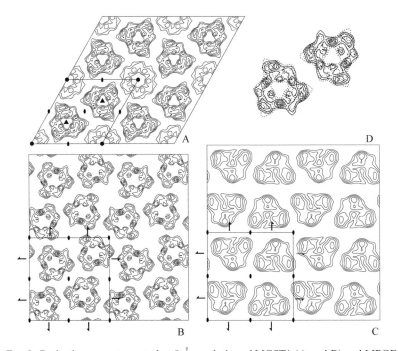

FIG. 2. Projection maps truncated at 8 Å resolution of MGST1 (A and B) and MPGES1 (C). The unit cells depicted have symmetries and parameters: p6, a = b = 81.8 Å, $\gamma = 120°$ (A), p22$_1$2$_1$, a = 91.9, b = 90.8 Å, $\gamma = 90°$ (B), and p22$_1$2$_1$, a = 97.0, b = 98.0 Å, $\gamma = 90°$ (C). The convention for the symmetry assignment of 2-D crystals is to specify the symmetry operation perpendicular to the layer, here the twofold and the sixfold axes, first. The contour lines were drawn at intervals of 0.5 σ. In (D) two adjacent trimers of MGST1 from the p6 (dashed contour lines) and the p22$_1$2$_1$ projection maps have been extracted and put in register.

orthorhombic crystal forms, MGST1 $p22_12_1$ and MPGES1 $p22_12_1$, they are local and may allow for slight asymmetry. A comparison of the packing arrangement of MGST1 in p6 and $p22_12_1$ crystals also suggests that inter-trimeric interactions exist leading to formation of trimer pairs (Schmidt-Krey et al., 1999) (Fig. 2D).

The lateral packing density of the proteins in the 2-D crystals ranges from 22.7–26.4 Da/Å^2. Similar values can be calculated for other well-ordered 2-D crystals of membrane proteins with most of their mass in the lipid bilayer (e.g., bacteriorhodopsin) (Henderson et al., 1990) and aquaporin-1 (Murata et al., 2000).

Fig. 3. Three-dimensional map of the MGST1 trimer truncated at 4.5 Å resolution obtained by merging data from the p6 and $p22_12_1$ crystals forms. The view is from the lumen side at a slight inclination from the membrane plane. A mask was applied to omit parts of the cytoplasmic domain to clearly visualize the orientations of the transmembrane helices. The most likely subdivision into a monomer is illustrated. (See color insert.)

Protein Fold

Even if density peaks in projection maps suggest positions for some transmembrane α-helices, it is difficult to draw any definite conclusions about structure, particularly if some of the helices are highly tilted relative to the normal of the membrane plane. Three-dimensional maps of MGST1 (Holm *et al.*, 2002; Schmidt-Krey *et al.*, 2000) demonstrate that a large proportion of the protein monomer forms a left-handed four-helix bundle (Fig. 3). Because the projection maps of all the analyzed MAPEG proteins are similar, it is very likely that the protein fold determined for MGST1 shows a common property among most, if not all, MAPEG proteins. The α-helices in MGST1 have overall inclination angles between 20.7° and 35.5° as measured to the membrane normal. The lengths of the helices are sufficient to span a lipid bilayer even at relatively high tilt. The clear asymmetry of protein mass in hydropathy plots of MGST1 together with density accumulation on one side of the 3-D maps define the luminal and cytoplasmic faces of the maps. On the lumen side, the helices end at approximately the same height, whereas they extend differently on the other side.

Acknowledgment

The structural work on MGST1 and MPGES1 has been supported by the Swedish Research Council.

References

Gonen, T., Sliz, P., Kistler, J., Cheng, Y., and Walz, T. (2004). Aquaporin-0 membrane junctions reveal the structure of a closed water pore. *Nature* **429**, 193–197.

Hebert, H. (1998). Two-dimensional crystals of membrane proteins. *In* "Biomembrane Structures" (D. Chapman and P. Haris, eds.), pp. 88–110. IOS Press, Amsterdam.

Hebert, H., Schmidt-Krey, I., and Morgenstern, R. (1995). The projection structure of microsomal glutathione transferase. *EMBO J.* **14**, 3864–3869.

Hebert, H., Schmidt-Krey, I., Morgenstern, R., Murata, M., Hirai, T., Mitsuoka, K., and Fujiyoshi, Y. (1997). The 3.0 Å projection structure of microsomal glutathione transferase as determined by electron crystallography of $p2_12_12$ two-dimensional crystals. *J. Mol. Biol.* **271**, 751–758.

Henderson, R., Baldwin, J. M., Ceska, T. A., Zemlin, F., Beckmann, E., and Downing, K. H. (1990). Model for the structure of bacteriorhodopsin based on high-resolution electron cryo-microscopy. *J. Mol Biol.* **213**, 899–929.

Henderson, R., Baldwin, J. M., Downing, K. H., Lepault, J., and Zemlin, F. (1986). Structure of purple membrane from *Halobacterium halobium*: Recording, measurement and evaluation of electron micrographs at 3.5 Å resolution. *Ultramicroscopy* **19**, 147–178.

Holm, P., Morgenstern, R., and Hebert, H. (2002). The 3-D structure of microsomal glutathione transferase 1 at 6 Å resolution as determined by electron crystallography of p22₁2₁ crystals. *Biochim. Biophys. Acta* **1594,** 276–285.

Holser, W. T. (1958). Point groups and plane groups in a two-sided plane and their subgroups. *Z. Kristallogr.* **110,** 266–281.

Kühlbrandt, W. (1992). Two-dimensional crystallization of membrane proteins. *Quart. Rev. Biophys.* **25,** 1–49.

Kühlbrandt, W., Wang, D. N., and Fujiyoshi, Y. (1994). Atomic model of plant light-harvesting complex by electron crystallography. *Nature* **367,** 614–621.

Lam, B. K., Penrose, J. F., Freeman, G. J., and Austen, K. F. (1994). Expression cloning of a cDNA for human leukotriene C4 synthase, an integral membrane protein conjugating reduced glutathione to leukotriene A4. *Proc. Natl. Acad. Sci. USA* **91,** 7663–7667.

Miyazawa, A., Fujiyoshi, Y., and Unwin, N. (2003). Structure and gating mechanism of the acetylcholine receptor pore. *Nature* **423,** 949–955.

Morgenstern, R., Guthenberg, C., and Depierre, J. W. (1982). Microsomal glutathione S-transferase. Purification, initial characterization and demonstration that it is not identical to the cytosolic glutathione S-transferases A, B and C. *Eur. J. Biochem.* **128,** 243–248.

Mosser, G. (2001). Two-dimensional crystallogenesis of transmembrane proteins. *Micron* **32,** 517–540.

Murata, K., Mitsuoka, K., Hirai, T., Walz, T., Agre, P., Heymann, J. B., Engel, A., and Fujiyoshi, Y. (2000). Structural determinants of water permeation through aquaporin-1. *Nature* **407,** 599–605.

Nogales, E., Wolf, S. G., and Downing, K. H. (1998). Structure of the alpha beta tubulin dimer by electron crystallography. *Nature* **391,** 199–203.

Schmidt-Krey, I., Kanaoka, Y., Mills, D. J., Irikura, D., Haase, W., Lam, B. K., Austen, K. F., and Kuhlbrandt, W. (2004). Human leukotriene C₄ synthase at 4.5 Å resolution in projection. *Structure* **12,** 2009–2014.

Schmidt-Krey, I., Lundqvist, G., Morgenstern, R., and Hebert, H. (1998). Parameters for the two-dimensional crystallization of the membrane protein microsomal glutathione transferase. *J. Struct. Biol.* **123,** 87–96.

Schmidt-Krey, I., Mitsuoka, K., Hirai, T., Murata, K., Cheng, Y., Fujiyoshi, Y., Morgenstern, R., and Hebert, H. (2000). The three-dimensional map of microsomal glutathione transferase 1 at 6 Å resolution. *EMBO J.* **19,** 6311–6316.

Schmidt-Krey, I., Murata, K., Hirai, T., Morgenstern, R., Fujiyoshi, Y., and Hebert, H. (1999). The projection structure of the membrane protein microsomal glutathione transferase at 3 Å resolution as determined from hexagonal two-dimensional crystals. *J. Mol. Biol.* **288,** 243–253.

Thorén, S., Weinander, R., Saha, S., Jegerschöld, C., Pettersson, P. L., Samuelsson, B., Hebert, H., Hamberg, M., Morgenstern, R., and Jakobsson, P. J. (2003). Human microsomal prostaglandin e synthase-1: Purification, functional characterization and projection structure determination. *J. Biol. Chem.* **278,** 22199–22209.

[11]   Plant Glutathione Transferases

By ROBERT EDWARDS and DAVID P. DIXON

Abstract

Soluble plant glutathione transferases (GSTs) consist of seven distinct classes, six of which have been functionally characterized. The phi and tau class GSTs are specific to plants and the most numerous and abundant of these enzymes. Both have classic conjugating activities toward a diverse range of xenobiotics, including pesticides, where they are major determinants of herbicide selectivity in crops and weeds. In contrast, the zeta and theta class GSTs are conserved in animals and plants and have very restricted activities toward xenobiotics. Theta GSTs function as glutathione peroxidases, reducing organic hydroperoxides produced during oxidative stress. Zeta GSTs act as glutathione-dependent isomerases, catalyzing the conversion of maleylacetoacetate to fumarylacetoacetate, the penultimate step in tyrosine degradation. The other two classes of plant GSTs, the dehydroascorbate reductases (DHARs) and lambda GSTs, differ from phi, tau, zeta, and theta enzymes in being monomers rather than dimers and possessing a catalytic cysteine rather than serine in the active site. Both can function as thioltransferases, with the DHARs having a specialized function in reducing dehydroascorbate to ascorbic acid. The determination of the diverse plant-specific functions of the differing GST classes is described.

Introduction

Plant GSTs were originally discovered just after their mammalian counterparts, but the major progress on their characterization, cloning, and classification has only occurred relatively recently (Dixon *et al.*, 2002b; Edwards *et al.*, 2000; Marrs, 1996). In common with the mammalian GSTs, the plant enzymes form a complex superfamily composed of a number of discrete classes. However, whereas mammals have a small number of GSTs per class, in plants classes can contain tens of members (Frova, 2003; McGonigle *et al.*, 2000). The presence of multiple closely related GSTs has greatly complicated the purification and study of individual isoenzymes isolated from plants. In the case of the dimeric GSTs, this situation is compounded by the participation of individual subunits in the formation of both homodimers and multiple heterodimers. It has,

METHODS IN ENZYMOLOGY, VOL. 401
Copyright 2005, Elsevier Inc. All rights reserved.
0076-6879/05 $35.00
DOI: 10.1016/S0076-6879(05)01011-6

therefore, proved to be preferable to characterize individual GSTs as the respective recombinant enzymes expressed in *Escherichia coli*, which in most cases reported results in good yields of active soluble proteins (Dixon *et al.*, 2002a; McGonigle *et al.*, 2000; Wagner *et al.*, 2002).

The GST classes present in plants are summarized in Table I along with key references describing their functional characterization and structural biology (where available). The plant-specific phi and tau GSTs represent the dominant classes and share many functional similarities with the drug-metabolizing GSTs in animals. Thus, they are dimeric and can catalyze the conjugation of a diverse range of xenobiotics, coming to prominence through their roles in detoxifying selective herbicides. The theta and zeta classes are also dimers and have close homologs in animals and fungi that

TABLE I
OVERVIEW OF THE CHARACTERISTICS OF THE DIFFERENT CLASSES OF PLANT GSTs

Class	Gene diversity (*A. thaliana*)[a]	Catalytic residue	Activity	Structure
Phi[b]	13	Serine[c]	Detox., ligandin[b]	Dimer, 3-D[c]
Tau[d]	28	Serine[e]	Detox., ligandin[b]	Dimer, 3-D[e]
Theta[f]	2	Serine[g]	GPOX[b]	Dimer, 3-D[g]
Zeta[h]	2	Serine[i]	MAAI, detox.[j]	Dimer[h], 3-D[i]
Lambda[k]	3	Cysteine[k]	Oxidoreductase?[k]	Monomer[k]
DHAR[k]	4	Cysteine[k]	DHAR[k]	Monomer[k]
TCHQD[a]	1	Serine?	?	?
Microsomal[l]	1	?	GPOX, detox., leukotriene synth (in animals).[m]	Trimer, 3-D[n]

[a] Number of genes for each class present in the *Arabidopsis* genome; http://www.arabidopsis.org/info/genefamily/gst.html.
[b] Edwards *et al.* (2000).
[c] Reinemer *et al.* (1996).
[d] Droog (1997).
[e] Thom *et al.* (2002).
[f] Dixon *et al.* (1999b).
[g] Board *et al.* (1995).
[h] Board *et al.* (1997).
[i] Thom *et al.* (2001).
[j] Dixon *et al.* (2000).
[k] Dixon *et al.* (2002a).
[l] Morgenstern *et al.* (1985).
[m] Jakobsson *et al.* (1999).
[n] Schmidt-Krey *et al.* (2000). Detox refers to a proven role in detoxifying herbicides or other xenobiotics, whereas ligandin indicates that members of this class have also been shown to bind natural products *in vivo*.

are presumed to indicate conserved functions across the phyla. Theta GSTs have limited transferase activity toward xenobiotics but are highly active glutathione-dependent peroxidases (GPOXs) that catalyze the reduction of lipid hydroperoxides to the respective monohydroxyalcohols. This is an important GST-associated antioxidant activity in plants that has received much less attention than that associated with the selenium-dependent GPOXs despite its demonstrated importance in stress tolerance (Roxas *et al.*, 2000). The zeta GSTs are also a near-ubiquitous class acting as glutathione-dependent maleylacetoacetate isomerases (MAAI) but also catalyzing glutathione-dependent dechlorination reactions of, for example, dichloroacetic acid. The most recently discovered lambda and dehydro-ascorbate reductase (DHAR) class enzymes are both plant specific. They differ from the other plant GSTs in being monomeric and act as glutathione-dependent oxidoreductases rather than conjugating enzymes (Dixon *et al.*, 2002a). A further gene related to GSTs can be identified in the genome of *Arabidopsis thaliana* (At1g77290). The functional characterization of this protein has not been reported to date, but the predicted protein most closely resembles the tetrachlorohydroquinone dehalogenase (TCHQD) enzymes from prokaryotes. Finally, plants also contain genes encoding microsomal GSTs, which although unrelated to the main GST superfamily, have similar glutathione-dependent activities. Microsomal GSTs are membrane-bound members of the MAPEG (membrane-associated proteins in eicosanoid and glutathione metabolism) superfamily (Jakobsson *et al.*, 1999). Based on sequence similarity, Arabidopsis contains a single microsomal GST gene, with other plant species containing one or more such genes. However, there is no published work on these enzymes in plants, and they are not considered further here.

Purification of Plant GSTs and Component Subunits

Because of the diversity within the plant GST superfamily, there are no generic purification protocols. Many of the phi and tau class enzymes can be successfully purified by affinity chromatography using glutathione-agarose (DeRidder *et al.*, 2002; Sappl *et al.*, 2004). However, this method does not work reliably with GSTs from other classes. Other useful affinity resins include S-hexylglutathione-agarose, which binds a number of GSTs, particularly those from the tau class (Dixon *et al.*, 1998) and bromosul-fophthalein-glutathione-agarose (Dixon *et al.*, 1999a) and Orange A agarose, which specifically retain phi-class enzymes isolated from maize (Dixon *et al.*, 1997). Although lambda GSTs and DHARs are not retained on conventional glutathione affinity columns, the presence of highly reactive cysteine residues in their active sites suggests that they would efficiently bind by means of disulfide interaction to activated thiol matrices such as activated thiol Sepharose or thiopropyl Sepharose as reported for other thioltransferases (reviewed in Sies and Packer, 2002). Where GSTs are retained on affinity supports of glutathione, or glutathione-conjugate affinity columns, these matrices can prove particularly invaluable in purifying the respective recombinant enzymes, because they allow easy purification of the protein without the presence of any peptide tag. However, when conjugates such as S-hexylglutathione are used as the eluting counter ligand, it is essential to ensure the final protein is extensively desalted to remove all traces of the glutathione conjugate because of their activity as competitive inhibitors of GST activity.

A useful method for purifying a wide range of plant GSTs and then subsequently resolving their component polypeptides is based on the relative hydrophobicities of these proteins. Protein preparations are adjusted to 1 M $(NH_4)_2SO_4$ by the addition of an equal volume of 2 M $(NH_4)_2SO_4$ and after clarifying by centrifugation, applied to a phenyl Sepharose column pre-equilibrated in buffer A (50 mM potassium phosphate, pH 7.2, containing 1 M $(NH_4)_2SO_4$ and 1 mM dithiothreitol (DTT). After washing with one column volume of buffer A, loosely bound proteins are recovered using a linearly decreasing concentration of salt over two column volumes ending with the column being washed with buffer B (50 mM potassium phosphate, 1 mM DTT, pH 7.2). After a further wash with one column volume equivalent of buffer B, the highly hydrophobic residual proteins, which are enriched in plant GSTs, are recovered with a mixture of buffer B: ethylene glycol (1:1 v/v). A minority of GSTs may be eluted in the low salt wash before the addition of ethylene glycol, and the fractionation of these enzymes based on hydrophobicity can greatly assist in determining the number and classes of GSTs present in whole plant extracts (Cummins et al., 1997). In each case, the purification of the GSTs is monitored by taking fractions and assaying for enzyme activities of interest (see later).

Because of the similarity in molecular masses of many plant GST subunits, it is often not possible to resolve individual polypeptides from one another using conventional sodium dodecyl sulfate-polyacrylamide gel electrophoresis (SDS-PAGE). Some improved resolution of GST polypeptides on the basis of their molecular mass can be achieved using specialized conditions for SDS-PAGE with tricine buffers (Schägger and von Jagow, 1987) or by using a high ratio of acrylamide/bis-acrylamide in the resolving gels (Mozer et al., 1983). An alternative higher resolution method is to use reversed-phase HPLC (Cummins et al., 1997). GST preparations are injected (200 μl) onto a C18 HPLC column (25 cm × 2.1 mm i.d., 5 μm packing, 300 Å pore size, Vydac) and eluted at 0.5 ml/min^{-1} with a linear gradient progressing from a ratio of solvent A/solvent B of 9:1 to a ratio of 2:8 over 50 min, where solvent A is 0.1% (v/v) aqueous trifluoroacetic acid (TFA) and solvent B is acetonitrile containing 0.1% (v/v) TFA. The elution of GST subunits is monitored from their absorbance at 280 nm, and the purified polypeptides then are further analyzed by SDS-PAGE or mass spectrometry.

Assay of Phi and Tau Class GSTs

Both phi and tau GSTs have broad ranging conjugating activity toward xenobiotics, and their presence can be readily determined using 1-chloro-2,4-dinitrobenzene (CDNB) and 4-nitrobenzylchloride (NBC). In addition, these plant GSTs also show activity to the other colorimetric substrates described elsewhere in this volume. However, in view of their importance in detoxifying pesticides, the assay for their conjugating activity toward herbicides is described here. In addition to conjugating herbicides, these plant GSTs have also been shown to catalyze glutathione-dependent isomerizations of thiadiazoline and isourazole herbicides (Iida et al., 1995; Nicolaus et al., 1996).

The diversity of herbicide chemistries undergoing GST-mediated glutathione conjugation in the course of their detoxification has been exhaustively reviewed (Edwards and Dixon, 2000). The importance of these reactions is that they frequently determine herbicide selectivity, because conjugation proceeds more readily in crops than in weed species. Key herbicide substrate classes (and notable examples) include the chloro-s-triazines (atrazine), chloroacetanilide (metolachlor), thiocarbamates (S-ethyl dipropylthiocarbamate sulfoxide), and the diphenyl ethers (fluorodifen). In addition, important individual compounds within the sulfonylurea (chlorimuron ethyl) and aryloxyphenoxypropionates (fenoxaprop ethyl) are detoxified directly by GSTs.

By obtaining those herbicides referred to in parentheses as analytical grade compounds, a good survey of the herbicide-conjugating activity of plant GSTs can be achieved, with some key examples shown (I), with sites of GSH substitution arrowed. GST activity toward these herbicide substrates is best achieved using HPLC-based assays to quantify the formation of conjugated products (Andrews et al., 1997; Hatton et al., 1996). This technology, although involved, can be readily adapted to high-throughput analysis (Clarke et al., 1998). However, with the diphenyl ether herbicide fluorodifen, GST activity can be determined directly with a simple spectrophotometric assay.

Atrazine   Metolachlor   Fluorodifen

Synthesis of Reference Glutathione Conjugates

Analytical grade herbicides (100 μmol) obtained from Greyhound Chemicals (ChemService, West Chester PA) or Fluka (Sigma-Aldrich, Gillingham Dorset, UK) are independently dissolved in 4 ml ethanol/acetonitrile (1:1 v/v) and mixed with 1 ml reduced glutathione (100 μmol) dissolved in double distilled water. The mixture is then adjusted to pH 9.5 with triethylamine, and the total volume made up to 6 ml with water. After incubating for 36 h at room temperature, 14 ml of acetone is added and after storage at $-20°$ for 24 h, the precipitate is collected on filter paper and washed with acetone to remove all traces of unreacted herbicide. The air-dried powder is then resuspended in methanol and a sample applied to an analytical TLC plate coated with silica containing fluorescent F_{254} indicator. The plate is developed with butan-1-ol/acetic acid/water (4:1:1 v/v) and glutathione-conjugates of the herbicides visualized using a combination of UV illumination and positive staining after a spray with ninhydrin dissolved in acetone (0.3% w/v) and heating to 110° in an oven. If further purification of the glutathione conjugate is required, this is best achieved using preparative HPLC using a C18 reverse phase column. Conjugates are eluted with a mixture of aqueous trifluoroacetic acid (0.5% v/v) and acetonitrile using a gradient of 10% acetonitrile to 80% acetonitrile to establish optimal conditions for isocratic elution while monitoring absorbance at 264 nm. The identity of conjugates can be confirmed by mass spectrometry after ionization by matrix-assisted laser desorption or electrospray (positive or negative mode) and the authenticated metabolites used to confirm the identity and quantity of reaction products derived from enzyme reactions.

Extraction of Plant GSTs

All steps are carried out at 4° unless otherwise specified. Frozen plant tissue in liquid nitrogen is ground to a powder using a pestle and mortar and then extracted in 0.1 M Tris-HCl, pH 7.5, containing 2 mM EDTA, 1 mM DTT, and 50 g/kg^{-1} polyvinylpolypyrrolidone. After straining through muslin followed by centrifugation (10,000g, 10 min), the supernatant is adjusted to 80% saturation with $(NH_4)_2SO_4$ and the protein pellet recovered by recentrifuging (13,000g, 20 min). Protein pellets will remain fully active for several months stored at $-20°$. The GSTs present can be assayed either directly or after purification by affinity or hydrophobic chromatography as described previously.

HPLC-Based Assays

GST preparations (plant extracts or recombinant enzymes) to be assayed are dissolved in 20 mM Tris-HCl, pH 7.5, and then desalted using Sephadex G25 columns before determining protein content. For comparative assays, protein content should be normalized to identical concentrations (1 mg/ml^{-1} for crude plant protein extracts). This can be important with pesticide substrates that are sparingly soluble in water, because binding to proteins influences the amount of herbicide bioavailable for conjugation. In particular, when using very low concentrations of pure recombinant protein, it may be beneficial to supplement the assays with bovine serum albumin (1 mg/ml^{-1}) to act as a carrier protein. The GST preparation (120 μl) is then added to either 50 μl 0.1 M potassium phosphate buffer, pH 6.8, or 50 μl 0.1 M Tris-HCl, pH 8.5, depending on the herbicide to be tested. Chloro-s-triazines (*e.g.*, atrazine) and chloroacetanilides (metolachlor) are assayed at pH 6.8, whereas diphenylethers (fluorodifen), aryloxyphenoxyprioionates (fenoxaprop ethyl), and sulfonylureas (chlorimuron ethyl) are assayed at pH 8.5. The mixture is then transferred to a water bath at 37° and 10 μl of a solution of the herbicide dissolved in acetone (10 mM) added, immediately followed by 20 μl of freshly prepared reduced glutathione (100 mM) adjusted to pH 7.0 with 0.1 M NaOH. At timed intervals up to 60 min, the reaction may be stopped by the addition of 10 μl HCl (3 M), and after standing on ice for 30 min, the protein precipitate is removed by centrifugation (12,000g, 5 min). Control incubations should consist of samples in which the glutathione is omitted to identify compounds that are not reaction products, as well as a boiled enzyme control to correct for the nonenzymic rate of conjugation. Aliquots of the sample (20–50 μl) are then injected onto a C18 reversed-phase HPLC column eluted using 1% (v/v) aqueous phosphoric acid and acetonitrile as the two mobile phase solvents starting at 20% acetonitrile and finishing at 80% to resolve the conjugates. The column should then be washed with 100% acetonitrile to remove unreacted parent herbicide. Conjugates may be identified and quantified using the reference compounds synthesized as described previously. Care should be taken to confirm that the production of conjugate is strictly dependent on time and protein content to ensure the assay is determining enzyme activity under conditions of first-order kinetics.

Spectrophotometric Assay

Unfortunately, the conjugation of most herbicides cannot be studied directly using spectrophotometric assays because of the relatively small changes in spectral characteristics resulting from conjugation. One exception to this is the conjugation of the diphenyl ether fluorodifen, which undergoes cleavage of the ether bond to release p-nitrophenol (I). The cleavage of diphenyl ethers is a reaction classically associated with tau class GSTs and can be usefully used to monitor their activity (Dixon et al., 2003). Fluorodifen is dissolved in ethanol (40 mM), and 10 μl is added to 50 mM glycine–NaOH, pH 9.5, (0.75 ml) and preincubated at 30° for 5 min before the addition of 0.1 ml GST preparation and 0.1 ml reduced glutathione (50 mM). At timed intervals over 60 min, the reaction is stopped with 10 M NaOH (50 μl) and after centrifuging (1,000g, 5 min), the absorbance of the sample determined at 400 nm and the amount of p-nitrophenol are calculated using a standard curve. Control incubations should include enzyme and the omission of glutathione to correct for the nonenzymic rate and any pigments present in the enzyme preparation.

Assay of Theta Class GSTs

Despite their widespread occurrence, plant theta class GSTs have received very little attention. These enzymes have little conjugating activity toward xenobiotics, with limited activity toward CDNB and NBC reported (Dixon et al., 1999b). Instead, these enzymes have very high GPOX activities directed toward organic hydroperoxides. Although these GPOX activities can be readily determined with synthetic cumene hydroperoxide (Edwards, 1996), it is preferable to assay their activity with naturally occurring fatty acid hydroperoxides.

Preparation of Fatty Acid Hydroperoxides

The method is applicable to a wide range of unsaturated fatty acids found in plants, notably linoleic (cis,cis-9,12-octadecadienoic acid), linolenic (cis,cis,cis-6,9,12octadecatrienoic acid), and ricinoleic (12-hydroxy- cis-octadecnoic acid) and is based on published methodology (Brash and Song, 1996). In the method described here, solid-phase extraction (SPE) cartridges are used in preference to solvent partitioning with diethyl ether or chloroform because of the higher recovery and overall yield of the fatty acid hydroperoxide product obtained by SPE. By scaling up the procedure, it is possible to produce milligram quantities of hydroperoxide.

Freshly prepared ammonium salts of the fatty acids (13.5 μmol) are added to 25 ml of 50 mM Tris-HCl, pH 9.0, that had been pre-aerated by bubbling air through for 30 min. The mixture, consisting of a dispersed emulsion, is then placed on an orbital shaker set at high speed to ensure continued thorough aeration. Samples of soybean lipoxidase (total 2 ml) prepared in 50 mM Tris-HCl, pH 9.0, to give approximately 100,000 enzyme units/ml^{-1} are added in 0.25-ml lots over 10 min, and the reaction is then allowed to go to completion after another 10 min of mixing. The reaction is then terminated with the addition of 6.5 ml of ethanol, and after cooling on ice, the reaction mix is adjusted to pH 3 with glacial acetic acid. After centrifugation to remove precipitated protein (17,000g, 30 min, 4°), the supernatant is applied to a C18 sample preparation column (6 ml). Each column is prepared for use by washing with ethanol (25 ml) followed by water (25 ml). The hydroperoxide preparation is applied in 5-ml portions to each pre-equilibrated column and the bound sample washed with 25 ml of 20% (v/v) ethanol followed by 50 ml water and finally 10 ml hexane. The hydroperoxide is then selectively recovered with methyl formate (10 ml) and concentrated to an oily residue under a stream of nitrogen before being resuspended in 0.5 ml methanol. The hydroperoxides can be stored in methanol at $-80°$ and quantified by performing a spectral scan (215–400 nm), with a typical molar extinction coefficient for the hydroperoxides being $\varepsilon = 23000\ M^{-1}/cm^{-1}$ at 235 nm.

GPOX Activity of Theta GSTs

To a plastic 1-ml cuvette add the following:

1. 500 μl 0.25 M potassium phosphate buffer, pH 7.0, containing 2.5 mM EDTA and 2.5 mM sodium azide
2. 100 μl glutathione reductase in 0.25 M potassium phosphate buffer pH 7 (6 units/ml^{-1})
3. 100 μl 10 mM reduced glutathione (adjusted to pH 7)
4. 100 μl 2.5 mM NADPH prepared in 0.1% w/v aqueous NaHCO$_3$

After pre-incubation at 37° for 10 min, the sample is placed in a spectrophotometer and the organic hydroperoxide substrate added. In the case of cumene hydroperoxide, 100 μl of an aqueous 12 mM solution is used. In the case of the fatty acid hydroperoxides synthesized previously, 50 μl of a 2 mM methanolic solution should be used because of the limited solubility of the substrates, and the buffer volumes in the cuvette should be adjusted accordingly. The enzyme extract (100 μl) is then added, and the change in absorbance at 340 nm is determined over 2 min. As a control, the change in absorbance should be determined with the enzyme omitted to correct for any changes caused by chemical oxidation of the glutathione or interference from opalescence caused by the fatty acid hydroperoxides. If the latter is a problem, the concentration of the substrate should be reduced. Results are expressed as nanomoles of product formed per second per milligram protein (nkat/mg^{-1}) using $\varepsilon = -6200\ M^{-1}/cm^{-1}$ for NADPH conversion.

Assay of the DHAR Class

Unlike the preceding GST classes, DHARs possess a cysteine residue at the active site and are unable to catalyze typical glutathione transferase reactions. Instead, the DHARs catalyze the glutathione-dependent reduction of dehydroascorbate to ascorbate, a vital step in the recycling of vitamin C. This activity is easily assayed because of an increase in absorbance at 265 nm as ascorbate is formed.

To assay for DHAR activity the following are mixed together at 30° in a quartz cuvette:

1. 900 μl 0.1 M potassium phosphate buffer, pH 6.5
2. 25 μl 20 mM dehydroascorbate in H_2O (freshly prepared)
3. 10 μl enzyme
4. 50 μl 100 mM reduced glutathione in H_2O (adjusted to pH 7.0 with 0.1 M NaOH)

After addition of glutathione, immediately transfer the cuvette to a spectrophotometer and measure the change in absorbance at 265 nm over 1 min. Ascorbate formation can then be calculated based on $\varepsilon = 14.0\ mM^{-1}/cm^{-1}$ at 265 nm. To correct for the nonenzymic rate of dehydroascorbate reduction, the incubation is repeated with 10 μl water replacing the enzyme.

This assay can be modified to measure the coupled NADPH-dependent reduction at 340 nm of the oxidized GSH formed by adding NADPH and glutathione reductase, as described for the GPOX and thioltransferase (see later) assays. This can be useful, for example, if measuring absorbance at 265 nm is impractical or if an accumulation of oxidized glutathione is suspected of causing enzyme inhibition.

Assay of Lambda Class GSTs

The function of lambda class GSTs is currently unknown, but based on the presence of a cysteine residue at the active site, it is highly likely that these enzymes catalyze a glutathione-dependent oxidoreductase reaction.

Until their true substrates are found, their activity can be monitored using a thiol transferase assay, with 2-hydroxyethyl disulfide (HED) as substrate (Dixon *et al.*, 2002a; Vlamis-Gardikas *et al.*, 1997). Disulfide exchange can then be monitored from the measurement of the NADPH-dependent reduction of oxidized glutathione formed during catalysis.

In a plastic cuvette, mix the following components:

1. 500 μl buffer (0.2 M Tris-Cl, pH 7.8, 4 mM EDTA)
2. 100 μl 2.5 mM NADPH in 0.1% NaHCO$_3$
3. 100 μl 10 mM reduced glutathione in H$_2$O (adjusted to pH 7.0 with NaOH)
4. 100 μl 6 U/ml glutathione reductase dissolved in assay in buffer
5. 100 μl 7 mM HED in H$_2$O

Incubate for 3 min at 30° after HED addition. This incubation increases enzyme activity, presumably because spontaneous disulfide exchange between HED and GSH gives a mixed disulfide of 2-mercaptoethanol-glutathione, which is a good substrate for the enzyme (II). As a result, this pre-incubation time must be kept constant between assays. Thioltransferase activity is then determined using the NADPH coupled assay by monitoring the change in absorbance at 340 nm after adding 100 μl enzyme according to the reaction.

The thioltransferase HED assay is also useful in assaying other GST superfamily members, including DHARs (Dixon *et al.*, 2002a), mammalian omega class GSTs (Board *et al.*, 2000), and bacterial beta class GSTs (Caccuri *et al.*, 2002) in addition to classic thioltransferases (Cho *et al.*, 1998, 1999).

Assay of Zeta Class GSTs

Zeta class GSTs are most easily assayed for their dechlorinating activity toward dichloroacetic acid. However, the natural function of this class is as maleylacetoacetate isomerase (MAAI; EC 5.2.1.2) catalyzing the isomerization of maleylacetoacetate to fumarylacetoacetate as part of the degradation of tyrosine (Fernández-Cañón and Peñalva, 1998). This isomerization reaction is very different from the dechlorinating activity, and its measurement is more involved. Maleylacetoacetate is unstable and not commercially available, and the chemical change effected by the reaction is very subtle and difficult to monitor. Two approaches have been used to allow the activity to be assayed. First, and less satisfactory, maleylacetone, an analog of maleylacetoacetate, can be chemically synthesized and used as substrate (Chen et al., 2003; Dixon et al., 2000; Fowler and Seltzer, 1970; Lantum et al., 2002). The use of this derivative assumes that the isomerase has sufficiently broad substrate specificity to accept this artificial substrate. The second approach uses the two enzymes coupled to MAAI in the tyrosine degradation pathway to first synthesize the required substrate and then the second to hydrolyze the product of the isomerization to produce a product that can be readily determined (III).

This reaction, therefore, requires a source of FAH and HDO, as well as the zeta class GST. As an example, we describe in the following the isolation and use of the respective recombinant enzymes from the model plant *Arabidopsis thaliana*. Similar assay methods have been described using enzymes purified from human tissue (Edwards and Knox, 1956) and extracts from the fungus *Aspergillus nidulans* (Fernández-Cañón and Peñalva, 1998).

Cloning and Expression of HDO and FAH

Both FAH and HDO can be successfully expressed as recombinant proteins in *E. coli* after their cloning into the expression vector pET24a, offering a C-terminal 6 x His tag. Polymerase chain reaction (PCR) products suitable for cloning into this vector can be generated from *Arabidopsis* cDNA using the oligonucleotide primers gcggcgcatatggaagagaagaagaaggagc and cgcgcggtcgacctccgaagctcctggttctttc for HDO and gcggcgcatatggcgttgctgaagtctttc and cgcgcgctcgagaggcggtgaaggaacaattttc for FAH. In each case, restriction digestion of the PCR product with *Nde*I and *Xho*I (*Sal*I for the HDO PCR product) gives a DNA fragment suitable for ligation directly into similarly digested pET24a. IPTG-induced expression of both constructs is best performed at 20° to minimize the amount of recombinant protein becoming insolublized in inclusion bodies. Crude protein extracts from *E. coli* expressing HDO and FAH can be used in the assays, or, alternately the recombinant proteins can be purified using Ni^{2+} immobilized metal affinity chromatography under standard conditions. However, both purified FAH and HDO are unstable in solution at 4° and become inactive once precipitated using $(NH_4)_2SO_4$ and are therefore best used as soon as possible after purification.

Assay of MAAI Activity

Maleylacetoacetate isomerase activity is measured using a coupled assay with excess fumarylacetoacetate hydrolase. Maleylacetoacetate is first enzymically synthesized by incubating at 30° 20 mM Tris-Cl, pH 7.8, containing 200 μM homogentisic acid with a crude centrifuged lysate

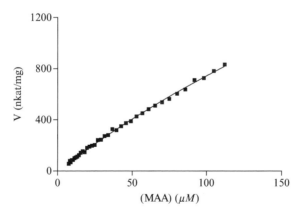

FIG. 1. Kinetic analysis of *At*GSTZ1-1 isomerase activity toward maleylacetoacetate (MAA) using data derived from a single assay. Calculated apparent K_m (MAA) = 500 μM V_{max} (MAA) = 4500 nkat/mg when assayed with 2.5 mM glutathione. Assays showed that *At*GSTZ1-1 had isomerase activity of 750 nkat/mg toward maleylacetoacetate (200 μM) and 185 nkat/mg toward maleylpyruvate (100 μM).

from sonicated bacteria expressing AtHDO. The conversion of homogentisate to maleylacetoacetate can be monitored at 330 nm and the reaction taken to completion. Next, add an excess of crude extract from bacteria expressing AtFAH, 1 mM glutathione (from 100 mM, pH 7.0, stock) and the zeta class GST preparation. The MAAI activity that is limiting in the coupled reaction can be monitored directly after the addition of the zeta GST by measuring the decrease in absorbance at 330 nm ($\varepsilon = -12,000$ M^{-1}/cm^{-1}).

Zeta GST–mediated isomerase activity toward maleylpyruvate can be similarly measured by replacing homogentisate with gentisate and HDO with a gentisate dioxygenase such as that from *Sphingomonas* sp. strain RW5 (Werwath *et al.*, 1998). In this case, the isomerase reaction results in a large enough decrease in absorbance at 330 nm ($\varepsilon = -5000$ M^{-1}/cm^{-1}; unpublished observation) that the coupled enzyme hydrolyzing fumarylpyruvate is not required. Additional assays should be performed with isomerase absent to correct for nonenzymic isomerization and other side reactions.

Notes

Add each of the following components in a minimal volume to minimize dilution of the assay mix.

1. Purified, recombinant his-tagged HDO (AtHDO) is able to cleave homogentisate with high activity (145 nkat/mg) but is unable to use gentisate.

2. HDO has previously been found to require ascorbate and Fe^{2+} for activity (Schmidt *et al.*, 1995). However, the activity of purified AtHDO was not decreased when assayed without these and actually increased approximately 30% when ascorbate was omitted. These reagents have, therefore, been omitted from the assay.

3. Purified, recombinant his-tagged AtFAH has a K_m toward FAA of 2.7 μM and a V_{max} of at least 40 nkat/mg.

4. If the resulting plot of absorbance vs. time is biphasic, this is an indication that the FAH activity is limiting in the first phase—either increase the concentration of FAH or reduce the MAAI concentration.

5. AtFAH has no detectable activity toward fumarylpyruvate.

6. Fumarylacetoacetate hydrolase activity can be assayed in a similar manner, except that excess isomerase enzyme should be used, and enzyme should be added after a short pre-incubation of the assay mix with glutathione and isomerase.

Because *At*FAH has a very low K_m for its substrate, the hydrolysis of fumarylacetoacetate will be sufficiently fast throughout the assay that it is safe to assume that no fumarylacetoacetate will accumulate and that the observed rate of hydrolysis is almost identical to the rate of isomerization. Also, the quantitative hydrolysis of fumarylacetoacetate means that MAAI is not subject to product inhibition. These factors mean that by following the reaction to completion, a single assay can yield rate data for a full range of substrate concentrations (Fig. 1).

References

Andrews, C. J., Skipsey, M., Townson, J. K., Morris, C., Jepson, I., and Edwards, R. (1997). Glutathione transferase activities toward herbicides used selectively in soybean. *Pestic. Sci.* **51,** 213–222.

Board, P. G., Baker, R. T., Chelvanayagam, G., and Jermiin, L. S. (1997). Zeta, a novel class of glutathione transferases in a range of species from plants to humans. *Biochem. J.* **328,** 929–935.

Board, P. G., Coggan, M., Chelvanayagam, G., Easteal, S., Jermiin, L. S., Schulte, G. K., Danley, D. E., Hoth, L. R., Griffor, M. C., Kamath, A. V., Rosner, M. H., Chrunyk, B. A., Perregaux, D. E., Gabel, C. A., Geoghegan, K. F., and Pandit, J. (2000). Identification, characterization, and crystal structure of the omega class glutathione transferases. *J. Biol. Chem.* **275,** 24798–24806.

Board, P. G., Coggan, M., Wilce, M. C. J., and Parker, M. W. (1995). Evidence for an essential serine residue in the active site of the theta class glutathione transferases. *Biochem. J.* **311,** 247–250.

Brash, A. R., and Song, W. (1996). Detection, assay, and isolation of allene oxide synthase. *In* "Cytochrome P450 Part B: RNA Polymerase and Associated Factors, Part A" (E. F. Johnson and M. R. Waterman, eds.), Vol. 272, pp. 250–259. Academic Press, San Diego.

Caccuri, A. M., Antonini, G., Allocati, N., Di Ilio, C., De Maria, F., Innocenti, F., Parker, M. W., Masulli, M., Lo Bello, M., Turella, P., Federici, G., and Ricci, G. (2002). GSTB1-1 from *Proteus mirabilis*—A snapshot of an enzyme in the evolutionary pathway from a redox enzyme to a conjugating enzyme. *J. Biol. Chem.* **277,** 18777–18784.

Chen, D., Kawarasaki, Y., Nakano, H., and Yamane, T. (2003). Cloning and *in vitro* and *in vivo* expression of plant glutathione *S*-transferase zeta class genes. *J. Biosci. Bioeng.* **95,** 594–600.

Cho, Y. W., Kim, J. C., Jin, C. D., Han, T. J., and Lim, C. J. (1998). Thioltransferase from *Arabidopsis thaliana* seed: Purification to homogeneity and characterization. *Mol. Cells* **8,** 550–555.

Cho, Y. W., Park, E. H., and Lim, C. J. (1999). A second thioltransferase from Chinese cabbage: Purification and characterization. *J. Biochem. Mol. Biol.* **32,** 133–139.

Clarke, E. D., Greenhow, D. T., and Adams, D. (1998). Metabolism-related assays and their application to agrochemical research: Reactivity of pesticides with glutathione and glutathione transferases. *Pestic. Sci.* **54,** 385–393.

Cummins, I., Cole, D. J., and Edwards, R. (1997). Purification of multiple glutathione transferases involved in herbicide detoxification from wheat (*Triticum aestivum* L.) treated with the safener fenchlorazole-ethyl. *Pestic. Biochem. Physiol.* **59,** 35–49.

DeRidder, B. P., Dixon, D. P., Beussman, D. J., Edwards, R., and Goldsbrough, P. B. (2002). Induction of glutathione *S*-transferases in *Arabidopsis* by herbicide safeners. *Plant Physiol.* **130**, 1497–1505.

Dixon, D., Cole, D. J., and Edwards, R. (1997). Characterisation of multiple glutathione transferases containing the GST I subunit with activities toward herbicide substrates in maize (*Zea mays*). *Pestic. Sci.* **50**, 72–82.

Dixon, D. P., Cole, D. J., and Edwards, R. (1998). Purification, regulation and cloning of a glutathione transferase (GST) from maize resembling the auxin-inducible type-III GSTs. *Plant Mol. Biol.* **36**, 75–87.

Dixon, D. P., Cole, D. J., and Edwards, R. (1999a). Dimerisation of maize glutathione transferases in recombinant bacteria. *Plant Mol. Biol.* **40**, 997–1008.

Dixon, D. P., Cole, D. J., and Edwards, R. (1999b). Identification and cloning of AtGST 10 (Accession Nos. AJ131580 and AJ132398), members of a novel type of plant glutathione transferases. *Plant Physiol.* **119**, 1568.

Dixon, D. P., Cole, D. J., and Edwards, R. (2000). Characterisation of a zeta class glutathione transferase from *Arabidopsis thaliana* with a putative role in tyrosine catabolism. *Arch. Biochem. Biophys.* **384**, 407–412.

Dixon, D. P., Davis, B. G., and Edwards, R. (2002a). Functional divergence in the glutathione transferase superfamily in plants—Identification of two classes with putative functions in redox homeostasis in *Arabidopsis thaliana*. *J. Biol. Chem.* **277**, 30859–30869.

Dixon, D. P., Lapthorn, A., and Edwards, R. (2002b). Plant glutathione transferases. *Genome Biol.* **3**, 3004.1–3004.10.

Dixon, D. P., McEwen, A. G., Lapthorn, A. J., and Edwards, R. (2003). Forced evolution of a herbicide detoxifying glutathione transferase. *J. Biol. Chem.* **278**, 23930–23935.

Droog, F. (1997). Plant glutathione *S*-transferases, a tale of theta and tau. *J. Plant Growth Regul.* **16**, 95–107.

Edwards, R. (1996). Characterisation of glutathione transferases and glutathione peroxidases in pea (*Pisum sativum*). *Physiol. Plant.* **98**, 594–604.

Edwards, R., and Dixon, D. P. (2000). The role of glutathione transferases in herbicide metabolism. *In* "Herbicides and Their Mechanisms of Action" (A. H. Cobb and R. C. Kirkwood, eds.), pp. 38–71. Sheffield Academic Press, Sheffield, UK.

Edwards, R., Dixon, D. P., and Walbot, V. (2000). Plant glutathione *S*-transferases: Enzymes with multiple functions in sickness and in health. *Trends Plant Sci.* **5**, 193–198.

Edwards, S. W., and Knox, W. E. (1956). Homogentisate metabolism: The isomerization of maleylacetoacetate by an enzyme which requires glutathione. *J. Biol. Chem.* **220**, 79–91.

Fernández-Cañón, J. M., and Peñalva, M. A. (1998). Characterization of a fungal maleylacetoacetate isomerase gene and identification of its human homologue. *J. Biol. Chem.* **273**, 329–337.

Fowler, J., and Seltzer, S. (1970). The synthesis of model compounds for maleylacetoacetic acid. Maleylacetone. *J. Org. Chem.* **35**, 3529–3532.

Frova, C. (2003). The plant glutathione transferase gene family: Genomic structure, functions, expression and evolution. *Physiol. Plant.* **119**, 469–479.

Hatton, P. J., Dixon, D., Cole, D. J., and Edwards, R. (1996). Glutathione transferase activities and herbicide selectivity in maize and associated weed species. *Pestic. Sci.* **46**, 267–275.

Iida, T., Senoo, S., Sato, Y., Nicolaus, B., Wakabayashi, K., and Böger, P. (1995). Isomerization and peroxidising phytotoxicity of thiadiazolidine-thione compounds. *Zeit. Naturforsch.* **50c**, 186–192.

Jakobsson, P. J., Morgenstern, R., Mancini, J., Ford-Hutchinson, A., and Persson, B. (1999). Common structural features of MAPEG—A widespread superfamily of membrane associated proteins with highly divergent functions in eicosanoid and glutathione metabolism. *Protein Sci.* **8**, 689–692.

Lantum, H. B. M., Liebler, D. C., Board, P. G., and Anders, M. W. (2002). Alkylation and inactivation of human glutathione transferase zeta (hGSTZ1-1) by maleylacetone and fumarylacetone. *Chem. Res. Toxicol.* **15**, 707–716.

Marrs, K. A. (1996). The functions and regulation of glutathione S-transferases in plants. *Annu. Rev. Plant Physiol. Plant Mol. Biol.* **47**, 127–158.

McGonigle, B., Keeler, S. J., Lau, S.-M. C., Koeppe, M. K., and O'Keefe, D. P. (2000). A genomics approach to the comprehensive analysis of the glutathione S-transferase gene family in soybean and maize. *Plant Physiol.* **124**, 1105–1120.

Morgenstern, R., De Pierre, J. W., and Jörnvall, H. (1985). Microsomal glutathione transferase: Primary structure. *J. Biol. Chem.* **260**, 13976–13983.

Mozer, T. J., Tiemeier, D. C., and Jaworski, E. G. (1983). Purification and characterisation of corn glutathione S-transferase. *Biochem.* **22**, 1068–1072.

Nicolaus, B., Sato, Y., Wakabayashi, K., and Böger, P. (1996). Isomerization of peroxidizing thiadiazolidine herbicides is catalyzed by glutathione S-transferase. *Zeit. Naturforsch.* **51c**, 342–354.

Reinemer, P., Prade, L., Hof, P., Neuefeind, T., Huber, R., Zettl, R., Palme, K., Schell, J., Koelln, I., Bartunik, H. D., and Bieseler, B. (1996). 3-dimensional structure of glutathione S-transferase from *Arabidopsis thaliana* at 2.2-angstrom resolution—structural characterization of herbicide-conjugating plant glutathione S-transferases and a novel active-site architecture. *J. Mol. Biol.* **255**, 289–309.

Roxas, V. P., Lodhi, S. A., Garrett, D. K., Mahan, J. R., and Allen, R. D. (2000). Stress tolerance in transgenic tobacco seedlings that overexpress glutathione S-transferase/ glutathione peroxidase. *Plant Cell Phys.* **41**, 1229–1234.

Sappl, P. G., Oñate-Sánchez, L., Singh, K. B., and Millar, A. H. (2004). Proteomic analysis of glutathione S-transferases of *Arabidopsis thaliana* reveals differential salicylic acid-induced expression of the plant-specific phi and tau classes. *Plant Mol. Biol.* **54**, 205–219.

Schägger, H., and von Jagow, G. (1987). Tricine sodium dodecyl sulfate-polyacrylamide gel electrophoresis for the separation of proteins in the range from 1 to 100 kDa. *Anal. Biochem.* **166**, 368–379.

Schmidt-Krey, I., Mitsuoka, K., Hirai, T., Murata, K., Cheng, Y., Fujiyoshi, Y., Morgenstern, R., and Hebert, H. (2000). The three-dimensional map of microsomal glutathione transferase 1 at 6 angstrom resolution. *EMBO J.* **19**, 6311–6316.

Schmidt, S. R., Müller, C. R., and Kress, W. (1995). Murine liver homogentisate 1,2-dioxygenase: Purification to homogeneity and novel biochemical properties. *Eur. J. Biochem.* **228**, 425–430.

Sies, H., and Packer, L. (2002). "Protein Sensors and Reactive Oxygen Species, Part B: Thiol Enzymes and Proteins." Academic Press, San Diego.

Thom, R., Cummins, I., Dixon, D. P., Edwards, R., Cole, D. J., and Lapthorn, A. J. (2002). Structure of a tau class glutathione S-transferase from wheat active in herbicide detoxification. *Biochem.* **41**, 7008–7020.

Thom, R., Dixon, D., Edwards, R., Cole, D., and Lapthorn, A. (2001). Structure determination of zeta class glutathione transferase from *Arabidopsis thaliana*. *Chem. Biol. Interact.* **133**, 53–54.

Vlamis-Gardikas, A., Åslund, F., Spyrou, G., Bergman, T., and Holmgren, A. (1997). Cloning, overexpression and characterization of glutaredoxin 2, an atypical glutaredoxin from *Escherichia coli. J. Biol. Chem.* **272**, 11236–11243.

Wagner, U., Edwards, R., Dixon, D. P., and Mauch, F. (2002). Probing the diversity of the *Arabidopsis* glutathione *S*-transferase gene family. *Plant Mol. Biol.* **49**, 515–532.

Werwath, J., Afrmann, H.-A., Pieper, D. H., Timmis, K. N., and Wittich, R.-M. (1998). Biochemical and genetic characterization of a gentisate 1,2-dioxygenase from *Sphingomonas* sp. strain RW5. *J. Bact.* **180**, 4171–4176.

[12] Phylogenies of Glutathione Transferase Families

By WILLIAM R. PEARSON

Abstract

The best known glutathione transferase family, with its class-alpha, -mu, -pi, -omega, -sigma, -theta, and -zeta subdivisions, is only one of four, or perhaps five, ancient protein families that conjugate glutathione or use a glutathione intermediate: (1) the cytoplasmic family, (2) the mitochondrial (kappa) family, (3) the microsomal (MAPEG) family, which may actually be two separate families, and (4) the fosphomycin/glyoxalase family. Although the cytoplasmic family is perhaps the most diverse, all four of these families have homologs in both prokaryotes and eukaryotes; it is striking that at least three, and perhaps as many as five, different protein folds capable of binding and positioning glutathione for a nucleophilic attack emerged more than 2 billion years ago. This chapter presents phylogenies for the four (or five) glutathione transferase families, focusing on the statistical evidence for homology (and non-homology).

Introduction

The phylogeny of glutathione transferase conjugation enzymes is complex; glutathione transferase activity has emerged independently at least four different times throughout evolutionary history, producing four different glutathione transferase families—cytoplasmic, microsomal, mitochondrial, and bacterial. Moreover, each of the four glutathione transferase protein families contains members that are clearly evolutionarily related, yet have different functions. Thus, there are at least four different glutathione transferase phylogenies, and several of these phylogenies include enzymes that do not conjugate glutathione. Because of the complex intersection of evolutionary history and biological function for these enzymes,

METHODS IN ENZYMOLOGY, VOL. 401   0076-6879/05 $35.00
Copyright 2005, Elsevier Inc. All rights reserved.   DOI: 10.1016/S0076-6879(05)01012-8

this review will consider each of the four families separately and will also discuss some proteins lacking glutathione conjugation activity in the analysis.

The Four Glutathione Transferase Families

Before discussing the evolutionary histories of the four glutathione transferase families, it is important to outline our criterion for inferring the non-homology of the four distinct glutathione transferase families and the homology of their family members. Although a recent review (Hayes et al., 2004) recognized the three families best known in eukaryotes—cytosolic, mitochondrial, and microsomal (referred to as MAPEG in Hayes et al., 2004)—earlier evolutionary classifications of some of the enzymes (e.g., the kappa class; Harris et al., 1991) have been mistaken. In this review, the inference of homology will be based on statistically significant sequence or structural similarity; if two proteins share significantly more similarity than is expected by chance, the most parsimonious explanation is that they descended from a common ancestor (they are homologous; Sierk and Pearson, 2004) and thus belong to the same family. In this chapter, stastical significance is reported as an expectation value; (e.g., E() $< 10^{-6}$) reports that the sequence (or structural) similarity would be expected by chance less than one time in 10^6 database searches. Unfortunately, although proteins that share statistically significant sequence similarity can be inferred to be homologous, the inverse is not true; proteins that do not share significant sequence similarity may still be inferred to be homologous if they share significant similarity in three-dimensional structure. However, if the proteins do not share *significant* structural similarity, then they could have arisen independently through convergent evolution. Fortunately, three-dimensional structures are available for members of all four glutathione transferase families; these provide the strongest evidence for classification.

High-resolution three-dimensional structures are available for three of the four glutathione transferase families: (1) the cytoplasmic glutathione transferases (at least 90 structures available), (2) the bacterial metallothiol transferases (19 structures), and (3) the mitochondrial kappa family (Ladner et al., 2004). A low-resolution structure is available for the microsomal family of glutathione transferases (MAPEG, Holm et al., 2002), which confirms that it does not share structural similarity with the other three families. It has been argued (Ladner et al., 2004) that the mitochondrial kappa family shares a common glutaredoxin/thioredoxin domain with the cytosolic enzymes, but the significance of the structural similarity is borderline. The deep phylogenetic relationships of the glutathione transferases will be discussed toward the end of the chapter.

Cytoplasmic Glutathione Transferases

Some of the phylogenetic complexity of the glutathione transferases was recognized almost 20 years ago (Mannervik *et al.*, 1985), when three classes of mammalian soluble enzymes were named, class-alpha, -mu, and -pi. The discovery of the class-theta proteins in plants and insects dramatically expanded the range of organisms with recognizable family members; insect class-theta proteins share strongly significant sequence similarity ($E() < 10^{-10}$) with proteins in vertebrates, invertebrates, plants, and bacteria. Indeed, the early division of the cytoplasmic enzymes into four classes, -alpha, -mu, -pi, and -theta, was somewhat misleading, because it suggested the classes were similar in scope. In fact, the -alpha, -mu, and -pi enzymes are much more closely related to each other, whereas the class-theta enzymes are much more phylogenetically diverse. Indeed, because of its great diversity, assignment to class-theta has little meaning.

Cytoplasmic glutathione transferases have been assigned to eight widely used classes, -alpha, -mu, -omega, -phi, -pi, -sigma, -theta, and -zeta. Of the eight, only class-phi is not found in mammals.

Because of their great diversity, it is difficult to produce a single evolutionary tree that includes all the cytoplasmic glutathione transferases. Figure 1 (class-alpha, -mu, -pi, and -sigma), Fig. 2 (class -phi, -theta, and -zeta), and Fig. 3 (class-omega, and others) show three overlapping parts of the phylogeny of cytosolic glutathione transferase family proteins.

The Metazoan-Specific Glutathione Transferases

Four cytoplasmic glutathione transferase classes are found exclusively in metazoa, the three original class-alpha, -mu, and -pi (Mannervik *et al.*, 1985), and the more recently recognized class-sigma (Jowsey *et al.*, 2001; Meyer and Thomas, 1995). A striking feature of the class-alpha, -mu, -pi, and -sigma proteins is their mixture of phylogenetic restriction and diversity. On the basis of the initial class-alpha, -mu, and -pi proteins, it is tempting to suggest that this branch of the glutathione transferase phylogeny is relatively young. The classical class-alpha, -mu, and -pi proteins are found only in vertebrates, suggesting that the common ancestor of these enzymes might have existed as little as 500 My ago. The vertebrate proteins on this tree are shaded in Fig. 1.

The other proteins in Fig. 1 produce a much more complex organismal phylogeny. Included near the mammalian class-mu proteins are a group of surface antigens and immunogens from flatworms (GT26_SCHJA, GT26_SCHMA, GT27_SCHMA, GT26–GT29_FASHE) and allergens from ticks and dust mites (GTM1_DERPT). The peculiar evolutionary placement of these proteins, and the lack of orthologs in other insects or

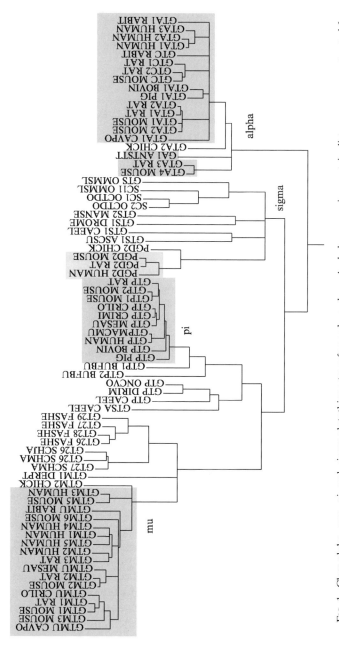

FIG. 1. Class-alpha, -mu, -pi, and -sigma glutathione transferase homologs. A phylogeny, using protein distances, was constructed for a set of class-alpha, -mu, -pi, and -sigma proteins from the Swiss-Prot protein sequence database (Bairoch and Apweiler, 1996). All of the proteins included in the tree share statistically significant similarity (E() < 10^{-3}) with GTM1_HUMAN (P09488). Sequences were aligned using the ClustalW program (Thompson *et al.*, 1994), and trees were built using the seqboot, protdist, and fitch programs from the PHYLIP v3.6 package (Felsenstein, 1989). The reliability of the branching order was estimated with 50 bootstrap samples of the multiple alignment; branches that occurred less than half the time were merged into polytomies. The tree was rooted using a *Drosophila* class theta sequence (not shown) as the out-group. Sets of mammalian proteins are shaded in gray.

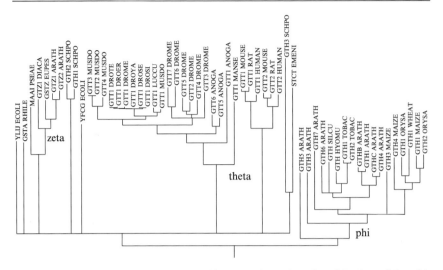

Fig. 2. Class-phi, -theta, and -zeta glutathione transferase homologs. Members of the -phi, -theta, and -zeta classes are shown. All of the proteins included in this tree share more than $E() < 10^{-6}$ similarity with GTH3_ARATH (P42761). Some homologs (elongation factor 1γ's) at this significance were excluded from the tree because of dramatically different lengths. The trees were built, and bootstraps performed, as described in Fig. 1. A mammalian class-mu glutathione transferase was used to root this tree.

simpler metazoans, together with their identification as immunogens, suggests that the flatworm and *acari* proteins arose through horizontal gene transfer. Thus, excluding horizontal gene transfer, one can argue that the class-mu enzymes arose concurrently with the emergence of vertebrates. Class-mu proteins have been identified in fish (*D. rerio*, NP_997841) and amphibians, as well as chickens (GTM2_CHICK) and mammals.

The evolutionary tree in Fig. 1, which was rooted using *Drosophila* class-theta sequence as the out-group, suggests that the class-alpha enzymes are the oldest members of this clade. The alpha class is striking, because there are no invertebrate members, unlike the close invertebrate homologs to class-mu, -pi, and -sigma. Likewise, most of the class-mu enzymes are vertebrate, except for the flatworm and tick proteins, which were probably acquired by horizontal gene transfer. Indeed, except for the tick/mite mu-like proteins, the only insect proteins from this branch of the tree belong in class-sigma. In contrast, *C. elegans* has dozens of proteins that are most closely related to class-pi and -sigma enzymes. Thus, if the class-alpha enzymes emerged first along this branch, they must have been lost from invertebrates, because invertebrate enzymes can be found in class-pi and -sigma. The analysis in Fig. 1 also suggests that the class-mu proteins arose most recently.

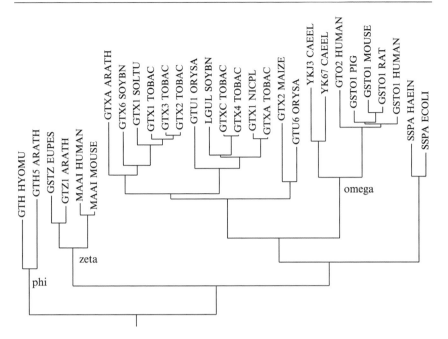

Fig. 3. Class-omega and other glutathione transferase homologs. Members of the -omega class, together with a sample of -phi and -zeta proteins, are shown. All of the proteins included in this tree share more than E() < 10^{-6} similarity with GSTO1_HUMAN (P78417). The trees were built, and bootstraps performed, as described in Fig. 1. A *Drosophila* class-theta sequence was used to root this tree.

Mammalian Orthologs and Paralogs

Among other activities, glutathione transferases detoxify chemical carcinogens, and dramatic increases in glutathione transferase expression are associated with the chemoprotective effects of antioxidants and other dietary supplements (Benson *et al.*, 1978; Ramos-Gomez *et al.*, 2001; Wattenberg, 1978). Direct molecular relationships between detoxification enzyme induction and diet are most easily demonstrated in rodents; to extend our understanding to humans, it is important to understand the evolutionary relationships between human and rodent detoxification genes.

All the proteins shown in Fig. 1 are clearly homologous; they all have statistical expectations with E() < 10^{-6} compared with human *GSTM1* (GTM1_HUMAN) and are generally more than 25% identical over their entire length. The mammalian proteins in the class-mu and class-alpha subfamilies are even more closely related; the most distantly related vertebrate class-mu protein, GTM1_CHICK, is 65% identical to GTM1_HUMAN, and the most divergent mammalian class-mu proteins (GTM3_HUMAN and GTM5_MOUSE) are more than 70% identical to GTM1_HUMAN. However, the detailed evolutionary relationships between the class-mu and class-alpha proteins are more ambiguous. Homologous proteins can be either *orthologous*—differing because of speciation events—or *paralogous*—the product of gene duplication events. Thus, most mammalian class-pi genes are orthologous (mouse is an exception, because of a gene duplication that seems to have occurred after the mouse/rat divergence); the class-pi section of the phylogeny in Fig. 1 presents a sensible phylogeny for mammals. The same is true for the mammalian class-sigma prostaglandin D_2 synthases.

The mammalian phylogenies for the class-alpha and class-mu enzymes are much more complex. In both the class-alpha and class-mu branches of the phylogeny in Fig. 1, glutathione transferase paralogs from the same taxonomic order (e.g., *Rodentia*) often cluster together, for example, the GTA1_HUMAN, GTA2_HUMAN, GTA3_HUMAN branch, which is separate from the GTA1_MOUSE, GTA2_MOUSE, GTA1_RAT, GTA2_RAT branch. For the class-alpha genes, there are no clear orthologs between human and rodents, and among the rodents, only the GTA3_RAT/GTA4_MOUSE and GTC_MOUSE/GTC2_RAT pairs seem clearly orthologous. For the class-mu genes, only the GTM3_HUMAN and GTM5_MOUSE (and also the rat M5, Rowe *et al.*, 1998, not shown in Fig. 1) are clearly orthologous between human and rodent. This orthology is supported by the unusual tail-to-tail arrangement of the human *GSTM3* gene (and rodent M5 genes) with respect to the cluster of head-to-head M1-like genes (Patskovsky *et al.*, 1999). It is perhaps not surprising that it is difficult to assign orthologies between human and mouse genes, because there are at least six mouse class-mu genes and only five human class-mu genes in their respective gene clusters.

Among the rodents, the GTM1_RAT/GTM1_MOUSE and GTM2_RAT/GTM2_MOUSE pairs are orthologous, and a rat ortholog of GTM3_MOUSE, rat Yb4 (*Gst4*, NP_065415) is known. The evolutionary history of the other paralogs suggests that additional duplications (or deletions) have taken place in the rodent line.

Because of the different numbers of genes, and the apparent gene conversion events that have occurred within the human class-mu gene cluster, it may not be possible to identify rodent/human class-mu or class-alpha orthologs other than the clear-cut GTM3_HUMAN/GTM3_MOUSE orthology. Unfortunately, this complicates the application of rodent models of class-mu and class-alpha glutathione transferase regulation to the human genes.

Class-Theta and Other Widely Distributed Glutathione Transferase Sub-Families

The phylogenetic tree shown in Fig. 2 contains the broadest range of glutathione transferases, which span the greatest evolutionary distance. Insect class-theta enzymes share statistically significant similarity with glutathione transferase homologs in mammals, plants, and bacteria. Class-theta enzymes were named with their discovery in mammals (Meyer *et al.*, 1991), but their relatively stronger similarity to the insect glutathione transferases caused the class to be extended to them as well. This is unfortunate, because the class-theta mammalian and insect glutathione transferases share less than 30% identity, and insect class-theta enzymes share no more similarity to the original human class-theta proteins than they do to class-zeta (but the zeta enzymes were not discovered until almost a decade later, using bioinformatics techniques).

In *Drosophila*, at least 27 of the 51 or so cytoplasmic glutathione transferases belong to class-theta, but the branching within the family is quite deep, similar to the deep branching shown in Fig. 2. *Drosophila* also has homologs that are most closely related to the sigma, omega, and zeta classes, and it seems to have an intracellular chloride channel homolog. In contrast, in *C. elegans*, there are no clearly class-theta enzymes, but there are more than 40 enzymes that are most closely related to enzymes from class-pi and -sigma (Fig. 1). *C. elegans* also has class-zeta and -omega homologs, as well as a chloride channel homolog.

Identifying Cytoplasmic Glutathione Transferase Homologs

Because of the very large number of known cytoplasmic glutathione transferase homologs, members of the cytoplasmic glutathione transferase protein family are easily identified with protein or translated-DNA sequence similarity searches using BLAST (Altschul *et al.*, 1997) or FASTA (Retief *et al.*, 1999). Indeed, new glutathione transferase classes (class-zeta, -omega, intracellular chloride channel) and class-members (mouse theta T3-3, Coggan *et al.*, 2002) were identified using bioinformatics techniques (Dulhunty *et al.*, 2001; Retief *et al.*, 1999). Today, the most distant cytoplasmic glutathione transferase homolog, the intracellular chloride channels (Dulhunty *et al.*, 2001), can be identified as glutathione transferase homologs readily after a few iterations of PSI-BLAST (Altschul *et al.*, 1997) or with the SSEARCH implementation of the Smith-Waterman algorithm (Pearson, 1998) using a significance threshold of E() < 0.002. Because glutathione transferases are soluble enzymes with "average" amino acid composition, statistical estimates in searches with cytoplasmic glutathione transferases tend to be very accurate, so that alignment scores with $E() < 0.01$ with SSEARCH, and $E() < 0.001$ with PSI-BLAST, reliably indicate homology.

Nonconjugating Glutathione Transferase Homologs

The cytoplasmic glutathione transferase family is very old; the proteins shown in Fig. 2 include enzymes from mammals, insects, and bacteria. Most of the proteins shown in Figs. 1–3 are likely to be glutathione transferase enzymes, although some, like maleylacetoacetate isomerase, have relatively low activities on traditional glutathione transferase substrates. Moreover, there are many cytoplasmic glutathione transferase family members that do not function as glutathione transferases. These include the cephalopod lens crystallins, which are clearly members of class-sigma (Fig. 1, Tomarev et al., 1991), bacterial stringent starvation proteins (Toung and Tu, 1992), and plant heat-shock proteins. All these proteins share strong similarity to cytoplasmic glutathione transferases and were quickly recognized as glutathione transferase homologs.

Mitochondrial Glutathione Transferases

In addition to the cytoplasmic glutathione transferases, three other glutathione transferase protein families have been identified—the mitochondrial family, also known as class-kappa, the microsomal enzymes, and the bacterial antibiotic resistance proteins. The mitochondrial glutathione transferases were originally named class-kappa because of a mistaken inference of homology to the cytoplasmic enzymes based on partial amino acid sequence similarity (Harris et al., 1991; Pemble et al., 1996), similar lengths, and dimeric structures. In fact, the mammalian "class-kappa" proteins do not show statistically significant sequence similarity to any members of the cytoplasmic glutathione transferase family; nor do any of the hundreds of cytoplasmic glutathione transferase family members share significant similarity to the mitochondrial "class-kappa" proteins. Instead, the mitochondrial enzymes share strong similarity with a family of prokaryotic 2-hydroxychromene-2-carboxylic acid isomerases (HCCI) involved in the catabolism of naphthalene (Eaton, 1994), dibenzyl compounds (Gregorio et al., 2004), and at least one DSBA oxidoreductase, CAE27624 from Rhodopseudomonas, Fig. 4). The complete lack of overlap between mitochondrial glutathione transferases/hydroxychromene-2-carboxylic acid isomerases and the cytoplasmic glutathione transferases strongly suggests an independent evolutionary origin.

The recently determined crystal structure of the mitochondrial class-kappa enzyme rGSTGK1-1 (Ladner et al., 2004) confirms the distinction

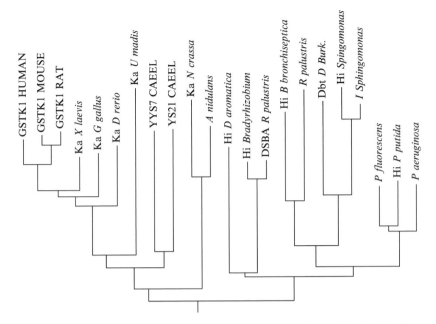

FIG. 4. Mitochondrial (kappa) glutathione transferase homologs. Mitochondrial-kappa homologs were identified with a search of the NCBI nr database using a human GSTK1 protein sequence (EAL23783). All of the sequences shown share statistically significant similarity (E() < 10^{-5}). To illustrate the types of functional annotations associated with the different sequences, proteins are either labeled using their Swiss-Prot name or using the annotated enzyme name (Ka for GST kappa), Hi for 2-hydroxychromene-2-carboxylic acid isomerases, or the gene name (DSBA, DbtD). The accession numbers for the non-SwissProt sequences are: Ka *X laevis*: AAH87819, Ka *G gallus*: XP_416525, Ka *D rerio*: NP_001002560, Ka *U madis*: EAK86081, *P fluorescens*: ZP_00262414, *A nidulans*: EAA59114, *P aeruginosa*: AAG03508, Hi *P putida*: NP_745026, Dbt *D Burk.*: AAK62352, DSBA *R palustris*: CAE27624, Ka *N crassa*: CAD21511, Hi *D aromatica*: ZP_00152853, *R palustris*: CAE26292, Hi *B bronchiseptica*: NP_887775, *I Sphingomonas*: CAG17583, Hi *Bradyrhizobium*: NP_772246, *Hi Sphingomonas*: AAD45416. Trees were built as described in Fig. 1, but no bootstrap sampling was done. The tree was rooted using mid-point rooting with retree.

between the mitochondrial family and the cytoplasmic glutathione trans-ferases. rGSTK1-1 (PDB 1R4W) shares weak structural similarity to *V. cholerae dsbA* disulfide oxidoreductase (1BED, VAST E() < 10^{-6}, 15% identical over 140 structurally aligned positions, DALI Lite Z = 12.3, E() < 0.0004, 15% identical over 143 structurally aligned residues). The best structural match in the cytoplasmic glutathione transferase family, 1YY7, has a lower level of structural similarity (VAST E() < 10^{-6}, 61 aligned residues; DALI-Lite Z < 3, not significant). As discussed by Ladner *et al.* (2004), the mitochondrial-kappa/HCCI enzymes share a different fold from the cytoplasmic enzymes, but one that may share an ancient relationship through the N-terminal glutathione-binding domain.

As shown in Fig. 4, the evolutionary history of the mitochondrial-kappa family is considerably less complex that the cytoplasmic glutathione transferases, although the mitochondrial family is also very ancient, with homologs in bacteria and eukaryotes. Current genome annotations on model organisms suggest that most vertebrate genomes contain a single orthologous mitochondrial-kappa protein; however, *Drosophila* does not have an annotated homolog, and *C. elegans* has two homologs. Although many members of this family are labeled hypothetical, the phylogeny in Fig. 4 suggests that all the eukaryotic enzymes are most closely related to mitochondrial-kappa, whereas the prokaryotic enzymes are HCCIs or DSBAs. Of course, many of the current annotations are based on sequence similarity, which may not reliably distinguish between these two possibilities.

Microsomal Glutathione Transferases

A third family of proteins—the microsomal enzymes, more recently termed the MAPEG family (membrane-associated proteins in eicosanoid and glutathione metabolism, Jakobsson *et al.*, 1999, 2000)—has glutathione transferase activity. The microsomal glutathione transferases differ substantially in size and structure from both the cytoplasmic and the mitochondrial-kappa glutathione transferase families; the microsomal enzymes are shorter, around 150–160 amino acids (compared with the 210–240 amino acids of the cytoplasmic enzymes), and some members of the family have been shown to be trimers (Jakobsson *et al.*, 1999). In humans, three different microsomal glutathione transferases have been described: MGST1, MGST2, and MGST3. MGST1 shares strong sequence similarity with prostaglandin E synthase (SwissProt PTGES_HUMAN, O14684), whereas the MGST2 proteins share strong sequence similarity with 5-lipoxygenase activating protein (FLAP_HUMAN, P20292) and leukotriene C4 synthase (LC4S_HUMAN, Q16873).

Although both the MGST1/PTGES and MGST2,3/FLAP/LC4S proteins are now classified as MAPEG family members in most protein and domain databases, the evidence for their homology is not strong. None of the more than 32 distinct proteins that share statistically significant similarity with MGST1/PTGES (referred to henceforth as the MAPEG1 family) share significant similarity with any protein that shares significant similarity with any of the 48 distinct members of the MGST2,3/FLAP/LC4S (MAPEG2) family currently available in the NCBI nr protein sequence

A MGST1/PTGES/MAPEG1 proteins

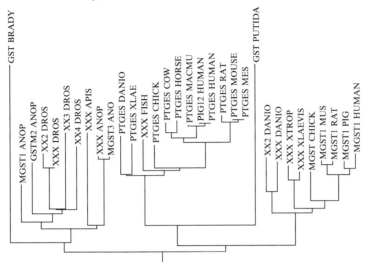

B MGST2/3/FLAP/LC4S/MAPEG2 proteins

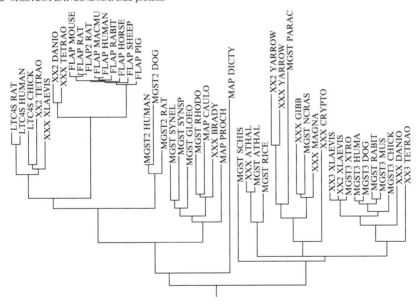

FIG. 5. Microsomal MAPEG1 and MAPEG2 glutathione transferase homologs. Two trees are shown; one composed of sequences that share statistically significant similarity with human MGST1 (A, MGST1_HUMAN, P10620) and the second with human MGSTM2 (B, MGST2_HUMAN, Q99735). Sequences sharing statistically significant similarity with

database (March, 2005) or vice versa. Moreover, a position-specific scoring matrix (PSSM) built from 48 distinct MAPEG2 family members identifies the MGST2/3/FLAP/LC4S proteins with $E() < 10^{-24}$, but does not find any of the MAPEG1 sequences with $E() < 0.001$ (a PTGE enzyme is found with $E() < 0.006$, and MGST1 with $E() < 0.04$, but these are very weak similarity scores for profile/PSSM searches, particularly with proteins containing multiple transmembrane domains). The absence of proteins that share significant similarity with both MAPEG1 and MAPEG2 family members is particularly surprising, because both families have clear homologs in the prokaryotes. MAPEG2 homologs are also found in plants, but MAPEG1 homologs are not. Although there is a strong chance that the MAPEG1 and MAPEG2 proteins share statistically significant structural similarity from which homology can be inferred, the two sets of proteins may be as distant as the cytoplasmic and mitochondrial-kappa enzymes and probably reflect parallel evolutionary paths.

The evolutionary histories of the MAPEG1 and MAPEG2 families are shown in Fig. 5. The phylogenies in Fig. 5 seem to place different enzyme activities on different deep branches of the trees, although it is difficult to be certain, because so many of the MAPEG1 and 2 family members have been annotated based on sequence similarity alone. In the MAPEG1 family, the PGTES orthologs are clearly separate from the MGST1 orthologs, which seem to be separate from the insect proteins. Two prokaryotic proteins are included in the family, one from the gamma-Proteobacteria *Pseudomonas putida*, a second from the alpha-Proteobacteria *Bradyrhizobium japonicum*. From the scarcity of MAPEG1 family members in prokaryotes, it seems likely that these bacteria obtained the protein through horizontal transfer. A similar phylogenetic structure is seen for the MAPEG2 family, with deep branching FLAP, LTC4S, MGSTM2, and MGST3 orthologs, where the MGST3 branch extends to fungi and plants. Many more prokaryotic MAPEG2 homologs can be

---

$E() < 10^{-6}$ are included in the two trees. SwissProt names are used when available, but when they were not available, XXX and an abbreviation for the species is used. When several homologs are available in the same species, XX2, XX3 was used. (A) The accession numbers for the non-SwissProt proteins are: XXX_XLAEV: AAH68870, XXX_FISH: CAF97117, XXX_DANIO: AAH83311, XXX_XTROP: AAH87821, XX2_DANIO: NP_001002215, XXX_DROS: NP_524696, XX2_DROS: AAN85305, XXX_ANOP: EAA14109, XXX_APIS XP: 394313, XX3_DROS: NP_788903, XX4_DROS: NP_788904. (B) The accession numbers for the non-SwissProt proteins are: XXX_XLAEV: AAH84377, XX2_XLAEV: AAH87365, XX3_XLAEV: AAH75123, XXX_TETRAO: CAG00412, XX2_TETRAO: CAG04538, XXX_DANIO: NP_998592, XX2_DANIO: NP_956355, XXX_BRADY: NP_767639, XX3_TE-TRA: CAG09920, XXX_ATHAL: AAF23833, XXX_CRYPTO: EAL18119, XXX_GIBB: EAA67799, XXX_MAGNA: EAA55250, XXX_YARROW: CAG79084, XX2_YARROW: CAG79755. Trees are shown using mid-point rooting.

detected, however, suggesting that this family may be considerably older than the MAPEG1 family.

Bacterial Antibiotic Resistance Proteins

A fourth (or fifth, if MAPEG1 and MAPEG2 are not homologous) family of glutathione transferase enzymes includes the metallothiol bacterial fosfomycin resistance proteins, which are related to glyoxylases, lactoglutathione lyases, and extradiol dioxygenases (Bernat *et al.*, 1997). X-ray crystal structures of these enzymes have been determined, showing a unique repeated domain structure that is distinct from either the cytoplasmic, mitochondrial-kappa, or MAPEG glutathione transferase families. Thus, it seems likely that glutathione transferase folds have arisen independently at least four times.

Like the other glutathione transferase families, the FOSA/glyoxalase/lactoglutathione lyase family contains a diverse set of proteins found in bacteria, plants, and animals. A phylogenetic tree of a sample of the family members is shown in Fig. 6. In addition to showing the age and diversity of this family, the tree also illustrates a functional overlap, which may lead to confusion, among the different glutathione transferase families. A set of lactoglutathione lyases (glyoxalases) from bacteria, animals, and plants belongs to this family, but a lactoglutathione lyase from soybean (LGUL_SOYBN, P46417) belongs to the omega branch of the cytoplasmic glutathione transferase family.

Thioredoxin and Glutathione Transferase Progenitors

Current domain and structural databases divide the cytoplasmic glutathione transferase structure into two domains, an N-terminal glutaredoxin-like domain and a C-terminal domain shared only by the cytoplasmic family. The NCBI Conserved Domain Database (CDD) distinguishes the N-terminal glutaredoxin domain from the much longer mitochondrial "class-kappa" domain structure, classifying the latter with protein disulfide isomerases. In contrast, both the CATH (Pearl *et al.*, 2005) and SCOP (Andreeva *et al.*, 2004) protein structure domain databases include both the cytoplasmic glutathione transferase structures (e.g., 1GTU, 1GUK, 4PGT; currently 92 cytoplasmic glutathione transferase structures are available) and the mitochondrial "class-kappa" (1R4W), which is grouped with *E. coli dsbA*.

Unfortunately, the statistics of structural similarity scores–the likelihood that a structural similarity might occur by chance–are poorly understood (Sierk and Pearson, 2004), so it is difficult to make a strong case that

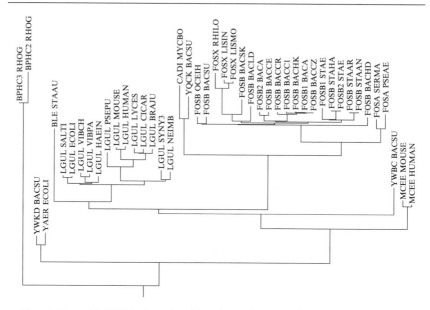

FIG. 6. Bacterial FOSA, lactoglutathione-lyase (glyoxalase), and related glutathione transferase homologs. Sequences from SwissProt sharing statistically significant similarity with a PSI-BLAST position specific scoring matrix built starting with FOSA_PSEAE (Q9I4K6) against the SwissProt protein database (a similar result would be generated starting with MCEE_HUMAN, Q96PE7) were aligned and used to build a tree as described in Fig. 1. Sequences with $E() < 10^{-6}$ are included in the tree. The tree was built as described in Fig. 1, without bootstrap sampling. Mid-point rooting was used.

the current cytoplasmic and mitochondrial enzymes share a common ancient ancestor with a thioredoxin fold, or, alternatively, whether these two similar folds arose independently through convergent evolution. Although the CATH and SCOP protein structure databases classify the cytoplasmic and mitochondrial structures as homologous, the PFAM domain database, which uses considerable structural information, classifies them separately.

However, the bacterial FOSA glutathione transferases clearly have a very distinct structure. SCOP places 1R9C, a fosfomycin resistance protein structure, in class alpha + beta, in contrast to the cytoplasmic and mitochondrial glutathione transferases, which are in class alpha/beta. CATH classifies the glyoxalase homologs of FOSA in the alpha/beta class, but with a roll architecture, rather than a alpha/beta/alpha sandwich for the thioredoxin-like fold of the cytoplasmic and mitochondrial proteins. Likewise, 1R9C does not share any significant similarity with the cytoplasmic or mitochondrial glutathione transferase structures. Thus, structures supporting glutathione transferase activity have arisen at least three times (cytoplasmic + mitochondrial, microsomal/MAPEG, and fosfomycin resistance) and may have arisen as many as five times (cytoplasmic, mitochondrial, MAPEG1, MAPEG2, and fosfomycin). Only the cytoplasmic and mitochondrial proteins share any similarity with the thioredoxin fold.

Conclusion

The glutathione transferase protein families show a remarkable range of evolutionary and functional diversity. Neither of these two aspects of glutathione transferases is unique in itself; there are hundreds of protein families that share homologs between mammals and bacteria, suggesting an evolutionary origin more than 2.5 billion years ago. Likewise, it is no longer surprising to find homologous proteins with dramatically different functions, although it is striking that in addition to the many catalytic activities and structural roles of cytoplasmic glutathione transferase homologs, members of the family also function as transcription factors (e.g., stringent starvation proteins) and as intracellular chloride transporters.

The remarkable feature of these families is their combination of phylogenetic ubiquity and functional diversity. One might certainly imagine that the emergence of the ancient cytoplasmic glutathione transferase fold, with its exceptional functional plasticity, might have precluded the independent emergence of other folds catalyzing similar chemistry. This is not the case; glutathione transferase activity has emerged, or perhaps was acquired, from at least three very different protein folds. This raises the intriguing possibility that the protein families that are currently recognized for their glutathione transferase activity originally emerged with different functions, which subsequently evolved to include glutathione chemistries.

There is strong evidence to suggest that all four of the glutathione transferase families are ancient–homologs are easily identified in bacteria. But in many cases, the ancient protein is not known to have glutathione transferase activity, which might have emerged later. It should be noted, however, that the emergence of different activities did not require substantial changes in domain architecture. The bacterial stringent starvation proteins, which are thought to be transcription factors, align with cytoplasmic class-theta proteins over 80% of their lengths, including both the N-terminal thioredoxin-like domain and the C-terminal glutathione transferase domain. Likewise, the mitochondrial kappa proteins align over the entire length of the hydroxychromeme-2-carboxylic acid protein, and the DSBA domain, and the mammalian MAPEG and FOSA homologs align across their prokaryotic homologs. Thus, very distinct folds have emerged with similar functions.

From a more practical perspective, the current class nomenclature for cytoplasmic glutathione transferases is most useful when classes are limited to proteins that share more than 50–60% identity. The -mu, -alpha, and -pi classes have been very helpful to investigators studying mammalian glutathione transferases, but the mammalian class-theta proteins are no more similar to the *Drosophila* class-thetas than they are to plant class-zeta proteins. Likewise, recent gene duplications in the mammalian class-mu and -alpha evolutionary histories can confuse identification of orthologs. Investigators must be careful when inferring similar functions or regulatory mechanisms among such an evolutionarily diverse group of protein families.

References

Altschul, S. F., Madden, T. L., Schaffer, A. A., Zhang, J., Zhang, Z., Miller, W., and Lipman, D. J. (1997). Gapped BLAST and PSI-BLAST: A new generation of protein database search programs. *Nucleic Acids Res.* **25**, 3389–3402.

Andreeva, A., Howorth, D., Brenner, S. E., Hubbard, T. J. P., Chothia, C., and Murzin, A. G. (2004). Scop database in 2004: Refinements integrate structure and sequence family data. *Nucleic Acids Res.* **32**, D226–D229.

Bairoch, A., and Apweiler, R. (1996). The Swiss-Prot protein sequence data bank and its new supplement TREMBL. *Nucleic Acids Res.* **24**, 21–25.

Benson, A. M., Batzinger, R. P., Ou, S.-Y., Bueding, E., Cha, Y.-N., and Talalay, P. (1978). Elevation of hepatic glutathione S-transferase activities and protection against mutagenic metabolites of benzo(a)pyrene by dietary antioxidants. *Cancer Res.* **38**, 4486–4495.

Bernat, B. A., Laughlin, L. T., and Armstrong, R. N. (1997). Fosfomycin resistance protein (FOSA) is a manganese metalloglutathione transferase related to glyoxalase I and the extradiol dioxygenases. *Biochemistry* **36**, 3050–3055.

Coggan, M., Flanagan, J. U., Parker, M. W., Vichai, V., Pearson, W. R., and Board, P. G. (2002). Identification and characterization of gstt3, a third murine theta class glutathione transferase. *Biochem. J.* **366**, 323–332.

Dulhunty, A., Gage, P., Curtis, S., Chelvanayagam, G., and Board, P. (2001). The glutathione transferase structural family includes a nuclear chloride channel and a ryanodine receptor calcium release channel modulator. *J. Biol. Chem.* **276**, 3319–3323.

Eaton, R. W. (1994). Organization and evolution of naphthalene catabolic pathways: Sequence of the DNA encoding 2-hydroxychromene-2-carboxylate isomerase and trans-O-hydroxybenzylidenepyruvate hydratase-aldolase from the NAH7 plasmid. *J. Bacteriol.* **176**, 7757–7762.

Felsenstein, J. (1989). Phylip—Phylogeny inference package (version 3.2). *Cladistics* **5**, 164–166.

Gregorio, S. D., Zocca, C., Sidler, S., Toffanin, A., Lizzari, D., and Vallini, G. (2004). Identification of two new sets of genes for dibenzothiophene transformation in *Burkholderia sp.* DBT1. *Biodegradation* **15**, 111–123.

Harris, J. M., Meyer, D. J., Coles, B., and Ketterer, B. (1991). A novel glutathione transferase (13-13) isolated from the matrix of rat liver mitochondria having structural similarity to class theta enzymes. *Biochem. J.* **278**, 137–141.

Hayes, J. D., Flanagan, J. U., and Jowsey, I. R. (2004). Glutathione transferases. *Annu. Rev. Pharmacol. Toxicol.* **45,** 51–88.

Holm, P. J., Morgenstern, R., and Hebert, H. (2002). The 3-D structure of microsomal glutathione transferase 1 at 6 Å resolution as determined by electron crystallography of p22(1)2(1) crystals. *Biochim. Biophys. Acta* **1594,** 276–285.

Jakobsson, P. J., Morgenstern, R., Mancini, J., Ford-Hutchinson, A., and Persson, B. (1999). Common structural features of MAPEG—a widespread superfamily of membrane associated proteins with highly divergent functions in eicosanoid and glutathione metabolism. *Protein Sci.* **8,** 689–692.

Jakobsson, P. J., Morgenstern, R., Mancini, J., Ford-Hutchinson, A., and Persson, B. (2000). Membrane-associated proteins in eicosanoid and glutathione metabolism (MAPEG) a widespread protein superfamily. *Am. J. Respir. Crit. Care Med.* **161,** S20–S24.

Jowsey, I. R., Thomson, A. M., Flanagan, J. U., Murdock, P. R., Moore, G. B., Meyer, D. J., Murphy, G. J., Smith, S. A., and Hayes, J. D. (2001). Mammalian class sigma glutathione S-transferases: Catalytic properties and tissue-specific expression of human and rat GSH-dependent prostaglandin D2 synthases. *Biochem. J.* **359,** 507–516.

Ladner, J. E., Parsons, J. F., Rife, C. L., Gilliland, G. L., and Armstrong, R. N. (2004). Parallel evolutionary pathways for glutathione transferases: Structure and mechanism of the mitochondrial class kappa enzyme rGSTK1-1. *Biochemistry* **43,** 352–361.

Mannervik, B., Alin, P., Guthenberg, C., Jensson, H., Tahir, M. K., Warholm, M., and Jornvall, H. (1985). Identification of three classes of cytosolic glutathione transferase common to several mammalian species: Correlation between structural data and enzymatic properties. *Proc. Natl. Acad. Sci. USA* **82,** 7202–7206.

Meyer, D. J., Coles, B., Pemble, S. E., Kilmore, K. S., Fraser, G. M., and Ketterer, B. (1991). Theta, a new class of glutathione transferases purified from rat and man. *Biochem. J.* **274,** 409–414.

Meyer, D. J., and Thomas, M. (1995). Characterization of rat spleen prostaglandin h d-isomerase as a sigma-class GSH transferase. *Biochem. J.* **311,** 739–742.

Patskovsky, Y. V., Huang, M., Takayama, T., Listowsky, I., and Pearson, W. R. (1999). Distinctive structure of the human GSTM3 gene-inverted orientation relative to the mu class glutathione transferase gene cluster. *Arch. Biochem. Biophys.* **361,** 85–93.

Pearl, F., Todd, A., Sillitoe, I., Dibley, M., Redfern, O., Lewis, T., Bennett, C., Marsden, R., Grant, A., Lee, D., Akpor, A., Maibaum, M., Harrison, A., Dallman, T., Reeves, G., Diboun, I., Addou, S., Lise, S., Johnston, C., Sillero, A., Thornton, J., and Orengo, C. (2005). The CATH domain structure database and related resources Gene3d and DHS provide comprehensive domain family information for genome analysis. *Nucleic Acids Res.* **33,** D247–D251.

Pearson, W. R. (1998). Empirical statistical estimates for sequence similarity searches. *J. Mol. Biol.* **276,** 71–84.

Pemble, S. E., Wardle, A. F., and Taylor, J. B. (1996). Glutathione S-transferase class kappa: Characterization by the cloning of rat mitochondrial GST and identification of a human homologue. *Biochem. J.* **319,** 749–754.

Ramos-Gomez, M., Kwak, M. K., Dolan, P. M., Itoh, K., Yamamoto, M., Talalay, P., and Kensler, T. W. (2001). Sensitivity to carcinogenesis is increased and chemoprotective efficacy of enzyme inducers is lost in NRF2 transcription factor-deficient mice. *Proc. Natl. Acad. Sci. USA* **98,** 3410–3415.

Retief, J. D., Lynch, K. R., and Pearson, W. R. (1999). Panning for genes—a visual strategy for identifying novel gene orthologs and paralogs. *Genome Res.* **9,** 373–382.

Rowe, J. D., Patskovsky, Y. V., Patskovska, L. N., Novikova, E., and Listowsky, I. (1998). Rationale for reclassification of a distinctive subdivision of mammalian class-mu glutathione S-transferases that are primarily expressed in testis. *J. Biol. Chem.* **273,** 9593–9601.

Sierk, M. L., and Pearson, W. R. (2004). Sensitivity and selectivity in protein structure comparison. *Protein Sci.* **13,** 773–785.

Thompson, J. D., Higgins, D. G., and Gibson, T. J. (1994). Clustal W: Improving the sensitivity of progressive multiple sequence alignment through sequence weighting, position-specific gap penalties and weight matrix choice. *Nucleic Acids Res.* **22,** 4673–4680.

Tomarev, S. I., Zinovieva, R. D., and Piatigorsky, J. (1991). Crystallins of the octopus lens. recruitment from detoxification enzymes. *J. Biol. Chem.* **266,** 24226–24231.

Toung, Y. P., and Tu, C. P. (1992). Drosophila glutathione S-transferases have sequence homology to the stringent starvation protein of *Escherichia coli. Biochem. Biophys. Res. Commun.* **182,** 355–360.

Wattenberg, L. W. (1978). Inhibitors of chemical carcinogenesis. *Adv. Cancer Res.* **26,** 197–226.

[13] *Drosophila* Glutathione S-Transferases

By CHEN-PEI D. TU AND BÜNYAMIN AKGÜL

Abstract

The *Drosophila* glutathione S-transferases (GSTs; EC2.5.1.18) comprise a host of cytosolic proteins that are encoded by a gene superfamily and a homolog of the human microsomal GST. Biochemical studies of certain recombinant GSTs have linked their enzymatic functions to important substrates such as the pesticide DDT and 4-hydroxynonenal, a reactive lipid metabolite. Moreover, a correspondence has been observed between resistance to insecticide substrates—such as DDT—and elevated enzyme levels in resistant strains. Such significant, recurring connections suggest that these *gst* genes may feature in a model for the development of insecticide resistance. We have amassed substantial biochemical support for relating the overexpression of a particular *gst* gene to insecticide resistance but are still short of solid genetic evidence to affirm a causal relationship.

With the *Drosophila* system, we have at our disposal genetic and molecular techniques such as *p*-element mutagenesis and excision, siRNA technology, and versatile transgenic techniques. We can use these methods to effect loss-of-function and gain-of-function conditions and, in these rendered contexts, study other potentially important functions of the *gst* gene superfamily. An immediate problem that comes to mind is the possible causal relationship between GST substrate specificity and chemical resistance phenotype(s).

In this chapter, we present an analysis of selected strategies and laboratory methods that may be useful in pursuing a variety of interesting problems. We will cover three kinds of approaches—biochemistry, genetics,

METHODS IN ENZYMOLOGY, VOL. 401
Copyright 2005, Elsevier Inc. All rights reserved.
0076-6879/05 $35.00
DOI: 10.1016/S0076-6879(05)01013-X

and genomics—as important instruments in a toolkit for studies of the *Drosophila gst* superfamily. We make the case that these approaches (biochemistry, genetics, and genomics) have helped us gain important insights and can continue to help the community gain a more complete understanding of the biological functions of GSTs. Such knowledge may be key in addressing questions about the detoxification of pesticides and how oxidative stresses affect life span. We hope that these techniques will prove fruitful in studying a host of other physiologic functions as well.

Introduction

Biochemical Function of GST

Glutathione S-transferases (GSTs, EC2.5.1.18) are a family of important detoxification enzymes that are active in phase II of drug metabolism (for reviews, see Mannervik and Danielson, 1988; Hayes and Pulford, 1995; Armstrong, 1997). In many organisms, *gsts* belong to one or the other of two different gene superfamilies—one of cytosolic isozymes (Hayes and Pulford, 1995; Lai and Tu, 1986), the other of microsomal isozymes (DeJong *et al.*, 1988; Jakobsson *et al.*, 1999, MAPEG). Most of the GSTs catalyze the conjugation of reduced glutathione (GSH) to some electrophilic substrate, such as 1-chloro-2, 4-dinitrobenzene (CDNB) (Habig and Jakoby, 1981). Some of the GSTs also have GSH-dependent peroxidase activities against lipid hydroperoxide and hydrogen peroxides (Mannervik and Danielson, 1988). Most of the GSTs have a broad substrate specificity against xenobiotic compounds. A few of them, however—usually the MAPEG family members—have very specific substrate associations (Jakobsson *et al.*, 1999).

Among members of the animal kingdom, species of the class *Insecta* have evolved to unmatched numbers and outstanding diversity. More than half a century ago, synthetic organochlorines and organophosphate insecticides were first introduced to support agriculture. But because these early forms of chemical "pest control" entered the environment, insect species have rapidly developed defenses against them. Their quickness to gain resistance has remained an important problem for agrochemical industries trying to combat hardy emerging species. Resistance caused by increased insecticide metabolism may involve one or a combination of the xenobiotic metabolizing enzymes such as cytochrome P450s, GSTs, carboxylesterases, and other phase I and phase II enzymes (see Clark [1990] for a brief review). The safe and sustainable control of various insect populations continues to have tremendous economic and biomedical importance. From safeguarding crops against hungry swarms to protecting humans from

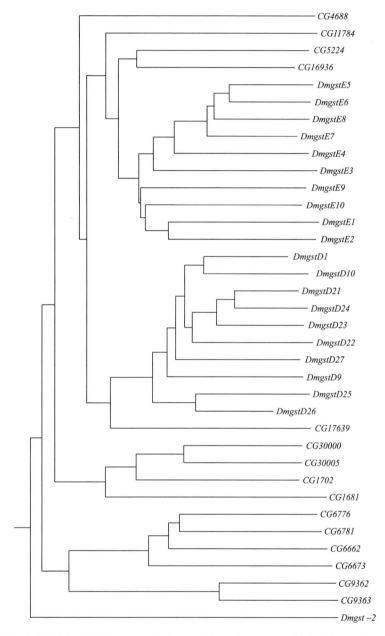

FIG. 1. Neighbor-joining tree of *Drosophila melanogaster* GSTs. The amino acid sequences of GSTs in Table I (except for GFZF and MGST-1) were aligned using ClustalW version 1.8 (Chenna *et al.*, 2003). The tree was constructed by the neighbor-joining method

disease vectors, many compelling causes are motivating our efforts to understand how insect systems respond to chemical substances.

As the complete genome sequences of *Drosophila* and mosquito species (Ding *et al.*, 2003; Enayati *et al.*, 2005) become more fully annotated, the diversity of GSTs will demand greater in-depth investigation, especially in the context of pesticide resistance. Now, with the advent of new technologies, we also have the unprecedented capacity to enhance our studies with the genomics approach to explore the physiological functions of the *Drosophila gst* gene superfamily.

Genetic Repertoire of Glutathione S-Transferases in *Drosophila melanogaster*

Annotation of the *Drosophila melanogaster* genome sequence revealed a complex gene superfamily behind the *Drosophila* GSTs (www.flybase. org). Among its varied members is a single gene with 45% amino acid sequence identity to the human microsomal *gst* gene (DeJong *et al.*, 1988). When this gene, known as the *Mgst-1*, is inactivated, the result is shortened lifespan in the flies (Toba and Aigaki, 2000; Toba *et al.*, 1999).

The cytosolic *gst* genes can be tentatively divided into six different classes (Fig. 1), according to sequence comparisons with known *gst*s. By this system, 11 genes at locus 87B would be assigned to the D or delta class (insect class I GSTs), and 10 genes at locus 55C6-7, plus 4 dispersed members would potentially fall into the epsilon class. The *GST-2* gene alone would represent the sigma class (Beall *et al.*, 1992; Singh *et al.*, 2001), and four genes clustered at locus 66D5 apparently would belong to the omega class. Two closely linked genes at locus 85D17 would constitute the zeta class, and two linked (locus 45F6) and two dispersed *gst* genes would round out the theta class (Table I).

A multiple zinc-finger-motif protein, GFZF, was found to have considerable homology to a bacterial GST (COG0625) in its C-terminal region. It was discovered that GFZF can be purified through GSH-agarose affinity matrix, but its GST activities have not yet been investigated (Dai *et al.*, 2004). Future finds may call for the designation of yet another class of *Drosophila* GSTs.

---

using Gonnet series of residue comparison matrix. The 36 putative and characterized GSTs are grouped according to classes epsilon (top, *CG4688, CG11784, CG5224, CG16936, E5, E6, E8, E7, E4, E3, E9, E10, E1, and E2*), delta (*D1, D10, D21[D2], D24[D5], D23[D4], D22 [D3], D27[D8], D9, D25[D6], D26[D7], CG17639*), theta (*CG30000, CG30005, CG1702, CG1681*), omega (*CG6776, CG6781, CG6662, CG6673*), zeta (*CG9362, CG9363*), and sigma (*gst-2* or *S1*, bottom). Dm, *Drosophila melanogaster*.

TABLE I

A Summary of *Drosophila melanogaster* GSTs

Gene	ID	Cytogenetic location	Other designation	Reference
Delta class				
*Gst*D1	CG10045	87B8	*gst*D1-1	Toung *et al.*, 1990
*Gst*D21	CG4181	87B8	*gst*D2	Toung *et al.*, 1993
*Gst*D22	CG4381	87B8	*gst*D3	Toung *et al.*, 1993
*Gst*D23	CG11512	87B8	*gst*D4	Toung *et al.*, 1993
*Gst*D24	CG12242	87B8	*gst*D5	Toung *et al.*, 1993
*Gst*D25	CG4423	87B8	*gst*D6	Toung *et al.*, 1993
*Gst*D26	CG4371	87B8	*gst*D7	Toung *et al.*, 1993
*Gst*D27	CG4421	87B8	*gst*D8	Toung *et al.*, 1993
*Gst*D9	CG10091	87B8		Sawicki *et al.*, 2003
*Gst*D10	CG18548	87B8		Sawicki *et al.*, 2003
—	*CG17639*	87B9		
Epsilon class				
*Gst*E1	CG5164	55C6	*gst*-3	Singh *et al.*, 2000
*Gst*E2	CG17523	55C6		Sawicki *et al.*, 2003
*Gst*E3	CG17524	55C7		Sawicki *et al.*, 2003
*Gst*E4	CG17525	55C7		Sawicki *et al.*, 2003
*Gst*E5	CG17527	55C7		Sawicki *et al.*, 2003
*Gst*E6	CG17530	55C7		Sawicki *et al.*, 2003
*Gst*E7	CG17531	55C7		Sawicki *et al.*, 2003
*Gst*E8	CG17533	55C7		Sawicki *et al.*, 2003
*Gst*E9	CG17534	55C8		Sawicki *et al.*, 2003
*Gst*E10	CG17522	55C6		Sawicki *et al.*, 2003
—	*CG4688*		49F12	
—	*CG5224*		55D1	
—	*CG16936*		60E1	
—	*CG11784*		45A9	
Sigma class				
Gst-2	CG8938	53F7-8	*gst*S1	Beall *et al.*, 1992
Putative *theta* class				
—	*CG1681*		11F4	
—	*CG1702*		19D1	
—	*CG30000*		45F6	
—	*CG30005*		45F6	
Putative *omega* class				
—	*CG6776*		66D5	
—	*CG6781*		66D5	
—	*CG6673*		66D5	
—	*CG6662*		66D5	
Putative *zeta* class				
—	*CG9362*		85D17	
—	*CG9363*		85D17	

(*continued*)

TABLE I (*continued*)

Gene	ID	Cytogenetic location	Other designation	Reference
Other *gst*-like genes				
Gfzf	CG10065	84C6		Dai *et al.*, 2004
Mgst-1				Toba *et al.*, 1999

Delta, epsilon and *sigma* class *gst*s were reported previously (Beall *et al.*, 1992; Sawicki *et al.*, 2003; Singh *et al.*, 2000; Toung *et al.*, 1990). The bioinformatics analysis of the annotated *Drosophila melanogaster* genes revealed the presence of more *gst*s. Classification was based on blasting the putative GST proteins, for which IDs were italicized, against the NCBI protein database. MGST-1 is a microsomal GST-like protein (Toba *et al.*, 1999), whereas GFZF is a GST-containing zinc finger protein (Dai *et al.*, 2004).

Biochemical Characterization of *Drosophila* GSTs

Purification of GSTs from the Organism and Cell Lines

Starting in the 1960s, a series of purification experiments and genetic analyses would reveal that insect GSTs closely resemble—both in size and structural complexity—those in mammals, including humans (Clark and Shamaan, 1984; Cochrane and LeBlanc, 1986; Cochrane *et al.*, 1987; Motoyama and Dauterman, 1975).

Our group later purified *Drosophila* GSTs containing subunits of 24 kDa and 28.5 kDa from Kc_0 cells. We used a two-step procedure starting with *S*-hexylGSH affinity chromatography, followed by FPLC on a MonoQ column (Tu *et al.*, 1990). We noted that a fraction (~1/3) of the CDNB conjugation activity failed to bind to the affinity matrix, which suggested the presence of additional forms of GST isozymes (Tu *et al.*, 1990). The GST-2 protein (now DmGSTS1-1), initially linked with flight muscle development and showing no CDNB activity (Beall *et al.*, 1992; Clayton *et al.*, 1998), is now known to be very active against 4-hydroxynonenal (4-HNE), one of the products of lipid peroxidation (Singh *et al.*, 2001). These findings would strongly suggest that *Drosophila* GSTs might be nearly as complex as the rat and human GSTs (Mannervik and Danielson, 1988). This conjecture was later substantiated by molecular genetic analyses of the *gst* genes and by bioinformatics analyses of the entire genome sequence (Toung *et al.*, 1990, 1993; http://www.flybase.org). Because of their genetic complexity, it has been difficult to purify GSTs from *Drosophila* specimens to the level of single isozymes.

Purifications of Drosophila *GSTs by the Recombinant DNA Approach*

We developed this approach for *Drosophila* GSTs after producing the first cDNA of a *Drosophila* GST (λGTDm1). The first step in the procedure was to raise antibodies against GSTs from Kc_0 cells; our aim was to isolate a clone (λGTDm1) from a λgt11 cDNA expression library (Toung *et al.*, 1990). Next, to authenticate the clone as a GST, we expressed the cDNA insert in *Escherichia coli,* then purified and characterized the recombinant protein against CDNB and several other common GST substrates (Habig and Jakoby, 1981). Homology analysis showed that the cDNA sequence, although very different from that of any mammalian GSTs, had 66% sequence identity with the maize GST III over a region of 44 amino acids (Toung *et al.*, 1990). The sequence also showed homology to the stringent starvation protein sequence from *E. coli* (Toung and Tu, 1992). After verifying its identity, we used the cDNA to isolate a pair of genomic clones (λGTDm101 and λGTDm102) and took the references from these to characterize most (8 of 11) of the D class genes (Tang and Tu, 1994; Toung *et al.*, 1990, 1993; *Delta* in Chelvanayagam *et al.*, 2001 and *class I insect GSTs* in Fournier *et al.*, 1992). This approach can be adapted and applied to any insect *gst* genes as long as their genomic sequences are known.

Recombinant DNA Method. We constructed expression clones for intronless *gst* coding sequences using direct polymerase chain reaction (PCR) amplification on *Drosophila* genomic DNA (Lee and Tu, 1995; Sawicki *et al.*, 2003; Tang and Tu, 1994). To optimize the initiation of target mRNA translation, translational regulatory sequences, such as the *E. coli* ribosome binding sequences, can be built into the forward PCR primer (Tang and Tu, 1994). This consideration may be less important in pET vector-based constructs, because the phage T7-promoter ensures exclusive transcription once the T7 RNA polymerase is induced (e.g., Sawicki *et al.*, 2003).

For coding sequences interrupted by one or more introns (e.g., the sigma class GSTS1-1), expression clones are best engineered from the mRNA templates by reverse transcriptase-polymerase chain reaction (RT-PCR) with properly designed primers (Singh *et al.*, 2001).

Synthetic Gene Method. If a particular spliced *gst* mRNA is rare, or if the developmental expression pattern is unknown for RNA isolation, a better strategy for expression is the synthetic gene method (Cheng *et al.*, 2001). This approach was used to investigate the rat GST M4-4 (Y_{b4}), whose existence was predicted from a genomic clone sequence (Lai *et al.*, 1988).

Next, the duplex gene sequences are divided into overlapping segments of large oligonucleotides. To ensure that sequences were of the correct lengths, oligonucleotides longer than 40-mer were gel-purified by their commercial suppliers. All of these, except for the two 5′ oligonucleotides (e.g., 0.8 nmol each), were phosphorylated separately by treatment with T4 polynucleotide kinase and ATP (1 μmol) in T4 ligase buffer (e.g., New England Biolabs, Beverly, MA). Two oligonucleotides at the 5′-ends of the duplex sequence were then dissolved (0.8 nmol each) in New England Biolabs T4 ligase buffer and combined with the other kinase reactions. These solutions were adjusted to 50 mM NaCl and boiled at 100° for 10 min and quickly cooled in ice water before ethanol precipitation.

The precipitated oligonucleotides were ligated and prepared for a proper restriction digest at 37° overnight. The reaction products were purified by one round of phenol extraction followed by two treatments of chloroform/isoamyl alcohol extraction and, finally, precipitated by ethanol. Equal molar amounts of the gene fragments and vector DNA (restriction enzyme-digested expression vector) were ligated and 50–100 ng of the ligated DNA were used to transform *E. coli* (DH5α or a proper host strain). After transformation, plasmid DNAs from selected transformants should be analyzed by PCR to detect the presence of a correct-sized insert. Several prospective clones may then be chosen and sequenced to find a correct clone or one with the fewest mutations. Insertions, deletions, and/or substitutions can be corrected by site-directed mutagenesis according to the Stratagene Quik Change protocol (La Jolla, CA). The final clone must be checked for expression levels before proceeding with large-scale purifications.

Advantages and Limitations. The synthetic gene method for purifying *Drosophila* GSTs is a straightforward and flexible method that takes advantage of existing, readily accessible resources for *Drosophila* research. In general, the coding sequence of a candidate *Drosophila gst* gene can be obtained from the centralized genome database (http://www.flybase.org). The potential splicing sites of a given gene can be estimated from the signature sequences and by assessing its homology to other known gene sequences (e.g., Cheng *et al.*, 2001). Another advantage of the synthetic gene method is that, with this technique, rare codons—which may adversely affect expression levels—can be replaced to increase protein expression levels.

As for the recombinant DNA approach, it offers a powerful means for purifying each individual GST. But certain kinds of information—for example, on the possible formation of specific heterodimers, a mechanism that generates additional diversity, and on functionally important post-translational modifications—require analysis (e.g., mass spectrometry) using GSTs purified directly from actual *Drosophila* specimens.

Substrate Specificities of Purified Drosophila *GSTs*

Insect molecular biologists have made great strides toward understanding metabolic resistance to insecticides by identifying candidate genes responsible for insecticide metabolism. Important insights have also come from association studies comparing the overexpression of xenobiotic metabolizing enzymes (e.g., GSTs or P450s) in resistant insect strains to low or lack of molecular expression in sensitive strain (e.g., Daborn *et al.*, 2002; Fournier *et al.*, 1992; Tang and Tu, 1994; Wei *et al.*, 2001). For example, DDT dehydrochlorinase activity was detected in DDT-resistant houseflies but was missing from extracts of sensitive control flies (Motoyama and Dauterman, 1975; Sternburg *et al.*, 1954). A number of laboratories have contributed to our collective knowledge of the biochemical and enzymatic behaviors of insect GSTs in the context of pesticide metabolism (Clark, 1990; Fournier *et al.*, 1992; Grant *et al.*, 1991; Ranson *et al.*, 2001; Tang and Tu, 1994; Wang *et al.*, 1991; Wei *et al.*, 2001).

It is a standard practice to test *Drosophila* GSTs for activity against three prospective substrates: the common substrates CDNB and cumene hydroperoxide (Habig and Jacoby, 1981), 4-hydroxynonenal (4-HNE), an α, β-unsaturated aldehyde produced in lipid peroxidation (Alin *et al.*, 1985; Sawicki *et al.*, 2003), and insecticide substrate(s) such as DDT (Clark and Shamaan, 1984; Sternburg *et al.*, 1954; Tang and Tu, 1994). A more complete list of assays conducted against pesticide substrates is available in a review by Clark (1990). A general caveat about these assays is that they often involve the use of highly toxic, organophosphorus compounds, which can be deadly (e.g., parathion) if not handled properly.

Catalytic Residue(s) for Drosophila *GSTs*

CDNB Plating Procedure. Board *et al.* (1995) determined that the *Lucilia cuprina* (Australian blowfly) GST has a Ser (Ser-9) as its catalytic residue. The hydroxyl group of Ser-9 contributes to the productive binding of GSH and to the stabilization of the ionized GSH (Caccuri *et al.*, 1997). This GST has been designated a theta class GST but is highly homologous to the previously published *Drosophila* GST D1-1 (Board *et al.*, 1994). Independent of the blowfly GST studies, we had isolated loss-of-activity mutants of *Drosophila* GST D1-1 using the CDNB plating procedure (Lee *et al.*, 1995). This technique assumes that *E. coli* growth will be inhibited by CDNB if a bacterial cell expresses an active GST—a likely result when GSH is depleted. We observed that extracts from one of the mutants that had a mutation at Ser-9 (GST D1-S9F) lost CDNB conjugation activity *in vitro* but still retained its binding affinity for the *S*-hexyl GSH matrix (J. Peel and C.P.D.T., unpublished). Similarly, we found that the S8A mutant of GST D27 would lose CDNB conjugation activity but retain the ability to bind to *S*-hexyl GSH-agarose. These behaviors are consistent with those seen from the mammalian active site mutants (e.g., Tyr to Phe mutants in the alpha, mu, pi classes of GSTs; see Armstrong, 1997).

Critical evidence for the functional significance of the *N*-terminal Ser residue would come through a third member of the D class GSTs, GST D21. GST D21 contains no Ser in its *N*-terminal region (the first 10 residues), having four glycines in a row (#7–10) there instead. In biochemical assay, it showed almost no activity against CDNB but would bind to glutathione conjugates such as *S*-hexyl GSH. Then, when a gain-of-activity mutant of GST D21 was isolated, it shored up existing indications that an *N*-terminal Ser was a key part of CDNB activity in D (delta) class GSTs. Results of mutagenesis experiments demonstrated that the G8S, and to a lesser extent the G9S mutants, were, in fact, acquiring the CDNB conjugation activity with the restoration of the Ser. However, the G8,9S double mutants only showed an additive effect in CDNB activity (J. Chapman and C.P.D.T., unpublished results). Thus, we have solid evidence that GST D21 may naturally be a GSH-conjugate binding protein (e.g., ligandin). The evolutionary wisdom of these GST-like ligandins is an intriguing mystery.

Chimeric GSTs

The standard GST-fold contains two domains: a GSH-binding domain in the *N*-terminal region and a xenobiotic substrate-binding domain (H-site) that resides mainly in the rest of the molecule (Armstrong, 1997; Mannervik and Danielson, 1988). Domain-switching experiments between two closely related members would, in theory, verify a close structural relationship between two GSTs. It was hypothesized that chimeric GSTs assembled by means of domain-switching between two closely related proteins would retain some or all of the parent molecules' properties.

Chimeric GSTs between GST D1 and GST D21-G8,9S, (a gain-of-activity mutant of D21) at position 66/65 showed a K_m^{GSH} characteristic of the parent GST. This result confirmed the independent domain I function of *Drosophila* D class GSTs. Binding of CDNB by the chimeric GSTs seems to be determined mainly by residues 67–170, as revealed by the switch at position 171/170. All four chimeric GSTs were measurably active against CDNB but none of them retained the catalytic efficiency of GST D1 (Jason Chapman, Hailing Cheng, Yen-Shing Loretta Tu, B. Akgül, Ming Tien, B.-C. Wang, and C.-P. D. Tu, unpublished result). Such hybrid constructs often give rise to novel phenotypes whose sources are difficult to pin down. For instance, we have seen chimeras between two closely related human alpha class GSTs (95% identity), A1-1 and A2-2, manifest a range of interesting properties (e.g., pH *vs* activity profiles) uncharacteristic of their parents (Corey Strickland and C.P.D.T., unpublished results). A point of interest is that identical amino acids can be as low as 38% and 28% in pairwise comparisons among the delta and epsilon classes of *Drosophila* GSTs, respectively (Figs. 2 and 3).

There remain many more questions to explore about the structure–function relationship of GSTs. Some members of the much-studied D class (D21, D22, and D26) are known to exhibit enzyme activities without the *N*-terminal Ser residue at position 8 or 9 (Fig. 2 and Sawicki *et al.*, 2003), supposedly key to their catalytic behaviors. Another interesting topic concerns GSTs of the epsilon class, which have a conserved Tyr at position 6 or 7, as well as a conserved Ser at position 11 (or 12 or 13). It is important to pin down the actual catalytic residue(s) with certainty, which would enable an informed class assignment of these dispersed *gst* genes.

Molecular Basis for Resistance to Insecticides

Background

Tang and Tu (1994) developed a highly DDT-resistant strain of *Drosophila melanogaster* (PSU-R). In PSU-R flies, GST D1-1, which has DDT dehydrochlorinase activity, as well as the *gstD1* mRNA level, is elevated twofold over levels in the sensitive parent strain. Over the past 20 years, this type of apparent relationship between GST overexpression and resistance to DDT (or other pesticides) has been explored further using purified GSTs to characterize their substrate specificities *in vitro* (Grant *et al.*, 1991; Ranson *et al.*, 2001; Tang and Tu, 1994; Wei *et al.*, 2001). Several laboratories have reported that mosquito (*Anopheles gambiae* and *Aedes aegypti*) GST(s) of the D (or delta) and epsilon classes exhibit DDT dehydrochlorinase activities (Ding *et al.*, 2003; Ranson *et al.*, 2001). Results obtained from these studies further support the association, but we are still short of the strict criteria for a causal relationship (Daborn *et al.*, 2002).

There are two major concerns about the hypothesized causal relationship that attributes DDT resistance in PSU-R flies to the overexpression of GST D1. The first concern is that *gstD1* mRNA is actually induced by DDT, so increased *gstD1* mRNA levels should contribute to a low level of DDT resistance in the PSU-R flies. It has been reported that the expression levels of several epsilon GSTs were elevated in a DDT-resistant strain of *Anopheles gambiae* mosquitoes, but only one of these GSTs is capable of converting DDT to DDE *in vitro* (Ding *et al.*, 2003; Grant *et al.*, 1991). This

```
DmgstD25   --MDLYNMSQSPSTRAVMMTAKAVGVEFNSI-QVNTFVGEQLEPWFVKINPQHTIPTLV  56
DmgstD26   MPNLDLYNFPMAPASRAIQMVAKAIGLELNSK-LINTMEGDQLKPEFVRINPQHTIPTLV  59
DmgstD21   --MDFYYMPGGGGGRTVIMVAKAIGLELNKK-LINTMEGEQLKPEFVKLNPQHTIPTLV  56
DmgstD24   --MDFYYSPRGSGGRTVIMVAKAIGVKLNMK-LINTLEKDQLKPEFVKLNPQHTIPTLV  56
DmgstD23   --MDFYYSPRSSGGRTIIMVAKAIGLELNKK-QLRITEGEHLKPEFLKLNPQHTIPTLV  56
DmgstD22   --------------MVGKAIGLEFNKK-IINTLKGEQMNPDFIKINPQHSIPTLV  40
DmgstD1    -MVDFYYLPGSSPCRSVIMTAKAVGVELNKK-LINLQAGEHLKPEFLKINPQHTIPTLV  57
DmgstD10   --MDLYYRPGSAPCRSVLMTAKAIGVEFDKKTIINTRAREQFTPEYLKINPQHTIPTLH  57
DmgstD27   --MDFYYHPCSAPCRSVIMTAKAIGVDLNMK-LIKVMDGEQLKPEFVKLNPQHCIPTLV  56
DmgstD9    -MIDFYYMLYSAPCRSILMTARAIGLELNKK-QVDLDAGEHLKPEFVKINPQHTIPTLV  57
                .           : *..:*:*:.:: :      ::: * :::*:****  ****
```

```
DmgstD25   DNLFVIWETRAIVVYLVEQYGKDDS-LYPKDPQKQALINQRLYFDMGTLYDGIAKYFFPL  115
DmgstD26   DNGFVIWESRAIAVYLVEKYGKPDSPLYPNDPQKRALINQRLYFDMGTLYDALTKYFFLI  119
DmgstD21   DNGFSIWESRAIAVYLVEKYGKDDY-LLENDPKKRAVINQRLYFDMGTLYESFAKYYYPL  115
DmgstD24   DNGFSIWESRAIAVYLVEKYGKDDT-LFEKDPKKQALVNQRLYFDMGTLYDSFAKYYYPL  115
DmgstD23   DNGFAIWESRAIAVYLVEKYGKDDS-LFENDPQKRALINQRLYFDMGTLHDSFMKYYYPF  115
DmgstD22   DNGFTIWESRAILVYLVEKYGKDDA-LYPKDIQKQAVINQRLYFDMALMYPTLANYYKA  99
DmgstD1    DNGFAIWESRAIQVYLVEKYGKTDS-LYPKCPKKRAVINQRLYFDMGTLYQSFANYYYPQ  116
DmgstD10   DHGFAIWESRAIMVYLVEKYGKDDK-LFEKDVQKQALINQRLYFDMGTLYKSFSEYYYPQ  116
DmgstD27   DDGFSIWESRAILIYLVEKYGADDS-LYPSDPQKKAVVNQRLYFDMGTLFQSFVEAIYPQ  115
DmgstD9    DDGFAIWESRAILIYLAEKYDKDGS-LYPKDPQQRAVINQRLFFDLSTLYQSYVYYYYPQ  116
           *. *  :**:*** :**.*:*.    * *. ::::*:*****:**:. :.       :
```

```
DmgstD25   LRT--GKPGTQENLEKLNAAFDLINNFLDGQDYVAGNQLSVADIVILATVSTTEMVDFDL  173
DmgstD26   FRT--GKFGDQEALDKVNSAFGFINTFLEGQDFVAGSQLTVADIVILATVSTVE-----  171
DmgstD21   FRT--GKPGSDEDLKRIETAFGFLDTFLEGQEYVAGDQLTVADIAILSTVSTFEVSEFDF  173
DmgstD24   FHT--GKPGSDEDFKKIESSFEYLNIFLEGQNYVAGDHLTVADIAILSTVSTFEIFDFDL  173
DmgstD23   IRT--GQLGNAENYKKVEAAFEFLDIFLVGQDYVAGSQLTVADIAILSSVSTFEVVEFDI  173
DmgstD22   FTT--GQFGSEEDYKKVQETFDFINTFLEGQDYAAGDSLTVADIAIVATVSTFEVAKFEI  157
DmgstD1    VFA--KAPADPEAFKKIEAAFEFINTFIEGQDYAAGDSLTVADIAILFLATVSTFDVAGFDF  174
DmgstD10   IFL--KKPANEENYKKIEVAFEFINTFLEGQTYSAGDYSLADIAFLATVSTFDVAGFDF  174
DmgstD27   IRN--NHPADPEAMQKVDSAFGHLDTFIEDQEYVAGDCLTIADIAILASVSTFEVVDFDI  173
DmgstD9    LFEDVKKPADPDNLKKIDDAFAMENTLLKGQYAAINKLTLADFAILATVSTFEISEYDF  176
             .  :  .::: :*  :::*  .* :* . :  :*:*:.::::.**. :     .
```

```
DmgstD25   KKFPNVDRWYKNAQKVTPGWDENLARIQSAKKFLAENLIEKL-  215
DmgstD26   --------------------------------  
DmgstD21   SKYSNVSRWYDNAKKVTPGWDENWEGLMAMKALFDARKLAAK-  215
DmgstD24   NKYPNVARWYANAKKVTPGWDENWKGAVELKGVFDARQAAAKQ  216
DmgstD23   SKYPNVARWYANAKKITPGWDENWKGLIQMKTMYEAQKASLK-  215
DmgstD22   SKYPNVARWYDHVKKITPGWEENWAGAIDVKKRIEEKQNAAK-  199
DmgstD1    SKYANVNRWYENAKKVTPGWEENWAGCLEFKKYFE------  209
DmgstD10   KRYANVARWYENAKKLTPGWEENWAGCQEFRKYFDN------  210
DmgstD27   AQYPNVASWYENAKEVTPGWEENWDGVQLIKKLVQ--ERNE--  212
DmgstD9    GKYPEVVRWYDNAKKVIPGWEENWEGCEYYKKLYLGAIINKQ-  218
             ..          .. ..  ..
```

FIG. 2. Amino acid alignment of *delta* class GSTs. The amino acid sequences were obtained from the flybase (www.flybase.org). The alignment was performed using ClustalW version 1.8 (Chenna *et al.*, 2003). *Dm, Drosophila melanogaster*; red, small and hydrophobic residues (AVFPMILW); blue, acidic residues (DE); magenta, basic residues (RHK); green, hydroxyl, amine and basic residues (STYHCNGQ). The columns identical in all sequences are labeled with a (*) sign. The columns that contain conserved and semiconserved substitutions are marked with a (:) and (.) sign, respectively. The Ser residues of the putative catalytic site are labeled by a (↑) sign. (See color insert.)

result is consistent with the notion that GSTs, regardless of their substrate specificity, were inducible by DDT in the resistant insects. But there is no evidence to ensure that the major source of resistance does not involve a change(s) in some other gene(s).

```
DmgstE1   MSSSGEVLYGTDLSPCVRTVKLTLKVINLDYEYKEVNLQAGEHLSEEYVKKNPQHTVPML 60
DmgstE2   MSDKLVLYGMDISPPVRACKLTLRAINLDYEYKEMDLLAGDHFKDAFIKKNPQHTVPLL 59
DmgstE10  -MANLILYGTESSPPVRAVILTLRAIQLDHEFHTLDMQAGDHLKPDMLRKNPQHTVPML 58
DmgstE9   -MGKLVLYGVEASPPVRACKLTLDAIGLQYEYRLVNLLAGEHKTKEFSLKNPQHTVPVL 58
DmgstE5   -MVKLTLYGVNPSPPVRAVKLTLAALQLPYEFVNVNISGQEQLSEEYLKKNPEHTVPTL 58
DmgstE6   -MVKLTLYGLDPSPPVRAVKLTLAALNLTYEYVNVDIVARAQLSPEYLEKNPQHTVPTL 58
DmgstE8   -MSKLILYGTEASPPVRAAKLTLAALGIPYEYVKINTLAKETLSPEFLRKNPQHTVPTL 58
DmgstE7   -MPKLILYGLEASPPVRAVKLTLAAALEVPYEFVEVNTRAKENFSEEFLKKNPQHTVPTL 58
DmgstE4   -MGKISLYGLDASPPTRACLITLKALDLPEEFVFWNLFEKENFSEDFSKKNPQHTVPLL 58
DmgstE3   -MGKLTLYGIDGSPPVRSVLLTLRAINLDFDYKIVNLMEKEHLKPEFLKINPLHTVPAL 58
          : *** : ** .*:  *** .* : .::   ::           ** **** *
```

```
DmgstE1   DDNGTFIWDSHAIAAYLVDKYAKSDELYPKDLAKRAIVNQRLFIDASVIYASIAN-VSRP 119
DmgstE2   EDNGALIWDSHAIVCYLVDKYANSDELYPRDLVIRAQVDQRLFFDASILFNSLRN-VSIP 118
DmgstE10  EDGESCIWDSHAIIGYLVNKYAQSDELYPKDPLKRAVVDQRLHFETGVLFHGIFKQLQRA 118
DmgstE9   EDDGKFIWESHAICAYLVRRYAKSDDLYPKDYFKRALVDQRLHFESGVLFQGCIRNIAIP 118
DmgstE5   EDDGNYIWDSHAIIAYLVSKYADSDALYPRDLLQRAVVDQRLHFETGVVFANGIKAITKP 118
DmgstE6   EDDGHYIWDSHAIIAYLVSKYADSDALYPKDPLKRAVVDQRLHFESGVVFANGIRSISKS 118
DmgstE8   EDDGHFIWDSHAISAYLVSKYGQSDTLYPKDLLQRAVVDQRLHFESGVVFWNGLRGITKP 118
DmgstE7   EDDGHYIWDSHAIIAYLVSKYGKTDSLYPKDLLQRAVVDQNQLTIADFSISTVSSLE-VFVKV 118
DmgstE4   QDDDACIWDSHAIMAYLVEKYAPSDELYPKDLLQRAKVDQLMHFESGVIFESALRRLTRP 118
DmgstE3   DDNGFYLADSHAINSYLVSKYGRNDSLYPKDLKKRAIVDQRLHYDSSVVTSTG-RAITFP 117
          :*.    : :**** *** :*. .* ***:*   ** *:* :.::::.:   . : .
```

```
DmgstE1   FWINGVTEVPQEKLDAVHQGLKLIETFIGNSPYLAGDSLTLADLSTGPTVSAVP-AAVDI 178
DmgstE2   YFLRQVSLVFKEKVDNIKDAYGHLENFIGDNPYLTGSQLTIADLCCGATASSLA-AVLDL 177
DmgstE10  LFKENATEVPKDRLAELKDAYALIEQFLAENPYVAGPQLTIADFSIVATVSTLHLSYCPV 178
DmgstE9   LFYKNITEVPRSQIDAIYEAYDFLEAFIGNQAYLCGPVITIADYSVVSSVSSLV-GLAAI 177
DmgstE5   LFFNGLNRIPKERYDAIVEIYDFVETFLAGHDYIAGDQLTIADFSLISITSLV-AFVEI 177
DmgstE6   VLFQGGQTKVPKERYDAIIEIYDFVETFLKGQDYIAGNQLTIADFSLVSSVASLE-AFVAL 177
DmgstE8   LFATGGQTTIPKERYDAVIEIYDFVETFLTGHDFIAGDQLTIADFSLITSITALA-VFVVI 177
DmgstE7   LFAGKQTMIPKERYDAIIEVYDFLEKFLAGNDYVAGNQLTIADFSISTVSSLE-VFVKV 177
DmgstE4   VLFFGEPTLPRNQVDHILQVYDFVETFLDDHDFVAGDQLTIADFSIVSTITSIG-VFLEL 177
DmgstE3   LFEWENKTEIPQARIDAIEGVYKSLNLFLENGNYLAGDNLTIADFHVIAGLTGFF-VFLPV 176
          :*: :   :     ::*:   ::*   :*:** .   :  .   :
```

```
DmgstE1   DPATYPKVTAWLDRLNKLPYYKEINEAPAQSYVAFIRSKWTKLGDK------------ 224
DmgstE2   DELKYPKVAAWFERLSKLPHYEEDNLRGLKKYINLLKPVIN-LEQ------------ 221
DmgstE10  DATKYPKLSAWLARISALPFYEEDNLRGARLLADKIRSKLPKQFDKLWQKAFEDIKSGAG 238
DmgstE9   DAKRYPKLNGWIDRMAAQPNYQSLNGNGAQMLIDMFSSKITKIV------------- 221
DmgstE5   DRLKYPRIIEWVRRLEKLPYYEEANAKGARELETILKSTNFTFAT------------- 222
DmgstE6   DTTKYPRIGAWIKKKLEQLPYYEEANGKGVRQLVAIKKTNFTFEA------------- 222
DmgstE8   DTVKYANITAWIKRIEELPYYEEACGKGARDLVTLLKKFNFTFST------------- 222
DmgstE7   DTTKYPRIAAWFKRLQKLPYYEEANGNGARTFESFIREYNFTFASN------------ 223
DmgstE4   DPAKYPKIAAWLERLKELPYYEEANGKGAAQFVELIRSKNFTIVS------------ 222
DmgstE3   DATKYPELAAWIKRIKELPYYEEANGSRAAQIIEFIKSKKFTIV------------- 220
          *    *..:  *. ::   * *:.                  :
```

```
DmgstE1   -
DmgstE2   -
DmgstE10  KQ 240
DmgstE9   -
DmgstE5   -
DmgstE6   -
DmgstE8   -
DmgstE7   -
DmgstE4   -
DmgstE3   -
```

FIG. 3. Amino acid alignment of *epsilon* class GSTs. The amino acid sequences were obtained from the flybase (www.flybase.org). The alignment was performed using ClustalW version 1.8 (Chenna *et al.*, 2003). *Dm, Drosophila melanogaster*; red, small and hydrophobic residues (AVFPMILW); blue, acidic residues (DE); magenta, basic residues (RHK); green, hydroxyl, amine and basic residues (STYHCNGQ). The columns identical in all sequences are labeled with a (*) sign. The columns that contain conserved and semi-conserved substitutions are marked with a (:) and (.) sign, respectively. (See color insert.)

The second major concern regards the possibility of multiple mutations giving rise to chemical resistance (Clark, 1990). Resistant insect strains, whether selected in the laboratory environment or captured in the wild, usually take many generations to develop. Thus, high-level resistance is likely the result of multiple genetic mutations, and elevated *gst* gene expression may be necessary but not sufficient to confer very high degrees of DDT resistance.

Daborn *et al.* (2002) reported that *Cyp6g1* was overexpressed in a highly DDT-resistant strain of *Drosophila* but that transgenic flies that over-transcribe *Cyp6g1* alone (100-fold) were only moderately resistant to DDT (10 μg DDT per vial). This study established a credible causal relationship between *Cyp6g1* over-transcription and a moderate level of DDT resistance, although the *Cyp6g1* gene product(s) has not yet been characterized *in vitro*. These findings do not exclude, however, contributions from GST D1 and, likely, other factors, culminating in very high resistance ($\gg$10 μg DDT per vial) in PSU-R flies and other DDT-resistant strains (A. H. Tang and CPDT, unpublished results).

Ideally, we would be able to compare between the complete genome sequences of the resistant and sensitive strains and between microarray expression profiles to identify precisely all candidate mutations with a potential role in the high DDT (or any other insecticide) resistance phenotypes. The extent to which overexpression of GST D1-1 in particular contributes to DDT resistance, and the mechanism by which D1 is over-expressed in PSU-R flies, will only become clear with more specific analyses. It will be necessary to use isogenic strains (or conditions) that enable us to compare effects with or without the *gst* gene overexpression, to ascertain the contributions of GST D1 itself in DDT resistance.

Transgenic Techniques for Drosophila *gsts*

A variety of transgenic techniques for studying gene function are available for *Drosophila* (Robertson *et al.*, 1988). Notable examples include the pCaSpeR transformation vectors (Thummel *et al.*, 1988), which provide heat shock–inducible general expression, and the Gal4-UAS enhancer trap systems, which enable transgene overexpression mediated by GAL4 drivers in different tissues or at different stages of development in the *Drosophila* life cycle (Brand and Perrimon, 1993; Rorth *et al.*, 1998; Toba *et al.*, 1999).

For investigating insecticide resistance, we have been using the heat shock–inducible *hsp70* promoter for regulating the expression of a *gst* transgene (Akgul and Tu, 2002, 2004; Thummel *et al.*, 1988). For entire flies or larvae, the heat-shock promoter provides a reliable switch for rapidly turning gene expression on and off (e.g., Akgul and Tu, 2002, 2004). In contrast, a xenobiotic compound (e.g., PB) consumed by flies will remain in their guts and continue to elevate gene expression until it peaks at 2–3 h time points (Akgül and Tu, 2002, 2004; Tang and Tu, 1995).

The heat shock–induced mRNA has the context of 5'-UTR(hsp70)-transgene sequence-3'-UTR (the trailer portion of act5C's 3'-UTR). If the target GST can be translated from this transgenic mRNA and remain stable, we can test the enzyme's effect on DDT resistance by comparing flies' resistance levels before and after the heat shock induction. So far, we have imbued transgenic strains with full-length gstD1 mRNA sequence and with the complete coding sequence, but none of these have shown high-level overexpression of the GST D1-1 protein (Akgül and Tu, 2002, 2004).

Although unexpected, this outcome falls in line with related observations. In the case of gstD21, a dramatic elevation of the mRNA levels (18-fold) by pentobarbital (PB)-induced condition fails to increase the production of GSTD21 protein (Akgül and Tu, 2002, 2004; Tang and Tu, 1995). These findings suggest that overexpression of the GST D proteins may be controlled at the translational level as well (Akgül and Tu, 2002, 2004; Tang and Tu, 1995).

In our work with the PSU-R flies and flies under PB induction, we detected no more than a twofold elevation of GST D1-1 in response to treatments (Tang and Tu, 1994, 1995). We know with certainty that GST D1–1 is a relatively abundant GST isozyme expressed at the major developmental stages of the Drosophila life cycle (Tang and Tu, 1995). Curiously, this protein shows a restricted expression pattern in embryos of stages 12–16 (Nakamura et al., 1999). Zou et al. (2000) found that in a microarray analysis survey of paraquat-treated and of aging flies, gstD1 mRNA was over-transcribed. Thus, the regulation of GST gene expression in the context of Drosophila development poses an interesting problem. Research on the expression of the Drosophila gst genes is only starting to acquire a broader functional perspective.

PB as a Model Chemical Inducer of Drosophila gsts

The compounds PB and phenobarbital (PhB) are known to regulate gst gene expression at both the transcription and posttranscription levels (Tang and Tu, 1995). We chose these compounds as model chemical inducers to perturb gst gene expression. We discuss the use of PB induction in this section.

The gstD21 mRNA, which is transcribed from an intronless gene, is uniquely suited for investigating drug (PB)-mediated stabilization of mRNA (Akgül and Tu, 2004; Tang and Tu, 1995). The gstD1 mRNA, known to be stable mRNA both in the presence and the absence of a drug (e.g., PB), may potentially be a good system for a variety of studies. The *gst*D1 mRNA is induced ~twofold by PB. A mechanism is needed to mediate the turnover of an otherwise stable mRNA when the inducer is no longer present (Akgül and Tu, 2002). It presents, for example, an opportunity to learn the details of positive mechanism for mRNA stabilization in a complex organism and to gain insight about the metabolism of a relatively stable mRNA after induction (Akgül and Tu, 2002). We describe in the following several procedures that could be useful for the readers in the insect GST community to investigate various aspects of *gst* gene expression. We use "Molecular Cloning: A Laboratory Manual" (Sambrook and Russell, 2003) as our general reference for molecular biology procedures.

Fly Sample Preparation, Collection, and Storage. Adult flies (2–3 days old) were distributed into clean milk bottles in similar numbers (~150– 200) for 5 h starvation at room temperature. A blotting paper strip (3 × 10 cm) saturated (but not dripping wet) with a 5% sucrose solution (control flies) or 200 mg/ml PB in 5% sucrose (PB-treated flies) was placed in each bottle for 2 h at room temperature.

Heat shock of the flies was carried out in clean milk bottles containing 5% sucrose-saturated paper strips normally at 35° for 1 h in a hybridization oven (Model 2000, Micro-hybridization incubator, Robbins Scientific Company, Sunnyvale, CA). It takes ~15 min for an empty milk bottle with a foam plug to warm to 35°. Once a warm bottle is removed from the 35° oven, the temperature inside the bottle drops to below 30° in 7.5 min.

For the combined induction of transgene (heat-shock inducible) and endogenous *gst*D21 (PB-inducible) mRNAs, the flies were treated with PB at room temperature for 1 h and then at 35° for the second hour (Akgül and Tu, 2002). The flies ingested more PB at 35°. For time course experiments of decay of induced mRNA, flies were allowed to recover after treatment at room temperature in milk bottles containing a paper strip soaked in 5% sucrose for varying durations. The flies were snap-frozen in liquid nitrogen (in 250-ml centrifuge bottles) and stored in 15-ml tubes at −70° until use.

Isolation of Total Fly RNAs by Guanidine Hydrochloride-CsCl Gradient Centrifugation. Although many rapid RNA isolation kits are available commercially, the procedure that follows consistently produces high-quality RNA in excellent yields from adults. Total RNAs from frozen adult flies were isolated according to Ullrich *et al.* (1977). Approximately 0.3–1 g flies were homogenized in 7 ml of the lysis buffer (4 *M* guanidinium hydrochloride, 1 *M* β-mercaptoethanol, 0.1 *M* NaOAc, and 0.01 *M* EDTA). The lysate was then adjusted for CsCl gradient centrifugation

(Beckman Ti70.1 rotor, 139,000g for 25 h). The quality of RNA prepared by this method survives long-term storage ($-70°$) without any degradation. For microarray analysis experiments, the quality of RNA was monitored by an Agilent 2100 bioanalyzer with the RNA 6000 Nano LabChip kit.

Analysis of Specific gst mRNAs by RPA Assays. To monitor changes in the expression of a specific *gst* mRNA among a multigene family, the method of choice is the ribonuclease protection assay (RPA, Calzone *et al.*, 1987; Friedberg *et al.*, 1990). Only the perfect complement to the radioactive cRNA probe will be protected from the subsequent RNase digestions. RPA assays can be carried out according to Ambion's instruction for RPA III. The power of this assay is extended by the ability to analyze simultaneously several RNAs if different lengths of cRNA probes can be designed (e.g., Akgül and Tu, 2002, 2004).

Investigation of Translational Regulation by Polysome Pattern Analysis. To investigate translational regulation of *gst*D21 mRNAs under various conditions, we examined their association with ribosomes to assess their translatability *in vivo*.

We isolated polysomes according to a previously published procedure with minor modifications (Bagni *et al.*, 2000). We homogenized 0.5-g flies in 3.5 ml of lysis buffer [100 mM NaCl, 10 mM MgCl$_2$, 30 mM Tris-HCl (pH 7.5), 0.1% Triton-X, and 100 μg/ml cycloheximide and 30 U/ml SUPERase-In RNase inhibitor (Ambion)] in a 10-ml Dounce homogenizer (pestle B). After 5 min of incubation on ice, the homogenates were centrifuged at 12,000g at 4° for 8 min; 1.5 ml of the resulting supernatant solution was sedimented in a 5%–70% (w/v) sucrose gradient prepared by a Teledyne ISCO's Programmable Density Gradient Fractionation System for cell organelle separation (Cat#679000177).

The gradients were centrifuged in a Beckman SW28 rotor at 27,000 rpm at 4° for 200 min. We then collected 12 fractions from the top of the gradient using an ISCO Tris pump while monitoring the absorbance at 254 nm with an ISCO UA-6 monitor. Peaks of ribosomal subunits, monoribosomes, and polyribosomes are clearly resolved by this system. Each fraction was extracted by RNase-free phenol/chloroform/IAA (Ambion) three times. The solution was made 0.5% SDS before the second extraction. We use one third of the total extracted RNA ($\sim$ 40–50 μg) for RPA assays according to Ambion's instructions (RPA III).

Although association with ribosomes is a prerequisite for translation *in vivo*, the translatability of isolated RNAs can be determined by the rabbit reticulocyte lysate system for *in vitro* translation. GSTs can be identified by immunoprecipitation of the product mixture with anti-GST antisera or by GSH- (or *S*-hexyl GSH-) agarose beads, followed by SDS polyacrylamide gel electrophoresis and autoradiography (Tu *et al.*, 1983).

Isolation of Nuclei and Transcription Rate Analysis. Isolated nuclei can be used to determine the possible changes of subcellular localization of *gst* mRNAs after chemical treatment (PB or insecticides). They can also be used for measuring changes in transcription rates by treatments of the *Drosophila* with insecticides or other chemicals (e.g., PB). The protocol we used for isolation of nuclei from entire flies was a slight modification of the procedure of Lee *et al.* (1992) as described in Tang and Tu (1995), which included the nuclear run-on data analysis. It is now feasible to use fluorescent nucleotides to label the nuclei RNA and hybridize to a special microarray chip of all the putative *Drosophila* (or any other insect) *gst* sequences and reference gene sequences.

Concluding Remarks and the Way Ahead

We have come far to learn the identities of each member of the *Drosophila gst* gene superfamily. Now, we are well prepared to work on characterizing all of them biochemically. The task of matching each isozyme to a signature substrate(s)—especially from among insecticides—is sure to be painstaking and long work, but with a solid knowledge base and powerful technologies, this goal is within our reach. In addition to very effective biochemical and molecular strategies, now we can also pursue advances in structural genomics (e.g., Agianian *et al.*, 2003) to validate known GST structures and perhaps find new variations on the theme of substrate binding among the GST fold (Armstrong, 1997). We can also imagine projects such as the construction of an atlas for the entire *gst* gene superfamily using monospecific antibody reagents and specific RNA probes.

A host of interesting problems exist, for which techniques in the *Drosophila* system are well suited. These include phenomena concerning the insect *Drosophila* specifically, as well as broader questions that involve all organisms and their environments. An immediate problem of economic and biomedical interest is the possible causal relationship between GST substrate specificity and chemical resistance phenotype(s). In addressing questions like these, studies of *gst* gene expression mechanisms at the molecular level will be critical in bridging the gap between biochemical properties *in vitro* and corresponding *in vivo* phenotype(s).

We have particularly gained informative insights from investigation of loss-of-function and gain-of-function mutants, by which we can construe the functions of each *gst* from a genetic perspective. The transposon insertion mutant technique, for example, has been adopted to probe for the function of the *Drosophila* microsomal *gst* gene. These studies would reveal that the inactivation of the gene leads to the shortened lifespan phenotype (Toba and Aigaki, 2000). Some of the *Drosophila gst* gene systems, like the divergently transcribed *gstD1* and *gstD21* mRNAs, also provide interesting contexts for investigating unique mechanisms of gene regulation.

But techniques initially developed for *Drosophila* are often versatile enough to transfer to studies of other insect *gst* genes. Many researchers may find it useful to test structure-function relationships and physiological functions of other *gst* sequences in the common, but heterospecific, environment of *Drosophila melanogaster*. The emerging genomics approach will encourage analysis from a more global perspective; for example, with microarray chips, we may better understand the development of insecticide resistance through considering genome-wide expression profiles.

We are optimistic that these three approaches (biochemistry, genetics, and genomics) will help to unlock a more complete understanding of the biological function of GSTs. Such knowledge may be key in addressing questions about the detoxification of pesticides and how oxidative stresses affect lifespan. Indeed, our experience assures us that much is possible on this front. But this is only one of many contexts in which these techniques can enable us to learn. We see great promise that these techniques will prove fruitful in studying a host of other physiological functions as well.

Acknowledgments

C.P.D.T. would like to thank Amy H. Tang, Yann-Ping Toung, and his former laboratory members for their various contributions to *Drosophila* glutathione *S*-transferases. We also thank Professor Tao-shih Hsieh of Duke University for timely assistance during the development of this project and Leslie Tu for assistance in the preparation of this manuscript. This project was supported by a grant from the National Institute of Environmental Health Sciences (ES02678) and a fellowship from the Turkish Government (to BA).

References

Agianian, B., Tucker, P. A., Schouten, A., Leonard, K., Bullard, B., and Gros, P. (2003). Structure of a *Drosophila* sigma class glutathione *S*-transferase reveals a novel active site topography suited for lipid peroxidation products. *J. Mol. Biol.* **326,** 151–165.

Akgül, B., and Tu, C.-P. D. (2004). Pentobarbital-mediated regulation of alternative polyadenylation in *Drosophila* glutathione *S*-transferase D21 mRNAs. *J. Biol. Chem.* **279,** 4027–4033.

Akgül, B., and Tu, C.-P. D. (2002). Evidence for a stabilizer element in the untranslated regions of *Drosophila* glutathione S-transferase D1 mRNA. *J. Biol. Chem.* **77,** 34700–34707.

Alin, P., Danielson, U. H., and Mannervik, B. (1985). 4-Hydroxyalk-2-enals substrates for glutathione transferase. *FEBS Lett.* **179,** 267–270.

Armstrong, R. N. (1997). Structure, catalytic mechanism and evolution of the glutathione transferases. *Chem. Res. Toxicol.* **10,** 2–18.

Bagni, C., Mannucci, L., Dotti, C. G., and Amaldi, F. (2000). Chemical stimulation of synaptosomes modulates alpha -Ca^{2+}/calmodulin-dependent protein kinase II mRNA association to polysomes. *J. Neuroscience* **20**, RC76.

Beall, C., Fyrberg, C., Song, S., and Fyrberg, E. (1992). Isolation of a *Drosophila* gene encoding glutathione *S*-transferase. *Biochem. Genet.* **30**, 515–527.

Board, P., Russell, R. J., Marano, R. J., and Oakeshott, J. G. (1994). Purification, molecular cloning and heterologous expression of a glutathione *S*-transferase from the Australian sheep blowfly (*Lucilia cuprina*). *Biochem. J.* **299**, 425–430.

Board, P. G., Coggan, M., Wilce, M. C., and Parker, M. W. (1995). Evidence for an essential serine residue in the active site of the Theta class glutathione transferases. *Biochem. J.* **311**, 247–250.

Brand, A. H., and Perrimon, N. (1993). Targeted gene expression as a means of altering cell fates and generating dominant phenotypes. *Development* **118**, 401–415.

Caccuri, A. M., Antonini, G., Nicotra, M., Battistoni, A., Bello, M. L., Boardi, P. G., Parker, M. W., and Ricci, G. (1997). Catalytic mechanism and role of hydroxyl residues in the active site of theta class glutathione S-transferases: Investigation of ser-9 and tyr-113 in a glutathione s-transferase from the Australian sheep blowfly, *Lucilia cuprina*. *J. Biol. Chem.* **272**, 29681–29686.

Calzone, F. J., Britten, R. S., and Davidson, E. H. (1987). Mapping of gene transcripts by nuclease protection assays and cDNA primer extension. *Meth. Enzymol.* **152**, 611–632.

Chelvanayagam, G., Parker, M. W., and Board, P. G. (2001). Fly fishing for GSTs: A unified nomenclature for mammalian and insect glutathione transferases. *Chem. Biol. Interact.* **133**, 256–260.

Cheng, H., Tchaikovskaya, T., Tu, Y. S., Chapman, J., Qian, B., Ching, W. M., Tien, M., Rowe, J. D., Patskovsky, Y. V., Listowsky, I., and Tu, T.-P. D. (2001). Rat glutathione S-transferase M4–4: An isoenzyme with unique structural features including a redox-reactive cysteine-115 residue that forms mixed disulphides with glutathione. *Biochem. J.* **356**, 403–414.

Chenna, R., Sugawara, H., Koike, T., Lopez, R., Gibson, T. J., Higgins, D. G., and Thompson, J. D. (2003). Multiple sequence alignment with the clustal series of programs. *Nucleic Acids Res.* **31**, 3497–3500.

Clark, A. G., and Shamaan, N. A. (1984). Evidence that DDT-dehydrochlorinase from the housefly is a glutathione *S*-transferase. *Pest. Biochem. Physiol.* **22**, 249–261.

Clark, A. G. (1990). The glutathione *S*-transferase and resistance to insecticide. *In* "Glutathione S-Transferases and Drug Resistance." (J. D. Hayes, C. B. Pickett, and T. J. Mantle, eds.), pp. 369–378. Taylor and Francis, London, New York, Philadelphia.

Clayton, J. D., Cripps, R. M., Sparrow, J. C., and Bullard, B. (1998). Interaction of troponin-H and glutathione *S*-transferase-2 in the indirect flight muscles of *Drosophila melanogaster*. *J. Muscle Res. Cell Motil.* **19**, 117–127.

Cochrane, B. J., and Le Blanc, G. A. (1986). Genetics of xenobiotics metabolism in *Drosophila*. I. Genetic environmental factors affecting glutathione-*S*-transferase in larvae. *Biochem. Pharmacol.* **35**, 1679–1684.

Cochrane, B. J., Morrissey, J. J., and Le Blanc, G. A. (1987). The genetics of xenobiotic metabolism in *Drosophila*. IV. Purification and characterization of the major glutathione-*S*-transferase. *Insect Biochem.* **17**, 731–738.

Daborn, P. J., Yen, J. L., Bogwitz, M. R., Le Goff, G., Feil, E., Jeffers, S., Tijet, N., Perry, T., Heckel, D., Batterham, P., Feyereisen, R., Wilson, T. G., and French-Constant, R. H. (2002). A single p450 allele associated with insecticide resistance in *Drosophila*. *Science* **297**, 2253–2256.

Dai, M. S., Sun, X. X., Qin, J., Smolik, S. M., and Lu, H. (2004). Identification and characterization of a novel *Drosophila melanogaster* glutathione S-transferase containing FLYWCH zinc finger protein. *Gene* **10**, 49–56.

De Jong, J. L., Morgenstern, R., Jornvall, H., De Pierre, J. W., and Tu, C.-P. D. (1988). Gene expression of rat and human microsomal glutathione *S*-transferases. *J. Biol. Chem.* **263**, 8430–8436.

Ding, Y., Ortelli, F., Rossiter, L. C., Hemingway, J., and Ranson, H. (2003). The *Anopheles gambiae* glutathione transferase supergene family: Annotation, phylogeny and expression profiles. *BMC Genomics* **4**, 35–50.

Enayati, A. A., Ranson, H., and Hemingway, J. (2005). Insect glutathione transferases and insecticide resistance. *Insect Mol. Biol.* **14**, 3–8.

Fournier, D., Bride, J. M., Poirie, M., Berge, J. B., and Plapp, F. W., Jr. (1992). Insect glutathione *S*-transferases. Biochemical characteristics of the major forms from houseflies susceptible and resistant to insecticides. *J. Biol. Chem.* **267**, 1840–1845.

Friedberg, T., Grassow, M. A., and Oesch, F. (1990). Selective detection of mRNA forms encoding the major phenobarbital inducible cytochromes P450 and other members of the P450IIB family by the RNases: A protection assay. *Archiv. Biochem. Biophys.* **279**, 167–173.

Grant, D. F., Dietze, E. C., and Hammock, B. D. (1991). Glutathione *S*-transferase isozymes in *Aedes aegypti*: Purification, characterization, and isozyme specific regulation. *Insect Biochem.* **4**, 421–433.

Habig, W. H., and Jakoby, W. B. (1981). Assays for differentiation of glutathione *S*-transferases. *Methods Enzymol.* **77**, 398–405.

Hayes, J. D., and Pulford, D. J. (1995). The glutathione *S*-transferase supergene family: Regulation of GST and the contribution of the isoenzymes to cancer chemoprotection and drug resistance. *Crit. Rev. Biochem. Mol. Biol.* **30**, 445–600.

Jakobsson, P. J., Morgenstern, R., Mancini, J., Ford-Hutchinson, A., and Persson, B. (1999). Common structural features of MAPEG—a widespread superfamily of membrane associated proteins with highly divergent functions in eicosanoid and glutathione metabolism. *Protein Sci.* **8**, 689–692.

Lai, H.-C. J., Qian, B., Grove, G., and Tu, C.-P. D. (1988). Gene expression of rat glutathione *S*-transferases. Evidence for gene conversion in the evolution of the Yb multigene family. *J. Biol. Chem.* **263**, 11389–11395.

Lai, H.-C. J., and Tu, C.-P. D. (1986). Rat glutathione *S*-transferases supergene family. Characterization of an anionic Yb subunit cDNA clone. *J. Biol. Chem.* **261**, 13793–13799.

Lee, H., Kraus, K. W., Wolfner, M. F., and Lis, J. T. (1992). DNA sequence requirements for generating pause polymerase at the start of *hsp70*. *Genes Dev.* **6**, 284–295.

Lee, H. C., Toung, Y.-P. S., Tu, Y.-S. L., and Tu, C.-P. D. (1995). A molecular genetic approach for the identification of essential residues in human glutathione S-transferase in *Escherichia coli*. *J. Biol. Chem.* **270**, 99–109.

Lee, H. C., and Tu, C.-P. D. (1995). *Drosophila* glutathione *S*-transferase D27: Functional analysis of two consecutive tyrosines near the *N*-terminus. *Biochem. Biophys. Res. Commun.* **209**, 327–334.

Mannervik, B., and Danielson, U. H. (1988). Glutathione transferases—structure and catalytic activity. *CRC Crit. Rev. Biochem.* **23**, 283–337.

Motoyama, N., and Dauterman, W. C. (1975). Interstrain comparison of glutathione-dependent reactions in susceptible and resistant strains of houseflies. *Pesti. Biochem. Physiol.* **5**, 489–495.

Nakamura, A., Yoshizaki, I., and Kobayashi, S. (1999). Spatial expression of *Drosophila* glutathione S-transferase D1 in the alimentary canal is regulated by the overlying visceral mesoderm. *Dev. Growth Differ.* **41**, 699–702.

Ranson, H., Rossiter, L., Ortelli, F., Jensen, B., Wang, X., Roth, C. W., Collins, F. H., and Hemingway, J. (2001). Identification of a novel class of insect glutathione *S*-transferases involved in resistance to DDT in the malaria vector *Anopheles gambiae*. *Biochem. J.* **359,** 295–304.

Robertson, H. M., Preston, C. R., Phillis, R. W., Johnson-Schlitz, D. M., Benz, W. K., and Engels, W. R. (1988). A stable genomic source of *P* element transposase in *Drosophila melanogaster*. *Genetics* **118,** 461–470.

Rorth, P., Szabo, K., Bailey, A., Laverty, T., Rehm, J., Rubin, G., Weigmann, K., Milan, M., Benes, V., Ansorge, W., and Cohen, S. (1998). Systematic gain-of-function genetics in *Drosophila*. *Development* **125,** 1049–1057.

Sambrook, J., and Russell, D. W. (2003). "Molecular Cloning: A Laboratory Manual," 3rd Ed., Vols. I–III. Cold Spring Harbor Laboratory Press, Cold Spring Harbor, NY.

Sawicki, R., Singh, S. P., Mondal, A. K., Benes, H., and Zimniak, P. (2003). Cloning, expression and biochemical characterization of one Epsilon-class (GST-3) and ten Delta-class (GST-1) glutathione S-transferases from *Drosophila melanogaster*, and identification of additional nice members of the Epsilon class. *Biochem. J.* **370,** 661–669.

Singh, M., Silva, E., Schulze, S., Sinclair, D. A., Fitzpatrick, K. A., and Honda, B. M. (2000). Cloning and characterization of a new theta-class glutathione *S*-transferase (GST) gene, *gst-3*, from *Drosophila melanogaster*. *Gene* **247,** 167–173.

Singh, S. P., Coronella, J. A., Benes, H., Cochrane, B. J., and Zimniak, P. (2001). Catalytic function of *Drosophila melanogaster* glutathione *S*-transferase DmGSTS1-1 (GST-2) in conjugation of lipid peroxidation. *Eur. J. Biochem.* **268,** 2912–2923.

Sternburg, J. G., Vinson, E., and Kearns, C. W. (1954). DDT-dehydrochlorinase, an enzyme found in DDT-resistant flies. *J. Agri. Food Chem.* **2,** 1125–1130.

Tang, A. H., and Tu, C.-P. D. (1995). Pentobarbital-induced changes in *Drosophila* glutathione *S*-transferase D21 mRNA stability. *J. Biol. Chem.* **270,** 13819–13825.

Tang, A. H., and Tu, C.-P. D. (1994). Biochemical characterization of *Drosophila* glutathione *S*-transferases D1 and D21. *J. Biol. Chem.* **269,** 27876–27884.

Thummel, C. S., Boulet, A. M., and Lipshitz, H. D. (1988). Vectors for *Drosophila* P-element-mediated transformation and tissue culture transfection. *Gene (Amst.)* **74,** 445–456.

Toba, G., Ohsako, T., Miyata, N., Ohtsuka, T., Seong, K. H., and Aigaki, T. (1999). The gene search system: A method for efficient detection and rapid molecular identification of genes in *Drosophila melanogaster*. *Genetics* **151,** 725–737.

Toba, G., and Aigaki, T. (2000). Disruption of the microsomal glutathione *S*-transferase-like gene reduces life span of *Drosophila melanogaster*. *Gene* **253,** 179–187.

Toung, Y.-P. S., Hsieh, T. S., and Tu, C.-P. D. (1990). *Drosophila* glutathione *S*-transferase 1-1 shares a region of sequence homology with the maize glutathione *S*-transferase III. *Proc. Natl. Acad. Sci. USA* **87,** 31–35.

Toung, Y.-P. S., and Tu, C.-P. D. (1992). *Drosophila* glutathione *S*-transferases have sequence homology to the stringent starvation protein of *Escherichia coli*. *Biochem. Biophys. Res. Commun.* **182,** 355–360.

Toung, Y.-P. S., Hsieh, T., and Tu, C.-P. D. (1993). The glutathione *S*-transferase D genes: A divergently organized, intronless gene family in *Drosophila melanogaster*. *J. Biol. Chem.* **268,** 9737–9746.

Tu, C.-P. D., Toung, Y.-P. S., Hsieh, T.-S., Simkovich, N. M., and Tu, Y.-S. L. (1990). Glutathione S-transferases from *Drosophila melanogaster*. *In* "Glutathione S-Transferases and Drug Resistance." (J. D. Hayes, C. B. Pickett, and T. J. Mantle, eds.), pp. 379–386. Taylor & Francis, London, New York, Philadelphia.

Tu, C.-P. D., Weiss, M. J., Li, N., and Reddy, C. C. (1983). Tissue-specific expression of the rat glutathione *S*-transferases. *J. Biol. Chem.* **258,** 4659–4662.

Ullrich, A., Shine, J., Chirgwin, J., Pictet, R., Tischer, E., Rutter, W. J., and Goodman, H. M. (1977). Rat insulin genes: Construction of plasmids containing the coding sequences. *Science* **196,** 1313–1319.

Wang, J. Y., McCommas, S., and Syvanen, M. (1991). Molecular cloning of a glutathione S-transferase overproduced in an insecticide-resistant strain of the housefly (*Musca domestica*). *Mol. Gen. Genet.* **227,** 260–266.

Wei, S. H., Clark, A. G., and Syvanen, M. (2001). Identification and cloning of a key insecticide-metabolizing glutathione S-transferase (MdGST-6A) from a hyper insecticide-resistant strain of the housefly *Musca domestica*. *Insect Biochem. Mol. Biol.* **31,** 1145–1153.

Zou, S., Meadows, S., Sharp, L., Jan, L. Y., and Jan, Y. N. (2000). Genome-wide study of aging and oxidative stress response in *Drosophila melanogaster*. *Proc. Natl. Acad. Sci. USA* **97,** 13726–13731.

[14] Mosquito Glutathione Transferases

By Hilary Ranson and Janet Hemingway

Abstract

The glutathione transferases (glutathione S-transferases, GSTs) are a diverse family of enzymes involved in a wide range of biological processes, many of which involve the conjugation of the tripeptide glutathione to an electrophilic substrate. Relatively little is known about the endogenous substrates of mosquito GSTs, and most studies have focused on their role in insecticide metabolism, because elevated levels of GST activity have been associated with resistance to all the major classes of insecticides. In addition, there is growing interest in the role of this enzyme family in maintaining the redox status of the mosquito cell, particularly in relation to vectorial capacity.

Most GSTs are cytosolic dimeric proteins, although a smaller class of microsomal GSTs exists in insects, mammals, and plants. Each GST subunit has a G site that binds glutathione and a substrate-binding site or H site. There are more than 30 GST genes in mosquitoes. Additional diversity is contributed by alternative splicing to produce GSTs with differing substrate specificities. In this review, we first discuss the diversity of insect GST enzymes and their mode of action before focusing on the various functions that have been attributed to specific mosquito GSTs.

Introduction

The glutathione S-transferases (GSTs) are important components of the detoxification pathway in almost all organisms. They act as phase II detoxifying enzymes, conjugating glutathione to products of metabolism or

Copyright 2005, Elsevier Inc. All rights reserved.
0076-6879/05 $35.00
DOI: 10.1016/S0076-6879(05)01014-1

xenobiotics, thereby increasing their solubility and aiding their excretion from the cell. The diverse ranges of substrates recognized by GSTs is partly due to the broad substrate specificities of some individual enzymes but is largely attributable to the extensive nature of this enzyme family in most eukaryotic organisms. The expansion in the number of GST subunits identified has been matched by a parallel expansion in the functions attributed to this enzyme family. GSTs are now recognized as playing important roles in protecting cells from the harmful effects of oxidative stress, cell signaling pathways, intracellular transport, and several biosynthetic pathways.

The GSTs from the major African malaria vector, *Anopheles gambiae* have been the most extensively studied of all the mosquito GST families. The complete draft genome sequence for this species was determined in 2002 (Holt *et al.*, 2002), enabling the full extent of the mosquito GST gene family to be recognized for the first time (Ranson *et al.*, 2002). At the time of writing, genome sequencing of the dengue vector, *Aedes aegypti*, is in progress and that for the lymphatic filariasis vector, *Culex quinquefasciatus*, is expected to follow shortly. Several GST genes have already been identified in the unassembled sequence data from *Ae. aegypti*, some of which have been further characterized (Lumjuan *et al.*, 2005). Initial observations on the similarities and differences between the GST supergene families in Anopheles and Aedes are discussed in the following. To date, little is known about the GSTs of *Cx quinquefasciatus*, but the ability of this species to survive in more highly polluted water than other mosquitoes may lead us to anticipate greater GST diversity.

Nomenclature

GSTs were first identified in rat livers in 1961 (Booth *et al.*, 1961; Combes *et al.*, 1961). It was quickly realized that multiple enzymes were present in rats and in other mammals, and different studies categorized GSTs according to their order of elution from affinity columns, substrate specificity, or immunological cross-reactivity. Eventually, in 1992, a unifying system of nomenclature was applied to the mammalian GSTs that classified each subunit to a different class, designated by a Greek letter, according to the degree of amino acid similarity (Mannervik *et al.*, 1992). As a general rule, GSTs sharing >40% amino acid similarity were assigned to the same class. Insect GSTs went through a similar transition of names. As the first insect GSTs were cloned and sequenced, two immunologically distinct classes were recognized, designated class I and class II (Fournier *et al.*, 1992). However, as we entered the genome era and searchable databases of expressed sequences became available, it was clear that additional classes of insect GSTs existed. In 2001, Chelvanayagam *et al.*, proposed a GST nomenclature for all organisms that closely paralleled the guidelines already in place for mammalian GSTs. Individual GST subunits are now assigned names indicating the species they were isolated from and the GST class. They are also given a number that may either reflect the order of discovery or the genomic organization. Thus, AgGSTe4 is the fourth member of the *A. gambiae* Epsilon class of GSTs to be identified. The proteins are represented by capital letters, are not italicized, and contain information on the composition of the dimer. So the designation AgGSTE4-4 represents a homodimer of two AgGSTE4 subunits, whereas AgGSTE3-4 would be heterodimer made up of a subunit of AgGSTE3 and a subunit of AgGSTE4. In practice, except where ambiguity may arise, the species prefix is usually omitted.

Classification of Mosquito GST Genes

At least six classes of soluble cytosolic insect GSTs have been identified in insects plus a single membrane-bound microsomal class (Ranson *et al.*, 2002). Criteria for inclusion in a particular class are primarily based on amino acid sequence identity and phylogenetic relationship, but chromosomal location and immunological properties, when known, are also taken into account. Thus, GSTE8 was assigned to the Epsilon class, despite sharing less than 28% amino acid identity with other members of this class, on the basis that its gene is located immediately adjacent to the seven Epsilon GSTs in *A. gambiae* and is immunologically related to this class (Ding *et al.*, 2003). Despite the availability of full genome sequences for two insect species (Adams *et al.*, 2000; Holt *et al.*, 2002), the possibility that further insect GST classes exist cannot be discounted. In fact, three of the cytosolic GST genes found in *A. gambiae* cannot be reliably assigned to any existing GST class. These are currently designated by the letter "u" for unclassified. Interestingly, as shown in Fig. 1, clear orthologs of these three genes are found in *Ae. Aegypti,* but similar sequences are not found in the fruit fly, *Drosophila melanogaster.*

Delta Class GSTs

The Delta class is the largest class of insect GSTs and was one of the first to be discovered. GSTs originally classified as belonging to insect class I have mostly been reclassified as Delta GSTs. The predominance of studies on Delta GSTs is partly a result of the ease of purification of this class. They are expressed at high levels in insects, especially in the immature larval stages, and most bind to glutathione-based affinity columns and

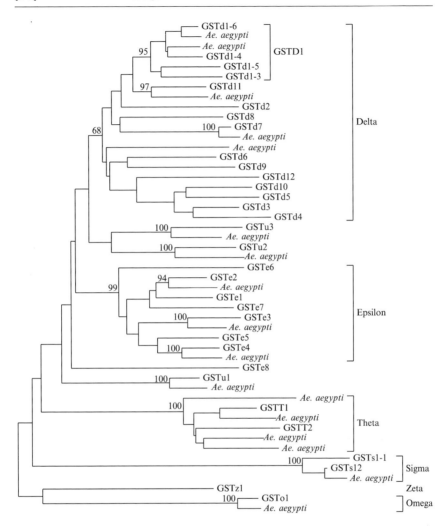

FIG. 1. Phylogenetic relationship of mosquito GSTs. A multiple alignment of 49 protein sequences was generated using ClusatlW and a distance tree was created using the neighbor-joining method. Some of the protein sequences from *Aedes aegypti* are predicted from genome sequence data. As gene names for the *Aedes* GSTs are tentative until the full subset of GSTs from this family has not been ascertained, they have been omitted. All named GSTs are from *Anopheles gambiae*. Bootstrap values (1000 replications) are shown as percent values for nodes defining a GST class or a putative orthologous pair between the two species.

hence are readily isolated from crude insect homogenates. There are 12 Delta GST genes in *A. gambiae* one of which, GSTd1, is alternatively spliced to produce four biochemically distinct subunits (Ranson *et al.*, 1998). Four Delta GST genes have been identified in *Ae. aegypti* so far, but partial purification and characterization suggests that additional Delta GSTs exist in this species. The expansion of the Delta GST class in insects occurred after the mosquito and fruit fly lineages split, approximately 250 million years ago (Ranson *et al.*, 2002). The relationship between the Delta GSTs in *Ae. aegypti* and *A. gambiae* cannot be fully resolved until the full complement of the Delta GST family has been identified in Aedes. However, it is already clear that this gene class is rapidly diverging, because a strict one-to-one orthology is not maintained within the mosquito Delta GST class from these two species (Fig. 1).

The independent expansion of the Delta class in different families of insects suggests that they are involved in adapting the insects to their particular ecological niches and are perhaps particularly important in the detoxification of environmental xenobiotics. This hypothesis has not been widely tested, but it is interesting to note that elevated levels of Delta class GSTs have been implicated in resistance to all the major classes of insecticide (Tang and Tu, 1994; Vontas *et al.*, 2002; Wang *et al.*, 1991).

Epsilon Class GSTs

The Epsilon class is also a large, insect-specific class. Eight members have been identified in *A. gambiae* and to date, three have been identified in the EST sequence database of *Ae. aegypti*. Most of the Epsilon GSTs characterized have low levels of activity with the model substrate, 1 chloro 2,4, dinitrobenzene (CDNB), which is usually used to detect GST activity, and many are not retained by glutathione-based affinity matrices. Both of these factors may explain the relatively recent discovery of this class. As with the Delta class, Epsilon GSTs have radiated independently in *Drosophila* and *Anopheles* (Ranson *et al.*, 2002). The fruit fly contains 10 Epsilon GSTs, but clear orthologs cannot be detected between these genes and those from the mosquito.

Members of the Epsilon class have also been implicated in the detoxification of insecticides (Huang *et al.*, 1998; Ortelli *et al.*, 2003; Wei *et al.*, 2001). In addition, some mosquito Epsilon GSTs have peroxidase activity and may be important in protection against secondary effects of oxidative stress (see the following) (Lumjuan *et al.*, 2005; Ortelli *et al.*, 2003).

Omega Class GSTs

The Omega GST class was first identified in humans, but members of this class also occur in nematodes, helminths, and insects (Board et al., 2000, Ranson et al., 2002). In most species, including A. gambiae, the Omega GSTs seem to be encoded by a single gene, but five putative Omega GST genes have been identified in D. melanogaster (Ding et al., 2003). The physiological function of Omega GSTs is unclear, but one suggestion is that they may play a housekeeping role in protecting against oxidative stress by removing S-thiol adducts from proteins (Board et al., 2000).

Sigma Class GSTs

A structural role has been suggested for the insect Sigma GSTs (formally known as the insect class II GSTs); as in D. melanogaster and Musca domestica, these proteins are found predominantly in the indirect flight muscles in close association with troponin-H. In both these species, a single Sigma GST exists with a proline/alanine rich–N-terminal extension that may aid attachment to the flight muscle. In contrast, the A. gambiae Sigma GST lacks this extension. Two alternative splice variants of AgGSTs1 have been detected in adult mosquitoes, but the tissue distribution of these has not yet been determined. Recently, insect Sigma GSTs (with or without the N-terminal extension) have been shown to be catalytically active (Singh et al., 2001). They have low levels of activity with typical GST substrates but a high affinity for the lipid peroxidation product, 4-hydroxynonenal (4-HNE). The high levels of Sigma GSTs found in the flight muscle may be necessary to protect this highly metabolically active tissue against by-products of oxidative stress rather than having a structural function.

Theta Class GSTs

The Theta GSTs are found in a diverse range of organisms and were originally postulated to be the progenitor of all GST classes. However, as more GSTs were identified, it became apparent that many GSTs, including the insect Delta class, were inappropriately assigned to this class, and these GSTs were subsequently renamed. Nevertheless, mosquitoes do possess several GST genes whose sequence identity and phylogeny warrant their inclusion in the Theta class. Interestingly, only two Theta GST genes have been identified in A. gambiae, but five putative Theta GSTs have already been identified in Ae. aegypti. These putative insect Theta GSTs have not yet been biochemically characterized, and their physiological role is unknown.

Zeta Class GSTs

Zeta GSTs are found in many different species, and their sequence is highly conserved, particularly at the N-terminus of the proteins where the SSCXWRVIAL motif is retained in plants, insects, and mammals (Board *et al.*, 1997). The highly conserved structure of this protein suggests it plays an essential housekeeping role, and in this regard, GSTZ1-1 catalyses an important step in the tyrosine degradation pathway. A single Zeta GST gene is present in *A. gambiae.*

Structure of Cytosolic GSTs

Cytosolic GSTs are dimeric proteins. Each subunit is composed of between 200 and 250 amino acid residues with typical molecular masses ranging from 20–28 kDa. Each GST subunit adopts a canonical GST fold of seven or eight alpha helices and four beta sheets to produce two distinct domains, the N- and C-terminal domains. Five insect GSTs have been crystallized, including three from mosquitoes (Chen *et al.*, 2003; Oakley *et al.*, 2001). The structure of a Delta class GST, bound to the inhibitor S-hexylglutathione, is shown in Fig. 2. The interface between the two subunits can be hydrophobic or hydrophilic, and interactions between residues in both subunits are essential for dimer stability. Incompatibility in interfacial residues prevents heterodimers forming between two GST subunits from different classes. However, within a class, the formation of heterodimers can expand the range of functional proteins produced. The prevalence of GST heterodimers has not been widely studied in insects, but in plants it has been shown that even within a class not all GSTs will dimerize with each other to form heterodimers (Dixon *et al.*, 1999).

GST subunits have two distinct binding sites: the G site that binds glutathione and the substrate binding or H site. The G site is mainly composed of amino acids in the N-terminal, including the active site residue that interacts with and activates the sulfhydryl group of glutathione to generate the catalytically active thiolate anion (Armstrong, 1991). In the Delta, Epsilon, Theta, and Zeta GSTs, the active site residue is a serine, and in Omega class GSTs, it is a cysteine; in Sigma GSTs (and most mammalian GSTs), this role is played by tyrosine (reviewed in Sheehan *et al.*, 2001). The H site is mainly found in the C-terminal and determines the substrate specificity of the GST. In *A. gambiae,* additional heterogeneity of the GST enzyme family is introduced by alternative splicing of two genes, GSTd1 and GSTs1. In both of these, a common 5′ exon, encoding the N-terminal domain can be spliced to alternative 3′ exons to generate

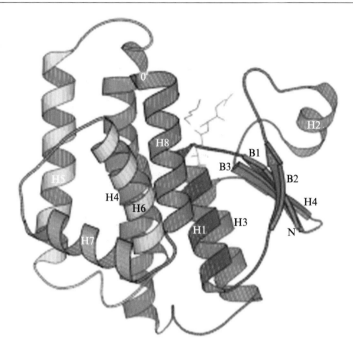

FIG. 2. The 3-D structure of AgGSTD1-6 bound to the inhibitor, S-hexylglutathione (shown in stick representation). Reproduced with permission from *Acta Crystallographica* with permission from Dr. Liqing Chen. (See color insert.)

subunits with differing biochemical properties (Fig. 3). Functional diversity is also maintained by allelic variation. Substitutions in a small number of amino acids can have a dramatic effect on the biochemical properties of the GST protein (Ketterman *et al.*, 2001). In some cases, exceptionally high levels of polymorphism in mosquito genes seem to be positively selected for. For example, two different alleles of AgGSTe1, which encode proteins that differ by 12 amino acids and have markedly different substrate specificities, are maintained in populations of *A. gambiae* throughout Africa (Ortelli *et al.*, 2003).

Microsomal GSTs

The microsomal GST class is evolutionarily and structurally distinct from the cytosolic class but catalyzes a similar range of reactions. Each subunit is composed of approximately 150 amino acids that assemble into four transmembrane alpha helices (Schmidt-Krey *et al.*, 2000). These form

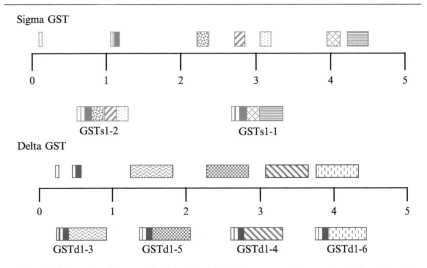

FIG. 3. Schematic diagram showing the alternative transcripts produced by alternative splicing of the *Anopheles gambiae* GSTd1 and GSTs1 genes. The genomic sequence is shown above the scale bar, transcripts are shown below. Empty rectangles indicate 5'UTR regions, solid or shaded rectangles represent coding regions. (See color insert.)

a trimeric protein with a molecular mass of approximately 50 KDa and a single substrate-binding site. Three members of this class have been identified in *A. gambiae*.

Functions of Mosquito GSTs

The reaction most commonly associated with the GST enzyme family is glutathione conjugation. This is a key stage in the conversion of lipophilic compounds to water-soluble metabolites that are more readily exported from the cell. Although it is well established that GSTs are involved in many other reactions besides conjugation, in reality, little is known about the endogenous substrates of mosquito GSTs. They presumably play a role in the metabolism of dietary components, such as plant allelochemicals and, as in other organisms, are important in intracellular transport, signaling pathways, and biosynthetic pathways, but most research on the mosquito GST family to date has focused on their importance in insecticide detoxification (see the following). However, recent advances in mosquito genomics mean that a functional genomics approach to explain the role of individual GST enzymes is now feasible. Mosquito research programs are primarily driven by the search for novel methods to kill the mosquito itself or the parasite or virus that it transmits. Manipulation of mosquito GST activity could provide novel opportunities to advance these goals. Some of the key areas of current research are outlined in the following.

GSTs and Insecticide Resistance

GSTs can catalyze the detoxification of several major classes of insecticide. For example, they catalyze the conjugation of glutathione to organophosphate insecticides such as tetrachlorvinphos and parathion, resulting in their O-dealkylation or O-dearylation, and they catalyze the dehydrochlorination of the organochlorine insecticide, DDT (Clark and Shamaan, 1984). GSTs with peroxidase activity may confer protection against the secondary effects of insecticide exposure, such as oxidative stress (Vontas et al., 2001, see the following). GSTs may also protect against insecticide exposure by acting as insecticide binding proteins, reducing the amount of insecticide that reaches the target (Kostaropoulos et al., 2001). Elevated GST activity is thus an important mechanism by which insects develop resistance to insecticides.

In mosquitoes, most research has focused on the role of GSTs in DDT resistance. Indoor residual house spraying with this insecticide was a central part of the malaria eradication program in many countries in the 1950s and 1960s. The campaign failed, in part, because of insecticide resistance and the inability of the public health infrastructure in many disease-endemic countries to deliver such a demanding program. Despite selection of resistance in many Anopheles populations, DDT continued to be the insecticide of choice for malaria control in many countries until the 1990s, with countries such as India arguing that resistance did not affect the repellent properties of DDT, which may be as important in reducing malaria transmission as the direct killing effects on the insects. The sole legitimate use for DDT through the recent Persistent Organic Pollutants Treaty remains indoor residual treatments for malaria control.

Several different approaches have been taken to identify the particular GSTs involved in DDT resistance in A. gambiae. From partial purification of mosquito GSTs and characterization of the different fractions' ability to dehydrochlorinate DDT, it became apparent that DDTase activity was not an inherent property of all GST enzymes. By a combination of different approaches, including genetic mapping, expression analysis, and in vitro expression of recombinant GST enzymes, the Epsilon GST, GSTE2, was identified as the enzyme responsible for DDT resistance in a strain of A. gambiae from East Africa. Homodimers of GSTE2 are extremely efficient at metabolizing DDT, and expression of this enzyme is elevated in DDT-resistant individuals (Ding et al., 2003).

Recent work has also implicated elevated activities of an Epsilon GST, which seems to be the ortholog of GSTE2, in conferring resistance to DDT in *Ae. aegypti*. This protein shares 71% amino acid identity with *A. gambiae* GSTE2, has high levels of DDTase activity, and is overexpressed in DDT-resistant mosquitoes. In contrast to its putative *A. gambiae* ortholog, the *Aedes* enzyme also has peroxidase activity, which may contribute to the high levels of pyrethroid resistance seen in this strain (Lumjuan *et al.*, 2005).

GSTs and the Processing of Odorant Signals

Olfaction plays an essential role in the reproduction and feeding behavior of mosquitoes. Odorant molecules enter olfactory sensilla located within the mosquito's antennae and interact with receptor molecules to trigger an appropriate response. The odorant molecules must ultimately be degraded to terminate the sensory response, and GSTs play an important role in this process. An antennal-specific GST involved in sex pheromone detection has been identified in the moth, *Manduca sexta* (Rogers *et al.*, 1999). As far as we are aware, there has been no systematic search for antennal-specific mosquito GSTs, but this is potentially an important area of future research. Unraveling the complex chemical cascade in the olfaction response could lead to a new generation of mosquito attractants and repellents.

GSTs as Protectors against Oxidative Stress

GSTs play a vital role in the inactivation of toxic products of oxygen metabolism. Reactive oxygen species (ROS), including hydrogen peroxide, superoxide anions, and hydroxyl radicals, are generated during aerobic respiration. Blood-feeding insects, such as mosquitoes, face a particular challenge from oxidative stress, because the digestion of hemoglobin results in a massive production of ROS. These ROS trigger a cascade of reactions that can be highly damaging to the cells. They can cause the inactivation of enzymes through conformational change, degrade DNA, and damage cellular membranes by the conversion of polyunsaturated fatty acids to lipid hydroperoxides. These, in turn, can give rise to highly reactive α,β-unsaturated aldehydes (e.g., 4-hydroxynonenal [4-HNE]) that, although essential components of signaling pathways, are toxic at high concentrations. GSTs have a protective effect at several different stages in this pathway. Peroxidase activity has been detected in insect GSTs from the Delta, Epsilon, and Sigma classes, using the model substrate cumene hydroperoxide (Ortelli *et al.*, 2003; Singh *et al.*, 2001; Vontas *et al.*, 2001). Other Delta and Sigma insect GSTs can detoxify 4-HNE by conjugation with glutathione (Sawicki *et al.*, 2003; Singh *et al.*, 2001).

More recently, the redox state of the cell has been shown to be an important determinant of parasite survival in the insect. Mosquito-borne parasites, such as the malaria protozoan, *Plasmodium*, have an obligate sexual reproductive stage in the mosquito midgut to produce a motile ookinete that penetrates the mosquito midgut wall. The mosquito immune response to parasites involves oxidative responses, and, indeed, some strains of mosquito that are in a permanent sate of chronic oxidative stress cannot support parasite development (Kumar *et al.*, 2003). This defensive mechanism against parasite invasion requires a balance of antioxidative enzymes to prevent self-damage. The role of GSTs in reducing the levels of ROS in the midgut and the significance of this for parasite transmission is unclear but of great interest. Could strains of mosquitoes with elevated levels of GST activity (perhaps selected for by insecticide exposure) be more efficient vectors of human pathogens?

Regulation of Mosquito GST Expression

Many GSTs are differentially regulated in response to various inducers or environmental signals, and some are expressed in a tissue- or developmental-specific manner. In *A. gambiae*, transcripts for all but one of the GST supergene family were detectable in 1-day-old adult mosquitoes by RT-PCR (Ding *et al.*, 2003), but preliminary data suggest that expression of some of these genes is restricted to certain tissues.

Some of the GST genes are tightly clustered within the mosquito genome. For example, six Delta class genes are located within 12 kb of chromosome arm 2L, and the eight Epsilon GSTs are found within 11 kb on chromosome 3R. In some cases, the intergenic distance between neighboring genes is less than 300 bp. Does a common promoter, upstream of the gene cluster, control expression of multiple GST genes, or does each GST gene contain its own basal promoter? To address this question, the putative promoter for GSTe2 was inserted upstream of a luciferase reporter gene and transfected into an *A. gambiae* cell line. The 300 bp between the end of the preceding gene, GSTe1, and the translation start point of GSTe2, was able to drive transcription, but, interestingly, inclusion of the 3'UTR of GSTe1 significantly increased promoter activity (Ding *et al.*, 2005).

Expression of *A. gambiae* Epsilon GSTs is induced by exposure to hydrogen peroxide, and several putative oxidant responsive elements are present in the intergenic regions of the Epsilon GST cluster (Ding *et al.*, 2005). In *D. melanogaster*, exposure to hydrogen peroxide and paraquat both had dramatic effects on the expression of multiple GSTs, but in addition to induction, some genes were repressed by oxidant exposure (Girardot *et al.*, 2004). Other factors that have been shown to induce

GST expression in insects include insecticides, fumigants, barbiturates, and plant allelochemicals.

The elevated levels of GSTs observed in insecticide-resistant strains of mosquitoes is a constitutive response caused by mutations in the regulatory regions of the GSTs. Both trans-acting and cis-acting factors have been implicated. For example, in *Ae. Aegypti*, a mutation in a trans-acting repressor element is the proposed mechanism for the enhanced expression of a Delta class GST in a DDT-resistant strain (Grant and Hammock, 1992). In *A. gambiae* overexpression of GSTe2 is associated with the deletion of two adenosine residues in the core promoter of this gene (Ding *et al.*, 2005).

Summary

In summary, mosquito GSTs form a complex, multi-class, family of enzymes that fulfil a raft of important housekeeping and protective roles within the insect. Their role in insecticide resistance is currently best understood, but our knowledge of the structure and function of this important enzyme class is expanding rapidly driven in part by recent advances in mosquito genome sequencing.

References

Adams, M.D, Clniker, S. E., Holt, R. A., Evans, C. A., Gocayne, J. D., *et al.* (2000). The genome sequence of *Drosophila melanogaster. Science* **287,** 2185–2195.

Armstrong, R. N. (1991). Glutathione S-transferases: Reaction mechanism, structure, and function. *Chem. Res. Toxicol.* **4,** 131–140.

Board, P. G., Baker, R. T., Chelvanayagam, G., and Jermiin, L. S. (1997). Zeta, a novel class of glutathione transferases in a range of species from plants to humans. *Biochem. J.* **328,** 929–935.

Board, P. G., Coggan, M., Chelvanayagam, G., Easteal, S., Jermiin, L. S., Schulte, G. K., Danley, D. E., Hoth, L. R., Griffor, M. C., Kamath, A. V., Rosner, M. H., Chrunyk, B. A., Perregaux, D. E., Gabel, C. A., Geoghegan, K. F., and Pandit, J. (2000). Identification, characterization, and crystal structure of the Omega class glutathione transferases. *J. Biol. Chem.* **275,** 24798–24806.

Booth, J., Boyland, E.., and Sims, P. (1961). An enzyme from rat liver catalyzing conjugation with glutathione. *Biochem. J.* **79,** 516–524.

Chelvanayagam, G., Parker, M..W., and Board, P. G. (2001). Fly fishing for GSTs: A unified nomenclature for mammalian and insect glutathione transferases. *Chem. Bio. Interact.* **133,** 256–260.

Chen, L., Hall, P. R., Zhou, X. E., Ranson, H., Hemingway, J., and Meehan, E. J. (2003). Structure of an insect Delta-class glutathione S-transferase from a DDT-resistant strain of the malaria vector *Anopheles gambiae. Acta Cryst.* **D59,** 2211–2217.

Clark, A. G., and Shamaan, N. A. (1984). Evidence that DDT dehydrochlorinase from the housefly is a glutathione S-transferase. *Pestic. Biochem. Physiol.* **22,** 249–261.

Combes, B., and Stakelum, G. S. (1961). A liver enzyme that conjugates sulfobromophthalein sodium with glutathione. *J. Clin. Invest.* **40**, 981–988.

Ding, Y., Ortelli, F., Rossiter, L., Hemingway, J., and Ranson, H. (2003). The *Anopheles gambiae* glutathione transferase family: Annotation, phylogeny and gene expression profiles. *BMC Genomics* **4**, 35.

Ding, Y., Hawkes, N., Meredith, J., Eggleston, P., Hemingway, J., and Ranson, H. (2005). Characterisation of the promoters of Epsilon glutathione transferases in the mosquito *Anopheles gambiae* and their response to oxidative stress. *Biochem J.* **387**, 879–888.

Dixon, D. P., Cole, D. J., and Edwards, R. (1999). Dimerisation of maize glutathione transferases in recombinant bacteria. *Plant Mol. Biol.* **40**, 997–1008.

Fournier, D., Bride, J. M., Poirie, M., Berge, J. B., and Plapp, F. W., Jr. (1992). Insect glutathione S-transferases. Biochemical characteristics of the major forms from houseflies susceptible and resistant to insecticides. *J. Biol. Chem.* **267**, 1840–1845.

Girardot, F., Monnier, V., and Tricoire, H. (2004). Genome wide analysis of common and specific stress responses in adult *Drosophila melanogaster. BMC Genomics* **5**, 74.

Grant, D. F., and Hammock, B. D. (1992). Genetic and molecular evidence for a trans-acting regulatory locus controlling glutathione S-transferase-2 expression in *Aedes aegypti. Mol. Gen. Genet.* **234**, 169–176.

Holt, R. A., Subramanian, G. M., Halpern, A., Sutton, G. G., Charlab, R., Nusskern, D. R., Wincker, P., Clark, A. G., Ribeiro, J. M., Wides, R., Salzberg, S. L., Loftus, B., Yandell, M., Majoros, W. H., Rusch, D. B., Lai, Z., Kraft, C. L., Abril, J. F., Anthouard, V., Arensburger, P., Atkinson, P. W., Baden, H., de, B. V., Baldwin, D., Benes, V., Biedler, J., Blass, C., Bolanos, R., Boscus, D., Barnstead, M., Cai, S., Center, A., Chatuverdi, K., Christophides, G. K., Chrystal, M. A., Clamp, M., Cravchik, A., Curwen, V., Dana, A., Delcher, A., Dew, I., Evans, C. A., Flanigan, M., Grundschober-Freimoser, A., Friedli, L., Gu, Z., Guan, P., Guigo, R., Hillenmeyer, M. E., Hladun, S. L., Hogan, J. R., Hong, Y. S., Hoover, J., Jaillon, O., Ke, Z., Kodira, C., Kokoza, E., Koutsos, A., Letunic, I., Levitsky, A., Liang, Y., Lin, J. J., Lobo, N. F., Lopez, J. R., Malek, J. A., McIntosh, T. C., Meister, S., Miller, J., Mobarry, C., Mongin, E., Murphy, S. D., O'Brochta, D. A., Pfannkoch, C., Qi, R., Regier, M. A., Remington, K., Shao, H., Sharakhova, M. V., Sitter, C. D., Shetty, J., Smith, T. J., Strong, R., Sun, J., Thomasova, D., Ton, L. Q., Topalis, P., Tu, Z., Unger, M. F., Walenz, B., Wang, A., Wang, J., Wang, M., Wang, X., Woodford, K. J., Wortman, J. R., Wu, M., Yao, A., Zdobnov, E. M., Zhang, H., Zhao, Q., Zhao, S., Zhu, S. C., Zhimulev, I., Coluzzi, M., della, T. A., Roth, C. W., Louis, C., Kalush, F., Mural, R. J., Myers, E. W., Adams, M..D., Smith, H. O., Broder, S., Gardner, M. J., Fraser, C. M., Birney, E., Bork, P., Brey, P. T., Venter, J. C., Weissenbach, J., Kafatos, F. C., Collins, F. H., and Hoffman, S. L. (2002). The genome sequence of the malaria mosquito *Anopheles gambiae. Science* **298**, 129–149.

Huang, H. S., Hu, N. T., Yao, Y. E., Wu, C. Y., Chiang, S. W., and Sun, C. N. (1998). Molecular cloning and heterologous expression of a glutathione S-transferase involved in insecticide resistance from the diamondback moth, *Plutella xylostella. Insect Biochem. Mol. Biol.* **28**, 651–658.

Ketterman, A. J., Prommeenate, P., Boonchauy, C., Chanama, U., Leetachewa, S., Promtet, N., and Prapanthadara, L. (2001). Single amino acid changes outside the active site significantly affect activity of glutathione S-transferases. *Insect Biochem. Mol. Biol.* **31**, 65–74.

Kostaropoulos, I., Papadopoulos, A. I., Metaxakis, A., Boukouvala, E., and Papadopoulou-Mourkidou, E. (2001). Glutathione S-transferase in the defence against pyrethroids in insects. *Insect Biochem. Mol. Biol.* **31**, 313–319.

Kumar, S., Christophides, G. K., Cantera, R., Charles, B., Han, Y. S., Meister, S., Dimopoulos, G., Kafatos, F. C., and Barillas-Mury, C. (2003). The role of reactive oxygen species on *Plasmodium* melanotic encapsulation in *Anopheles gambiae. Proc. Natl. Acad. Sci. USA* **100,** 14139–14144.

Lumjuan, N., McCarroll, L., Prapanthadara, L., Hemingway, J. and Ranson, H. (2005). The conserved role of GSTe2 in metabolically-based resistance to DDT between Anopheles and *Aedes* mosquitoes. *Insect. Biochem. Mol. Biol.* **35,** 861–871.

Mannervik, B., Awasthi, Y. C., Board, P. G., Hayes, J. D., Di Ilio, C., Ketterer, B., Listowsky, I., Morgenstern, R., Muramatsu, M., Pearson, W. R., Pickett, C. B., Sato, K., Widerston, M., and Wolf, C. R. (1992). Nomenclature for human glutathione S-transferases. *Biochem. J.* **282,** 305–306.

Ortelli, F., Rossiter, L. C., Vontas, J., Ranson, H., and Hemingway, J. (2003). Heterologous expression of four glutathione transferase genes genetically linked to a major insecticide resistance locus, from the malaria vector *Anopheles gambiae. Biochem. J.* **373,** 957–963.

Ranson, H., Collins, F., and Hemingway, J. (1998). The role of alternative mRNA splicing in generating heterogeneity within the *Anopheles gambiae* class I glutathione S-transferase family. *Proc. Natl. Acad. Sci. USA* **95,** 14284–14289.

Ranson, H., Claudianos, C., Ortelli, F., Abgrall, C., Hemingway, J., Sharakhova, M. V., Unger, M. F., Collins, F. H., and Feyereisen, R. (2002). Evolution of supergene families associated with insecticide resistance. *Science* **298,** 179–181.

Rogers, M..E., Jani, M..K., and Vogt, R. G. (1999). An olfactory-specific glutathione-S-transferase in the sphinx moth *Manduca sexta. J. Exp. Biol.* **202**(Pt. 12), 1625–1637.

Sawicki, R., Singh, S. P., Mondal, A. K., Benes, H., and Zimniak, P. (2003). Cloning, expression and biochemical characterization of one Epsilon-class (GST-3) and ten Delta-class (GST-1) glutathione S-transferases from *Drosophila melanogaster*, and identification of additional nine members of the Epsilon class. *Biochem. J.* **370,** 661–669.

Schmidt-Krey, I., Murata, K., Hirai, T., Mitsuoka, K., Cheng, Y., Morgenstern, R., Fujiyoshi, Y., and Hebert, H. (1999). The projection structure of the membrane protein microsomal glutathione transferase at 3 A resolution as determined from two-dimensional hexagonal crystals. *J. Mol. Biol.* **288,** 243–253.

Sheehan, D., Meade, G., Foley, V. M., and Dowd, C. A. (2001). Structure, function and evolution of glutathione transferases: Implications for classification of non-mammalian members of an ancient enzyme superfamily. *Biochem. J.* **360,** 1–16.

Singh, S. P., Coronella, J. A., Benes, H., Cochrane, B. J., and Zimniak, P. (2001). Catalytic function of *Drosophila melanogaster* glutathione S-transferase DmGSTS1-1 (GST-2) in conjugation of lipid peroxidation end products. *Eur. J. Biochem.* **268,** 2912–2923.

Tang, A. H., and Tu, C. P. (1994). Biochemical characterization of *Drosophila* glutathione S-transferases D1 and D21. *J. Biol. Chem.* **269,** 27876–27884.

Vontas, J. G., Small, G. J., and Hemingway, J. (2001). Glutathione S-transferases as antioxidant defence agents confer pyrethroid resistance in *Nilaparvata lugens. Biochem. J.* **357,** 65–72.

Vontas, J. G., Small, G. J., Nikou, D. C., Ranson, H., and Hemingway, J. (2002). Purification, molecular cloning and heterologous expression of a glutathione S-transferase involved in insecticide resistance from the rice brown planthopper, *Nilaparvata lugens. Biochem. J.* **362,** 329–337.

Wang, J..Y., McCommas, S., and Syvanen, M. (1991). Molecular cloning of a glutathione S-transferase overproduced in an insecticide-resistant strain of the housefly (*Musca domestica*). *Mol. Gen. Genet.* **227,** 260–266.

Wei, S..H., Clark, A..G., and Syvanen, M. (2001). Identification and cloning of a key insecticide-metabolizing glutathione S-transferase (MdGST-6A) from a hyper insecticide-resistant strain of the housefly Musca domestica. *Insect Biochem. Mol. Biol.* **31**, 1145–1153.

[15] Glutathione S-transferase from Malarial Parasites: Structural and Functional Aspects

By MARCEL DEPONTE and KATJA BECKER

Abstract

Malaria represents an emerging disease because of increasing parasite resistance against available drugs and because of increasing geographical distribution of the causative agent, *Plasmodium falciparum*. The complete genome of *Plasmodium* was sequenced recently, revealing that the parasite harbors only one glutathione S-transferase (PfGST). This observation was of particular interest: First, certain antimalarial drugs such as chloroquine and methylene blue presumably influence the glutathione metabolism in which PfGST is involved. Second, PfGST might play a significant role in drug resistance. PfGST was studied in parasite extracts and as recombinant protein, and its x-ray structure has been solved. The available data indicate that the homodimeric PfGST cannot be assigned to any of the previously known GST classes. PfGST exhibits significant structural differences to human GSTs, particularly at the so-called hydrophobic binding pocket (H-site) where the second substrate binds. Inhibition of PfGST is expected to act at different vulnerable metabolic sites of the parasite in parallel; it is likely to disturb GSH-dependent detoxification processes, to increase the levels of cytotoxic peroxides, and possibly to increase the concentration of toxic hemin. In this chapter, we summarize the current knowledge on PfGST, including aspects of structure, function, and future drug development.

Functions of Parasite GSTs

Glutathione S-transferases (GSTs) of protozoan and metazoan parasites have gained increasing attention in the past few years. Because these enzymes might be involved in drug resistance and the removal of endogenous and exogenous cytotoxic metabolites, GSTs from parasitic nematodes, trematodes, ticks, and malarial parasites have been studied. Some

METHODS IN ENZYMOLOGY, VOL. 401
Copyright 2005, Elsevier Inc. All rights reserved.
0076-6879/05 $35.00
DOI: 10.1016/S0076-6879(05)01015-3

of these proteins might be exploited as drug target or might be good candidates for vaccine development (Fritz-Wolf *et al.*, 2003; Johnson *et al.*, 2003; McTigue *et al.*, 1995; Ouaissi *et al.*, 2002; Rossjohn *et al.*, 1997). Indeed, identification of the *Schistosoma japonicum* GST as a potential vaccine candidate "accidentally" lead to the development of a GST fusion protein system that is, for example, extensively used to analyze protein–protein interactions by pull-down assays (Smith and Johnson, 1988; Smith *et al.*, 1987).

The functions of GSTs from different parasites include (1) the more or less specific nucleophilic addition of GSH to electrophils, (2) the GSH-dependent reduction of hydroperoxides, and (3) the binding of ligands. However, to which degree all of these functions play a significant role *in vivo* has to be studied for each parasite and protein. Some parasite GSTs might also catalyze the GSH-dependent isomerization of metabolites such as prostaglandins, and as a result, the parasite might modulate the immune system of the host (Angeli *et al.*, 2001; Johnson *et al.*, 2003; Ouaissi *et al.*, 2002).

Several parasitic worms possess more than one GST (Liebau *et al.*, 1996), whereas, for example, *Plasmodium falciparum* seems to have only one classical GST (Fritz-Wolf *et al.*, 2003). Not surprisingly, parasite GSTs differ significantly with respect to their functions: GST3 from *Onchocerca volvulus* has been shown to increase resistance of transgenic *Caenorhabditis elegans* to internal and external oxidative stress. In addition, expression of the respective gene is inducible by oxidative stress. These results suggest that GST3 could be involved in the defense against reactive oxygen species (ROS) resulting from cellular metabolism and against ROS derived from the host's immune system (Kampkotter *et al.*, 2003). A GST from *Haemonchus contortus* shows only limited activity with classical GST substrates such as 1-chloro-2,4-dinitrobenzene (CDNB). However, the protein effectively binds hematin in contrast to a 60% identical GST from the closely related nonparasitic nematode *C. elegans,* suggesting that the high-affinity binding of hematin as a ligand may represent a parasite adaptation to blood feeding (van Rossum *et al.*, 2004). GSTs from the parasitic nematodes *Ascaris suum* and *Onchocerca volvulus* efficiently use CDNB, unsaturated carbonyl compounds such as *trans*-2-nonenal, and hydroperoxides such as cumene hydroperoxide as substrates *in vitro*. Tissue distribution and localization of these nematode GSTs differ, depending on the protein and organism. Thus, some of the GSTs (but not all of them) are acting at the host–parasite interface and are, indeed, accessible to inhibitors (Liebau *et al.*, 1996, 1997; Wildenburg *et al.*, 1998). As reported by Rao *et al.* (2000), *Brugia*, a pathogenic filarial nematode, may modify the host's defense mechanisms by a detoxification process involving GSTs. Determination of GST activity in soluble parasite extracts and in excretory-secretory products of B. malayi suggests that GST is secreted *in vivo* and cross-reacts immunologically with the GSTs from other filarial nematodes. GST inhibitors reduced the viability and motility of microfilariae, third-stage larvae, and adult worms.

GST Activity in Malarial Parasites

GST activity has been detected in all *Plasmodium* species studied so far, as well as in all intraerythrocytic stages of the parasite (Dubois *et al.*, 1995; Harwaldt *et al.*, 2002; Liebau *et al.*, 2002; Srivastava *et al.*, 1999). At 37°, GST activities between 20 and 95 mU/mg total protein were determined in *P. berghei*, *P. yoelii*, and *P. falciparum*, whereas *P. knowlesi* showed lower activities around 5 mU/mg. Stage-specific studies revealed that there is a decreasing activity from schizonts to rings to trophozoites (Srivastava *et al.*, 1999). In accordance with these results, GST activities determined in a different study in extracts from isolated trophozoites of eight *P. falciparum* strains indicated values between 5.6 and 22 mU/mg at 37° (Harwaldt *et al.*, 2002). As further reported by Srivastava *et al.* (1999), GST activity increases significantly in chloroquine-resistant strains. This increase in enzyme activity was directly related to drug pressure of resistant *P. berghei*.

Because strategies directed against the malaria vector *Anopheles* are of major interest in malaria control, it is worth mentioning here that DDT resistance in both adults and larvae of *Anopheles gambiae* is mediated by stage-specific insect class I GSTs (Ranson *et al.*, 1997). A DDT-metabolizing GST has also been characterized in *Anopheles dirus* (Prapanthadara *et al.*, 1996; for review see Eaton, 2000; see also Chapter 14).

Biochemical Properties of PfGST

The gene encoding PfGST has been cloned, and the corresponding recombinant protein has been purified and characterized independently by two groups (Harwaldt *et al.*, 2002; Liebau *et al.*, 2002). The gene structure, comprising two exons, and the amino acid sequence of PfGST (accession number AY014840) are in full agreement with the *P. falciparum* genomic database (Kissinger *et al.*, 2002; http://www.plasmodb.org). The activity and Michaelis–Menten parameters of recombinant PfGST were determined in a standard GSH-dependent CDNB-assay (Harwaldt *et al.*, 2002): 1 ml assay mixture containing 1 mM GSH and PfGST in 100 mM Hepes, 1 mM EDTA, pH 6.5, was equilibrated to 25°. The reaction was started by the addition of 0.5 mM CDNB, and the formation of

S-(2,4-dinitrophenyl)glutathione was monitored spectrophotometrically at 340 nm ($\varepsilon_{340\ nm} = 9.6$ mM^{-1}cm^{-1}). Alternatively, the reaction can be started by the addition of the enzyme. The specific activity of recombinant PfGST in this assay is 0.20 U/mg (5.2 U/μmol). By varying the concentration of GSH, the K_m for GSH was determined by Harwaldt et al. (2002) and Liebau et al. (2002) to be 164 ± 20 μM and 156 ± 13 μM, respectively. Thus, under physiological conditions (for review see Becker et al., 2003), PfGST can be saturated by GSH in vivo. For CDNB, a K_m value of >2 mM was estimated. It has to be kept in mind that the spontaneous background reaction between GSH and CDNB increases with increasing pH or increasing substrate concentrations. Thus, if necessary, the measured activities have to be corrected.

PfGST has a pH optimum of 8.1, and its activity is sensitive to NaCl and potassium phosphate (Harwaldt et al., 2002): comparison with 100 mM potassium phosphate, 1 mM EDTA, 100 mM Tris/HCl, and 1 mM EDTA at the same pH showed a decrease in enzyme activity to 15% and 90% of the activity in 100 mM Hepes buffer, respectively. Addition of NaCl to the assay system strongly decreased enzyme activity (83% residual activity at 50 mM NaCl, 66% at 100 mM NaCl, 19% at 200 mM NaCl). Recombinant N-terminally His-tagged enzyme is very stable. The K_m value for GSH and the specific activity with GSH and CDNB as substrates were not significantly changed after more than 1 year (unpublished). However, precipitating recombinant PfGST with ammonium sulfate (Liebau et al., 2002) or dialyzing the protein resulted in partial loss of activity.

Apart from CDNB, 1,2-dichloro-4-nitrobenzene, bromosulfophthalein, and ethacrynic acid have been tested as electrophilic substrates. Glutathione consumption was determined with 5,5'-dithio-bis-(2-nitrobenzoic acid) (DTNB) measuring the formation of 5-thio-2-nitrobenzoate at 412 nm ($\varepsilon_{412\ nm} = 13.6$ mM^{-1}cm^{-1}, for details see Harwaldt et al., 2002). The specific activity for 1 mM ethacrynic acid and 1 mM o-nitrophenyl acetate is 0.19 U/mg (5.0 U/μmol) and 0.06 U/mg (1.5 U/μmol), respectively, and the K_m for ethacrynic acid is approximately 0.5–1.5 mM (Harwaldt et al., 2002; Liebau et al., 2002).

Because malarial parasites digest huge amounts of hemoglobin, and because they are exposed to oxidative stress inside the erythrocyte (for review see Becker et al., 2004), detoxification of peroxidases and binding of heme-containing compounds as ligands might be further functions of PfGST in vivo. Indeed, PfGST binds hemin/ferriprotoporphyrin IX in the lower micromolar range and shows slight peroxidase activity with GSH and cumene hydroperoxide as substrates. By measuring the GST activity in extracts from isolated trophozoites of eight different P. falciparum strains, the amount of PfGST can be estimated to be >1% of cellular protein

(Harwaldt *et al.*, 2002). In a different experimental approach, the concentration was estimated by Liebau *et al.* (2002) to be 0.1% of the total protein content in trophozoites. Depending on the true cellular concentration, PfGST might protect the parasite against oxidative stress or might act as a buffer for heme-containing compounds *in vivo* (Harwaldt *et al.*, 2002; Liebau *et al.*, 2002). Thus, further work is needed to decipher the function(s) of PfGST *in vivo*.

Tertiary and Quaternary Structure of PfGST

Recently, the crystal structure of PfGST was solved independently by two groups at 1.9 and 2.2 Å resolution (Burmeister *et al.*, 2003; Fritz-Wolf *et al.*, 2003; Perbandt *et al.*, 2004). Apart from GST from other *Plasmodium* species, PfGST shares highest sequence identities with the pi-class GSTs from *Dirofilaria immitis* and *O. volvolus* (approximately 35% identity). Structural alignment of PfGST with members of the alpha-, mu-, and pi-classes indicated an rms deviation of at least 1.2 Å, which is significantly higher than expected for members within the same class (<0.7 Å). Because PfGST adopts the canonical GST fold but cannot be assigned to any of the known GST classes, comparisons with human GSTs provide highly valuable data that could help to develop specific and potent inhibitors of the parasite protein. These inhibitors might be suited as lead compounds for the development of new antimalarial drugs.

PfGST possesses a shorter C-terminal section, resulting in a more amphiphilic and more solvent-accessible hydrophobic binding pocket (H-site) compared with other GST structures (Fritz-Wolf *et al.*, 2003). On binding of the inhibitor *S*-hexyl-glutathione, the glutathione binding site (G-site) and the overall tertiary structure of PfGST remain almost unchanged with exception of the flexible C-terminus (Perbandt *et al.*, 2004). Detection of PfGST in *P. falciparum* homogenates using native polyacrylamide gel electrophoresis followed by western blot analysis suggested that the protein exists as a dimer *in vivo* (Liebau *et al.*, 2002). Using gel filtration, we also determined recombinant PfGST to be dimeric in the presence of 2 mM GSH. However, in the absence of GSH, the MRGSH$_6$GS-tagged protein forms tetrameric (98%) and octameric (2%) oligomers under nonreducing and reducing (2 mM 1,4-dithiothreitol) conditions (unpublished). The crystal structure of PfGST in the absence of glutathione reveals two crystallographically independent monomers in the asymmetric unit leading to two types of homodimers (A, A′, and B, B′). By comparison with known GST dimer structures, monomers A and B do not form the enzymatically active dimer, but surprisingly a loop between the α-helices 3 and 4 of monomer B interacts with the H-site of the active site of monomer A (Fritz-Wolf *et al.*, 2003). Considering the protein species with an apparent molecular weight corresponding to a tetrameric structure observed in our gel filtration experiments, the crystal structure can be interpreted as the interaction of two dimers forming a tetramer, with one dimer occupying the active site of the second dimer in the absence of GSH. Whether changes of the quaternary structure have any relevance *in vivo* (for example, regulating enzyme activity) might be interesting to investigate.

PfGST and Drug Resistance

Glutathione *S*-transferase conjugates GSH to toxic electrophiles predisposing them to export by specific membrane pumps, also known as multidrug resistance–associated proteins (MRPs; for reviews on glutathione and redox metabolism see Becker *et al.*, 2003, 2004). Many cells also display a GSSG pump that overlaps to some extent the action of MRP. This overall picture also seems to be present in *Plasmodium falciparum*. As reported by Klokouzas *et al.* (2004), *Plasmodium falciparum* expresses a multidrug resistance–associated protein. All other components of the GSH system, except for glutathione peroxidases, have been identified in *P. falciparum* (Becker *et al.*, 2003, 2004). The development of drug resistance often encountered in cancer therapy and the therapy of parasitic diseases has been related to the function of GST, which can contribute to drug clearance (Salinas and Wong, 1999). There are reports indicating that in chloroquine-resistant parasites, GST activity is related to drug pressure (Dubois *et al.*, 1995; Srivastava *et al.*, 1999). This topic is, however, controversially discussed (Ferreira *et al.*, 2004; Harwaldt *et al.*, 2002).

A model for the antimalarial action of chloroquine, one of the most important antimalarial drugs, has recently been proposed: Intraerythrocytic malarial parasites digest the hemoglobin of their host cell, yielding globin-derived amino acids and potentially toxic heme. Apart from heme polymerization as a strategy for detoxifying heme, GSH is capable of modifying or degrading it. Loading of intact erythrocytes with heme is followed by acceleration of the pentose phosphate pathway because of the production of H_2O_2, the oxidation of intracellular GSH, and redox-cycling of iron (Becker *et al.*, 2004; Famin *et al.*, 1999; Platel *et al.*, 1999). In agreement with this mechanism of action, increasing the cellular levels of glutathione leads to increased resistance to chloroquine and *vice versa* (Dubois *et al.*, 1995). The quantitative contribution of GSH-dependent heme degradation to the total heme degradation remains, however, to be established.

In the study by Dubois *et al.* (1995), the relationship between the intracellular GSH levels as well as glutathione *S*-transferase, glutathione reductase (GR), and glutathione peroxidase (GPx) activities and how they relate to *Plasmodium berghei* resistance to chloroquine was investigated. Interestingly, marked increases in GSH levels and GST activity were observed in resistant parasites when directly comparing them with sensitive strains. On the other hand, GR and GPx activities were found to be similar. Treatment with chloroquine did not influence the intracellular level of GSH, but it was found to significantly decrease GR activity. Intracellular depletion of GSH, by buthionine sulfoximine, sensitized the resistant parasites to chloroquine. These results suggest that the *P. berghei* resistance results from altered GSH and GST levels and activity, which enable the detoxification of chloroquine in resistant parasites. The data of Dubois *et al.* do contrast with values determined by Harwaldt *et al.* (2002). In this latter study, GST activities determined in extracts from eight *P. falciparum* strains of different chloroquine sensitivity did not differ significantly between sensitive (14.5 mU/mg protein) and resistant strains (10.3 mU/mg protein).

These data are supported by the recent studies from Ferreira *et al.* (2004). Comparing the sequences of *P. chabaudi* genes encoding enzymes involved in glutathione metabolism (namely γ-glutamylcysteine synthetase, glutathione synthetase, glutathione peroxidase, glutathione reductase, and gst) with those of chloroquine-resistant *P. falciparum* gene orthologs revealed no point mutations in the resistant parasites. A real-time PCR approach furthermore indicated that transcription levels of the genes were not changed between chloroquine-sensitive and chloroquine-resistant parasite clones and that treatment with chloroquine did not induce an alteration in the expression of these genes in sensitive or resistant parasites. The authors thus conclude that chloroquine resistance in this species is determined by a mechanism that is independent of these genes, and most likely, of GSH metabolism.

A strategy to overcome multidrug resistance in cancer cells, and thus also parasites that share many features with cancer cells, involves treatment with a combination of the antineoplastic agent and a chemomodulator that inhibits the activity of the resistance-causing protein. In cases in which GSTs are thought to play a role in drug resistance, chemomodulation might be achieved by using inhibitors of glutathione synthesis or by using GST inhibitors. To address this problem, Mukanganyama *et al.* (2002) investigated the effects of antimalarial drugs on human recombinant GSTs. GST A1-1 activity was inhibited by artemisinin with an IC_{50} of 6 μM, whilst GST M1-1 was inhibited by quinidine and its diastereoisomer quinine with IC_{50} values of 12 μM and 17 μM, respectively. GST P1-1 was the most susceptible enzyme to inhibition by antimalarials with IC_{50} values of 1, 2, 1, 4, and 13 μM for pyrimethamine, artemisinin, quinidine, quinine, and tetracycline, respectively. The IC_{50} values obtained for artemisinin, quinine, quinidine, and tetracycline are below peak plasma concentrations obtained during therapy of malaria with these drugs. The authors concluded that the addition of nontoxic reversing agents such as antimalarials could enhance the antineoplastic efficacy of a variety of alkylating agents.

The antimalarial drug artemisinin is a sesquiterpene lactone containing an endoperoxide bridge. It is particularly useful against the drug-resistant strains of *Plasmodium falciparum*. Mukanganyama *et al.* (2001) have thus investigated whether the drug is broken down by a typical reductive reaction in the presence of human GSTs. Artemisinin was shown to stimulate NADPH oxidation in cytosolic cell extracts, as well as in assays containing human recombinant GSTs. The authors thus hypothesized that GSTs may contribute to the metabolism of artemisinin and proposed a model based on the known reactions of GSTs and sesquiterpenes, in which (1) artemisinin reacts with GSH resulting in oxidized glutathione; (2) the oxidized glutathione is then converted to reduced glutathione by means of glutathione reductase; and (3) the latter reaction may then result in the depletion of NADPH by means of glutathione reductase. Although it cannot be excluded that this reaction sequence plays a role in some (patho)physiological contexts, it should be mentioned here that we were not able to reproduce these results with human placenta GST. PfGST, which became available in recombinant form directly after the study by Mukanganyama *et al.*, did not show any reaction under the published assay conditions either. One of the hindrances to a successful assay was the fact that artemisinin is not soluble in the millimolar concentrations reported to be used in the assays by Mukanganyama *et al.* To exclude that the low solubility of artemisinin led to the negative result, we then used artesunate, a water-soluble derivative of artemisinin, as substrate of PfGST and human placenta GST. However, activity was not detected either (unpublished results).

The Potential of PfGST as Drug Target

Malaria represents an increasing threat to human health and welfare; 300–500 million people per year are infected, and more than 2 million people die of malaria annually (Greenwood and Mutabingwa, 2002). One of the major reasons for this situation is the emergence of drug resistance to currently used drugs. Thus, new drugs directed against novel targets are urgently and continuously required (see also Schirmer *et al.*, 1995). As rapidly growing and multiplying cells, malarial parasites do depend on a functional GST. This assumption is supported by the first results obtained from knock out studies (unpublished data), as well as by the fact that PfGST is the only canonical GST present in *P. falciparum*.

Inhibition of PfGST is expected to act at different vulnerable metabolic sites of the parasite in parallel, which further supports its promising potential as drug target. PfGST inhibition is likely to disturb GSH-dependent detoxification processes, to enhance the levels of cytotoxic peroxides, and possibly to increase the concentration of toxic hemin. It has been shown that in the presence of GSH the parasitotoxic hemin inhibits PfGST in the lower micromolar range, indicating that free hemin might be buffered by the enzyme *in vivo* (Harwaldt *et al.*, 2002; Liebau *et al.*, 2002). Furthermore, it has been shown that chloroquine inhibits hemin catabolism, leading to intracellular hemin accumulation (Famin *et al.*, 1999). It might thus be speculated that PfGST inhibitors act synergistically with chloroquine.

As delineated previously, PfGST represents a novel GST-isoform, and its H-site differs significantly from the ones of the human counterparts (Fritz-Wolf *et al.*, 2003). In contrast to all other GSTs, PfGST contains only five residues following helix $\alpha 8$, which is too short to form a wall (mu- or pi-class) or an α-helix (alpha-class). This leads to a more solvent-accessible H-site in PfGST than in the other classes, which suggests that the substrate spectrum of PfGST is broader, includes amphiphilic compounds, and is accessible to amphiphilic inhibitors that are not able to enter the H-site of the human isoforms. In the structure reported by Fritz-Wolf *et al.* (2003), the region of the H-site is occupied by the peptide segment Asn B112 to Thr B121 of the non-crystallographically related monomer B. The corresponding synthesized undecapeptide Asn-Asn-Thr-Asn-Leu-Phe-Lys-Asn-Asn-Ala-Thr inhibits PfGST with an IC_{50} of 300 μM. This inhibition was competitive with GSH, the calculated K_i being 115 μM. For human placenta GST, the IC_{50} value was found to be >1 mM. These data provide valuable hints for interactions of inhibitors with H-site residues, which are currently exploited for the design of more potent inhibitors.

Apart from peptide inhibitors, hemin and protoporphyrin IX ($K_i = 10$ μM), cibacron blue ($K_i = 0.5$ μM), ethacrynic acid ($IC_{50} = 30$ μM), and S-hexylglutathione ($K_i = 35$ μM) have been characterized as PfGST inhibitors (Harwaldt *et al.*, 2002). The x-ray structure of PfGST in complex with *S*-hexyl-glutathione has been reported by Perbandt *et al.* (2004). Because PfGST has been discussed over the past years as a potential target of clinically used antimalarial drugs, the effects of chloroquine, primaquine, quinine, methylene blue, and artemisinin on PfGST were tested at a final concentration of 50 μM in the CDNB assay. However, under these conditions, only chloroquine inhibited PfGST by 20% (Harwaldt et al., 2002). In partial contrast to these data obtained for the isolated enzyme, Srivastava et al. (1999) reported a stage-dependent inhibition of PfGST in parasite extracts by 20 μM chloroquine (59–100% inhibition), artemisinin (20–41%), and primaquine (32–62%) in P. knowlesi. Further studies will have to reveal whether these differences in results are based on artefacts produced by the heterogeneity of the cell extract or on differences between the two Plasmodium species.

As PfGST inhibitors directed against the G-site, we tested—apart from S-hexylglutathione—a number of different S-substituted glutathione derivatives in the CDNB assay (100 mM Hepes, 1 mM EDTA, pH 6.5, 0.5 mM CDNB, 1 mM GSH, 25°). The compounds were active with the following IC_{50} values given in parentheses: S-propylglutathione ($IC_{50} = 500 \mu M$), S-(p-nitrobenzyl)glutathione ($IC_{50} = 170 \mu M$), S-(N-hydroxy-N-phenylcarbamoyl)glutathione ($IC_{50} = 190 \mu M$), S-(N-hydroxy-N-chlorophenylcarbamoyl)glutathione ($IC_{50} = 40 \mu M$), S-(N-hydroxy-N-4-bromophenylcarbamoyl)glutathione ($IC_{50} = 32 \mu M$) (unpublished results).

Further inhibitors that might be considered and exploited for PfGST inhibition studies include the phenoxyacid herbicides, 2,4-dichlorophenoxyacetate and 2,4,5-trichlorophenoxyacetate. These compounds inhibit all known isoenzymes of human liver and erythrocyte GSTs. However, the maximal inhibition and the IC_{50} values vary significantly for different isoenzymes (Singh et al., 1985). Glutathione S-sulfonate (GS-SO$_3^-$) was found to be a potent competitive inhibitor of rat lung GST ($K_i = 9 \mu M$) and human lung tumor cell GST ($K_i = 4 \mu M$) (Leung et al., 1985). Di-n-butyltin dichloride (DBTC) and tricyclohexyltin chloride (TCHTC) (Henninghausen and Merkord, 1985), as well as different metals including mercury, copper (II), and cadmium (Dierickx, 1982; Reddy et al., 1981) have also been found to act as GST inhibitors. Furthermore, γ-L-Glu-L-SerGly (GOH), as well as γ-L-Glu-L-AlaGly (GH) have been designed as dead-end inhibitors of different GSTs with K_i values of 13 and 116 μM, respectively (Chen et al., 1985). Differential inhibition of GSTs has also been reported for quercetin, alizarin, purpurogallin, and ellagic acid (Das et al., 1986). For a review on GST inhibitors, please see Mannervik and Danielson (1988).

The explanation of further PfGST–inhibitor interactions using steady-state kinetics, site-directed mutagenesis, and x-ray crystallography will allow for optimization of G-site and H-site inhibitors. These compounds could then be even coupled, resulting in highly specific double-headed drugs.

Acknowledgments

The authors wish to thank Nicole Hiller and Dr. Karin Fritz-Wolf for helpful discussions and Marina Fischer for her excellent technical assistance. Our studies on PfGST are supported by the Deutsche Forschungsgemeinschaft (DFG) BE 2554/1–1.

References

Angeli, V., Faveeuw, C., Roye, O., Fontaine, J., Teissier, E., Capron, A., Wolowczuk, I., Capron, M., and Trottein, F. (2001). Role of the parasite-derived prostaglandin D2 in the inhibition of epidermal Langerhans cell migration during schistosomiasis infection. *J. Exp. Med.* **193**, 1135–1147.

Becker, K., Rahlfs, S., Nickel, C., and Schirmer, R. H. (2003). Glutathione – functions and metabolism in the malarial parasite *Plasmodium falciparum*. *Biol. Chem.* **384**, 551–566.

Becker, K., Tilley, L., Vennerstrom, J. L., Roberts, D., Rogerson, S., and Ginsburg, H. (2004). Oxidative stress in malaria parasite-infected erythrocytes: Host-parasite interactions. *Int. J. Parasitol.* **34**, 163–189.

Burmeister, C., Perbandt, M., Betzel, Ch., Walter, R. D., and Liebau, E. (2003). Crystallization and preliminary X-ray diffraction studies of the glutathione *S*-transferase from *Plasmodium falciparum*. *Acta Crystallogr. D Biol. Crystallogr.* **59**, 1469–1471.

Chen, W. J., Boehlert, C. C., Rider, K., and Armstrong, R. (1985). Synthesis and characterization of the oxygen and desthio analogues of glutathione as dead-end inhibitors of glutathione *S*-transferase. *Biochem. Biophys. Res. Commun.* **16**, 233–240.

Das, M., Singh, S. V., Mukhatar, H., and Asasthi, I. C. (1986). Differential inhibition of rat and human glutathione *S*-transferase isoenzymes by plant phenols. *Biochem. Biophys. Res. Commun.* **30**, 1170–1176.

Dierickx, P. J. (1982). *In vitro* inhibition of the soluble glutathione *S*-transferases from rat liver by heavy metals. *Enzyme* **27**, 25–32.

Dubois, V. L., Platel, D. F., Pauly, G., and Tribouley-Duret, J. (1995). *Plasmodium berghei*: Implication of intracellular glutathione and its related enzyme in chloroquine resistance *in vivo*. *Exp. Parasitol.* **81**, 117–124.

Eaton, D. L. (2000). Biotransformation enzyme polymorphism and pesticide susceptibility. *Neurotoxicology* **21**, 101–111.

Famin, O., Krugliak, M., and Ginsburg, H. (1999). Kinetics of inhibition of glutathione-mediated degradation of ferriprotoporphyrin IX by antimalarial drugs. *Biochem. Pharmacol.* **58**, 59–68.

Ferreira, I. D., Nogueira, F., Borges, S. T., do Rosario, V. E., and Cravo, P. (2004). Is the expression of genes encoding enzymes of glutathione (GSH) metabolism involved in chloroquine resistance in *Plasmodium chabaudi* parasites? *Mol. Biochem. Parasitol.* **136**, 43–50.

Fritz-Wolf, K., Becker, A., Rahlfs, S., Harwaldt, P., Schirmer, R. H., Kabsch, W., and Becker, K. (2003). X-ray structure of glutathione *S*-transferase from the malarial parasite *Plasmodium falciparum*. *Proc. Natl. Acad. Sci. USA* **100**, 13821–13826.

Greenwood, B., and Mutabingwa, T. (2002). Malaria in 2002. *Nature* **415**, 670–672.

Harwaldt, P., Rahlfs, S., and Becker, K. (2002). Glutathione *S*-transferase of the malarial parasite *Plasmodium falciparum* characterization of a potential drug target. *Biol. Chem.* **383**, 821–830.

Henninghausen, G., and Merkord, J. (1985). Meso-2,3-dimercaptosuccinic acid increases the inhibition of glutathione S-transferase activity from rat liver cytosol supernatants by di-n-butyltin dichloride. *Arch. Toxicol.* **57**, 67–68.

Johnson, K. A., Angelucci, F., Bellelli, A., Herve, M., Fontaine, J., Tsernoglou, D., Capron, A., Trottein, F., and Brunori, M. (2003). Crystal structure of the 28 kDa glutathione S-transferase from *Schistosoma haematobium. Biochemistry* **42**, 10084–10094.

Kampkotter, A., Volkmann, T. E., de Castro, S. H., Leiers, B., Klotz, L. O., Johnson, T. E., Link, C. D., and Henkle-Duhrsen, K. (2003). Functional analysis of the glutathione S-transferase 3 from *Onchocerca volvulus* (Ov-GST-3): A parasite GST confers increased resistance to oxidative stress in *Caenorhabditis elegans. J. Mol. Biol.* **325**, 25–37.

Kissinger, J. C., Brunk, B. P., Crabtree, J., Fraunholz, M. J., Gajria, B., Milgram, A. J., Pearson, D. S., Schug, J., Bahl, A., Diskin, S. J., Ginsburg, H., Grant, G. R., Gupta, D., Labo, P., Li, L., Mailman, M. D., McWeeney, S. K., Whetzel, P., Stoeckert, C. J., and Roos, D. S. (2002). The *Plasmodium* genome database. *Nature* **419**, 490–492.

Klokouzas, A., Tiffert, T., van Schalkwyk, D., Wu, C. P., van Veen, H. W., Barrand, M. A., and Hladky, S. B. (2004). *Plasmodium falciparum* expresses a multidrug resistance-associated protein. *Biochem. Biophys. Res. Commun.* **321**, 197–201.

Leung, K. H., Post, G. B., and Menzel, D. B. (1985). Glutathione S-sulfonate, a sulfur dioxide metabolite, as a competitive inhibitor of glutathione S-transferase, and its reduction by glutathione reductase. *Toxicol. Appl. Pharmacol.* **15**, 388–394.

Liebau, E., Bergmann, B., Campbell, A. M., Teesdale-Spittle, P., Brophy, P. M., Luersen, K., and Walter, R. D. (2002). The glutathione S-transferase from *Plasmodium falciparum. Mol. Biochem. Parasitol.* **124**, 85–90.

Liebau, E., Eckelt, V. H., Wildenburg, G., Teesdale-Spittle, P., Brophy, P. M., Walter, R. D., and Henkle-Duhrsen, K. (1997). Structural and functional analysis of a glutathione S-transferase from *Ascaris suum. Biochem. J.* **324**, 659–666.

Liebau, E., Wildenburg, G., Brophy, P. M., Walter, R. D., and Henkle-Duhrsen, K. (1996). Biochemical analysis, gene structure and localization of the 24 kDa glutathione S-transferase from *Onchocerca volvulus. Mol. Biochem. Parasitol.* **80**, 27–39.

Mannervik, B., and Danielson, U. H. (1988). Glutathione transferases—structure and catalytic activity. *CRC Crit. Rev. Biochem.* **23**, 283–337.

McTigue, M. A., Williams, D. R., and Tainer, J. A. (1995). Crystal structures of a schistosomal drug and vaccine target: Glutathione S-transferase from *Schistosoma japonica* and its complex with the leading antischistosomal drug praziquantel. *J. Mol. Biol.* **246**, 21–27.

Mukanganyama, S., Naik, Y. S., Widersten, M., Mannervik, B., and Hasler, J. A. (2001). Proposed reductive metabolism of artemisinin by glutathione transferases *in vitro. Free Radic. Res.* **35**, 427–434.

Mukanganyama, S., Widersten, M., Naik, Y. S., Mannervik, B., and Hasler, J. A. (2002). Inhibition of glutathione S-transferases by antimalarial drugs: Possible implications for circumventing anticancer drug resistance. *Int. J. Cancer* **97**, 700–705.

Ouaissi, A., Ouaissi, M., and Sereno, D. (2002). Glutathione S-transferases and related proteins from pathogenic human parasites behave as immunomodulatory factors [review]. *Immunol. Lett.* **81**, 159–164.

Perbandt, M., Burmeister, C., Walter, R. D., Betzel, C., and Liebau, E. (2004). Native and inhibited structure of a Mu class-related glutathione S-transferase from *Plasmodium falciparum. J. Biol. Chem.* **279**, 1336–1342.

Platel, D. F., Mangou, F., and Tribouley-Duret, J. (1999). Role of glutathione in the detoxification of ferriprotoporphyrin IX in chloroquine resistant *Plasmodium berghei. Mol. Biochem. Parasitol.* **98**, 215–223.

Prapanthadara, L.-A., Koottathep, S., Promtet, N., Hemingway, J., and Kettermann, A. J. (1996). Purification and characterization of a major glutathione *S*-transferase from the mosquito *Anopheles dirus* (species B). *Insect Biochem. Mol. Biol.* **26,** 277–285.

Ranson, H., Cornel, A. J., Fournier, D., Vaughan, A., Collins, F. H., and Hemingway, J. (1997). Cloning and localization of a glutathione *S*-transferase class I gene from *Anopheles gambiae. J. Biol. Chem.* **272,** 5464–5468.

Rao, U. R., Salinas, G., Mehta, K., and Klei, T. R. (2000). Identification and localization of glutathione S-transferase as a potential target enzyme in *Brugia* species. *Parasitol. Res.* **86,** 908–915.

Reddy, C. C., Scholz, R. W., and Massaro, E. J. (1981). Cadmium, methylmercury, mercury, and lead inhibition of calf liver glutathione *S*-transferase exhibiting selenium-independent glutathione peroxidase activity. *Toxicol. Appl. Pharmacol.* **61,** 460–468.

Rossjohn, J., Feil, S. C., Wilce, M. C., Sexton, J. L., Spithill, T. W., and Parker, M. W. (1997). Crystallization, structural determination and analysis of a novel parasite vaccine candidate: *Fasciola hepatica* glutathione *S*-transferase. *J. Mol. Biol.* **273,** 857–872.

Salinas, A. E., and Wong, M. G. (1999). Glutathione *S*-transferases—a review. *Curr. Med. Chem.* **6,** 279–309.

Schirmer, R. H., Müller, J. G., and Krauth-Siegel, L. (1995). Disulfide-reductase inhibitors as chemotherapeutic agents: The design of drugs for trypanosomiasis and malaria. *Angew. Chem. Int. Ed. Engl.* **34,** 141–154.

Singh, S. V., and Awasthi, Y. C. (1985). Inhibition of human glutathione *S*-transferases by 2,4-dichlorophenoxyacetate (2,4-D) and 2,4,5-trichlorophenoxyacetate (2,4,5-T). *Toxicol. Appl. Pharmacol.* **81,** 328–336.

Smith, D. B., and Johnson, K. S. (1988). Single-step purification of polypeptides expressed in *Escherichia coli* as fusions with glutathione S-transferase. *Gene* **67,** 31–40.

Smith, D. B., Davern, K. M., Board, P. G., Tiu, W. U., Garcia, E. G., and Mitchell, G. F. (1987). Mr 26,000 antigen of *Schistosoma japonicum* recognized by resistant WEHI 129/J mice is a parasite glutathione *S*-transferase. *Proc. Natl. Acad. Sci. USA* **83,** 8703–8707.

Srivastava, P., Puri, S. K., Kamboj, K. K., and Pandey, V. C. (1999). Glutathione-*S*-transferase activity in malarial parasites. *Trop. Med. Int. Health* **4,** 251–254.

van Rossum, A. J., Jefferies, J. R., Rijsewijk, F. A., LaCourse, E. J., Teesdale-Spittle, P., Barrett, J., Tait, A., and Brophy, P. M. (2004). Binding of hematin by a new class of glutathione transferase from the blood-feeding parasitic nematode *Haemonchus contortus. Infect. Immun.* **72,** 2780–2790.

Wildenburg, G., Liebau, E., and Henkle-Duhrsen, K. (1998). *Onchocerca volvulus*: Ultrastructural localization of two glutathione S-transferases. *Exp. Parasitol.* **88,** 34–42.

[16] Optimizing the Heterologous Expression of Glutathione Transferase

By BENGT MANNERVIK

Abstract

The heterologous expression of a protein may be enhanced by silent mutations in the coding region of its corresponding DNA. This simple approach has been successfully used for optimized production of a number of glutathione-linked enzymes. For example, the yield of human glutathione transferase M2-2 was elevated by 140-fold in a clone isolated by immunoscreening of a library of plasmids with randomized synonymous codons in the $5'$-segment of the region encoding the enzyme.

Background

Rapid advances have been made in the study of protein structure and function as a result of the development of recombinant DNA techniques for their heterologous expression (Sørensen and Mortensen, 2005). Recombinant proteins can often be obtained in the high purity required for accurate structural and spectroscopic studies, which, complemented by mutagenesis, can give insights into functionally important building blocks and their relevance to biological activities. Proteins expressible in the bacterium *Escherichia coli* can frequently be produced in large quantities with limited efforts and at reasonable costs. However, an important limitation is that many proteins in their native state are posttranslationally modified in fashions that cannot be reproduced by the bacterial expression system. This applies to, for example, glycosylation, which characterizes many eukaryotic proteins.

Another constraint on bacterial protein production is that heterologous DNA sequences may need decoding that is not optimal for high-level expression. The readout of the genetic code in *E. coli* is translated by the ribosome into 21 amino acids (including selenocysteine, specified by the stop codon UGA [Chambers *et al.*, 1986]). However, with the exception of Met, Trp, and selenocysteine, all amino acids are represented by more than one codon; Ser, Leu, and Arg are represented by six codons each. These synonymous codons for a given amino acid are not used to the same extent in different organisms or even for different proteins in the same genome (Kane, 1995). A rare codon may require a cognate tRNA in

0076-6879/05 $35.00
DOI: 10.1016/S0076-6879(05)01016-5

correspondingly short supply. The codon bias may consequently be detrimental to the heterologous production of some proteins. Incorporation of erroneous amino acids, translational frameshifts, and lack of protein product are possible outcomes resulting from of the presence of rare codons.

To overcome the possible problems associated with rare codons, attempts have been made to synthesize DNA templates containing codons that are used naturally in the host organism for its production of abundant proteins (www.kazusa.or.jp/codon). However, it seems that the occurrence of rare codons may not necessarily be the main limiting factor for high-level expression (Wu et al., 2004). Instead, mRNA instability, interfering secondary RNA structure formation, and unfavorable folding kinetics of the nascent polypeptide chain can be more important problems. Changing the strain of host cells or adjustment of growth conditions, such as temperature, have also been tried to circumvent poor yields and optimize expression.

Optimization of Heterologous Glutathione Transferase Expression in *Escherichia coli*

Choice of Expressed Form of the Recombinant Protein

In the design of the expression strategy, it is important to consider the desired final form of the protein and the purification procedure to be used. Established methods for production of fusion proteins are generally robust and have high success rates. The most frequently used is probably the one based on a plasmid vector containing a GST sequence from *Schistosoma mansoni* (Smith and Johnson, 1988), which is commercially available. Affinity chromatography on immobilized glutathione (Simons and Vander Jagt, 1981) is generally the purification method of choice.

For expression of other GSTs, a fusion with the *Schistosoma* GST may not be desirable. A ubiquitin-fusion system has been developed and applied to the expression of mammalian GSTs (Tan and Board, 1996). For some applications, the ballast of the fusion partner may be retained and even used to advantage for affinity purification and detection. If the native protein is the goal, it can usually be released by selective proteolysis of a designed scissile bond in the fusion protein.

Many proteins can be purified by immobilized metal ion affinity chromatography (Porath *et al.*, 1975) and the use of Fe(II)–IMAC is exemplified with rat GST T2-2 (Jemth *et al.*, 1996). The presence of a sterically accessible cluster of at least two His residues in the protein enables purification by Ni (II)–IMAC (Chaga *et al.*, 1994; Yilmaz *et al.*, 1995). Most applications of the Ni(II)–IMAC technique have been based on introduction of a terminal hexa-His sequence in the protein, and in the case of mammalian GSTs, this modification seems to have a minor effect on the catalytic functions of the enzyme (Gustafsson and Mannervik, 1999).

However, this chapter will focus on the expression of the native protein directly from a plasmid vector in *E. coli*, with the assumption that post-translational modifications are not required. The only well-established modification of GSTs is the N-terminal acetylation of some, but far from all, mature mammalian enzymes (Ålin *et al.*, 1986; Rowe *et al.*, 1997). Many GSTs can be affinity purified using immobilized S-hexylglutathione (Mannervik and Guthenberg, 1981), like glyoxalase I (Aronsson and Mannervik, 1977), or immobilized glutathione (Simons and Vander Jagt, 1981) directly from a particle-free bacterial lysate. In some cases, when the expression level is high, GSTs can be isolated by a single step of ion-exchange chromatography (Gustafsson and Mannervik, 1999).

Optimization of the Codon Usage by Silent Mutations in the 5'-Region of the Coding Sequence

Enhanced expression of several human proteins in *E. coli* has been accomplished by exploiting the degeneracy of the genetic code. The principle is to create a library of DNA sequences with silent mutations in the 5'-end of the coding region and screen for clones with elevated expression levels of the desired protein (e.g., by immunodetection with antibodies directed to the protein). Most amino acids are encoded by more than one codon and by use of synthetic oligodeoxyribonucleotides synonymous codons can be introduced in chosen positions by means of the polymerase chain reaction (PCR). If the number of positions is limited to approximately 10 and rare codons are avoided, the library is not larger than what can be examined without the use of high-throughput screening. Obviously, the amino acid residues in the expressed protein will be the same for all variant sequences, and the differences only come into play at the nucleic acid level, where they may influence transcription and translation. For illustration, the optimization of human GST M2-2 expression in *E. coli* will be described (Johansson *et al.*, 1999). In this case, an improvement of up to 140-fold was noted as compared to expression from the natural DNA sequence.

Construction of a GST M2-2 Expression Library

Alternative nucleotides were introduced in the third ("wobble") position of codons 2, 4–7, and 10–14 of cDNA encoding GST M2-2. To limit the resulting library to a size that could be screened with good coverage, the choice of alternative codons was restricted to those that are used in *E. coli* genes expressed at high levels (Bulmer, 1988). The alternative nucleotides and the corresponding wild-type sequence are shown in Table I.

The library theoretically comprised 3,456 different 5′—end sequences in the coding region. The construction of the library also involved a substitution of the TAG termination codon present in the wild-type cDNA by TAA and the introduction of a supplementary adjacent TGA stop codon. The alteration was made because the efficiency of mRNA translation termination in *E. coli* is strongly influenced by the identity of the stop codon and shows a high preference for UAA (Sharp and Bulmer, 1988). In addition, the translation termination efficiency is further improved if the nucleotide immediately after the stop codon is U (Poole *et al.*, 1995). A second reason for replacing TAG is that some *E. coli* strains used for heterologous expression (such as XL1-Blue) carry the supE mutation, which may cause suppression of the translation termination efficiency of the UAG stop codon.

Construction of a Partially Codon-Randomized cDNA Library Encoding Human GST M2-2

PCR primers were designed on the basis of the previously published human GST M2-2 cDNA sequence (Vorachek *et al.*, 1991). The 5′-primer was constructed to introduce random mutations into silent positions in codons 2, 4–7, and 10–14. The endonuclease recognition site for *Eco*RI (italicized) was included in the degenerate 5′—primer, 5′—AG GAG AGA *GAA TTC* ATG CCK ATG ACY CTB GGY TAY TGG AAC ATY CGY GGY CTB GCD CAT TCC ATC CGC-3′ (where K is G/T, Y is C/T, B is C/G/T, and D is A/G/T). The 3′-primer was designed to replace the GST M2-2 wild-type stop codon (TAG) with two sequential stop codons (TAA TGA) and to insert the endonuclease recognition site for *Sal*I (italicized): 5′—GAG GAG *GTC GAC* TCA TTA CTT GTT GCC CCA GAC AGC CAT-3′. The PCR was performed in a reaction system of 100 μl containing 1 μM of each primer, 1.5 mM MgCl$_2$, 0.2 mM dNTPs, and 10 units/ml Taq DNA polymerase (Boehringer Mannheim, Mannheim, Germany). The DNA template was 10 ng/ml pGEM-3Zf(+)GSTM2, containing GST M2-2 cDNA that had been previously isolated and cloned from a human substantia nigra cDNA library (Baez *et al.*, 1997). Thirty cycles of the PCR were carried out with denaturation at 94° for 1 min, annealing at 60° for 2 min, and polymerization at 72° for 2 min. The PCR product was subjected to chloroform extraction and ethanol precipitation followed by digestion with the restriction enzymes *Eco*RI and *Sal*I. The library of fragments with variant 5′—end coding regions was purified using Gene-Clean II (BIO 101, La Jolla, CA) and inserted into the similarly digested and purified pKK-D expression vector (Björnestedt *et al.*, 1992)

TABLE I

COMPARISON OF 5'-END SEQUENCES OF WILD-TYPE GST M2-2, MUTANT LIBRARY, AND HIGH-LEVEL-EXPRESSING GST M1-1cDNA WITH THE SEQUENCE OF THE SELECTED HUMAN GST M2-2 CLONE PROVIDING HIGH-LEVEL EXPRESSION IN *ESCHERICHIA COLI*

N-terminal amino acid sequence	Met	Pro	Met	Thr	Leu	Gly	Tyr	Trp	Asn	Ile	Arg	Gly	Leu	Ala
No.	1	2	3	4	5	6	7	8	9	10	11	12	13	14
Wild-type GST M2-2 cDNA	ATG	CCC	ATG	ACA	CTG	GGG	TAC	TGG	AAC	ATC	CGC	GGG	CTG	GCC
Degeneracy introduced in library	--[a]	--G / T	---	--C / T	--C / G / T	--C / T	--C / T	---	---	--C / T	--C / T	--C / T	--C / G / T	--A / G / T
High-level-expressing GST M2-2 cDNA (pKHXhGM2)	---	--T	---	--C	--T	--T	--C	---	---	--C	--C	--T	--C	--G
High-level expressing GST M1-1 cDNA[b]	---	--T	---	ATA[c]	--T	--T	--C	---	GAC[c]	--T	--T	--C	--G	--G
Preference in *E. coli*[d]	---	--G	---	--C	--G	--T	--C	---	---	--C	--T	--T	--G	--T

[a] A dash indicates nucleotide identical to that of wild-type GST M2-2.

[b] Widersten *et al.* (1996).

[c] In GST M1-1 codons 4 and 9 specify Ile and Asp, respectively.

[d] The codon use most preferred for proteins expressed at high levels (Bulmer, 1988).

using T4 DNA ligase (Boehringer Mannheim). The resulting library was used to transform electrocompetent *E. coli* XL1-Blue cells (Stratagene, La Jolla, CA) by electroporation. Transformed cells were allowed to recover for 1 h at 37° with shaking in 2 ml 2TY medium (1.6 % [w/v] tryptone, 1% [w/v] yeast extract, 0.5% [w/v] NaCl) before being plated onto LB–ampicillin plates (1% [w/v] tryptone, 0.5% [w/v] yeast extract, 1% [w/v] NaCl, 1.5% [w/v] Bacto-agar and 100 mg/ml ampicillin).

Immunoscreening

Individual colonies from the partially randomized cDNA library encoding human GST M2-2 were transferred onto fresh LB plates containing 50 μg/ml ampicillin by dotting with sterile toothpicks. In this way, 94 individual clones could be screened on each plate. Cells transformed with the pKK-D vector lacking an insert and cells transformed with the pKK-D vector containing the unmodified GST M2-2 cDNA were also subcultured on each plate as controls. After overnight growth at 37° the colonies were lifted onto a nitrocellulose membrane presoaked in 2TY medium containing 1 mM isopropyl-β-D-thio-galactopyranoside (IPTG). The membranes were incubated at 37° for 1 h to enable protein expression. The cells were then lysed by placing the membranes on filter paper soaked with 5% sodium dodecyl sulfate (SDS) followed by heating at 100° for 15 min. The expressed proteins were electrophoretically transferred to a nitrocellulose membrane (150 mA for 1 h) as described by Towbin *et al.* (1979). The relative level of human GST M2-2 expression was determined for each colony by means of immunoscreening, essentially as described by Stanley (1983) using antiserum raised in rabbits against human GST M2-2. Nonspecific protein binding was blocked using 3% (w/v) bovine serum albumin (BSA), 0.02% (w/v) sodium azide, and 0.1 mg/ml DNase (Boehringer Mannheim) in phosphate-buffered saline (PBS). After washing in 0.2% (w/v) BSA, 0.02% (w/v) SDS, 0.1% (v/v) Triton X-100, and 0.02% (w/v) thimerosal in PBS, the membranes were incubated with anti-GST M2-2 antiserum. The secondary antibody used was horseradish peroxidase–linked donkey anti-rabbit IgG (Amersham Life Science, Buckinghamshire, UK). Immunoreactive protein was detected with 3,3′-di-aminobenzidine tetrahydrochloride and 0.03% (v/v) H_2O_2 and by using ECL Western blotting detection reagents (Amersham Life Science). The GST M2-2 expression relative to the controls was scored by eye.

Small-Scale Expression and Activity Screening

Single colonies judged to display high expression were subcultured into 2 ml 2TY containing 100 μg/ml ampicillin and grown overnight at 37° with shaking. Overnight cultures were then diluted 200-fold into 20 ml

2TY containing 100 μg/ml ampicillin and allowed to grow at 37° with shaking until reaching the state of exponential growth (OD_{600} ~0.3). Expression was then induced through addition of IPTG (0.2 mM final concentration), and the cells were further grown overnight with shaking at 37°. All subsequent procedures were carried out at 4°. Cells were harvested by centrifugation at 5000g for 5 min, and the bacterial pellet was resuspended in 0.1 M sodium phosphate buffer, pH 7.0, containing 0.02% (w/v) sodium azide and 50 mM phenylmethanesulfonyl fluoride (PMSF). Cells were lysed by means of sonication and insoluble debris pelleted by centrifugation at 8,000g for 20 min. Lysates were assayed for GST activity in 0.1 M sodium phosphate, pH 6.5, at 30° using 1 mM glutathione (GSH) and 1 mM 1-chloro-2,4-dinitrobenzene (CDNB) (Habig et al., 1974).

Large-Scale Expression and Purification

The clone that was shown by small-scale expression and immuno-screening to express the highest yield of GST M2-2 was given the name pKHXhGM2 (pKK-D High-level–expressing human GST M2-2). This clone was used for large-scale expression and purification. GST M2-2 was expressed overnight from pKHXhGM2 in a 500-ml culture as described for the small-scale expression. The cells were harvested by centrifugation at 5000g for 10 min, and the bacterial pellet was resuspended in 10 mM Tris-HCl, pH 7.8, containing 50 mM PMSF. The cells were lysed by sonication. Insoluble cell debris was pelleted by means of centrifugation at 12,000g for 20 min and the lysate collected. The remaining pellet was resuspended and subjected to a further two rounds of sonication and centrifugation, and the three resulting supernatant fractions were pooled. GST M2-2 was separated from other cytosolic proteins using affinity chro-matography on glutathione Sepharose, prepared as described by Simons and Vander Jagt (1981). (The glutathione-affinity matrix is also available commercially.) The pooled lysate was mixed with the affinity matrix in a glass beaker, and protein binding was allowed to proceed with gentle agitation for 2 h. The gel was then poured into a chromatography column, and nonspecifically bound proteins were removed by excessive washing with 10 mM Tris-HCl, pH 7.8, containing 0.2 M NaCl. Bound GST M2-2 was eluted with 50 mM glycine-NaOH, pH 10.0, and the eluate neutra-lized by addition to 1:10 volume of 2 M Tris-HCl, pH 7.2. The purified enzyme was stored at 4° in the presence of 50 mM PMSF and 0.02% (w/v) sodium azide. The purification of GST M2-2 was monitored by enzyme activity measurements.

Results of the Chosen Example

High-Level Expression Clones

Approximately 1600 individual clones from the cDNA library were subjected to immunoscreening to identify the clones that displayed the highest expression level of GST M2-2. Twenty clones showing high-level expression of the enzyme compared with controls were selected for further analysis of the expression level. The crude lysates from small-scale expressions of these clones were screened for enzyme activity as measured with GSH and CDNB. Activities per milliliter of the crude lysates from the 20 selected clones relative to the activity of the lysate obtained from cells that contained the original GST M2-2 cDNA construct are shown in Fig. 1. Nineteen of the 20 selected clones exhibited higher specific activity than the unmodified cDNA clone of GST M2-2. The relative GST activity in the crude lysates of the investigated clones varied from 1–146. However, most of the selected clones displayed relative activity values that correspond to a 1- to 30-fold improvement of the expression level. The purification of GST M2-2 expressed from pKHXhGM2 could be made with a 90% recovery from the bacterial lysates and a total yield of approximately 190 mg/l of bacterial culture. This corresponds to 12% of the total amount of soluble protein in the cells.

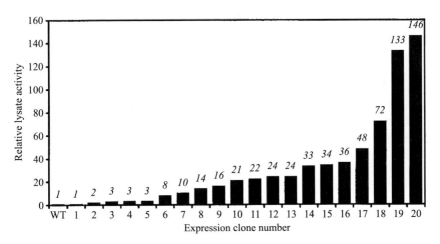

FIG. 1. Relative enzyme activities of 20 silent-mutation clones showing increased GSTM2-2 expression in bacterial lysates. (From Johansson *et al.*, 1999, with permission).

Nucleotide Sequence Analysis of the High-Level-Expressing Clone

The DNA sequence of the high-level-expressing GST M2-2 cDNA was determined to enable comparison of the codon use with the original cDNA clone. It was also confirmed that no unwanted mutations had been introduced during the construction of the library. The sequence of the selected cDNA clone, pKHXhGM2, is shown in Table I. The third nucleotide in the codons of the modified region was altered in 7 of 10 positions in the high-level-expressing GST M2-2 construct. However, five of these alterations were inevitable, because the wild-type codon was not present in the primer that was used to construct the library. The selected codons at positions 4, 6, 7, 10, and 12 in the pKHXhGM2 clone are the most frequently used codons in highly biased genes in *E. coli* (Bulmer, 1988). These selections contrast with those at positions 5 and 13, which encode leucine. In *E. coli*, the GST M2-2 wild-type codon CTG is normally strongly preferred, but in the high-level-expressing clone two different codons, which are significantly less frequent in highly biased genes in *E. coli*, were observed. Also, codon 2, which encodes proline, is less frequently used in proteins expressed at high levels in *E. coli*. These unexpected selections all entail pyrimidine bases.

General Comments

Many factors influence the efficiency of recombinant protein production, and suitable optimization may enhance the level of gene expression. Vector design as well as stability of the expressed protein should be considered (Bass and Yang, 1998; Makrides, 1996). Heterologous protein expression is also influenced by codon use, as demonstrated here with cDNA encoding human GST M2-2. A comparison of the selected codons in pKHXhGM2 and the codon preference in *E. coli* (Table I) shows, in agreement with previous findings, that the presence only of the most preferred codon in *E. coli* does not necessarily produce increased or maximal efficiency of protein expression (Makrides, 1996). Analysis of a cDNA clone that was obtained in a similar manner for high-level expression of GST M1-1 (Widersten *et al.*, 1996), an enzyme closely related to GST M2-2, shows that identical codons have been selected in the first four of the nine common codons that were subjected to modification (Table I). Furthermore, these findings together with those of Widersten *et al.* (1996) indicate that a high pyrimidine (C/T) content in the 5'—end of the DNA encoding human Mu class GSTs increases the efficiency of protein expression. It would seem that translation efficiency is strongly influenced by the codon use in the region immediately after the start codon.

This highlights the problems faced when trying to engineer high-level–expressing clones based solely on the established codon bias in the host cells used for expression (Kolm *et al.*, 1995). Thus, the approach of introducing random silent mutations into nucleotide sequences may often prove more successful for obtaining a more efficient expression clone. An early example of this approach was the optimizing of human glutathione reductase expression in *E. coli* (Bücheler *et al.*, 1990). Recently, an extensive study of a large number of mRNA sequences essentially confirmed the importance of the 5'-codon region for the translation efficiency (Voges *et al.*, 2004).

By use of the same general approach of creating libraries of expression clones with silent mutations in the 5'-end of the coding region, the expression of several GSTs (Johansson *et al.*, 1999; Nilsson and Mannervik, 2001; Widersten *et al.*, 1996), glyoxalase I (Ridderström and Mannervik, 1996), glyoxalase II (Ridderström *et al.*, 1996), glutathione reductase (Jiang and Mannervik, 1999), and other proteins (Wisén *et al.*, 1999) has been improved in our laboratory. It should be emphasized that despite the mutations introduced into the DNA, the amino acid sequence of the protein is identical to the wild-type sequence. The approach is relatively simple and efficient for increasing heterologous protein production and is limited only by the need for a method for specific screening of the protein expression.

Acknowledgments

This work was funded by grants from the Swedish Cancer Society and the Swedish Research Council. The work on GST M2-2 was carried out in our laboratory by Drs. Robyn Bolton-Grob and Ann-Sofie Johansson.

References

Ålin, P., Mannervik, B., and Jörnvall, H. (1986). Cytosolic rat liver glutathione transferase 4-4. Primary structure of the protein reveals extensive differences between homologous glutathione transferases of classes Alpha and Mu. *Eur. J. Biochem.* **156,** 343–350.

Aronsson, A.-C., and Mannervik, B. (1977). Characterization of glyoxalase I purified from pig erythrocytes by affinity chromatography. *Biochem. J.* **165,** 503–509.

Baez, S., Segura-Aguilar, J., Widersten, M., Johansson, A.-S., and Mannervik, B. (1997). Glutathione transferases catalyse the detoxication of oxidized metabolites (o-quinones) of catecholamines and may serve as an antioxidant system preventing degenerative cellular processes. *Biochem. J.* **324,** 25–28.

Bass, S., and Yang, M. (1998). Expressing cloned genes in *Escherichia coli.* *In* "Protein Function. A Practical Approach" (T.E. Creighton, ed.), pp. 29–55. IRL Press, Oxford.

Björnestedt, R., Widersten, M., Board, P. G., and Mannervik, B. (1992). Design of two chimaeric human–rat class Alpha glutathione transferases for probing the contribution of C-terminal segments of protein structure to the catalytic properties. *Biochem. J.* **282,** 505–510.

Bücheler, U. S., Werner, D., and Schirmer, R. H. (1990). Random silent mutagenesis in the initial triplets of the coding region: A technique for adapting human glutathione reductase-encoding cDNA to expression *in Escherichia coli. Gene* **96,** 271–276.

Bulmer, M. (1988). Are codon usage patterns in unicellular organisms determined by selection–mutation balance? *J. Evol. Biol.* **1**, 15–26.

Chaga, G., Widersten, M., Andersson, L., Porath, J., Danielson, U. H., and Mannervik, B. (1994). Engineering of a metal coordinating site into human glutathione transferase M1-1 based on immobilized metal ion affinity chromatography of homologous rat enzymes. *Protein Engineer* **7**, 1115–1119.

Chambers, I., Frampton, J., Goldfarb, P., Affara, N., McBain, W., and Harrison, P. R. (1986). The structure of the mouse glutathione peroxidase gene: The selenocysteine in the active site is encoded by the 'termination' codon, TGA. *EMBO J.* **5**, 1221–1227.

Gustafsson, A., and Mannervik, B. (1999). Benzoic acid derivatives induce recovery of catalytic activity in the partially inactive Met208Lys mutant of human glutathione transferase A1-1. *J. Mol. Biol.* **288**, 787–800.

Habig, W. H., Pabst, M. J., and Jakoby, W. B. (1974). Glutathione S-transferases. The first enzymatic step in mercapturic acid formation. *J. Biol. Chem.* **249**, 7130–7139.

Jemth, P., Stenberg, G., Chaga, G., and Mannervik, B. (1996). Heterologous expression, purification and characterization of rat class Theta glutathione transferase T2-2. *Biochem. J.* **316**, 131–136.

Jiang, F., and Mannervik, B. (1999). Optimized heterologous expression of glutathione reductase from cyanobacterium Anabaena PCC 7120 and characterization of the recombinant protein. *Protein Expression Purif.* **15**, 92–98.

Johansson, A.-S., Bolton-Grob, R., and Mannervik, B. (1999). Use of silent mutations in cDNA encoding human glutathione transferase M2-2 for optimized expression in *Escherichia coli*. *Protein Expression Purif.* **17**, 105–112.

Kane, J. F. (1995). Effects of rare codon clusters on high-level expression of heterologous proteins in *Escherichia coli*. *Curr. Opin. Biotechnol.* **6**, 494–500.

Kolm, R. H., Stenberg, G., Widersten, M., and Mannervik, B. (1995). High-level bacterial expression of human glutathione transferase P1-1 encoded by semisynthetic DNA. *Protein Expression Purif.* **6**, 265–271.

Mannervik, B., and Guthenberg, C. (1981). Glutathione transferase (human placenta). *Meth. Enzymol.* **77**, 231–235.

Makrides, S. C. (1996). Strategies for achieving high-level expression of genes in *Escherichia coli*. *Microbiol. Rev.* **60**, 512–538.

Nilsson, L. O., and Mannervik, B. (2001). Improved heterologous expression of human glutathione transferase A4-4 by random silent mutagenesis of codons in the 5′-region. *Biochim. Biophys. Acta* **1528**, 101–106.

Poole, E. S., Brown, C. M., and Tate, W. P. (1995). The identity of the base following the stop codon determines the efficiency of *in vivo* translational termination in *Escherichia coli*. *EMBO J.* **14**, 151–158.

Porath, J., Carlsson, J., Olsson, I., and Belfrage, G. (1975). Metal chelate affinity chromatography, a new approach to protein fractionation. *Nature* **258**, 598–599.

Ridderström, M., and Mannervik, B. (1996). Optimized heterologous expression of the human zinc enzyme glyoxalase I. *Biochem. J.* **314**, 463–467.

Ridderström, M., Saccucci, F., Hellman, U., Bergman, T., Principato, G., and Mannervik, B. (1996). Molecular cloning, heterologous expression and characterization of human glyoxalase II. *J. Biol. Chem.* **271**, 319–323.

Rowe, J.D, Nieves, E., and Listowsky, I. (1997). Subunit diversity and tissue distribution of human glutathione S-transferases: Interpretations based on electrospray ionization-MS and peptide sequence-specific antisera. *Biochem. J.* **325**, 481–486.

Sharp, P. M., and Bulmer, M. (1988). Selective differences among translation termination codons. *Gene* **63**, 141–145.

Simons, P. C., and Vander Jagt, D. L. (1981). Purification of glutathione S-transferases by glutathione affinity chromatography. *Methods Enzymol.* **77,** 235–237.

Smith, D. B., and Johnson, K. S. (1988). Single-step purification of polypeptides expressed in *Escherichia coli* as fusions with glutathione S-transferase. *Gene* **67,** 31–40.

Sørensen, H. P., and Mortensen, K. K. (2005). Advanced genetic strategies for recombinant protein expression in *Escherichia coli. J. Biotechnol.* **115,** 113–128.

Stanley, K. K. (1983). Solubilization and immune-detection of β-galactosidase hybrid proteins carrying foreign antigenic determinants. *Nucleic Acids Res.* **11,** 4077–4092.

Tan, K. L., and Board, P. G. (1996). Purification and characterization of a recombinant human Theta-class glutathione transferase (GSTT2-2). *Biochem. J.* **315,** 727–732.

Towbin, H., Staehelín, T., and Gordon, T. (1979). Electrophoretic transfer of proteins from polyacrylamide gels to mitocellulose sheets: Procedure and some applications. *Proc. Natl. Acad. Sci. USA* **76,** 4350–4354.

Voges, D., Watzele, M., Nemetz, C., Wizemann, S., and Buchberger, B. (2004). Analyzing and enhancing mRNA translation efficiency in an *Escherichia coli in vitro* expression system. *Biochem. Biophys. Res. Commun.* **318,** 601–614.

Vorachek, W. R., Pearson, W. R., and Rule, G. S. (1991). Cloning, expression, and characterization of a class-Mu glutathione transferase from human muscle, the product of the GST4 locus. *Proc. Natl. Acad. Sci. USA* **88,** 4443–4447.

Widersten, M., Huang, M., and Mannervik, B. (1996). Optimized heterologous expression of the polymorphic human glutathione transferase M1-1 based on silent mutations in the corresponding cDNA. *Protein Expression Purif.* **7,** 367–372.

Wisén, S., Jiang, F., Bergman, B., and Mannervik, B. (1999). Expression and purification of the transcription factor NtcA from the cyanobacterium Anabaena PCC 7120. *Protein Expression Purif.* **17,** 351–357.

Wu, X., Jörnvall, H., *et al.* (2004). Codon optimization reveals critical factors for high level expression of two rare codon genes in *Escherichia coli*: RNA stability and secondary structure but not tRNA abundance. *Biochem. Biophys. Res. Commun.* **313,** 89–96.

Yilmaz, S., Widersten, M., Emahazion, T., and Mannervik, B. (1995). Generation of a Ni(II) binding site by introduction of a histidine cluster in the structure of human glutathione transferase A1-1. *Prot. Engineer* **8,** 1163–1169.

[17] Human Glutathione Transferase A3-3 Active as Steroid Double-Bond Isomerase

By FRANÇOISE RAFFALLI-MATHIEU and BENGT MANNERVIK

Abstract

Glutathione transferases (GSTs) constitute a superfamily of detoxifying enzymes with a major role in protecting cellular macromolecules from reactive electrophilic compounds. A growing body of evidence suggests, however, that at least certain glutathione transferases are involved in other essential cellular processes, such as cellular signaling and anabolic

METHODS IN ENZYMOLOGY, VOL. 401   0076-6879/05 $35.00
Copyright 2005, Elsevier Inc. All rights reserved.   DOI: 10.1016/S0076-6879(05)01017-7

pathways. One of them is the human GST A3-3, which is selectively expressed in steroidogenic organs and which efficiently catalyzes the obligatory isomerization of the Δ^5-ketosteroid precursors in the biosynthesis of progesterone and testosterone. In this chapter, we summarize the current knowledge on human GST A3-3 and describe methods for heterologous expression and functional characterization of the enzyme.

Background

Biosynthesis of Steroid Hormones

All steroid hormones in humans originate from cholesterol. The first and rate-limiting step of steroid hormone biosynthesis is the transfer of cholesterol from the outer to the inner mitochondrial membrane in a process mediated by the steroidogenic acute regulatory (StAR) protein. As depicted in Fig. 1, numerous enzymatic reactions catalyzed by various cytochrome P450 (CYP) enzymes and the 3β- and 17β-steroid dehydrogenases are involved in the multistep conversion of cholesterol to active steroid hormones (Payne and Hales, 2004). Although the individual enzymes catalyzing each step of steroid hormone synthesis have been known for many years, recent studies indicate a role of glutathione transferases in the biosynthetic pathway of testosterone and progesterone (Johansson and Mannervik, 2001, 2002), as described in this chapter.

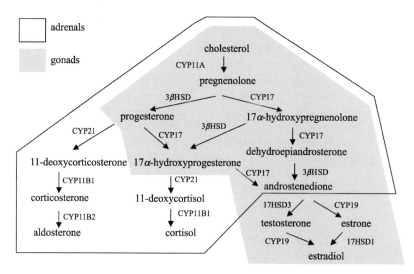

FIG. 1. Biosynthesis of steroid hormones in major steroidogenic organs, the gonads (gray shading) and the adrenals (boxed). Modified from Payne and Hales (2004).

Glutathione Transferases

Glutathione transferases (GSTs) are classically considered contributors to the detoxification processes. They form a superfamily of enzymes catalyzing the nucleophilic attack of the tripeptide glutathione (γ-Glu-Cys-Gly) on electrophilic compounds, thereby increasing their water solubility and facilitating their elimination from the organism. Besides, GSTs have been shown to participate in other biological processes, such as cellular signaling, intracellular binding of various organic molecules, and synthesis of endogenous molecules (Hayes et al., 2005).

Discovery of the Isomerase Activity of Certain Glutathione Transferases

A glutathione-stimulated ketosteroid isomerase activity was discovered in a rat liver cytosol fraction by Benson and Talalay (1976), who later demonstrated that this activity was linked to one of the major soluble GSTs (Benson et al., 1977). Whereas GSTA1-1 was the first human GST identified as an efficient Δ^5-Δ^4-steroid double-bond isomerase (Pettersson and Mannervik, 2001), another member of the Alpha class, GST A3-3, was subsequently found to exhibit 20 times higher catalytic efficiency than GST A1-1 toward Δ^5-ketosteroids (Johansson and Mannervik, 2001). GST A3-3 is currently the most efficient Δ^5-Δ^4-steroid isomerase known in human tissues (Fig. 2).

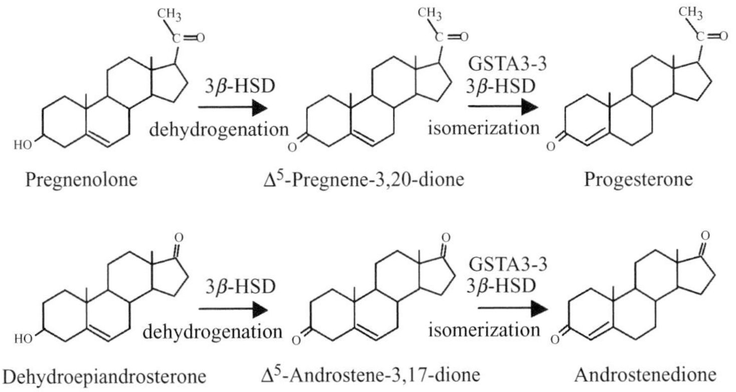

FIG. 2. Double-bond isomerizations catalyzed by human GST A3-3. The sequence of reactions shown is classically attributed to the dehydrogenase and isomerase activities of the 3β-hydroxysteroid dehydrogenase (3β-HSD). Modified from Payne and Hales (2004).

GST A3-3

Although the human *GSTA3* gene was described in 1993 (Suzuki *et al.*, 1993), a full-length cDNA was isolated from a placental cDNA library in 2001 (Johansson and Mannervik, 2001).

GSTA3 *Gene*

The gene is localized on chromosome 6, within the GST Alpha gene cluster, which consists of five genes and seven pseudogenes (Morel *et al.*, 2002) (Fig. 3). The *GSTA3* gene comprises seven exons, spanning 13 kb.

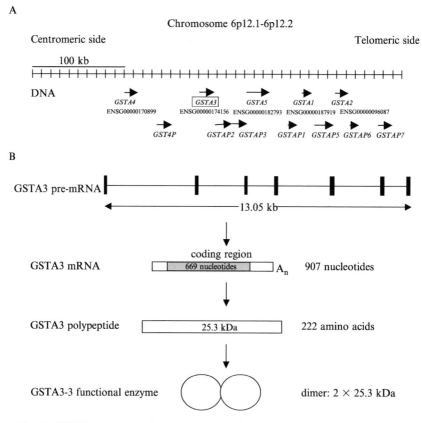

FIG. 3. GSTA3 gene, transcript and protein. (A) Organization of the GST Alpha gene cluster on human chromosome 6. Five genes (GSTA1→GSTA5; Ensemble data base reference numbers indicated) and seven pseudogenes (GSTA1P→GSTA7P) are present. Arrows indicate length and orientation of the genes and pseudogenes. Expression of GSTA5 has not been demonstrated. (B) Structure of the GSTA3 pre-mRNA, mRNA, and protein. Exons are shown as black boxes, introns as a line.

Expression In Vivo

Tissue-specific Expression. Reverse transcriptase-polymerase chain reaction (RT-PCR) analysis (Johansson and Mannervik, 2001; Morel *et al.*, 2002) reveals that *GSTA3* expression is restricted mainly to steroidogenic tissues (adrenal gland, ovary, placenta, and testis, see Fig. 4). An immuno-histochemical study shows that Alpha-class GSTs are found in the steroidogenic cells of the human ovary (Tiltman and Haffajee, 1999). Even though GST A3-3 cannot be distinguished from other Alpha-class GSTs by the immunohistochemical methods used, the observation that Alpha-GSTs proteins are confined to steroidogenic cells of the ovary, together with the specific detection of GSTA3 message in the ovary and other organs producing steroid hormones, strongly supports the notion that GST A3-3 is present predominantly in cells that actively synthesize steroids.

Splice Variants of GSTA3. The *GSTA3* gene is subject to alternative splicing, and shorter splice variants of the transcript have been detected in tissues where the full length GSTA3 mRNA is found (Board, 1998; Johansson and Mannervik, 2001; Morel *et al.*, 2002).

As shown in Fig. 5, splice variant 1 lacks exon 3, which results in a +1 frame shift, leading to a premature stop codon after the 32nd amino acid. A second splice variant lacks both exons 1 and 2, whereas 26 nucleotides apparently unrelated to any known GST sequence are attached to the 5′ end of exon 3. The third variant has lost exons 2 and 3. Both variants 2 and 3 are short of the start codon used in the full length GSTA3, because it is located within exon 2.

The proportions of the three RNA species vary among different tissues. It is not yet known whether the truncated splice variants are translated or if they have any biological activity, as regulatory transcripts, for instance. It is

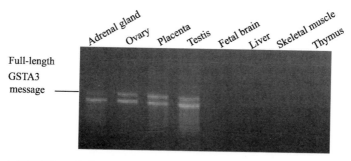

FIG. 4. RT-PCR analysis of the expression of GSTA3 mRNA in steroidogenic tissues. Modified from Johansson and Mannervik (2001).

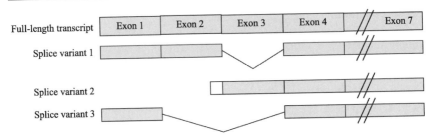

FIG. 5. GSTA3 splice variants. Variant 1 has lost exon 3. Variant 2 has lost exons 1 and 2 and starts with 26 nucleotides not coding for any known GST sequence (white box). Variant 3 has lost exons 2 and 3.

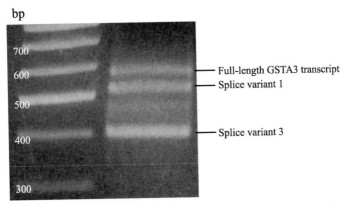

FIG. 6. Electrophoresis on 1.5% agarose gel of RT-PCR products (visualized under UV light after ethidium bromide staining) obtained using the primers described (Morel *et al.*, 2002). The full-length GSTA3 transcript and splice variants 1 and 3 are detected in JEG-3 choriocarcinoma cells.

noteworthy that in certain tissues, such as adrenal gland and testis, splice variant 1 seems to be expressed at even higher levels than the full-length transcript.

The full-length GSTA3 transcript and variants 1 and 3 can be detected by RT-PCR (Fig. 6).

DNase-digested total RNA (1.5 μg) is reverse-transcribed using the AMV reverse transcriptase, and 5 μl of the cDNA is used in a PCR reaction as described by Morel *et al.* (2002).

The primers used are the following:

Sense: 5′-CGGAGACCGGCTAGACTTTA-3′ corresponding to nucleotide assignment 34–53 (GeneBank NM_000847). Antisense:

5'-TGGAGTCAAGCTCTTCCACA-3' corresponding to nucleotide assignment 577–596 (GeneBank NM_000847).

The PCR cycles are as follows: 94°, 2 min, followed by 35 cycles consisting of denaturation 30 s at 94°, annealing 30 s at 56°, extension 45 s at 72°. Extension of the amplicons is completed by 8 min at 72°.

GST A3-3 Enzyme

Cloning of Human GST A3-3. Two procedures of cloning and expression of the recombinant GST A3-3 enzyme are described in the following. The first one details cloning into the expression vector pKK-D (Björnestedt *et al.*, 1992), which is robust and easy to use. However, this vector may produce relatively smaller amounts of protein compared with the expression vector pET-21(a)+ (Novagen) used in cloning procedure II. The pET construct is less stable but usually affords higher yields of recombinant protein.

CLONING PROCEDURE I. Gene Pool cDNA derived from normal human placenta (Invitrogen) is used to specifically amplify the coding region of GSTA3 by PCR performed with the *Pfu* DNA polymerase. The PCR primers are as follows:
Sense:

start
codon

5'-AATAATGAATTCATGGCAGGGAAGCCAAGCTT-3'

EcoRI site

Nucleotide assignment 80–100
GenBank NM_000847

Antisense:

5'-AATAATGGATCCTTCTTAGCCTCCATGGCTGCT-3'

BamHI site

Nucleotide assignment 752–772
GenBankNM_000847

The PCR cycles are as follows: 94°, 5 min, followed by 35 cycles consisting of denaturation 1 min at 94°, annealing 2 min at 54°, and extension 2 min at 72°. Extension of the amplicons is completed by 7 min at 72°.

The purified PCR product cloned into the pGEM^R-3Z vector (Promega) is propagated in *Escherichia coli* XL1-Blue cells (Stratagene). The absence of mutations is checked by sequencing.

Subcloning into the pKK-D expression vector: An undesired *Sal*I restriction site within GSTA3 cDNA is removed through an A→T silent mutation (boxed) by inverted PCR, using the following primers:

Sense: 5′-CCTTATGTCG[T]CCTGAGAAAAAGAT-3′

Antisense: 5′-GCAGAAGAAGGATCATTTCATTCAAAT-3′

PCR protocol: 95°, 10 min, followed by 35 cycles consisting of denaturation 1 min at 95°, annealing 1 min at 54°, extension 10 min at 72°. Extension of the amplicons is completed by 30 min at 72°.

The resulting mutated cDNA is subcloned into the expression vector pKK-D. The entire coding region of the resulting expression clone pKK-DGSTA3 is sequenced.

CLONING PROCEDURE II. GSTA3 cDNA is amplified from the pKK-DGSTA3 clone by PCR using the following primers:

Sense:

start
codon

5′-ATATGAATTCATATGGCAGG[T]AAGCCCAAGCTTCAC-3′.
       *Eco*RI   *Nde*I sites

This primer introduces a silent G → T mutation (boxed) in the third codon to avoid a seldom-used codon in *E. coli*.

Antisense:

5′AATAAT**GTCGAC**TTGTTAGCCTGGATGGCTGCT-3′.
         *Sal*I site

The PCR product is subcloned into the pGEM expression vector using *Eco*RI and *Sal*I. After sequencing and propagation, the GSTA3 cDNA is subcloned into the pET-21a(+) expression vector.

NOTE: In this second procedure, amplification of the GSTA3 coding sequence is performed using the pKK-D clone. However, using these primers GSTA3 may be amplified directly from a cDNA library.

Expression of Recombinant GST A3-3

Using the pKK-D Plasmid. *E. coli* XL-1 Blue cells carrying the pKK-DGSTA3 plasmid are grown in 2TY medium at 37° overnight and diluted 200 times before growing to an OD_{600} of 0.35. IPTG (0.2 mM) is then used to induce protein expression for 16 h. After ultrasonication of the cellular suspension, GST A3-3 is purified from the lysate by affinity chromatography (Simons and Vander Jagt, 1981) using glutathione-Sepharose (GE Healthcare). Elution is performed using 10 mM glutathione.

Normally, 2.5 mg pure protein is obtained from 1 l of bacterial culture. *Using the pET-21a(+) Plasmid. E. coli* BL-21(DE3) cells are used to express GST A3-3 cloned into the pET-21a(+) vector. The cells are grown to an OD_{600} of 0.7, and expression is induced by IPTG (1 mM) for 4 h. After ultrasonication of the cells, the lysate is desalted on a PD-10 gel filtration column (GE Healthcare). The proteins are eluted in 20 mM sodium phosphate, pH 7.0, and subsequently loaded on a Hi-Trap SP cation exchanger (GE Healthcare). The proteins are eluted using a salt gradient. The yield is typically 100 mg pure protein from 1 l culture.

Protein purity is determined by sodium dodecyl sulfate–polyacrylamide gel electrophoresis (SDS-PAGE) followed by Coomassie Brilliant Blue staining.

Measurement of Enzymatic Activities

Classical GST Substrates. Functional characterization of GSTs is achieved by measuring the catalytic activities of the enzymes toward a range of standard substrates, for which convenient spectrophotometric assays are available (Mannervik and Widersten, 1995). The specific activities of GST A3-3 determined with some of these substrates are compared with those of the other Alpha-class GSTs (Table I).

Steroid Substrates. The heterologously expressed GST A3-3, besides acting on typical GST substrates, possesses an exceptionally high steroid

TABLE I
SPECIFIC ACTIVITIES OF HUMAN ALPHA GSTs TOWARD TYPICAL GST SUBSTRATES

Substrate	Substrate mM	GSH mM	λ nm	Extinction coefficient mM^{-1}.cm^{-1}	Specific activity μmol.mg^{-1}.min^{-1} A3-3	A1-1	A2-2	A4-4
1-Chloro-2,4-dinitrobenzene	1	1	340	9.6	23	80	80	7.5
Nonenal	0.1	0.5	225	13.75	2.2	0.8	nd	205
Ethacrynic acid	0.2	0.25	270	5	0.17	0.2	0.1	1.9
Phenethyliso-thiocyanate	0.1	1	274	8.89	4.1	1.7	nd	0.2
Sulforaphane	0.4	1	274	8.0	4.0	1.9	nd	nd
Cumene hydroperoxide	1.5	1	340	−6.2	2.6	10	10	1

All measurements were performed at 30°, with 0.1 M sodium phosphate buffer. From Johansson and Mannervik (2001) Mannervik and Widersten (1995).
nd, not determined.

TABLE II
SPECIFIC ACTIVITIES OF HUMAN ALPHA GSTs TOWARD STEROID SUBSTRATES

Substrate	Substrate mM	GSH mM	λ nm	Extinction coefficient mM^{-1}.cm^{-1}	Specific activity μmol.mg^{-1}.min^{-1}			
					A3-3	A1-1	A2-2	A4-4
Δ^{5}androstene-3,17-dione	0.1	1	248	16.3	197	40	0.2	0.03
Δ^5-pregnene-3,20-dione	0.01	1	248	17	37	3.2	nd	nd

Measurements were performed at 30°, in PBS buffer. From Johansson and Mannervik (2001) and Mannervik and Widersten (1995).
nd, not determined.

double-bond isomerase activity with Δ^5-pregnene-3,20-dione and Δ^5-androstene-3,17-dione. These steroids are intermediates in the biosynthetic pathways of progesterone and testosterone, respectively, as presented in Fig. 2. These 3-oxosteroids are formed from corresponding 3-hydroxysteroids by the catalytic action of 3β-hydroxysteroid dehydrogenase (3βHSD) (Fig. 2). The dehydrogenase has an associated double-bond isomerase activity, but GST A3-3 displays a catalytic efficiency (k_{cat}/Km) up to 230 times higher than that of the 3β-hydroxysteroid dehydrogenase (3βHSD) toward the precursors of testosterone and progesterone (Table II).

Amino Acid Residues Contributing to High Steroid Isomerase Activity

Five amino acids of GST A3-3 (Fig. 7 and Table III) have been shown by site-directed mutagenesis to be crucial to the high steroid isomerase activity (Johansson and Mannervik, 2002; Pettersson et al., 2002).

Possible Reaction Mechanism

A reaction mechanism, in which the thiolate of glutathione serves as a base, was proposed for GST A3-3 on the basis of the kinetics and pH dependence of the double-bond isomerization of Δ^5-androstene-3,17-dione (Fig. 8).

The crystal structure of GST A3-3 has recently been determined in the absence of a ligand (Gu et al., 2004). Modeling of the ternary complex enzyme/glutathione/Δ^5-androstenedione supports the role of glutathione as a base in the catalytic mechanism deduced from the kinetic studies.

MAGKPKLHY**F**[10] N**G**[12]RGRMEPIR WLLAAAGVEF EEKFIGSAED LGKLRNDGSL MFQQVPMVEI
DGMKLVQTRA ILNYIASKYN LYGKDIKERA LIDMYTEGMA DLNEMILLLP **L**[111]CRPEEKDAK
IALIKEKTKS RYFPAFEKVL QSHGQDYLVG NKLSRADISL VELLYYVEEL DSSLISNFPL
LKALKTRISN LPTVKKFLQP GSPRKPP **A**[208]DA KALEE **A**[216]RKIF RF

FIG. 7. Amino acid sequence of the human GST A3–3 enzyme. Amino acids important for high steroid isomerase activity are indicated with their respective positions.

TABLE III

AMINO ACIDS IMPORTANT FOR HIGH STEROID ISOMERASE ACTIVITY IN HUMAN GST
A3-3: COMPARISON WITH OTHER HUMAN ALPHA-CLASS GSTs

	Amino acid residue				
GST	10	12	111	208	216
A3-3	F	G	L	A	A
A1-1	F	A	V	M	A
A2-2	S	I	F	M	S
A4-4	P	G	F	P	V

From Johansson and Mannervik (2001).

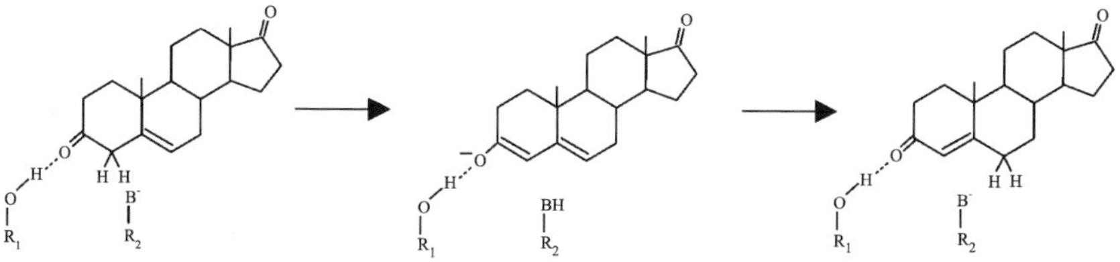

Δ^5-androstenedione   dienolate Δ^4-androstenedione

FIG. 8. Proposed reaction mechanism of the Δ^5-Δ^4 isomerization of the testosterone precursor Δ^5-androstene, 3,17-dione. The base B⁻ is identified with the thiolate of glutathione. Modified from Johansson and Mannervik (2002).

Other Mammalian GSTs with High Sequence Similarity to Human GST A3-3

Three Alpha-class GSTs (GST A1, GST A2, GST A3) have been isolated from a hepatic library from the primate *Macaca fascicularis*. These enzymes show high sequence similarity to the human GST A3-3 protein (Wang *et al.*, 2002).

Two glutathione transferases of the Alpha-class (GST A1, GST A2) are expressed at high levels in bovine ovarian follicles, particularly in

TABLE IV
AMINO ACIDS IMPORTANT FOR HIGH STEROID ISOMERASE ACTIVITY. COMPARISON OF THE
HUMAN GSTA3-3 ENZYME (HA3-3) WITH *MACACA FASCICULARIS* (MFA) AND
BOVINE (B) ALPHA-CLASS ENZYMES

GST	Amino acid residue				
	10	12	111	208	216
hA3-3	F	G	L	A	A
Mfa A1-1	F	A	I	M	A
Mfa A2-2	F	A	I	M	A
Mfa A3-3	F	A	I	M	A
bA1-1	F	G	L	T	A
bA2-2	F	G	L	M	A

From Johansson and Mannervik (2001), Rabahi *et al.* (1999), and Wang *et al.* (2002).

steroidogenically active cells (Rabahi *et al.*, 1999), and these isoforms have been shown to be regulated by gonadotropins. The highest amounts of GSTA1 and GSTA2 transcripts are found in bovine steroidogenic organs and in liver. Interestingly, both enzymes seem to be up-regulated in dominant follicles selected during the estrous cycle (Fayad *et al.*, 2004). Of five amino acids known in human GST A3-3 to be essential for high steroid isomerase activity, four residues are conserved in bovine GST A1 and GST A2 (Table IV).

In porcine ovary, four Alpha-class GSTs have been found. The Alpha class is the most abundantly expressed GST class in porcine corpus lutea. Moreover Alpha-class GSTs are specifically expressed in Leydig and Sertoli cells in porcine testis (Eliasson *et al.*, 1999). It remains to be found whether any of the homologous GSTs show the same high steroid isomerase activity as human GST A3-3.

Acknowledgments

The research from the authors' laboratory was supported by the Swedish Cancer Society and the Swedish Research Council. The original work on GST A3-3 was performed by Ann-Sophie Johansson. We thank Carolina Orre for valuable suggestions on the manuscript.

References

Benson, A. M., and Talalay, P. (1976). Role of reduced glutathione in the Δ^5-3-ketosteroid isomerase reaction of liver. *Biochem. Biophys. Res. Commun.* **69,** 1073–1079.

Benson, A. M., Talalay, P., Keen, J. H., and Jakoby, W. B. (1977). Relationship between the soluble glutathione-dependent Δ^5-3-ketosteroid isomerase and the glutathione S-transferases of the liver. *Proc. Natl. Acad. Sci. USA* **74**, 158–162.

Björnestedt, R., Widersten, M., Board, P. G., and Mannervik, B. (1992). Design of two chimaeric human-rat class Alpha glutathione transferases for probing the contribution of C-terminal segments of protein structure to the catalytic properties. *Biochem. J.* **282**, 505–510.

Board, P. G. (1998). Identification of cDNAs encoding two human Alpha class glutathione transferases (GSTA3 and GSTA4) and the heterologous expression of GSTA4-4. *Biochem. J.* **330**, 827–831.

Eliasson, M., Stark, T., and DePierre, J. W. (1999). Expression of glutathione transferase isoenzymes in the porcine ovary in relationship to follicular maturation and luteinization. *Chem. Biol. Interact.* **117**, 35–48.

Fayad, T., Levesque, V., Sirois, J., Silversides, D. W., and Lussier, J. G. (2004). Gene expression profiling of differentially expressed genes in granulosa cells of bovine dominant follicles using suppression subtractive hybridization. *Biol. Reprod.* **70**, 523–533.

Gu, Y., Guo, J., Pal, A., Pan, S.-S., Zimniak, P., Singh, S. V., and Ji, X. (2004). Crystal structure of human glutathione S-transferase A3-3 and mechanistic implications for its high steroid isomerase activity. *Biochemistry* **43**, 15673–15679.

Hayes, J. D., Flanagan, J. U., and Jowsey, I. R. (2005). Glutathione transferases. *Annu. Rev. Pharmacol. Toxicol.* **45**, 51–88.

Johansson, A.-S., and Mannervik, B. (2001). Human glutathione transferase A3-3, a highly efficient catalyst of double-bond isomerization in the biosynthetic pathway of steroid hormones. *J. Biol. Chem.* **276**, 33061–33065.

Johansson, A.-S., and Mannervik, B. (2002). Active-site residues governing high steroid isomerase activity in human glutathione transferase A3-3. *J. Biol. Chem.* **277**, 16648–16654.

Mannervik, B., and Widersten, M. (1995). Glutathione transferases. *In* "Advances in Drug Metabolism in Man" (G. M. Pacifici and G. N. Fracchia, eds.), pp. 407–459. European Commission, Luxembourg.

Morel, F., Rauch, C., Coles, B., Le Ferrec, E., and Guillouzo, A. (2002). The human glutathione transferase alpha locus: Genomic organization of the gene cluster and functional characterization of the genetic polymorphism in the hGSTA1 promoter. *Pharmacogenetics* **12**, 277–286.

Payne, A. H., and Hales, D. B. (2004). Overview of steroidogenic enzymes in the pathway from cholesterol to active steroid hormones. *Endocr. Rev.* **25**, 947–970.

Pettersson, P. L., and Mannervik, B. (2001). The role of glutathione in the isomerization of Δ^5-androstene- 3,17-dione catalyzed by human glutathione transferase A1-1. *J. Biol. Chem.* **276**, 11698–11704.

Pettersson, P. L., Johansson, A.-S., and Mannervik, B. (2002). Transmutation of human glutathione transferase A2-2 with peroxidase activity into an efficient steroid isomerase. *J. Biol. Chem.* **277**, 30019–30022.

Rabahi, F., Brule, S., Sirois, J., Beckers, J. F., Silversides, D. W., and Lussier, J. G. (1999). High expression of bovine alpha glutathione S-transferase (GSTA1, GSTA2) subunits is mainly associated with steroidogenically active cells and regulated by gonadotropins in bovine ovarian follicles. *Endocrinology* **140**, 3507–3517.

Simons, P. C., and Vander Jagt, D. L. (1981). Purification of glutathione S-transferases by glutathione-affinity chromatography. *Methods Enzymol.* **77**, 235–237.

Suzuki, T., Johnston, P. N., and Board, P. G. (1993). Structure and organization of the human alpha class glutathione S-transferase genes and related pseudogenes. *Genomics* **18,** 680–686.

Tiltman, A. J., and Haffajee, Z. (1999). Distribution of glutathione S-transferases in the human ovary: An immunohistochemical study. *Gynecol. Obstet. Invest.* **47,** 247–250.

Wang, C., Bammler, T. K., and Eaton, D. L. (2002). Complementary DNA cloning, protein expression, and characterization of Alpha-class GSTs from Macaca fascicularis liver. *Toxicol. Sci.* **70,** 20–26.

[18] A Subclass of Mu Glutathione S-Transferases Selectively Expressed in Testis and Brain

By IRVING LISTOWSKY

Abstract

A subclass of glutathione S-transferases (GSTs), exemplified by the human hGSTM3 and rodent GSTM5 subunits, has properties that distinguish it from other Mu class GSTs. Thus, they originate from single copy genes that are in an inverted order and, apart from the coding regions, share little sequence homology relative to the others in the Mu cluster. The genes for this M3/M5 subgroup encode for proteins that are in many ways unique, including their extended lengths with key amino acid substitutions. The M3/M5 subclass is selectively expressed in testis and brain and could function differently from the other GSTs.

Introduction

Mammalian glutathione S-transferases (GSTs) are products of gene superfamilies. At least eight categories of soluble GSTs have been described in the literature (Hayes and Pulford, 1995; Mannervik and Danielson, 1988), of those the Mu, Alpha, and Pi forms have been studied most extensively. Multiple genes from within each class probably originated from gene duplication or conversion events. Campbell *et al.* (1990) first cloned a gene from a human testis cDNA library that was categorized as a class Mu GST and recognized that the gene differed from other GSTs in this class from many standpoints. For instance, the protein product, designated as hGSTM3, was not detected in human liver, but high levels were found in testis and brain. Subsequently, counterparts of hGSTM3 in

METHODS IN ENZYMOLOGY, VOL. 401
Copyright 2005, Elsevier Inc. All rights reserved.
0076-6879/05 $35.00
DOI: 10.1016/S0076-6879(05)01018-9

mouse (mGSTM5) and rat (rGSTM5) were cloned (Fulcher *et al.*, 1995; Rowe *et al.*, 1998). Although orthologs of this subclass of GST can readily be identified in different species, unfortunately, the numerical assignments adopted (i.e., human M3 and rodent M5) are based on their original designations in the literature (Mannervik *et al.*, 1992) and thus are not consistent with one another.

GSTs of this human GSTM3/rodent GSTM5 subclass are often referred to as *M3-like, brain-testis-specific* or *germ-cell-specific*, because they are preferentially expressed in those cell types; however, lower levels are also detected in other organs. The class will also be referred to as the *M3/M5 subgroup* in this chapter. These forms of GST are encoded by single copy genes that are most distantly related to the other Mu class GSTs of each species. Whereas GST isoenzymes within a class usually are characterized by extensive sequence homology, subunits of that subclass usually exhibit less than 70% identity to the other Mu forms. Thus, it was recognized that they constitute a distinct group and were regarded as a distinctive subset of the Mu class GST (Rowe *et al.*, 1998).

Properties of the M3/M5 GST Subclass

The Genes

The Mu GST clusters map to human chromosome 1, mouse chromosome 3, and rat chromosome 2. The *M3-like* subclass apparently appeared in evolution after mammals diverged from birds. These single copy genes are in an inverted tail-to-tail orientation relative to the other genes in the Mu cluster (Patskovsky *et al.*, 1999). The genes contain eight exons and are considerably shorter than the others (particularly intron 8). The Mu subclass is also unique in that apart from their coding regions, they do not share sequence homology with the other Mu genes.

The human *hGSTM3* gene exhibits several polymorphisms. For instance, a polymorphic 3-base pair deletion in intron 6 yields an apparent YY recognition sequence (Inskip *et al.*, 1995). A fairly common functional polymorphism of the *hGSTM3* gene is a variant at the penultimate residue at the C-terminus (V224I), which exhibited an increased catalytic activity (Tetlow *et al.*, 2004). On the other hand, when Patskovsky *et al.* (1999) deleted the PVC C-terminal extension of hGSTM3, the resultant protein had very similar catalytic activity to that of the wild-type enzyme, even though the mutant lacking the C-terminal cysteine residue was found to be more stable. Methods for expression of some hGSTM3 variants are presented by Patskovsky *et al.* (1999) and Tetlow *et al.* (2004).

The Proteins

There are some common and some distinguishing structural features and properties of the proteins of this human M3/rodent M5 subdivision compared with other GSTs. Amino acid sequences for this subclass of GST of five different species are shown in Fig. 1. The extensive sequence identity of members of this class of GST among the different species is extraordinary, particularly in view of their limited homology (usually approximately 70% identity) relative to other Mu class GSTs even from within the same species. Thus, the chimpanzee and macaque orthologs are, respectively, 99% and 97% identical to the human hGSTM3 counterpart. Even the mouse and rat sequences, which are almost identical to each other, are approximately 90% homologous to that of the human hGSTM3.

Unlike other Mu GSTs, members of this M3/M5 subclass have blocked N-terminal serine residues (N-acetylated). The proteins are cysteine rich with conserved C-terminal Cys residues (Fig. 1). Other distinguishing structural features of this GST Mu subclass include C- and N-terminal

```
14250650_human    1-MSCESSMVLGYWDIRGLAHAIRLLLEFTDTSYEEKRYTCGEAPDYDRSQWLDVKFKLDL
55587744_chimp     -MSCESSMVLGYWDIRGLAHAIRLLLEFTDTSYEEKRYTCGEAPDYDRSQWLDVKFKLDL
13516449_macaca    -MSCESSMVLGYWDIRGLAHAIRLLLEFTDTSYEEKRYTCGEAPDYDRSQWLDVKFKLDL
25282395_rat       -MSCSKSMVLGYWDIRGLAHAIRMLLEFTDTSYEEKQYTCGEAPDYDRSQWLDVKFKLDL
6754086_mouse     --MSSKSMVLGYWDIRGLAHAIRMLLEFTDTSYEEKRYICGEAPDYDKSQWLDVKFKLDL
57088159_dog       MSLSKSNMVLGYWDIRGLAHAIRMLLEFTDTCYEERRYTCGEAPDYDKSQWLDVKFKLDL
                    ....**************:.*******.***::*.********:*:***********

14250650_human    60 DFPNLPYLLDGKNKITQSNAILRYIARKHNMCGETEEEKIRVDIIENQVMDFRTQLIRLC
55587744_chimp       DFPNLPYLLDGKNKITQSNAILRYIARKHNMCGETEEEKIRVDIIENQVMDFRTQLIRLC
13516449_macaca      DFPNLPYLMDGKNKITQSNAILRYIARKHNMCGETEEEKIRVDIIENQVMDFRTQLIRLC
25282395_rat         DFPNLPYLMDGKNKITQSNAILRYIARKHNMCGDTEEEKIRVDIMENQIMDFRMQLVRLC
6754086_mouse        DFPNLPYLMDGKNKITQSNAILRYIARKHNMCGDTEEEKIRVDIMENQIMDFRMQLVRLC
57088159_dog         DFPNLPYLMDGKNKITQSNAILRYIARKHNMCGETEEEKIRVDIMENQIMDFRIQLVQLC
                     ********:*********************:****:*****:***:**** **::**

14250650_human   120 YSSDHEKLKPQYLEELPGQLKQFSMFLGKFSWFAGEKLTFVDFLTYDILDQNRIFDPKCL
55587744_chimp       YSSDHEKLKPQYLEELPGQLKQFSMFLGKFSWFAGEKLTFVDFLTYDILDQNRIFDPKCL
13516449_macaca      YSSDHEKLKPQYLEELPGQLKQFSVFLGKFSWFAGEKLTFVDFLTYDILDQNRIFEPKCL
25282395_rat         YNSNHESLKPQYLEQLPAQLKQFSLFLGKFTWFAGEKLTFVDFLTYDVLDQNRMFEPKCL
6754086_mouse        YNSNHENLKPQYLEQLPAQLKQFSLFLGKFTWFAGEKLTFVDFLTYDVLDQNRMFEPKCL
57088159_dog         YSPDLEKLKPRYLEQLPGQLKQFSLFLGKFSWFAGEKLTFVDFLTYDVLDQNRMFEPKCL
                     *..:.*.***:***:**.*******:*****:***************:*****:*:****

14250650_human   180 DEFPNLKAFMCRFEALEKIAAYLQSDQFCKMPINNKMAQWGNKPVC  225
55587744_chimp       DEFPNLKAFMCRFEALEKIAAYLQSDQFFKMPINNKMAQWGNKPVC
13516449_macaca      DEFPNLKAFMCRFEALEKIAAYIQSDQFFKMPINNKMAQWGNKPVC
25282395_rat         DEFPNLKAFMCRFEALEKIAAFLQSDRCFKMPINNKMAKWGNKSIC
6754086_mouse        DEFPNLKAFMCRFEALEKIAAFLQSDRFFKMPINNKMAKWGNKCLC
57088159_dog         DEFPNLKAFMCRFEALEKIANYMQSDRFLKMPINNKMALWGNKRIC
                     *******************  ::.***:  ********.****.:*
```

FIG. 1. Sequence alignment of GSTs from the M3/M5 subclass in different species. The asterisks indicate identities in all five species. Residues are numbered including the initiation methionine. The unique Asn212 and Asn213 residues conserved at the H-site only for members of this subclass (see text) are also indicated in bold letters. Individual GI accession numbers are shown.

peptide extensions; most other Mu class GSTs have 217—218 residue lengths. The N-terminal extended sequences in particular are more divergent than the rest of the polypeptide chains (depicted in bold letters in Fig. 1). Thus, the human, chimpanzee, macaque, and rat forms have six residue N-terminal extensions, yet the mouse has five residues, and the dog has seven. The functional significance of the extra lengths of this subclass has not been determined, even though the rare human genetic polymorphism recently discovered (encoding PIC instead of PVC at the C-terminus–see previously) was reported to produce an enzyme that showed increased catalytic activity (Tetlow *et al.*, 2004). Heterodimeric combinations of hGSTM3 subunits with hGSTM2 subunits have been detected (Hussey and Hayes, 1993); however, no hGSTM3 heterodimers with other Mu subunits have been reported.

Although their primary structures are divergent relative to the other class Mu GSTs, the three-dimensional folds of this M3/M5 subclass are very similar to those of the other class Mu proteins (e.g., PDB 3GTU [Patskovsky *et al.*, 1999]). In particular, the architecture and amino acid side chains involved in GSH binding domains (G-sites) are conserved. However, key features of the second substrate-binding site (H-site) differ from those of other Mu class GSTs. In particular, in the three-dimensional structures, the Asn212 residue of the testis–brain subclass (M3/M5) (shown in bold in Fig. 1) corresponds to a conserved Phe208 residue that is present in other Mu class GSTs. The Asn213 residue (Fig. 1) is also close to the active site pocket. Replacement of the aromatic residue with a polar asparagine at position 212 introduces some instability in a cluster of hydrophobic side chains and increases the flexibility of this segment of the proteins of this subclass. This circumstance accounts for the relatively low turnover numbers for nucleophilic aromatic substitution reactions catalyzed by this subclass (see Table I). In fact, a mutated Asn212Phe form of the enzyme substantially increased the catalytic efficiency of the enzyme (Patskovsky *et al.*, 1999).

Methods and Protocols

Resolution of GST Subunits by HPLC

Cytosolic GSTs from tissue, cells, or recombinant proteins are routinely purified by a single step of epoxy-linked GSH affinity chromatography. Subunits of the GST preparations are resolved by analytical or preparative HPLC methods to yield characteristic profiles from the different sources. High-resolution subunit profiles can be achieved using reversed-phase C4 columns; analytical columns (4.6 mm ID × 250 mm L and 5-μm bore) are

TABLE I
KINETIC CONSTANTS FOR HUMAN, RAT, AND MOUSE FORMS OF THE SELECT
MU–GST SUBCLASS

Enzyme[a]	K_m^{CDNB} (mM)	K_m^{GSH} (mM)	k_{cat} (s^{-1})	k_{cat}/K_m^{CDNB} ($\times 10^6$ M^{-1} min^{-1})	k_{cat}/K_m^{GSH} ($\times 10^6$ M^{-1} min^{-1})
hGSTM3-3	2.80	0.08	5.0	0.11	3.6
rGSTM5-5	1.08	0.14	7.2	0.4	3.1
mGSTM5-5	1.10	0.14	6.9	0.38	3.0

[a] Recombinant proteins were used with the 1-chloro-2,4-dinitrobenzene (CDNB) and GSH substrates (see Rowe et al., 1998).

used to achieve separations such as those shown in Fig. 2. For most separations, linear binary gradients of water containing 0.1% trifluoroacetic acid (TFA) (solvent A) and acetonitrile containing 0.1% TFA (solvent B) will suffice to resolve a full complement of soluble GSTs. Most GST subunits elute between 40–55% of solvent B. In some cases, flow rates may need to be adjusted to resolve minor subunits from more abundant forms with similar retention times. For high resolution of samples with low concentrations of protein, microbore technology (C4 columns with 1.0-mm ID) are used with a Hewlett Packard HP 1100 HPLC unit.

Mouse testicular cytosolic GST subunits are displayed in the HPLC profiles shown in Fig. 2. The mGSTM5 subunit, which is the major GST at puberty (Fig. 2A), has a characteristic retention time such that it is well resolved from the other subunits. It is noteworthy that GSTs of this subclass from different species (mGSTM5, rGSTM5, and hGSTM3) have similar retention times under the same conditions. The data in Fig. 2 show that mGSTM5 is by far the predominant subunit at puberty; testis is the only tissue in which this class predominates. Subunits from this subclass of GST also have significantly greater molecular masses that distinguish them from most other GSTs.

Expression of the M3/M5 *GST Subclass*

Total RNA is purified from freshly removed tissues by use of an RNA purification kit (Qiagen). Primers were designed on the basis of comparison of unique sequences of the *M3-like* subclass. To synthesize cDNAs, reverse transcriptase (*RAV-2*, Amersham Pharmacia Biotech) and the reverse primers were used according to published methods. An aliquot (2 μl) of the cDNA solution was used as a template for PCR amplification. The reaction mixture contains 0.5 mM each of dNTP, 4 mM MgCl$_2$,

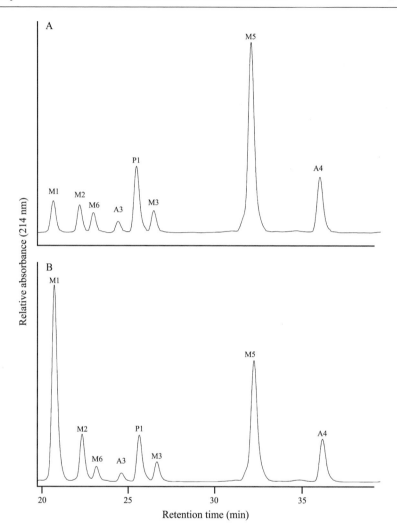

FIG. 2. HPLC profiles of GSTs from mouse testis. Purified mouse testicular GSTs (150 μg) obtained from GSH–agarose affinity columns for each treatment were resolved by HPLC on an analytical C4 (0.46 × 25 cm) reversed phase column (Grace Vydac, Hesperia, CA) using an HP-1090 instrument (Hewlett Packard, Palo Alto, CA). Proteins were eluted with linear gradients of buffer B (0.1% trifluoroacetic acid TFA in acetonitrile) at flow rates of 0.7 ml/min in two phases (2%/min, 0–10 min; 0.5%/min, 10–50 min). Fractions were collected and peaks detected by absorbance at 214 nm and were classified by peptide-specific antibody recognition and identified by electrospray ionization mass spectrometry (ESI-MS) using a quadrupole ion-trap mass spectrometer (LCQ-Finnigan Corp., San Jose, CA). Molecular masses of mouse GSTs are given in Table III. Quantitative estimates of relative amounts of each subunit can be obtained by measuring the areas under their peaks relative to the total area and known amounts of loaded protein. (A) GSTs from 5-week-old C57/BL mouse testis. (B) Testicular GSTs from a 13-week-old mouse. Subunits abbreviations are designated mGSTM5 is labeled as M5, etc.

TABLE II
PCR Primer Sequences Used to Express the Distinct Subclass of GST

hGSTM3-sense	5'-AGC CCG TCC ATA TGT CGT GCG AGT CGT CTA-3';
hGSTM3-antisense	5'-TGC AAG TCT GGA TCC TGA TCA GCA TAC AGG-3';
rGSTM5-sense	5'-AAG ATC CCC CCA TAT GTC GTG CTC C-3';
rGSTM5-antisense	5'-AGA GCA GCG GGA TCC AGC TCA GCA-3'
mGSTM5-sense	5'-GCC AAG ATC GCC CCA TAT GTG ATC CAA G-3'
mGSTM5-antisense	5'-CTC AGC AGC AGC GGG ATC CGG CTC AGC A-3'

All oligonucleotides contain the underlined NdeI or BamHI restriction sites that are used for cloning into expression vectors. The sense primers contain the ATG initiation codons of the respective mRNAs. PCR reactions are performed in the presence of the sets of sense–antisense primers and cDNA templates used according to conditions outlined in Rowe *et al.* (1998).

0.10–0.2 μmol of primers, 5 units of *Taq* DNA polymerase in a total volume of 100 μl. The PCR reaction is performed in 30–40 cycles, each consisting of steps for denaturation (1 min, 94°), annealing (1 min, 55°), and elongation (1 min, 72°),followed by a final extension step (10 min, 72°). The product is purified by agarose gel electrophoresis and cloned into pCRII TA-vector (Invitrogen) according to the procedure described in the instruction manual. DNA sequencing confirms the fidelity of the products, and the expressed proteins are authenticated by ESI-MS. Suggested primers for use to express the *M3-like* subclass of GST are shown in Table II.

Peptide Sequence-Specific Antisera

On the basis of the unique sequences near the C-terminal domains of the *M3/M5* subclass (see Fig. 1), oligomeric peptides corresponding to those C-terminal sequences can be used to generate antibodies highly specific for this Mu subclass (Rowe *et al.*, 1997). The peptide sequences underlined in Fig. 1 have been used to raise antibodies against the mouse mGSTM5 and human hGSTM3 subunits. The peptide sequence-specific antisera recognize *only* these proteins on immunoblots (Rowe *et al.*, 1998) and tissue specimens by immunohistochemical means (Rowe *et al.*, 1998).

Questions To Be Resolved

The existence of this distinctive GST M3/M5 subclass that is primarily expressed in postmeiotic germ cells and certain neurons in brain points to new potential areas of study in the GST field. Members of this subgroup of GST are cysteine-rich and thereby could function in a redox capacity in the brain, the reproductive tract, and elsewhere. Their localization in the

TABLE III
Murine GST Subunits, Resolved by HPLC and Identified by ESI MS

| Mouse GST subunit | Molecular mass | |
	Experimental[a]	Database (accession no.)
M1	25,839	25,839 (P10649)
M4	25,571	25,570 (P19639)
M2	25,586	25,585 (P15626)
M6	25,494	25,497 (O35660)
A3	25,271[b]	25,229 (P30115)
P2	23,403	23,406 (P46425)
P1	23,477	23,478 (P19157)
M3	25,579	25,579 (AK002213)
A5	25,491	No entry
A1	25,517[b]	25,477 (P13745)
A2	25,450[b]	25,402 (P10648)
M5	26,546[b]	26,504 (P48774)
M7	25,384	25,388 (AAM67419)
A4	25,478	25,477 (Q9CQ81)

[a] Observed mass ± 3 a.m.u. Results were obtained for at least three different preparations of the GST subunit.
[b] These subunits are N-acetylated at their N-termini.

fibrous sheath of flagellum of spermatozoa and in some neuronal processes indicates that the protein may be involved in sperm motility and can also serve as structural components. The observation that this subclass is primarily a neuronal GST, whereas other brain GSTs are mainly found in glial cells, also underscores its special nature. A recent report showing that hGSTM3 is localized in microglia, plaque, and tangles in brains of individuals with Alzheimer's disease (Tchaikovskaya et al., 2005), suggests that this type of GST could function in capacities other than as detoxicants. Indeed, their catalytic efficiencies for conjugation of nucleophilic aromatic substitution reactions with some common electrophilic substrates of GSTs are relatively poor (see Table II for example).

The genes for this subclass differ from those of other GSTs in regard to their orientation in the *Mu GST* clusters and in that they are single copy versions. There are no data thus far on the regulation of expression of their genes; this GST subclass apparently has no functional ARE elements (Hayes et al., 2004) and does not respond to the traditional inducers of other GSTs. Neither the functions nor the basis for this highly selective cell-type expression of this GST subclass are known. A targeted disruption of the *mGSTM5* gene in the mouse has been produced. The *mGSTM5* null animals are viable and produce progeny. Although their reproductive functions apparently are intact, the *mGSTM5* deletion renders the animals more susceptible to some oxidants (unpublished). The null and transgenic models should provide information about the functions and mechanisms of cell type–specific regulation of expression of the M3/M5 type genes.

References

Campbell, E., Takahashi, Y., Abramovitz, M., Peretz, M., and Listowsky, I. (1990). A distinct human testis and brain Mu-class glutathione S-transferase. Molecular cloning and characterization of a form present even in individuals lacking hepatic type Mu isoenzymes. *J. Biol. Chem.* **265**, 9188–9193.

Fulcher, K. D., Welch, J. E., Klapper, D. G., O'Brien, D. A., and Eddy, E. M. (1995). Identification of a unique Mu-class glutathione S-transferase in mouse spermatogenic cells. *Mol. Reprod. Dev.* **42**, 415–424.

Hayes, J. D., Flanagan, J. U., and Jowsey, I. R. (2005). Glutathione transferases. *Annu. Rev. Pharmacol. Toxicol.* **45**, 51–88.

Hayes, J. D., and Pulford, D. J. (1995). The glutathione S-transferase supergene family: Regulation of GST and the contribution of the isoenzymes to cancer chemoprotection and drug resistance. *Crit. Rev. Biochem. Mol. Biol.* **30**, 445–600.

Hussey, A. J., and Hayes, J. D. (1993). Human Mu-class glutathione S-transferases present in liver, skeletal muscle and testicular tissue. *Biochim. Biophys. Acta* **1203**, 131–141.

Inskip, A., Elexperu-Camiruaga, J., Buxton, N., Dias, P. S., MacIntosh, J., Campbell, D., Jones, P. W., Yengi, L., Talbot, J. A., Strange, R. C., and Fryer, A. A. (1995). Identification of polymorphism at the glutathione S-transferase, GSTM3 locus: Evidence for linkage with GSTM1*A. *Biochem. J.* **312**(Pt. 3), 713–716.

Mannervik, B., Awasthi, Y. C., Board, P. G., Hayes, J. D., Di Ilio, C., Ketterer, B., Listowsky, I., Morgenstern, R., Muramatsu, M., Pearson, W. R., Pickett, C. B., Sato, K., Widersten, M., and Wolf, C. R. (1992). Nomenclature for human glutathione transferases. *Biochem. J.* **282**(Pt. 1), 305–306.

Mannervik, B., and Danielson, U. H. (1988). Glutathione transferases—structure and catalytic activity. *CRC Crit. Rev. Biochem.* **23**, 283–337.

Patskovsky, Y. V., Huang, M. Q., Takayama, T., Listowsky, I., and Pearson, W. R. (1999). Distinctive structure of the human GSTM3 gene-inverted orientation relative to the Mu class glutathione transferase gene cluster. *Arch. Biochem. Biophys.* **361**, 85–93.

Patskovsky, Y. V., Patskovska, L. N., and Listowsky, I. (1999). An asparagine-phenylalanine substitution accounts for catalytic differences between hGSTM3-3 and other human class Mu glutathione S-transferases. *Biochemistry* **38**, 16187–16194.

Rowe, J. D., Nieves, E., and Listowsky, I. (1997). Subunit diversity and tissue distribution of human glutathione S-transferases: Interpretations based on electrospray ionization-MS and peptide sequence-specific antisera. *Biochem. J.* **325**(Pt. 2), 481–486.

Rowe, J. D., Patskovsky, Y. V., Patskovska, L. N., Novikova, E., and Listowsky, I. (1998). Rationale for reclassification of a distinctive subdivision of mammalian class Mu glutathione S-transferases that are primarily expressed in testis. *J. Biol. Chem.* **273**, 9593–9601.

Rowe, J. D., Tchaikovskaya, T., Shintani, N., and Listowsky, I. (1998). Selective expression of a glutathione S-transferase subclass during spermatogenesis. *J. Androl.* **19**, 558–567.

Tchaikovskaya, T., Fraifeld, V., Urphanishvili, T., Andorfer, J. H., Davies, P., and Listowsky, I. (2005). Glutathione S-transferase hGSTM3 and ageing-associated neurodegeneration: Relationship to Alzheimer's disease. *Mech. Ageing Dev.* **126,** 309–315.
Tetlow, N., Robinson, A., Mantle, T., and Board, P. (2004). Polymorphism of human Mu class glutathione transferases. *Pharmacogenetics* **14,** 359–368.

[19] Glutathione S-Transferases as Regulators of Kinase Pathways and Anticancer Drug Targets

By Danyelle M. Townsend, Victoria L. Findlay, and Kenneth D. Tew

Abstract

Anticancer drug development using the platform of glutathione (GSH), glutathione S-transferases (GST) and pathways that maintain thiol homeostasis has recently produced a number of lead compounds. GSTπ is a prevalent protein in many solid tumors and is overexpressed in cancers resistant to drugs. It has proved to be a viable target for pro-drug activation with at least one candidate in late-stage clinical development. In addition, GSTπ possesses noncatalytic ligand-binding properties important in the direct regulation of kinase pathways. This has led to the development and testing of agents that bind to GSTπ and interfere with protein–protein interactions, with the phase II clinical testing of one such drug. Attachment of glutathione to acceptor cysteine residues (glutathionylation) is a posttranslational modification that can alter the structure and function of proteins. Two agents in preclinical development (PABA/NO, releasing nitric oxide on GST activation, and NOV-002, a pharmacologically stabilized pharmaceutical form of GSSG) can lead to glutathionylation of a number of cellular proteins. The biological significance of these modifications is linked with the mechanism of action of these drugs. In the short term, glutathione-based systems should continue to provide viable targets and a platform for the development of novel cancer drugs.

Introduction

Drug discovery in cancer has evolved significantly in the past few years. High-throughput screening and cancer-specific target discrimination have essentially supplanted the classical synthetic chemistry structure–activity approaches to identify new lead compounds. Pathways that involve

METHODS IN ENZYMOLOGY, VOL. 401   0076-6879/05 $35.00
Copyright 2005, Elsevier Inc. All rights reserved.   DOI: 10.1016/S0076-6879(05)01019-0

proteins aberrantly expressed in cancer cells are optimal as targets for drug intervention. Increased expression of the GSTπ isozyme (the most ubiquitous and prevalent GST in nonhepatic tissues) has been linked to both drug resistance and the malignant phenotype of many solid tumors (Tew, 1994). In addition, GSTπ has been found to be an endogenous regulator of c-Jun NH$_2$-terminal kinase (JNK) (Adler *et al.*, 1999a). On the basis of such differential expression and unusual signaling function in tumors, it was concluded that GSTπ might be an opportunistic drug target that could provide for an enhanced therapeutic index in the treatment of cancer.

The K$_{cat}$ values for GSTπ-mediated GSH conjugation reactions with a number of anticancer drug substrates have been measured and are not impressive. Although rate and extent of conjugation for some alkylating drugs can be enhanced by GSTπ catalysis (Ciaccio *et al.*, 1991), it has never seemed unreasonable to presume that some GSTπ functionality other than catalysis may be of consequence to the biological importance of the protein. The recent description of protein–protein interactions between GSTπ and JNK serve to extend the basic principles of the ligand-binding properties of GST isozymes. Indeed, early characterization of the GSTs centered on their capacity to act as a ligand in association with other proteins, particularly nonsubstrate ligands such as heme and bilirubin (Litwack *et al.*, 1971). Although comparatively large macromolecules, they do not encompass the dimensions of a kinase such as JNK, and, not surprisingly, interactions between these two proteins are complex, with apparent association constants in the nanomolar range (Wang *et al.*, 2001). The following sections describe how the evolving understanding of GST functions in cells can be adapted to the principles of drug design and development.

GSH and GST in Cell Signaling

Adding a further layer of complexity to the understanding of GST function is the fact that proteins rarely act in isolation in a cellular milieu. Rather, essential protein–protein interactions govern how cellular events unfold (Golemis *et al.*, 2002). This process has proved to be significant to the regulation of JNK signaling by GSTπ (Adler *et al.*, 1999a; Wang *et al.*, 2001). This same paradigm seems to hold for additional redox proteins and associated kinases such as thioredoxin and apoptosis signal-regulating kinase, ASK1 (Saitoh *et al.*, 1998), implying the possible existence of a general regulatory mechanism for kinases that may involve GSH and associated pathways (Adler *et al.*, 1999b; Davis *et al.*, 2001). In addition, emergent literature suggests that direct glutathionylation of critical signaling molecules may also serve as a trigger for cellular events that are influenced by oxidative stress (Adachi *et al.*, 2004; Cross and Templeton,

2004). It would seem that GSH and GSTs have roles that extend much further than simple detoxification reactions.

Other small redox-active proteins such as peroxiredoxin have the potential to heterodimerize with GSTπ. Oxidative stress-mediated signaling events can be activated by GST-mediated catalytic addition of GSH to a critical cysteine residue in a sterically protected region of peroxiredoxin (Manevich *et al.*, 2004). This observation broadens the functional importance of GSTπ into yet another arena, emphasizing how redox-active proteins have roles that are more than just removal of reactive oxygen species but are central to the signaling processes required in the cell's response to stress. Changes in redox conditions can trigger cellular responses through a number of different pathways. The nature and extent of the ROS insult may determine the threshold of the cellular response manifest as proliferation, stress response and damage repair or apoptosis. With further understanding, the link between thiol-active proteins, GSTs, and stress-activated protein kinases exemplified by JNK and ASK may become an expansive series of interconnected pathways. In an unstressed cellular environment, JNK is kept in an inactive mode by the presence of one or more repressors. Under conditions of oxidative stress, GSTπ dissociates from JNK and forms dimers and/or multimeric complexes (Adler *et al.*, 1999a). Meanwhile, the liberated JNK regains its functional capacity to be phosphorylated and to phosphorylate c-Jun. This process can activate the stress cascade involving the numerous sequential downstream kinases. However, in GSTπ-overexpressing cells, constitutively active MEKK1 effectively phosphorylated both MKK4 (immediately upstream of JNK) and JNK but did not result in jun phosphorylation, confirming the specificity of the GSTπ/JNK association (Yin *et al.*, 2000). Mouse embryo fibroblast cell lines (MEF) from mice (Henderson *et al.*, 1998) engineered to be null for GSTπ expression have high basal levels of JNK activity that can be reduced if these cells are transfected with GSTπ cDNA. In addition, treatment of GSTπ wild-type cells (but not null cells) with a specific GSTπ inhibitor, TLK199, causes activation of JNK activity. Also, human HL60 cells chronically exposed to this inhibitor develop tolerance to the drug and also overexpress JNK, presumably as a means of compensating for the constancy of GSTπ inhibition and the perceived chronic stress (Ruscoe *et al.*, 2001).

These combined data provide evidence that GSTπ has a nonenzymatic, regulatory role in controlling cellular response to external stimuli. MEFs from GST$^{-/-}$ mice have a 24-h doubling time compared with 36 h for wild type (Ruscoe *et al.*, 2001). TLK199, although designed to be a modulator of drug resistance through preferential inhibition of GSTπ (Morgan *et al.*, 1996), was found to possess additional pharmacological activity as a myeloproliferative agent (Gate *et al.*, 2004). Each study provides support for the concept that GSTπ has a role in regulation of proliferative pathways. Although GSTπ regulates JNK activity through protein–protein interactions, the influence of GST on GSH–GSSG homeostasis could also be a contributory factor. For example, the GSH binding site of GSTs (G-site) may be an important sequestration site for cellular GSH with concomitant impact on cellular redox status.

There are indications that GSH and associated enzymes play a role in cellular immunity. For example, GSH levels in antigen-presenting cells determine whether a TH1 or TH2 pattern of response predominates (Peterson *et al.*, 1998). The TH1 response is characterized by production of interleukin-12 (IL-12) and interferon α and the enhancement of delayed hypersensitivity response; TH2 by IL-4 and IL-10 production and up-regulation of a number of antibody responses. The molecular basis for this difference is not known, but it is also significant that patients with HIV receiving n-acetylcysteine (a bioavailable precursor of GSH biosynthesis) have longer survival times than untreated controls. In diseases such as HIV, TH2 predominance is a critical component of immune response and, thus, GSH levels in antigen-presenting cells may play an integral role in determining disease progression (Herzenberg *et al.*, 1997). The question now extends to a possible link between JNK and immune response/myeloproliferation. To this end, T cells from a JNK1 null mouse hyperproliferated, exhibited decreased activation, and induced cell death and preferentially differentiated into TH2 cells (Dong *et al.*, 1998). Despite the redundancy accorded to this system by the continued production of JNK2, it seems reasonable to conclude that the JNK1 signaling pathway may be playing a role in T-cell receptor–initiated TH cell proliferation, differentiation, and apoptosis. By association, GSTπ must also be involved.

There are other examples where ROS or electrophilic insults stimulate stress response pathways. Thioredoxins are a family of redox proteins of approximately 12 kDa responsible for mediating numerous cytoplasmic functions largely influenced by the Cys 73 residue of the monomeric protein. Dimerization at this site mitigates many of the redox-dependent functions of the protein. Recent data implicate a secreted form of thioredoxin in control of cell growth, where the redox function is essential for growth stimulation (Powis *et al.*, 1998). Tumor cells transfected with thioredoxin demonstrate increased growth and decreased sensitivity to drug-induced apoptosis. As mentioned earlier, thioredoxin has also been shown to bind to ASK1 to inhibit its activity as a kinase (Saitoh *et al.*, 1998). This inhibition is attenuated by ROS, which causes dimerization of thioredoxin (Gotoh and Cooper, 1998). Thioredoxin regulation of the glucocorticoid receptor through association at the DNA-binding domain has also been reported (Makino *et al.*, 1999), and this too is influenced by ROS. These provide examples analogous to the GSTπ/JNK association and perhaps suggest a more broad ranging role for thiol–disulfide regulation of proliferation.

In summary, alterations in redox balance by exposure to ROS cause dose-dependent changes in GSH/GSSG ratios that can potentially influence a number of target proteins by causing oxidation and disulfide exchange reactions at specific cysteine residues. Specifically for GSTπ, the conversion from monomer to dimer (or multimers) causes dissociation from JNK with resultant activation of the kinase. With subsequent phosphorylation of c-Jun and enhanced transcription of AP-1–responsive genes, the stage is set for signal transduction for stress response, proliferation, or apoptosis. Cysteine residues of GSTπ have been shown to be sensitive to oxidation by H_2O_2 (Shen *et al.*, 1993). The model shown in Fig. 1 illustrates

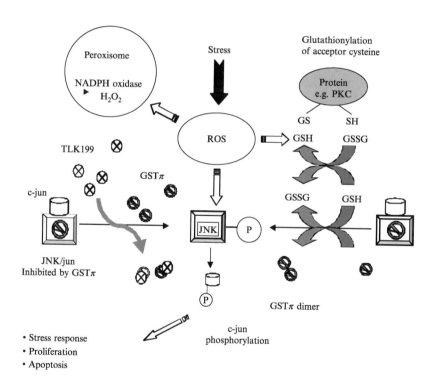

FIG. 1. Model scheme showing a means by which oxidative stress can be transmitted through a GST "switch" connecting to kinase cascades influencing cell signaling. In the example shown, TLK199 can cause a disassociation of the GST–JNK complex, activating JNK. PKC is an example of a protein susceptible to glutathionylation.

how oxidative stress can transmit through a GST "switch" into the kinase cascade of pathways. Of the numerous literature reports of drug-induced perturbations in GSH/GST levels, few consider whether the biological implications extend further than an enhanced rate of catalytic thioether product formation with a prescribed electrophilic center of an administered drug. The early nomenclature of "ligandin" (Litwack et al., 1971) may prove to have been prescient, particularly regarding the interaction with kinases. This property forms the basis for much of the rationale for drug design in directly inhibiting GSTπ and is described in the following section.

GSTπ as a Drug Target

In the past, modulation by inhibition of GST has been attempted as a means to improve response to cancer drugs. Use of, for example, ethacrynic acid, although effective in its experimental effects on various GST isozymes, was not successful enough in the clinic to merit continued development. The dose-limiting toxicity of fluid imbalance was triggered by the diuretic properties of the drug (O'Dwyer et al., 1991). However, one consequence of this approach was the conceptual design of a peptidomimetic inhibitor of GSTπ, TLK199 (γ-glutamyl-S-(benzyl)cysteinyl-R(-)phenyl glycine diethyl ester; Fig. 2). In concept, the drug was envisioned as a plausible means to sensitize drug-resistant tumors that overexpress GSTπ (Morgan et al., 1996). However, preclinical and mechanism of action studies with this agent revealed an unexpected effect in animals, namely that the drug caused myeloproliferative activity (Gate et al., 2004; Ruscoe et al., 2001). Given the link between GSTπ and the kinase pathways, a model for how TLK199 can produce proliferative effects in the marrow compartment has been proposed (Gate et al., 2004). Initial clues were provided from the observation of increased myeloproliferation in GSTπ-deficient mice compared with wild-type animals (Ruscoe et al., 2001). A general increase in white blood cell counts in GSTπ$^{-/-}$ mice but no change in leukocyte composition was observed. Spleen cell counts were higher in knock out animals, and this was associated with a twofold increase in B lymphocytes, whereas T lymphocytes and NK cell counts were similar in both animal strains. In contrast, no difference in thymocyte counts and thymus subset composition was observed. Red blood cell and platelet counts were also higher in GSTπ$^{-/-}$ mice. Taken together, these data inferred that the absence of GSTπ expression potentiates hematopoiesis by influencing the proliferation and/or the differentiation of hematopoietic cells.

In in vitro hematopoiesis experiments, IL-3, GM-CSF, or G-CSF induced more colonies in GSTπ$^{-/-}$ cells than in wild type. TLK199 stimulated colony formation in wild-type, but not in knock out, animals. The JNK

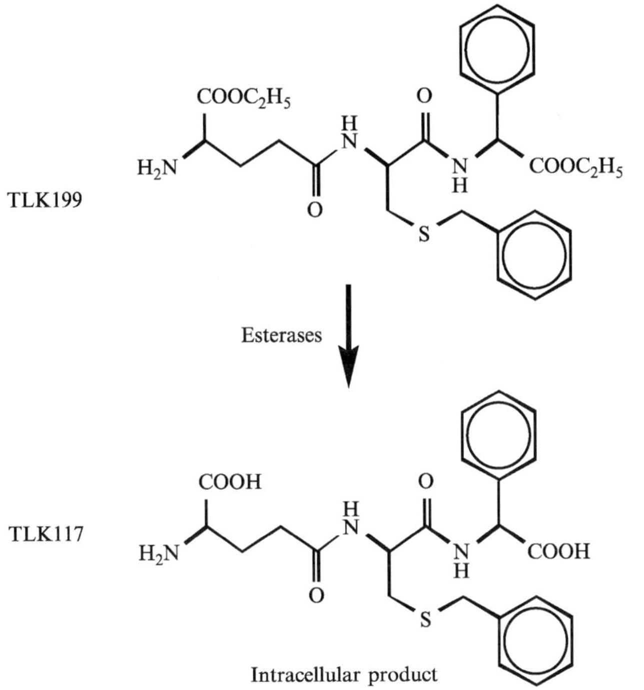

Fig. 2. Structure of a peptidomimetic inhibitor of GSTπ, TLK199 (γ-glutamyl-S-(benzyl) cysteinyl-R(-)phenyl glycine diethyl ester) and the de-esterified active inhibitor of GSTπ, TLK117.

inhibitor SP600125 decreased marrow colonies produced by cytokine treatment of knock out animals, and JNK phosphorylation was endogenously elevated in bone marrow cells from GSTπ$^{-/-}$ animals. These data are consistent with GSTπ acting as a physiological inhibitor of JNK, where TLK199 disassociates GSTπ from JNK, allowing kinase phosphorylation and subsequent activation of the kinase cascade (see Fig. 1; Adler et al., 1999a; Ruscoe et al., 2001). Inhibition of JNK, abrogates the increased phosphorylation of this kinase observed in the presence of TLK199 and consequently reduces the myeloproliferative effect of the drug. Because the cell compositions of the colonies were similar in both mouse strains after exposure to cytokines, the increase is a function of proliferation, rather than differentiation, of hematopoietic cells. Taken together, these data suggest that JNK plays an integral role in the elevated myeloproliferation observed in GSTπ-deficient mice and the myelostimulant properties of TLK199. This is consistent with reports indicating a possible role for JNK in control of proliferation. Yang and colleagues have shown that the inhibition of JNK1 or JNK2 expression in human prostate carcinoma was associated with a decrease in cell proliferation (Yang et al., 2003). Similarly, it has been observed that JNK activity was required for rat liver regeneration by increasing cyclin D1 expression and allowing G0–G1 transition (Schwabe et al., 2003). In addition, JNK phosphorylates and induces the transactivation of the transcription factors c-Jun, JunB, and JunD (Kallunki et al., 1996). Overexpression or activation of Jun and JunD has been linked with cell proliferation and transformation (Shaulian and Karin, 2001), and JNK activation has been linked to induction of apoptosis (Tournier et al., 2000). More recent data have suggested that, after UV, JNK1 was more likely to be proapoptotic, whereas JNK2 was associated with survival (Hochedlinger et al., 2002). The discrimination between the survival and apoptotic functions of JNK also seems to correlate with the level and duration of the enzyme activation. A strong and sustained activation is associated with apoptosis, whereas a weaker and transient phosphorylation is correlated with proliferation (Shaulian and Karin, 2001). For example, in mouse hematopoietic BaF3 cells, JNK activity was three times lower when cells were exposed to mitogenic concentrations of IL-3 than to cytotoxic concentrations of anisomycin (Terada et al., 1997). From such reports, it seems likely that regulation of JNK activity by GSTπ should be a viable target for drug intervention. As a consequence, company-sponsored (Telik Inc., Palo Alto, CA) clinical trials of TLK199 (now named *Telintra*) in patients with myelodysplastic (MDS) syndrome have now been instigated. An ongoing multicenter phase II trial has produced interim results as of the last quarter of 2004. For 34 patients with MDS, clinically significant improvement in one or more blood cell lineages was observed in 61.5% of patients from all major FAB subtypes (Callander, 2004). Responses were associated with decreased requirements for red blood cell, platelet, and growth factor support, in some cases leading to transfusion independence. Although these results are preliminary in nature, they do provide encouragement for the translational relevance of targeting GSTπ and for the mechanism of action data for TLK199.

GST-Activated Prodrugs

TLK286 (Telcyta)

The latent prodrug, TLK286 [γ-glutamyl-α-amino-β(2-ethyl-N,N,N′,N′-tetrakis (2-chloroethyl)phosphorodiamidate)-sulfonyl-propionyl-(R)-(-) phenylglycine] was synthesized as the lead candidate from a group of rationally designed glutathione analogs designed, once again, to exploit high

GSTπ levels associated with malignancy, poor prognosis, and the development of drug resistance (Hayes and Pulford, 1995; Tew, 1994). Thus, selective targeting of susceptible tumor phenotypes is a strategy that should result in the "release" of more active drug in malignant cells compared with normal tissue, thereby achieving an improved therapeutic index.

In TLK286, the sulfhydryl of a glutathione conjugate has been oxidized to a sulfone. The tyrosine-7 in GSTπ promotes a β-elimination reaction that cleaves the compound (see Fig. 3). The cleavage products are a

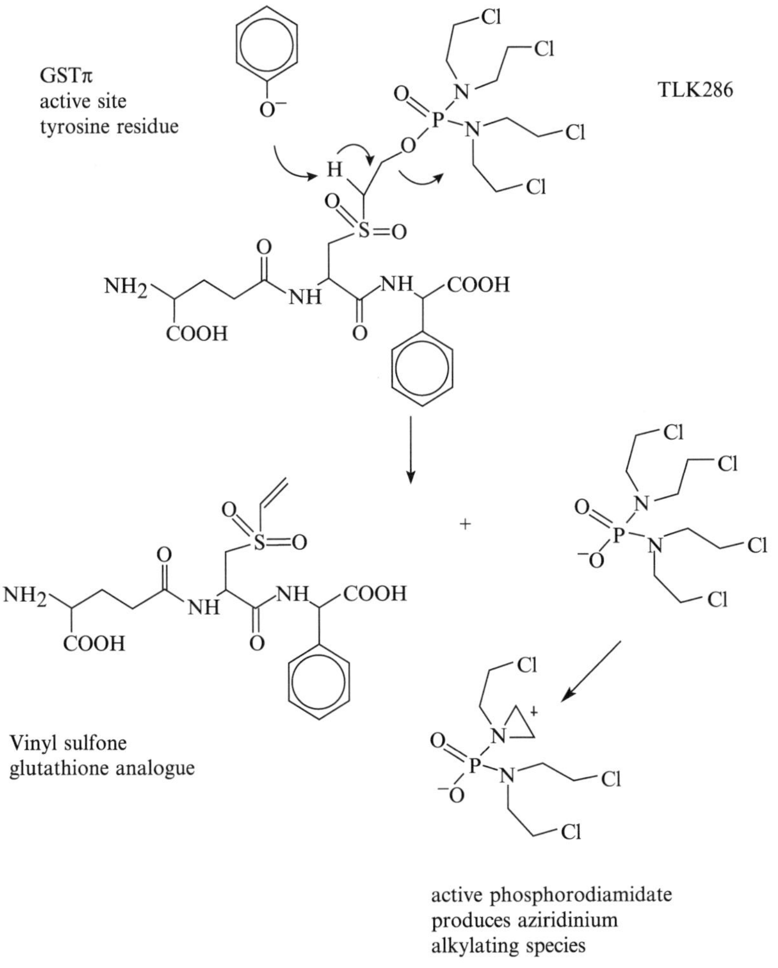

FIG. 3. Structure of TLK286 and its activation by GSTπ.

glutathione analog and a phosphorodiamidate, which in turn spontaneously forms aziridinium species, the actual alkylating moieties. The cytotoxic moiety has tetrafunctional alkylating properties, similar in concept to bifunctional nitrogen mustards. Each can react with cellular nucleophiles with a short half-life (Lyttle *et al.*, 1994; Satyam *et al.*, 1996). The other part of the molecule contains the glutathione backbone and an electrophilic vinyl sulfone moiety. The contribution of this component to drug efficacy or toxicity is not entirely clear. Glutathione conjugates of the sulfone are possible, and these could be substrates for transporters such as MRP1. In addition, the vinyl sulfone could be a factor in chain reactions leading to lipid peroxidation and even production of hydrogen peroxide (Comporti, 1989), considerations that could explain the enhanced expression of catalase in the HL60 cell line selected for resistance to TLK286 (Rosario *et al.*, 2000).

TLK286 exhibits activity against a variety of tumors and tumor cell lines. *In vitro* studies have shown that elevated GSTπ in stably transfected cell lines correlates with increased sensitivity to TLK286 cells. Similarly, drug-resistant cell lines that overexpress GSTπ are more sensitive to TLK286 (Morgan *et al.*, 1996). In an *ex vivo* clonogenic assay against human solid tumors, TLK286 showed activity against 15 of 21 lung tumors and 11 of 20 breast tumors tested. In addition, effective antitumor activity was found *in vivo* using xenografted human tumors in nude mice, with only mild bone marrow toxicity (Morgan *et al.*, 1998).

Three distinct model systems have been used to study the pharmacology of TLK286 with the ultimate goal of gaining proof-of-principle with respect to mechanism of action. The first required establishment of a TLK286 acquired–resistant cell line, a task that proved more difficult to achieve than would normally have been expected. Although cells frequently survived the initial low concentration–selecting treatment and partially repopulated the culture, recovery to full viability was difficult to attain. Considering the relative ease with which HL60 cells can usually be made resistant to anticancer drugs, this result was somewhat surprising. This may indicate that resistance to TLK286 is governed by multiple factors or that survival response pathways are not readily invoked after chronic drug exposure. A characteristic of GSTπ as a GSH-conjugating enzyme is the generally low catalytic efficiency with broad substrate "specificity." The observation that GSTπ is directly involved in the regulation of JNK-mediated stress response emphasizes the ligand binding, noncatalytic function for the protein. In turn, this may provide a partial explanation for the high GSTπ levels seen in many tumors, where kinase cascade pathways involving JNK may be imbalanced. Although for TLK286 the β-elimination reaction catalyzed by GSTπ does not inactivate the protein, it may serve to compartmentalize it away from the JNK ligand–binding function. This may influence the stoichiometry that controls kinase-mediated prolif-erative/apoptotic pathways and may be a contributory factor in the diffi-culty experienced in establishing a TLK286 resistant cell line. With perseverance, eventually a fivefold resistant cell line was established. Phe-notypically, decreased expression of GSTπ in resistant cells supported a mechanism of action based on the rational design of the drug (Rosario *et al.*, 2000). Two other experimental systems served to corroborate these results. For example, increased resistance to TLK286 in the MEF cell lines derived from GSTP1-1$^{-/-}$mice (Henderson *et al.*, 1998) was consistent with reduced activation of the drug. In addition, increased sensitivity in a NIH3T3 cell line transfected to overexpress GSTπ also confirmed a direct involvement of the isozyme in determining cytotoxicity.

The resistant cell line also expressed increased glutathione levels, a mechanism commonly associated with resistance to a range of alkylating agents (Tew, 1994). Enhanced GSH levels were not a consequence of induced overexpression of the primary enzymes responsible for *de novo* (γ-glutamyl cysteine synthetase; γ) or salvage (γ-transpeptidase; γ-GT) synthesis of the tripeptide. Similarly, MRP expression was unaltered in resistant cells. A co-coordinated, increased expression of γ-GCS, GSTP1-1, and MRP1 has been shown in cells selected for resistance to ethacrynic acid, a drug with Michael addition properties (Ciaccio *et al.*, 1996). TLK286 produces aziridinium moieties characteristic of other nitrogen mustards but distinct in electrophilic properties from Michael acceptors. This difference may account for the absence of evidence for inducible expression of the cadre of GSH-related detoxification gene products in HL60/TLK286-resis-tant cells. In considering the other major metabolites of TLK286, it is less clear how the release of the vinyl sulfone impacts on the pharmacology of TLK286. Its electrophilic characteristics would predict reactivity with cel-lular nucleophiles, and because GS conjugates are primary substrates for MRP (Keppler *et al.*, 1997), a GS–vinyl sulfone could prove to be an effective substrate for MRP. Thus, although the vinyl sulfone component of the drug is unlikely to be important to the therapeutic alkylating activity, it may prove to have some pharmacological significance.

Approximately 8 years encompassed the design, synthesis, and pretest-ing of TLK286. As such, many of the preclinical studies discussed previ-ously were instrumental in leading to clinical trial design. Perhaps the most informative progress to report is the recent clinical results presented at the 2004 American Society of Clinical Oncology meeting. In a series of ab-stracts, a 46% objective response rate was reported for a combination of Telcyta and liposomal doxorubicin in patients with platinum-refractory ovarian cancer; a 56% objective response rate for Telcyta and Carboplatin in platinum-resistant ovarian cancer; and a 27% objective response rate for combinations of Telcyta and Docetaxel in platinum-resistant non-small cell lung cancer (NSCLC). Telcyta is also under active testing in a clinical phase III setting for a number of disease states, including non-small cell, ovarian, and colon cancers.

PABA-NO

High GSTπ expression levels in human cancers and drug-resistant disease, together with the knowledge that nitric oxide (NO) has therapeutic potential, also provided the rationale for the design of the NO-releasing GST-activated pro-drug, O^2-[2,4-dinitro-5-(N-methyl-N-4-carboxyphenyla-mino)phenyl] 1-N,N-dimethylamino)diazen-1-ium-1,2-diolate; THF, tetra-hydrofuran (PABA/NO; (Saavedra *et al.*, 2004)). GSTπ-catalyzed conjugation of GSH to PABA/NO releases a diazeniumdiolate ion, with subsequent release of NO (Fig. 4). Other NO donors of the diazenium-diolate class are known to release NO in an enzyme-catalyzed manner, with activation by cytochrome P450's (Saavedra *et al.*, 1997) or esterases (Saavedra *et al.*, 2000).

To confirm the GSTπ activation requirements of PABA/NO, a number of model systems have been used. For example, mouse embryo fibroblasts from GST$\pi^{-/-}$ mice showed a decreased sensitivity to PABA/NO. These data are consistent with those for activation of TLK286 (Rosario *et al.*, 2000). The quantitative effect on cytotoxicity of eliminating GST$\pi^{-/-}$ is influenced by the expression of other GST isoforms that can also activate PABA/NO at different rates. In addition, slow spontaneous activation of PABA/NO can occur through noncatalytic GSH conjugation. Other components involved in metabolism and detoxification pathways were analyzed in a model system with stably transfected GSTπ, γGCS, and MRP1. A role for GST and GSH in the activation of PABA/NO was confirmed (Findlay *et al.*, 2004). Forced expression of γGCS and MRP1 also provided insight into cellular resistance mechanisms towards PABA/NO. Unlike TLK286, increased resistance to PABA/NO was conferred by the overexpression of MRP1, supporting the concept that this transporter (an efflux pump with affinity for GS conjugates) is involved in the removal of PABA/NO and/or active metabolite(s). Inhibition of LTC$_4$ transport (a known substrate for MRP1) was observed in the presence of both PABA/NO and GSH but not with either alone. Furthermore, GSTπ stimulated PABA/NO-induced in-hibition of LTC$_4$ transport, suggesting that metabolites of PABA/NO may be substrates for, and effluxed by, the transporter (Findlay *et al.*, 2004).

The *in vivo* efficacy of PABA/NO has been shown in tumor-bearing animals (Findlay *et al.*, 2004; Shami *et al.*, 2003). For example, nontoxic

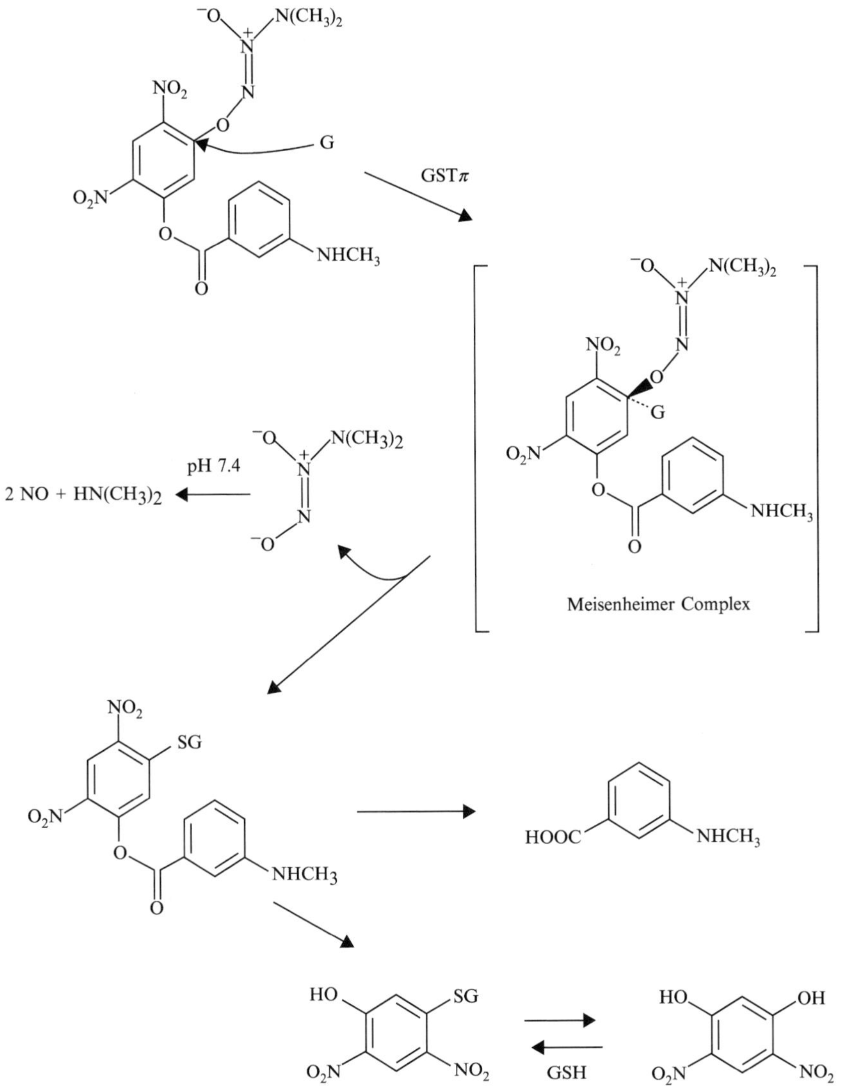

FIG. 4. Structure of an NO-releasing GST-activated pro-drug, O^2-[2,4-dinitro-5-(N-methyl-N-4-carboxyphenylamino)phenyl] 1-N,N-dimethylamino)diazen-1-ium-1,2-diolate; THF, tetrahydrofuran (PABA/NO). Metabolism and release of nitric oxide are based on the schema outlined elsewhere (Keefer, 2005).

doses of PABA/NO led to a significant growth delay in a human ovarian cancer model in SCID mice, with results comparable to those seen with cisplatin (the standard of care for management of ovarian cancer). Of interest, selective GST activation of PABA/NO may produce a therapeutic advantage, because GSTπ overexpression has been associated (although not necessarily causally linked) with resistance to platinum drugs (Townsend and Tew, 2003). PABA/NO exerts an apoptotic effect by induction of p38 and JNK. As noted earlier, this observation carries greater significance because GSTπ functions as an endogenous negative regulatory switch for these same regulatory kinase pathways (Adler et al., 1999a). NO has diverse roles in a variety of physiological processes, and augmentation of endogenous with exogenous NO provides the foundation for a broad range of therapeutic applications (Keefer, 2003; Napoli and Ignarro, 2003). NO has been reported to modulate the apoptotic process in a number of cell types. Bifurcating pathways of inhibition or induction of cell death seem to be tissue specific and to depend on the amount, duration, and site of NO production (Umansky and Schirrmacher, 2001). In support of both cellular antioxidant and prooxidant actions of NO in vivo, it has been reported that low doses of NO protect cells against peroxide-induced death, whereas higher doses result in increased killing (Joshi et al., 1999). PABA-NO has a dose- and time-dependent effect for drug-induced activation of the kinases, and MRP1 abrogates and/or delays the effect, primarily as a consequence of reducing the effective intracellular concentration of the drug (Findlay et al., 2004). Whether kinase activation occurs as a result of direct NO interaction (e.g., nitrosylation/nitration of residues in JNK) or by the impact of PABA/NO (or metabolites including GSNO) effects on the GSTπ–JNK complex remains to be shown. PABA/NO remains in early preclinical testing but has the potential disadvantage of poor solubility and stability in aqueous solution (Srinivasan, 2005). However, reasonable in vivo antitumor data suggest that it is a good lead compound for further structure activity and drug discovery efforts (Keefer, 2003).

Modulation of Glutathione and Glutathionylation

One of the more interesting conundrums to emerge from the completion of the genome project is the realization that humans are a composite of <30,000 genes, and yet the complexity of protein structure/function seems distinctly more layered. In the burgeoning era of proteomics, it becomes clear that the central dogma of genetic determinism can be influenced by a number of processes that include polymorphic variants, gene-splicing events, exon shuffling, protein domain rearrangements, and the large number of posttranslational modifications that contribute to alterations in tertiary and quaternary protein structure. Among these, phosphorylation, glycosylation, methylation, and acetylation can account for a large proportion of modifications, and indeed, there is evidence to suggest that GSTs may be subject to either glycosylation (Kuzmich *et al.*, 1991) or phosphorylation (Lo *et al.*, 2004). More recently, the addition of glutathione to available cysteine residues (glutathionylation; Fig. 5) has been shown to be of consequence to protein function. The importance of modifying cysteine residues is not necessarily restricted to redox regulation but now seems to be a plausible event that can lead to changes in signaling processes, particularly as a response to a divergent number of stress responses. By adding the GSH tripeptide to a target protein, an additional negative charge is introduced (as a consequence of the glu residue), and a change in protein conformation is made likely. The implication from this analysis is that cells actively participate in the stochastic production of multiple protein building blocks with the

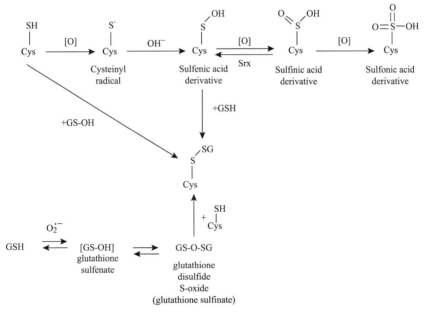

FIG. 5. Possible mechanisms of ROS-induced protein glutathionylation. ROS may induce glutathionylation of protein thiols by many different routes. Those highlighted here include the direct oxidation of protein cysteines to generate reactive protein thiol intermediates such as the reactive cysteinyl radical or sulfenic acid, which can further react with GSH to form a mixed disulfide. Alternately, a mixed disulfide is formed through reaction with oxidized forms of GSH (i.e., GS-OH or GS[O]SG).

intent of realizing functional nonredundancy. A recent report (Manevich *et al.*, 2004) showed that GSTπ is capable of acting in a catalytic manner to glutathionylate peroxiredoxin, a non-selenium–dependent lipid peroxidase that converts lipid hydroperoxides to corresponding alcohols. The catalytically important cysteine residue on peroxiredoxin is sterically inaccessible within the globular dimeric complex, and GSTπ facilitates GSH transfer to this site. The resultant enzyme activation serves a cellular regulatory role, particularly with respect to response to oxidative stress. It will be interesting for future research to determine how widespread the phenomenon of GSTπ catalyzed glutathionylation may be.

In the past, modulation of GSH and GST has been attempted as a means to improve response to cancer drugs. Use of, for example, BSO and ethacrynic acid, although effective in their experimental effects on each system, was not successful enough in the clinic to merit continued development (Bailey *et al.*, 1997; O'Dwyer *et al.*, 1991). One consequence of these approaches was the conceptual design of a peptidomimetic inhibitor of GSTπ discussed earlier. More recently, Novelos Inc. has developed a platinum-stabilized coordination complex of oxidized glutathione called NOV-002 (Fig. 6). It is oxidized glutathione complexed with cis-platinum at an approximate 1000:1 ratio. Standard animal and patient dosing with NOV-002 results in a cumulative total of cis-platinum that is equivalent to <2% of a typical standard of care in oncology *single*-dose cis-platinum. As such, it seems unlikely that the platinum component contributes substantially to the pharmacology of the compound. In some manner, the platinum (and perhaps the cis-amine groups) serves to stabilize the GSSG. In rodents, the serum and tissue levels of GSSG are profoundly affected by this stabilizing influence where standard doses will routinely produce a

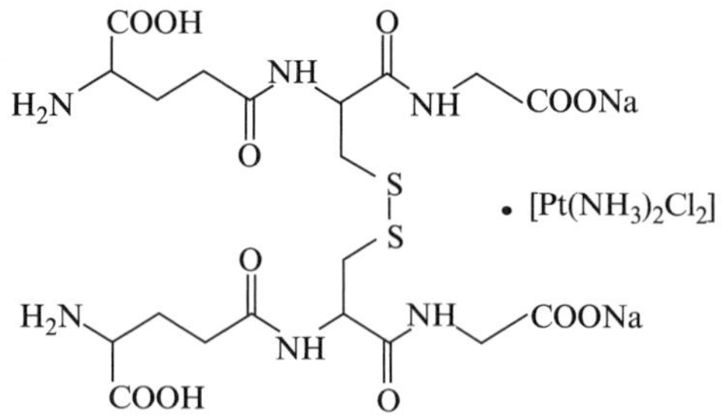

FIG. 6. Structure of a platinum-stabilized coordination complex of oxidized glutathione, NOV-002.

measurable elevation in circulating GSSG. Clinical data have been already been generated in Russia. Evidence of efficacy has been reported in 340 patients with diseases such as NSCLC, colorectal, pancreatic, and breast cancer. In particular, a 55% 1-year survival rate in patients with NSCLC (17% in control arm) improved the Karnofsky score for quality of life, and improved hematopoietic parameters (with results not dissimilar to those generated for TLK199) have been demonstrated. Lymphocyte counts were elevated in the NOV-002–treated animals, where the numbers of both $CD25^+$ and $CD16^+/CD56^+$ cells increased approximately fourfold after long-term treatment (60 days) with the drug.

Concentrations of NOV-002 between 10–100 mM induce glutathionylation of proteins in target cells. Although this concentration range may seem high, it is achievable in the plasma and tissues of rodents and has no obvious toxicities in the animals. NOV-002 affects the phosphorylation of both ERK and p38, two kinases with critical regulatory roles in governing cell proliferation and apoptotic pathways. *In vitro* data for NOV-002 seem to mimic those where GSSG is used, emphasizing that this moiety is, indeed, the active component of the drug. Once again, the similarity between NOV-002 and the data for TLK199 seem to indicate the general importance of GSH and GST pathways in governing proliferation in normal tissues and apoptosis in cancer cells.

NOV-002 has undergone significant clinical testing in Russia, and these trials are now being repeated in the United States. Treatment was associated with increases in circulating lymphocyte, monocyte, T-cell, and NK cell counts, and patient response rates were affected through reduced morbidity and the capacity to tolerate longer periods of standard chemotherapy. Quite independent of the clinical data, there are important implications for mechanism of action for NOV-002 that stem from the possibility that GSSG can act as a donor of glutathione (Fig. 5) in glutathionylation of receptive cysteine residues in target proteins.

References

Adachi, T., Pimentel, D. R., Heibeck, T., Hou, X., Lee, Y. J., Jiang, B., Ido, Y., and Cohen, R. A. (2004). S-glutathiolation of Ras mediates redox-sensitive signaling by angiotensin II in vascular smooth muscle cells. *J. Biol. Chem.* **279**, 29857–29862.

Adler, V., Yin, Z., Fuchs, S. Y., Benezra, M., Rosario, L., Tew, K. D., Pincus, M. R., Sardana, M., Henderson, C. J., Wolf, C. R., Davis, R. J., and Ronai, Z. (1999a). Regulation of JNK signaling by GSTp. *EMBO J.* **18**, 1321–1334.

Adler, V., Yin, Z., Tew, K. D., and Ronai, Z. (1999b). Role of redox potential and reactive oxygen species in stress signaling. *Oncogene* **18**, 6104–6111.

Bailey, H. H., Ripple, G., Tutsch, K. D., Arzoomanian, R. Z., Alberti, D., Feierabend, C., Mahvi, D., Schink, J., Pomplun, M., Mulcahy, R. T., and Wilding, G. (1997). Phase I study of continuous-infusion L-S,R-buthionine sulfoximine with intravenous melphalan. *J. Natl. Cancer Inst.* **89,** 1789–1796.

Callander, N., Ochoa-Bayona, J., Piro, L., Guba, S., Shapiro, G., Williams, S., Oliff, I., Burris, H., Jameson, A., Patel, K., Brown, G., FAderl, S., Estrov, Z., and Emanuel, P. (2004). Hematologic improvement following treatment with TLK199 (Telintra™), a novel glutathione analog inhibitor of GST P1–1, in myelodysplastic syndrome (MDS): Interim results of a dose-ranging phase 2a study. Session Type: Poster Session 582-I. *American Society Hematology Abstract* no. 1428.

Ciaccio, P. J., Shen, H., Kruh, G. D., and Tew, K. D. (1996). Effects of chronic ethacrynic acid exposure on glutathione conjugation and MRP expression in human colon tumor cells. *Biochem. Biophys. Res. Commun.* **222,** 111–115.

Ciaccio, P. J., Tew, K. D., and LaCreta, F. P. (1991). Enzymatic conjugation of chlorambucil with glutathione by human glutathione S-transferases and inhibition by ethacrynic acid. *Biochem. Pharmacol.* **42,** 1504–1507.

Comporti, M. (1989). Three models of free radical-induced cell injury. *Chem. Biol. Interact.* **72,** 1–56.

Cross, J. V., and Templeton, D. J. (2004). Oxidative stress inhibits MEKK1 by site-specific glutathionylation in the ATP-binding domain. *Biochem. J.* **381,** 675–683.

Davis, W., Jr., Ronai, Z., and Tew, K. D. (2001). Cellular thiols and reactive oxygen species in drug-induced apoptosis. *J. Pharmacol. Exp. Ther.* **296,** 1–6.

Dong, C., Yang, D. D., Wysk, M., Whitmarsh, A. J., Davis, R. J., and Flavell, R. A. (1998). Defective T cell differentiation in the absence of Jnk1. *Science* **282,** 2092–2095.

Findlay, V. J., Townsend, D. M., Saavedra, J. E., Buzard, G. S., Citro, M. L., Keefer, L. K., Ji, X., and Tew, K. D. (2004). Tumor cell responses to a novel glutathione S-transferase-activated nitric oxide-releasing prodrug. *Mol. Pharmacol.* **65,** 1070–1079.

Gate, L., Majumdar, R. S., Lunk, A., and Tew, K. D. (2004). Increased myeloproliferation in glutathione S-transferase pi-deficient mice is associated with a deregulation of JNK and Janus kinase/STAT pathways. *J. Biol. Chem.* **279,** 8608–8616.

Golemis, E. A., Tew, K. D., and Dadke, D. (2002). Protein interaction-targeted drug discovery: Evaluating critical issues. *Biotechniques* **32,** 636–638, 640, 642 passim.

Gotoh, Y., and Cooper, J. A. (1998). Reactive oxygen species- and dimerization-induced activation of apoptosis signal-regulating kinase 1 in tumor necrosis factor-alpha signal transduction. *J. Biol. Chem.* **273,** 17477–17482.

Hayes, J. D., and Pulford, D. J. (1995). The glutathione S-transferase supergene family: Regulation of GST and the contribution of the isoenzymes to cancer chemoprotection and drug resistance. *Crit. Rev. Biochem. Mol. Biol.* **30,** 445–600.

Henderson, C. J., Smith, A. G., Ure, J., Brown, K., Bacon, E. J., and Wolf, C. R. (1998). Increased skin tumorigenesis in mice lacking pi class glutathione S-transferases. *Proc. Natl. Acad. Sci. USA* **95,** 5275–5280.

Herzenberg, L. A., De Rosa, S. C., Dubs, J. G., Roederer, M., Anderson, M. T., Ela, S. W., and Deresinski, S. C. (1997). Glutathione deficiency is associated with impaired survival in HIV disease. *Proc. Natl. Acad. Sci. USA* **94,** 1967–1972.

Hochedlinger, K., Wagner, E. F., and Sabapathy, K. (2002). Differential effects of JNK1 and JNK2 on signal specific induction of apoptosis. *Oncogene* **21,** 2441–2445.

Joshi, M. S., Ponthier, J. L., and Lancaster, J. R., Jr. (1999). Cellular antioxidant and pro-oxidant actions of nitric oxide. *Free Radic. Biol. Med.* **27,** 1357–1366.

Kallunki, T., Deng, T., Hibi, M., and Karin, M. (1996). c-Jun can recruit JNK to phosphorylate dimerization partners via specific docking interactions. *Cell* **87**, 929–939.

Keefer, L. K. (2003). Progress toward clinical application of the nitric oxide-releasing diazeniumdiolates. *Annu. Rev. Pharmacol. Toxicol.* **43**, 585–607.

Keefer, L. K. (2005). Nitric Oxide (NO)- and Nitroxyl (HNO)-Generating Diazeniumdiolates (NONOates): Emerging Commercial Opportunities. *Curr. Top. Med. Chem.* **5**, 625–636.

Keppler, D., Leier, I., and Jedlitschky, G. (1997). Transport of glutathione conjugates and glucuronides by the multidrug resistance proteins MRP1 and MRP2. *Biol. Chem.* **378**, 787–791.

Kuzmich, S., Vanderveer, L. A., and Tew, K. D. (1991). Evidence for a glycoconjugate form of glutathione S-transferase pI. *Int. J. Pept. Protein Res.* **37**, 565–571.

Litwack, G., Ketterer, B., and Arias, I. M. (1971). Ligandin: A hepatic protein which binds steroids, bilirubin, carcinogens and a number of exogenous organic anions. *Nature* **234**, 466–467.

Lo, H. W., Antoun, G. R., and Ali-Osman, F. (2004). The human glutathione S-transferase P1 protein is phosphorylated and its metabolic function enhanced by the Ser/Thr protein kinases, cAMP-dependent protein kinase and protein kinase C, in glioblastoma cells. *Cancer Res.* **64**, 9131–9138.

Lyttle, M. H., Satyam, A., Hocker, M. D., Bauer, K. E., Caldwell, C. G., Hui, H. C., Morgan, A. S., Mergia, A., and Kauvar, L. M. (1994). Glutathione-S-transferase activates novel alkylating agents. *J. Med. Chem.* **37**, 1501–1507.

Makino, Y., Yoshikawa, N., Okamoto, K., Hirota, K., Yodoi, J., Makino, I., and Tanaka, H. (1999). Direct association with thioredoxin allows redox regulation of glucocorticoid receptor function. *J. Biol. Chem.* **274**, 3182–3188.

Manevich, Y., Feinstein, S. I., and Fisher, A. B. (2004). Activation of the antioxidant enzyme 1-CYS peroxiredoxin requires glutathionylation mediated by heterodimerization with pi GST. *Proc. Natl. Acad. Sci. USA* **101**, 3780–3785.

Morgan, A. S., Ciaccio, P. J., Tew, K. D., and Kauvar, L. M. (1996). Isozyme-specific glutathione S-transferase inhibitors potentiate drug sensitivity in cultured human tumor cell lines. *Cancer Chemother. Pharmacol.* **37**, 363–370.

Morgan, A. S., Sanderson, P. E., Borch, R. F., Tew, K. D., Niitsu, Y., Takayama, T., Von Hoff, D. D., Izbicka, E., Mangold, G., Paul, C., Broberg, U., Mannervik, B., Henner, W. D., and Kauvar, L. M. (1998). Tumor efficacy and bone marrow-sparing properties of TER286, a cytotoxin activated by glutathione S-transferase. *Cancer Res.* **58**, 2568–2575.

Napoli, C., and Ignarro, L. J. (2003). Nitric oxide-releasing drugs. *Annu. Rev. Pharmacol. Toxicol.* **43**, 97–123.

O'Dwyer, P. J., LaCreta, F., Nash, S., Tinsley, P. W., Schilder, R., Clapper, M. L., Tew, K. D., Panting, L., Litwin, S., Comis, R. L., and Ozols, R. F. (1991). Phase I study of thiotepa in combination with the glutathione transferase inhibitor ethacrynic acid. *Cancer Res.* **51**, 6059–6065.

Peterson, J. D., Herzenberg, L. A., Vasquez, K., and Waltenbaugh, C. (1998). Glutathione levels in antigen-presenting cells modulate Th1 versus Th2 response patterns. *Proc. Natl. Acad. Sci. USA* **95**, 3071–3076.

Powis, G., Kirkpatrick, D. L., Angulo, M., and Baker, A. (1998). Thioredoxin redox control of cell growth and death and the effects of inhibitors. *Chem. Biol. Interact.* **111–112**, 23–34.

Rosario, L. A., O'Brien, M. L., Henderson, C. J., Wolf, C. R., and Tew, K. D. (2000). Cellular response to a glutathione S-transferase P1-1 activated prodrug. *Mol. Pharmacol.* **58**, 167–174.

Ruscoe, J. E., Rosario, L. A., Wang, T., Gate, L., Arifoglu, P., Wolf, C. R., Henderson, C. J., Ronai, Z., and Tew, K. D. (2001). Pharmacologic or genetic manipulation of glutathione S-transferase P1-1 (GStpi) influences cell proliferation pathways. *J. Pharmacol. Exp. Ther.* **298**, 339–345.

Saavedra, J. E., Billiar, T. R., Williams, D. L., Kim, Y. M., Watkins, S. C., and Keefer, L. K. (1997). Targeting nitric oxide (NO) delivery *in vivo*. Design of a liver-selective NO donor prodrug that blocks tumor necrosis factor-alpha-induced apoptosis and toxicity in the liver. *J. Med. Chem.* **40**, 1947–1954.

Saavedra, J. E., Bohle, D. S., Smith, K. N., George, C., Deschamps, J. R., Parrish, D., Ivanic, J., Wang, Y. N., Citro, M. L., and Keefer, L. K. (2004). Chemistry of the diazeniumdiolates. O- versus N-alkylation of the RNH[N(O)NO](-) ion. *J. Am. Chem. Soc.* **126**, 12880–12887.

Saavedra, J. E., Shami, P. J., Wang, L. Y., Davies, K. M., Booth, M. N., Citro, M. L., and Keefer, L. K. (2000). Esterase-sensitive nitric oxide donors of the diazeniumdiolate family: *In vitro* antileukemic activity. *J. Med. Chem.* **43**, 261–269.

Saitoh, M., Nishitoh, H., Fujii, M., Takeda, K., Tobiume, K., Sawada, Y., Kawabata, M., Miyazono, K., and Ichijo, H. (1998). Mammalian thioredoxin is a direct inhibitor of apoptosis signal-regulating kinase (ASK) 1. *EMBO J.* **17**, 2596–2606.

Satyam, A., Hocker, M. D., Kane-Maguire, K. A., Morgan, A. S., Villar, H. O., and Lyttle, M. H. (1996). Design, synthesis, and evaluation of latent alkylating agents activated by glutathione S-transferase. *J. Med. Chem.* **39**, 1736–1747.

Schwabe, R. F., Bradham, C. A., Uehara, T., Hatano, E., Bennett, B. L., Schoonhoven, R., and Brenner, D. A. (2003). c-Jun-N-terminal kinase drives cyclin D1 expression and proliferation during liver regeneration. *Hepatology* **37**, 824–832.

Shami, P. J., Saavedra, J. E., Wang, L. Y., Bonifant, C. L., Diwan, B. A., Singh, S. V., Gu, Y., Fox, S. D., Buzard, G. S., Citro, M. L., Waterhouse, D. J., Davies, K. M., Ji, X., and Keefer, L. K. (2003). JS-K, a glutathione/glutathione S-transferase-activated nitric oxide donor of the diazeniumdiolate class with potent antineoplastic activity. *Mol. Cancer Ther.* **2**, 409–417.

Shaulian, E., and Karin, M. (2001). AP-1 in cell proliferation and survival. *Oncogene* **20**, 2390–2400.

Shen, H., Tsuchida, S., Tamai, K., and Sato, K. (1993). Identification of cysteine residues involved in disulfide formation in the inactivation of glutathione transferase P-form by hydrogen peroxide. *Arch. Biochem. Biophys.* **300**, 137–141.

Terada, K., Kaziro, Y., and Satoh, T. (1997). Ras-dependent activation of c-Jun N-terminal kinase/stress-activated protein kinase in response to interleukin-3 stimulation in hematopoietic BaF3 cells. *J. Biol. Chem.* **272**, 4544–4548.

Tew, K. D. (1994). Glutathione-associated enzymes in anticancer drug resistance. *Cancer Res.* **54**, 4313–4520.

Tournier, C., Hess, P., Yang, D. D., Xu, J., Turner, T. K., Nimnual, A., Bar-Sagi, D., Jones, S. N., Flavell, R. A., and Davis, R. J. (2000). Requirement of JNK for stress-induced activation of the cytochrome c-mediated death pathway. *Science* **288**, 870–874.

Townsend, D., and Tew, K. (2003). Cancer drugs, genetic variation and the glutathione-S-transferase gene family. *Am. J. Pharmacogenomics* **3**, 157–172.

Umansky, V., and Schirrmacher, V. (2001). Nitric oxide-induced apoptosis in tumor cells. *Adv. Cancer Res.* **82**, 107–131.

Wang, T., Arifoglu, P., Ronai, Z., and Tew, K. D. (2001). Glutathione S-transferase P1-1 (GSTP1-1) inhibits c-Jun N-terminal kinase (JNK1) signaling through interaction with the C terminus. *J. Biol. Chem.* **276**, 20999–21003.

Yang, Y. M., Bost, F., Charbono, W., Dean, N., McKay, R., Rhim, J. S., Depatie, C., and Mercola, D. (2003). C-Jun NH(2)-terminal kinase mediates proliferation and tumor growth of human prostate carcinoma. *Clin. Cancer Res.* **9**, 391–401.

Yin, Z., Ivanov, V. N., Habelhah, H., Tew, K., and Ronai, Z. (2000). Glutathione S-transferase p elicits protection against H_2O_2-induced cell death via coordinated regulation of stress kinases. *Cancer Res.* **60**, 4053–4057.

[20] Modification of N-Acetyltransferases and Glutathione S-Transferases by Coffee Components: Possible Relevance for Cancer Risk

By WOLFGANG W. HUBER and WOLFRAM PARZEFALL

Abstract

Enzymes of xenobiotic metabolism are involved in the activation and detoxification of carcinogens and can play a pivotal role in the susceptibility of individuals toward chemically induced cancer. Differences in such susceptibility are often related to genetically predetermined enzyme polymorphisms but may also be caused by enzyme induction or inhibition through environmental factors or in the frame of chemopreventive intervention. In this context, coffee consumption, as an important lifestyle factor, has been under thorough investigation. Whereas the data on a potential procarcinogenic effect in some organs remained inconclusive, epidemiology has clearly revealed coffee drinkers to be at a lower risk of developing cancers of the colon and the liver and possibly of several other organs. The underlying mechanisms of such chemoprotection, modifications of xenobiotic metabolism in particular, were further investigated in rodent and *in vitro* models, as a result of which several individual chemoprotectants out of the >1000 constituents of coffee were identified as well as some strongly metabolized individual carcinogens against which they specifically protected. This chapter discusses the chemoprotective effects of several coffee components and whole coffee in association with modifications of the usually protective glutathione-S-transferase (GST) and the more ambivalent N-acetyltransferase (NAT). A key role is played by kahweol and cafestol (K/C), two diterpenic constituents of the unfiltered beverage that were found to reduce mutagenesis/tumorigenesis by strongly metabolized compounds, such as 2-amino-1-methyl-6-phenylimidazo-[4,5-*b*]pyridine, 7,12-dimethyl-benz[a]anthracene, and aflatoxin B_1, and to cause various modifications of

METHODS IN ENZYMOLOGY, VOL. 401
Copyright 2005, Elsevier Inc. All rights reserved.
0076-6879/05 $35.00
DOI: 10.1016/S0076-6879(05)01020-7

xenobiotic metabolism that were overwhelmingly beneficial, including induction of GST and inhibition of NAT. Other coffee components such as polyphenols and K/C-free coffee are also capable of increasing GST and partially of inhibiting NAT, although to a somewhat lesser extent.

Introduction: Xenobiotic Metabolism and Cancer

The presence or absence of activating or detoxifying enzymes of xenobiotic metabolism and variation in the activity of these enzymes are of utmost importance for the sensitivity of an individual to toxicants. This is of particular relevance for chemical carcinogens that often require metabolic activation. Consequently, epidemiological investigations have shown that polymorphic distribution of certain enzymes of xenobiotic metabolism may be associated with differences in cancer risk, usually under conditions of known exposure to metabolized carcinogens (reviewed e.g., in Thier *et al.* [2003]). In this context, great relevance has to be attributed to polymorphisms of glutathione *S*-transferases (GSTs) and *N*-acetyltransferases (NATs), two important conjugating (phase II) enzyme systems of xenobiotic metabolism. GSTs are predominantly involved in carcinogen-detoxification, and carriers of functionally deficient genotypes (e.g., of the subtypes GSTA1, GSTM1, GSTP1, GSTT1, or combinations) have consequently been reported in numerous studies to be at an increased risk of developing cancers of, for example, lung (Hou *et al.*, 2000; Malats *et al.*, 2000; Miller *et al.*, 2002, 2003; Saarikoski *et al.*, 1998; Stucker *et al.*, 1999, 2002), pleura (Hirvonen *et al.*, 1995), bladder (Johns and Houlston, 2000; Steinhoff *et al.*, 2000), prostate (Steinhoff *et al.*, 2000), testis (Harries *et al.*, 1997), breast, (Mitrunen *et al.*, 2001) or colon/rectum (Coles *et al.*, 2001; Sweeney *et al.*, 2002; Zhong *et al.*, 1993) and/or to accumulate a greater amount of genotoxic damage (Knudsen *et al.*, 1999; Scarpato *et al.*, 1997; Schröder *et al.*, 1995). Frequently, such findings were associated with specific exposures for example through tobacco smoke (Miller *et al.*, 2003), well-done meat (Sweeney *et al.*, 2002), air pollution (Knudsen *et al.*, 1999), and occupation through such substances as aromatic amines and asbestos (Hirvonen *et al.*, 1995, 1996). However, certain carcinogens such as dihaloalkanes and trichloroethene are also activated by GSTs (Sherratt *et al.*, 1998; Thier *et al.*, 1993, 1996), which may explain why functional GSTM1 and GSTT1 status was more frequently observed in patients with renal cell cancer who had been exposed to trichloroethene (Bruning *et al.*, 1997). It also has to be kept in mind that many anticancer drugs are inactivated by GSTs, so that such therapies would require low rather than high activity of GSTs associated with maximum toxicity in the cancer cells (Coles and Kadlubar, 2003).

NATs are generally more ambivalent than GSTs, because they were reported to activate food-borne heterocyclic amines leading to colon cancer (Hein et al., 2000; Lang et al., 1994; Minchin et al., 1993) but to detoxify industrial aromatic amines such as benzidine by which they prevent cancers of the bladder (Golka et al., 2002; Kadlubar and Badawi, 1995). Thus, the classical polymorphism of NAT2 implicates an overrepresentation of rapid over slow acetylators among patients with colon cancer, the opposite being true for those who have bladder cancer. The ambivalence of NAT was underlined by further studies that associated fast acetylation with greater susceptibility toward cancers of lung (Cascorbi et al., 1996) and prostate (Fukutome et al., 1999), whereas slow acetylators displayed a higher background level of chromosomal aberrations in lymphocytes (Norppa, 2001).

Xenobiotic Metabolism and Chemoprevention

Apart from mere genetic predetermination, a considerable number of enzymes of xenobiotic metabolism, including GSTs and, to a lesser extent, NATs, may be induced or inhibited by externally applied substances. Consequently, such modifications may influence the susceptibility of the exposed individual to toxic effects and must, therefore, be considered in the corresponding risk estimates (Coles and Kadlubar, 2003). Importantly, inhibitors of carcinogen activation and inducers of carcinogen detoxification have become a major focus in the development of strategies of chemoprevention. Both potentially protective GST induction and anti-mutagenic/anticarcinogenic effects in a putatively causal relationship have, for instance, been observed in vivo with the antischistosomal test drug oltipraz (Buetler et al., 1995; Kensler, 1997), the food additives butylated hydroxyanisole (BHA) (Buetler et al., 1995; Lam and Zhang, 1991; Reddy and Maeura, 1984) and ethoxyquin (Buetler et al., 1995; Kensler et al., 1985), as well as with the plant components diallyl sulfide (Le Bon et al., 1997; Sumiyoshi and Wargovich, 1990; Wargovich, 1987), and α-angelicalactone (Nijhoff and Peters, 1994; Wattenberg et al., 1979). Other than GSTs that may be regarded as one of the classical inducible enzyme systems, NATs had, until recently, been considered as being exclusively under genetic regulation (Lampe et al., 2000). This opinion was apparently confirmed by the observed high degree of agreement between NAT genotypes and phenotypes (Gross et al., 1999; Le Marchand et al., 1996). However, decreases in acetylated urinary and/or fecal metabolites of indicator compounds were observed in rats pretreated with BHA and butylated hydroxytoluene (Chung, 1999) and in humans receiving the over-the-counter drug acetaminophen (Rothen et al., 1998). Moreover, a series of further compounds (e.g., diallylsufide or vitamin C) caused a drop in acetylation products in several *in vitro* systems that included human-derived cell lines from liver (Wu and Chung, 1998; Wu *et al.*, 2001) and colon (Chang *et al.*, 2001; Chen *et al.*, 1998). Aloe-emodin, berberin, curcurmin, and paclitaxel caused an *in vitro* decrease of NAT mRNAs, indicating direct interference with synthesis regulation at least in the extrahepatic and extracolonic models concerned (Chen *et al.*, 2003; Chung *et al.*, 2003; Wang *et al.*, 2002; Yang *et al.*, 2003). Thus, despite some open questions regarding these data, modification of NAT activity might become a promising and fairly novel means of chemopreventive intervention.

As far as chemoprotection in general is concerned, great interest is being dedicated to the enzyme-modifying potential of so-called lifestyle-related chemicals (i.e., substances to which millions of people are voluntarily exposed in large quantities, mostly on a daily basis). Such a definition is certainly valid for the components of coffee that are ingested with one of the most popular beverages in the world. Thus, this chapter will primarily deal with the effects of whole coffee and its most important constituents on GSTs and NATs and with the resulting possible implications on cancer risk and prevention.

Coffee and Cancer: Epidemiology

On the one hand, factors of lifestyle may cause exposure to numerous carcinogenic chemicals, for instance, through tobacco smoke or unhealthy food characteristics that are often related to Western lifestyle. On the other hand, aspects of lifestyle have been found to play a role in the prevention of cancer. Therefore, as an extremely widespread lifestyle factor, coffee consumption has been thoroughly investigated for procarcinogenic and anticarcinogenic effects. A potential procarcinogenic effect of coffee has been discussed for tumors of bladder, ovary, and pancreas, but the data remained inconclusive and/or the modifications extraordinarily modest (Weiderpass *et al.*, 1998; Isaksson *et al.*, 2002; Jacobsen and Heuch, 2000; Lin *et al.*, 2002; Michaud *et al.*, 2001; Porta *et al.*, 2000; Silverman *et al.*, 1998; Slebos *et al.*, 2000; Tavani and La Vecchia, 2000; Tavani *et al.*, 2001; Villeneuve *et al.*, 2000; Zeegers *et al.*, 2001a,b). A rather clear protection of coffee drinkers was, however, seen against carcinogenesis of the colon (Giovannucci, 1998; IARC, 1991; Tavani and La Vecchia, 2000, 2004) and, more recently, the liver (Gallus *et al.*, 2002a; Inoue *et al.*, 2005), the latter being in agreement with the protection against liver cirrhosis that coffee consumption had also been reported to provide (Gallus *et al.*, 2002b; Klatsky and Armstrong, 1992; Tverdal and Skurtveit, 2003). Additional although not always reconfirmed epidemiological studies have, moreover, suggested inhibition of malignomas in the following extrahepatic and extracolonic organs: skin (melanoma) (Veierod *et al.*, 1997); gallbladder (Ghadirian *et al.*, 1993); breast (in lean women) (Vatten *et al.*, 1990); endometrium (Petridou *et al.*, 2002; Terry *et al.*, 2002); mouth, tongue, and pharynx (Franceschi *et al.*, 1992; Rodriguez *et al.*, 2004; Tavani *et al.*, 2003); esophagus (Tavani *et al.*, 2003), lung (in female smokers) (Kubik *et al.*, 2004); ovary (Jordan *et al.*, 2004); and thyroid (Takezaki *et al.*, 1996). Thus, in summary, the available data suggest a predominantly beneficial effect of coffee with regard to cancer risk, particularly when considering colon and liver cancer.

Individual Chemoprotective Coffee Constituents and Susceptible Carcinogenic Exposures: Investigation in Animal and *In Vitro* Studies

General Principles and Chemoprotection by Whole Coffee

Animal and *in vitro* studies were carried out to further elucidate the potential chemoprotection by coffee and the underlying mechanisms. In agreement with human epidemiology, preparations of whole coffee and coffee beans, applied either in the feed or drinking fluid, were found to exert antitumorigenic effects in several organs of rodents (Gershbein, 1994; Miller *et al.*, 1988; Stalder *et al.*, 1990; Tanaka *et al.*, 1990; Wattenberg, 1983; Woutersen *et al.*, 1989) and whole coffee preparations were likewise shown to be antimutagenic in *in vivo* (Abraham, 1991; Abraham *et al.*, 1998; Turesky *et al.*, 2003) and *in vitro* (Abraham and Stopper, 2004) mammalian assays. However, coffee is a mixture of more than 1000 individual substances whose final pattern in the consumed beverage is, moreover, greatly influenced by the method of preparation (e.g., by filtering), which would explain why coffee studies may occasionally also reveal promutagenic (Aeschbacher *et al.*, 1989; Nagao *et al.*, 1986) and protumorigenic (Saroja *et al.*, 2001) effects that are not necessarily reflected in human epidemiology. Thus, to evaluate the chemoprotective potential of coffee, it became inevitable to identify and to further investigate the individual protective constituents or constituent combinations. Furthermore, work was dedicated to single out carcinogenic exposures that are potentially susceptible to such coffee-related chemoprevention. Because of the scope of this chapter, we confine ourselves to reviewing those coffee constituents that were found to be capable of modifying GST and/or NAT (i.e., where the induced or inhibited enzymes may at least in part explain the accompanying antimutagenic/antitumorigenic effects that would consequently have to occur during the tumor initiation phase).

Chemoprotection by Kahweol and Cafestol (K/C)

Among coffee constituents, a relevant protective potential was revealed for kahweol and cafestol (K/C; the abbreviation usually refers to a 1:1 mixture of either the palmitates or equimolar amounts of the free alcohols) (for formulas, see Fig. 1), two diterpenes that account for 1–2% of the *Coffea Arabica* coffee bean and of which up to 90 mg each may be found in a liter of unfiltered coffee (e.g., in the varieties prepared in the Turkish or Scandinavian style) (Gross *et al.*, 1997; Urgert *et al.*, 1995). However, only traces remain in filtered or instant coffee, whereas intermediate amounts are encountered in espresso. Beans from *Coffea Arabica* contain more of both substances than those from *Coffea Canephora* (*Robusta*) (Mensink *et al.*, 1995). Chemoprotection by K/C was observed against several important carcinogens/mutagens. For instance, hamsters on a diet containing 0.2% K/C were found to develop fewer buccal pouch tumors upon painting with 7,12-dimethylbenz[a]anthracene (Miller *et al.*, 1991). A 10-day pretreatment of male F344 rats with the same food concentration of K/C decreased by approximately 50% the formation of colonic DNA adducts that followed a challenge with the heterocyclic amine 2-amino-1-methyl-6-phenylimidazo[4,5-b]-pyridine (PhIP) (Huber *et al.*, 1997), a so-called cooked food mutagen to which humans are strongly exposed and to which a role in the development of human colon cancer is attributed (Adamson *et al.*, 1996; Nowell *et al.*, 2002). Furthermore, subcellular liver fractions of K/C-treated rats were capable of inhibiting the *in vitro* covalent DNA-binding of aflatoxin B_1 (AFB1) (Cavin *et al.*, 1998). Decreases in DNA adducts of AFB1, PhIP, N-nitrosodimethylamine, and benzo[a]pyrene were also observed in K/C-exposed primary rat hepatocyte cultures and, importantly, in human-derived hepatic or pulmonary cell lines (Cavin *et al.*, 2001, 2003; Majer *et al.*, 2005). Notably, chemoprotection by K/C seemed to be predominantly effective against carcinogens that are strongly metabolized.

Chemoprotection by Polyphenols

Another group of coffee components that raised interest with regard to chemoprotection are polyphenols such as chlorogenic (CGA), ferulic (FA), caffeic (CA), p-coumaric (p-CMA), and tannic (TA) acids (for formulas, see Fig.1) (Chung *et al.*, 1998; Mori *et al.*, 2000). The term "CGA" may, at times, be used for all esters of phenolic acids and the aliphatic alcohols of quinic acid, but according to IUPAC, it should refer only to 5-caffeoylquinic acid that represents approximately 60% of total CGAs in green coffee beans (Illy and Viani, 1995; Farah *et al.*, 2005), the remaining part consisting of further monoesters and diesters of CA and of esters of FA (5–6%)

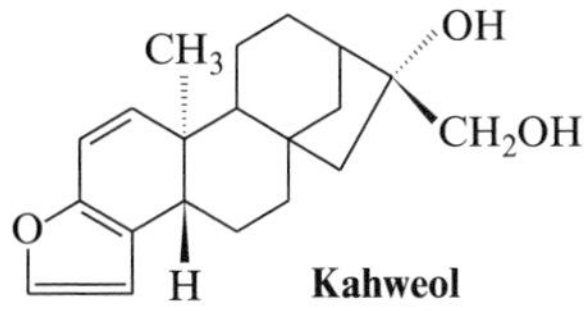

Kahweol

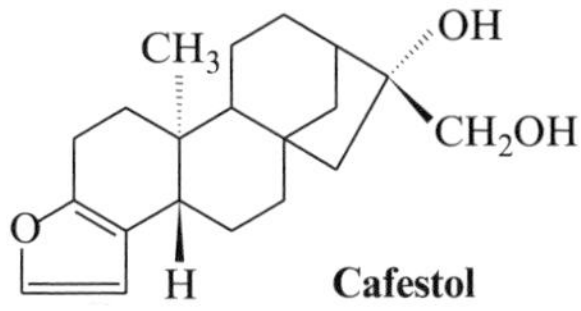

Cafestol

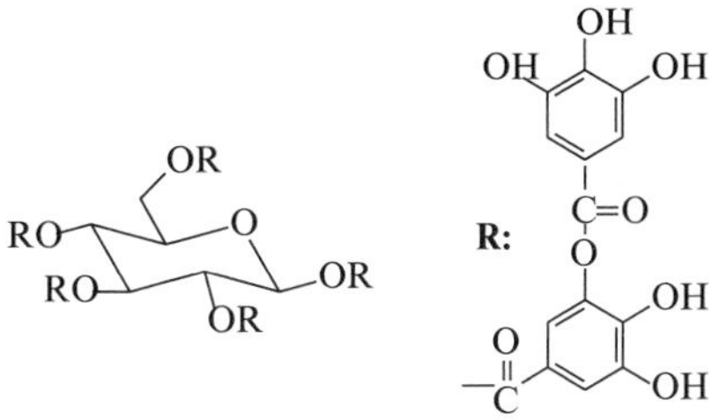

Tannic acid (TA)

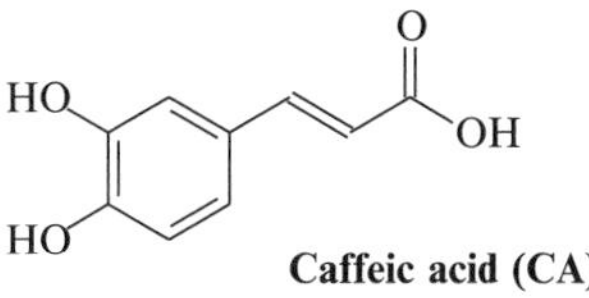

R:

Caffeic acid (CA)

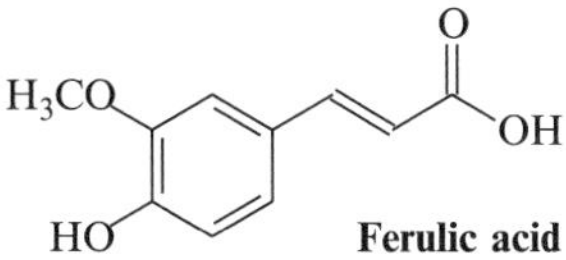

Ferulic acid (FA)

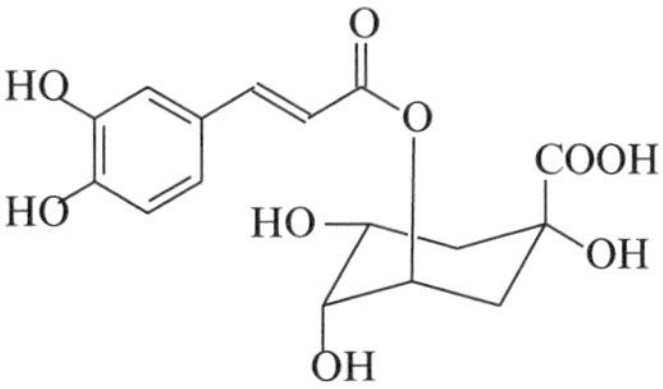

Chlorogenic acid
(5-caffeoylquinic acid; CGA)

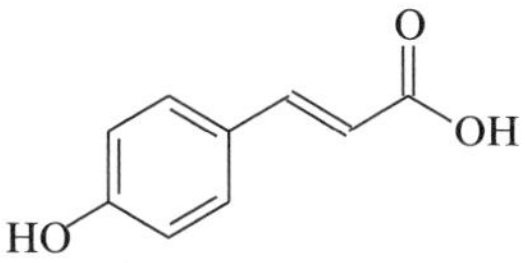

p-coumaric acid (p-CMA)

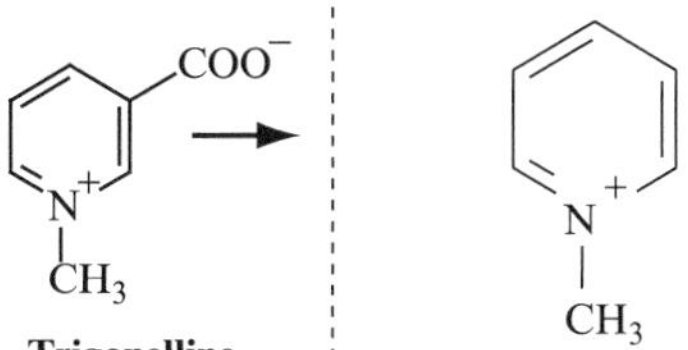

Trigonelline

N-methylpyridinium
ion

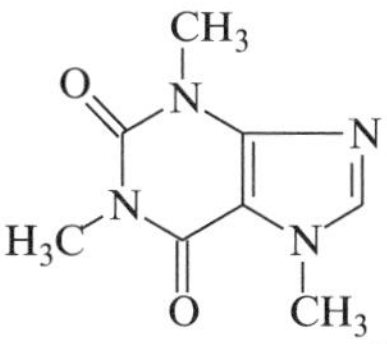

Caffeine

and p-CMA (8–9%). Total CGAs were reported to make up 6–7% of *Arabica* and 10% of *Robusta* beans, but roasting reduced the content to 2.5 and 3.8%, respectively (Illy and Viani, 1995). The opposite effect of roasting was observed for TA equivalents in coffee, which increased from 0.6–0.7% to 1.7–1.8% (Savolainen, 1992).

CGA, FA, and CA in the diet inhibited the development of tongue carcinogenesis in rats given 4-nitroquinoline-1-oxide (Tanaka *et al.*, 1993). In the same species, feeding with CA reduced by almost 50% the number and the area of spontaneously occurring preneoplasia (GSTP + foci) in the liver (Hagiwara *et al.*, 1996), and FA in the food for 5 weeks protected clearly against the formation of colonic aberrant crypt foci by azoxymethane given by subcutaneous injections during this period (Kawabata *et al.*, 2000; Wargovich *et al.*, 2000). Apart from protective effects upon topical application to the skin (Mukhtar *et al.*, 1988), which is not representative of coffee consumption, TA was shown to inhibit benzo[a] pyrene-induced DNA adduct formation and tumorigenesis in the lungs and forestomachs of mice when given in the diet (Athar *et al.*, 1989). Polyphenols are well-known as antioxidants (Chung *et al.*, 1998; Guglielmi *et al.*, 2003; Nardini *et al.*, 1997; Khan *et al.*, 2000; Kono *et al.*, 1997) and, as further potentially chemoprotective mechanisms, they have, for instance, been reported to inhibit nitric oxide synthetase (Srivastava *et al.*, 2000), peroxy-nitrite-dependent nitration (Pannala *et al.*, 1998), nitrosation (Kono *et al.*, 1995; Rundlof *et al.*, 2000), cell proliferation (Hudson *et al.*, 2000; Iwai *et al.*, 2004), AP-1 transcriptional activity (Maggi-Capeyron *et al.*, 2001), and leukotriene C4 synthesis (Leung, 1986). On the other hand, they were also capable of inducing prooxidant, promutagenic, and/or protumorigenic effects under particular circumstances, which may be partially related to higher doses or differing time points of application (Hagiwara *et al.*, 1991; Hirose *et al.*, 1991, 1992, 1999; Huber *et al.*, 1997; Lutz *et al.*, 1997; Yamanaka *et al.*, 1997).

FIG. 1. Chemical structures and GST-enhancing capacity of coffee components discussed in this chapter. GST-enhancing capacity is indicated by the margins of the boxes; thick, solid line, strong capacity, greater than twofold maximally observed enhancement of GST-CDNB *in vivo* (any organ); thin solid line, intermediate capacity, 1.5- to 2-fold maximally observed enhancement of GST-CDNB *in vivo* (any organ); dashed line, moderate capacity, less than 1.5-fold maximal enhancement only observed *in vitro* or of specific parameter(s) other than GST-CDNB; dash-dotted line, questionable capacity, only limited indirect evidence alongside of several negative results.

Chemoprotection by Caffeine

Despite being regarded almost as a synonym of coffee by the public, caffeine, which is contained at approximately 1–2% in the coffee bean and rather resistant to roasting (Illy and Viani, 1995) (for formula, see Fig. 1), has thus far not played more than a moderate role in considerations on coffee-related and also tea-related chemoprotection against cancer. This may be partially due to the widely known adverse effects on heart, blood circulation, and nervous system. Moreover, interaction between caffeine and the carcinogenic process has turned out to be extraordinarily complex, which has led to both anticarcinogenic and procarcinogenic study outcomes (reviewed in Conney [2003]). Protective effects against carcinogen exposure of animals *in vivo* were seen in skin (Lou *et al.*, 1999), lung (Nomura, 1980, 1983), and breast (Hagiwara *et al.*, 1999; Hirose *et al.*, 2002; Petrek *et al.*, 1985; Welsch *et al.*, 1988), the latter being in agreement with a human study in which an inverse relationship between methylxanthine exposure and breast cancer was observed (Lubin *et al.*, 1985). However, other investigations showed caffeine-related stimulation of parameters of tumorigenesis in breast, skin, and colon (Conney, 2003; Hagiwara *et al.*, 1999; Tsuda *et al.*, 1999).

Chemoprotection by Other Coffee Components

Melanoidins are a chemically heterogeneous group of compounds that are created during the roasting procedure by the Maillard reaction and are associated with the brown color of coffee and other alimentary products (Anese and Nicoli, 2003; Borrelli *et al.*, 2003). Thus far, their chemoprotective potential has been little investigated, and the associated studies have focused on their antioxidant capacity (Daglia *et al.*, 2004). Furthermore, a role in chemoprotection was attributed to coffee fiber. Possible involvement of the contained polyphenols and melanoidins as well as the capacity of the fibers to adsorb carcinogens are discussed in connection with this effect (Borrelli *et al.*, 2004; Kato *et al.*, 1991; Rao *et al.*, 1998).

Summary of Chemoprotectants in Coffee

All chemoprotective components specifically mentioned in this chapter, with the possible exception of caffeine, have been found to interfere with xenobiotic metabolism by GST and partially NAT, albeit both to a varying degree. As will be described in detail later, the most convincing effects with regard to both NAT and GST have been identified with K/C.

Therefore, K/C will be the focus of the following sections. Also, other than whole coffee, caffeine, and polyphenols, K/C have thus far not been associated with any promutagenic or protumorigenic effects. Negative results were obtained in two genotoxicity studies in *Salmonella* (Pezzuto *et al.*, 1986) and HepG2 (Majer *et al.*, 2005), and no indications of early tumor formation emerged after 90 days of treatment (Miller *et al.*, 1991; Schilter *et al.*, 1996). Although K/C was capable of enhancing hepatic GST-π and γ-glutamyl transpeptidase (GGT, see later), which may serve as indicators of preneoplasia, the required focal distribution of this enzyme induction was missing (Huber *et al.*, 2002a,b; Schilter *et al.*, 1996). Moreover, *in vivo* and *in vitro* indicators of oxidative stress remained unchanged or were even somewhat reduced by K/C (Huber *et al.*, 2002b; Scharf *et al.*, 2003). Still, caution has to be exercised with regard to the potential of K/C to raise blood cholesterol in humans (Urgert and Katan, 1997; Weusten-Van der Wouw *et al.*, 1994).

Effects of K/C on NAT, GST, and GST-Related Parameters
(see also Fig. 2)

NAT

The dose of 0.2% K/C in the diet of male F344 rats for 10 days caused an 80% decrease in the NAT-dependent activation of N-OH-PhIP to DNA-binding species by the hepatic cytosols of these animals in an *in vitro* assay (Huber *et al.*, 2004). The effect was dose-dependent, and a statistically significant decrease by 13% occurred even at 0.02% K/C in the diet. In agreement, NAT mRNAs in rat liver were decreased (unpublished observation), and both kahweol and cafestol palmitates inhibited by 80% the *in vitro* acetylation of 2-aminofluorene in primary cultures of rat hepatocytes (Huber *et al.*, 2004). However, in the human-derived hepatoma cell line HepG2, NAT activity, as measured by the metabolism of p-aminobenzoic acid, was reported to remain unmodified (Majer *et al.*, 2005). K/C *in vivo* was not found to decrease NAT activity in the cytosols of the colon, although the same treatment reduced the formation of PhIP-DNA adducts in this organ (Huber *et al.*, 1997, 2004). This discrepancy might be explained by the observation that N-acetoxy-PhIP was sufficiently stable to reach extrahepatic organs (Lin *et al.*, 1994), which makes it conceivable that diminished formation of N-acetoxy-PhIP in the liver would also lead to reduced amounts of this activated mutagen in other parts of the body. Thus, it is suggested that K/C is, indeed, capable of *in vivo* NAT inhibition and that this effect may contribute to the body's defense against heterocyclic amines such as PhIP. This conversion of fast to slow acetylator phenotype in rats seems to be a rather unique element of chemoprevention, because, in the same study, it was not observed with black tea and benzyl isothiocyanate, although these treatments caused a similar decrease in PhIP-DNA adducts as K/C (Huber *et al.*, 1997, 2004). On the other hand, it cannot entirely be ruled out that K/C might also be capable of inhibiting NAT-related detoxification of aromatic amines, possibly raising the risk of bladder cancer. Interestingly, in some studies, coffee consumption was, indeed, associated with a marginal increase of cancer at this site (Zeegers *et al.*, 2001b). However, the extraordinarily small degree of the reported increase may indicate that other beneficial effects of K/C (see later) may have essentially neutralized this potential hazard.

GST

K/C were shown to be efficient and versatile *in vivo* inducers of GSTs in the rat (Cavin *et al.*, 1998; Huber *et al.*, 1997; 2002a, 2004; Schilter *et al.*, 1996), as well as in the mouse (Di Simplicio *et al.*, 1989; Lam *et al.*, 1982, 1987; McMahon *et al.*, 2001). For instance, overall hepatic GST activity, as determined by the metabolism of 1-chloro-2,4-dinitrobenzene (GST-CDNB), was increased threefold to fourfold when rats were exposed to a dietary concentration of 0.2% K/C for 10 days (Huber *et al.*, 2004). The effect was dose dependent, and enhancements in the range of 30% were still observed at one-tenth the dietary concentration (i.e., 0.02%), although the changes at this dose were not always statistically significant. The induction of GST-CDNB by K/C was found to be strongly organ-specific. As far as extrahepatic organs of the rat are concerned, the increase was considerable in kidney; weaker in lung; marginal in colon; and absent in pancreas, spleen, salivary gland, heart, and testis (Huber *et al.*, 2002a). Investigation of enzyme proteins, mRNAs, and/or the activity toward subtype-specific substrates revealed that hepatic GST-induction by K/C in the rat concerned all four major subtypes (α, μ, π, and θ) (Cavin *et al.*, 1998; Huber *et al.*, 1997; 2002a, 2004; Schilter *et al.*, 1996). The changes ranged from a barely twofold increase in GST-θ–specific metabolism of 1,2-epoxy-3-(p-nitrophenoxy)propane to an up to 23-fold higher activity toward the GST-π–specific substrate 4-vinylpyridine, the latter, however, compared with the low control levels characteristic of GST-π in the liver. As an apparent exception to these inductions, the protein of the α-subunit rGSTA1 remained unmodified, thus displaying a sharp contrast to the protein of rGSTA2 that was increased sevenfold in the same animals (Huber et al., 2004). These in vivo data were in agreement with studies in vitro that revealed strong increases in GST-π and the AFB1-metaboliz-ing GST-α subunit rGSTA5 in primary rat hepatocytes (Cavin et al., 2001, 2003), as well as significantly elevated GST-CDNB in the human-derived hepatoma cell line HepG2 (Majer et al., 2005). Increases in GST-μ protein, but not in GST-α, GST-π, and GST-CDNB, were observed in the human-derived liver epithelial cell line THLE (Cavin et al., 2001), and a moderate enhancement of GST-π occurred in Beas-2B cells, a nonhepatic (i.e., bronchial) cell line of human origin (Cavin et al., 2003). In in vivo studies in the mouse (male NMRI and/or female ICH/Ha) that investigated cafes-tol and kahweol treatments separately, hepatic GST-CDNB displayed twofold and threefold increases, respectively, which was also reflected by a 3.5-fold elevation of specific cytosolic GST content in the cafestol-treated animals (Di Simplicio et al., 1989; Lam et al., 1987). Other than in rat, the murine increases in these studies concerned primarily the metabolism by GST-μ, whereas the metabolisms of GST-α– and GST-π–specific substrates remained unmodified or were slightly depressed. Even stronger elevations (fourfold to sevenfold) in GST-CDNB than in the liver were found in the small intestines of these mice, whereas increases appeared weaker in the forestomach (1.2–twofold) and were not observed in the lung. The strong inducibility of GST-CDNB in the murine small intestine was con-firmed by a more recent study that showed a 2.5-fold increase, this time accompanied by elevations of both GST-μ and GST-α proteins (McMahon et al., 2001).

Glutathione, γ-Glutamylcysteine Synthetase, and GGT (see also Fig. 2)

Apart from inducing GST, 0.2% dietary K/C given to the rat in vivo were also found to enhance up to threefold the hepatic levels of the corresponding cofactor glutathione (GSH), an elevation that was to a more moderate degree also observed in kidney, lung, and colon and was likely due to the concomitant induction of γ-glutamylcysteine synthetase (GCS), the rate-limiting enzyme of glutathione synthesis (Huber et al., 2002b). Exposure to K/C increased GCS activity as well as the mRNAs of the light and heavy subunits of the enzyme. Both enhanced GCS activity and GSH elevation were qualitatively reproduced in KC-treated HepG2, suggesting human relevance, and, in the same study, it was shown that the increase in GCS activity preceded that of GSH level, which is supportive of the assumed causal relationship (Scharf et al., 2003). An induction of GCS by K/C was also found in the small intestine of the mouse that was apparently associated with the Nrf2 transcription factor to which importance in the

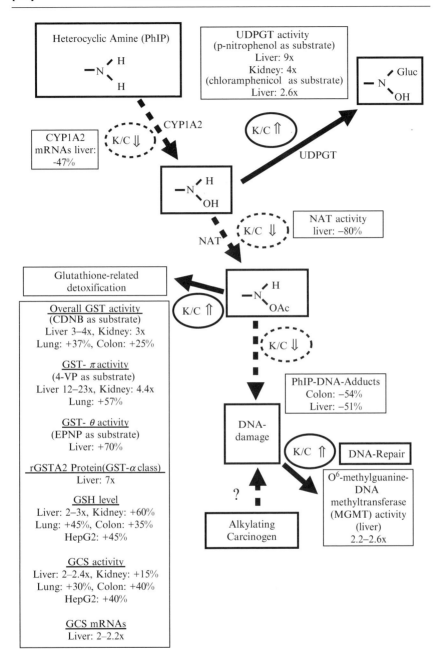

regulation of antioxidative response element (ARE)–driven genes is attributed (McMahon *et al.*, 2001). In rat liver, but not in HepG2, K/C enhanced the activity of GGT, an enzyme that is involved in the degradation of GSH and GSH conjugates and in substrate supply for GSH synthesis (Huber *et al.*, 2002b; Scharf *et al.*, 2003). Thus, K/C seem to stimulate a wide range of elements of GSH-related xenobiotic metabolism concerning GSTs, GSH, GCS, and GGT, and all these effects may greatly contribute to carcinogen detoxification.

Interpretation of Time Course and Dose Requirement of the Modifications of GST and NAT in Light of Chemoprevention by Human Coffee Consumption

Upon K/C treatment, the time courses of the inhibition of NAT activity and the induction of GST-CDNB were inversely correlated (Huber *et al.*, 2004). Both effects developed after 2–5 days and were only reversed after more than 5 days of K/C withdrawal. The induction of GST-CDNB was not further increased even when the K/C treatment was continued for 90 days (Schilter *et al.*, 1996). Thus, the rapid onset of the effects and their persistence would be in agreement with the concept of possible chemoprevention under the time course of regular coffee drinking.

However, some reservation with respect to transferability of the beneficial effects of K/C to the human coffee drinker is tied to dose. The lowest observed effect level of enzyme modification by K/C in the rat was found to be in the range of a dietary concentration of 0.02%. This corresponds to a daily intake of 10–20 mg K/C per kg body weight, of which only one eighth to one quarter was calculated to be transferred to a 70-kg human with a liter of unfiltered coffee (Gross *et al.*, 1997; Huber *et al.*, 2004). Thus, before an actual role in chemoprotection by coffee consumption can be attributed to K/C-related enzyme modifications, additional questions have to

FIG. 2. Potentially protective effects of K/C in the rat and HepG2 in connection with the metabolism of heterocyclic amines and the repair of DNA damage by alkylating carcinogens. Large arrows with solid line, double-lined arrow pointing up and solid margin in oval fields surrounding "K/C", increase by K/C, large arrows with dashed line; double-lined arrow pointing down and dashed margin in oval fields surrounding "K/C", decrease by K/C. Quantification of increases, percentage preceded by plus sign or numbers followed by "x" indicating "-fold." Quantification of decreases, percentage preceded by minus sign; question mark, remains to be explained. In rats, K/C were given in the diet for 10 days at 0.2% of K/C palmitates or as equimolar amount of the free alcohols. In HepG2, K/C were applied at 62 μM. Data are taken from Huber *et al.* (1997, 2002a,b, 2003, 2004), Scharf *et al.* (2003), and manuscript in preparation.

answered. These would concern the comparison of sensitivities in rat and human (see also later), the possible existence of high-sensitivity groups within humans, and the degree of variations in the K/C content of coffees (for instance, one article reported on a coffee containing 1.25 g/l, which was approximately sevenfold in excess of the maximum in other studies that we used for our calculations (Ruiz del Castillo et al., 1999). Moreover, K/C-based chemoprotective strategies using other means of application than coffee (e.g., as food or beverage additives) may be promising as well.

Effects of K/C on UDP-Glucuronosyltransferase and Other Parameters (see also Fig. 2)

The potentially chemoprotective effects of K/C on phase II xenobiotic metabolism are by no means limited to NATs and GSTs. The diterpenes were also found to enhance UDP-glucuronosyltransferase (UDPGT) activity as assayed with p-nitrophenol (UDPGT-pNP) and chloramphenicol (UDPGT-CA), two substrates that are representative of the UDPGT subfamilies UGT1 and UGT2, respectively (Huber et al., 2002a). The dietary concentration of 0.2% K/C, given to male F344 rats for 10 days, which increased hepatic GST-CDNB threefold to fourfold (see earlier), caused a ninefold elevation of UDPGT-pNP in the same organ, and 0.02% K/C were still associated with a twofold increase in enzyme activity (Huber et al., 2003). Just as with GST-CDNB, stimulation of UDPGT-pNP by K/C was organ-specific with a clear fourfold increase in renal activity at the high dose but no apparent changes in lung, colon, salivary gland, spleen, and testis. A significant increase in the UDPGT-related metabolism of methylumbelliferone was also found in HepG2 (Majer et al., 2005).

In several in vivo and in vitro studies, K/C were shown to inhibit cytochrome P450 isozymes (CYP450s) involved in the activation of relevant carcinogens such as PhIP and AFB1 (CYP1A2, CYP3A2, CYP2C11), an effect that may thus contribute considerably to chemoprotection by K/C (Cavin et al., 1998, 2001, 2003; Huber et al., manuscript in preparation; Majer et al., 2005). This inhibitory effect of K/C is in contrast with the induction of CYP450s that is seen with whole coffee, the latter probably being associated with the presence of caffeine (Turesky et al., 2003). Furthermore, K/C led to an almost twofold induction of NADPH/quinone oxidoreductase in the small intestine of the mouse (McMahon et al., 2001).

Apart from enzymes of xenobiotic metabolism, K/C enhanced approximately 2.5-fold the activity of O^6-methylguanine-DNA methyltransferase, a protein that is capable of repairing DNA damage by alkylating agents (Huber et al., 2003). To date, it remains to be elucidated whether this effect also implicates an inhibition of the mutagenic/tumorigenic damage that occurs on actual challenge with such agents. Finally, kahweol and cafestol both suppressed cyclooxygenase-2 (Kim *et al.*, 2004a), and kahweol the inducible nitric oxide synthetase (Kim *et al.*, 2004b), in a murine macrophage cell line, the latter probably being mediated by inhibition of Nf-κB functions.

Effects of Whole Coffee on NAT and GST

Animals

   In light of the strong effects of K/C on NAT and GST, the question arises to what extent contributions may be made by other components of coffee. Studies on preparations of the whole beverage or the beans have provided useful information in this context. Mice that were exposed for 2 weeks to a diet that contained 5% of a powder of whole green coffee beans, which may have been equivalent to 0.05–0.1 % K/C, developed strong increases in GST-CDNB (i.e., 2.6-fold in the liver and 5.7-fold in intestinal mucosa; Sparnins *et al.*, 1982). Raising the concentration of coffee beans to 20%, likely corresponding to 0.2–0.4% K/C, elevated hepatic and intestinal GST-CDNB further to 5- and 11-fold the control values, respectively. On the basis of these early mouse data, K/C were calculated to be responsible for approximately 50% of the increase in GST-CDNB (Wattenberg, 1983). Two additional mouse studies revealed only modest increases in hepatic GST-CDNB and/or GSH by 20–30%, either after a single gavage of 500 mg (Abraham *et al.*, 1998) or after 10 daily gavages of 140 mg coffee/kg body weight (Abraham and Singh, 1999). However, these results are not in contradiction with the strong enhancements described previously, because even the 500 mg/kg gavage provided only a dose that corresponded to 1 day on a chow with 0.5% coffee or with <0.01% K/C.

   Rats that received Turkish coffee as the drinking fluid *ad libitum* for 10 days displayed significantly increased GST-CDNB, GSH levels, and GCS activity in their livers (Huber *et al.*, 2003). Extension of the treatment period to 20 days did not lead to any further strengthening of these effects. The comparatively moderate degree of these increases (by <30%) corresponded rather well with the doses of K/C that the animals were estimated to have ingested with the coffee, being comparable to the consumption of food containing 0.02–0.04% K/C. However, in another study (Turesky *et al.*, 2003), rats received a coffee that had been deprived of virtually all of its K/C content at 5% in their chow and still displayed a 1.6-fold elevation of hepatic GST-CDNB and 1.4- and 2.6-fold higher expressions of GSTA1 and GSTA3 protein, respectively. Notably, NAT activity in this 2-week study was not significantly modified. An overall quantitative comparison between the two studies and the different pathways of application is difficult to make because of divergent gastrointestinal absorption and behavior during beverage preparation of the numerous components. However, when using the published standard values as a basis for calculation, *ad libitum* consumption of food with 5% unaltered coffee bean powder would lead to a twofold to threefold higher ingestion of K/C than *ad libitum* consumption of Turkish coffee as drinking fluid. Thus, a higher intake of unremoved components other than K/C in the feed study may, as well, be conceivable. Interestingly, in a further investigation, a similar dietary content of 4.5% of a lyophilized coffee beverage made from decaffeinated filtered coffee, thus probably not containing more than traces of K/C either, led to only slight and statistically nonsignificant increases in hepatic and renal GST-CDNB in rats after 15 days of exposure (Somoza *et al.*, 2003). Then again, it has to be noted that the same preparation was capable of elevating GST-CDNB in human intestinal CaCo-2 cells at the concentrations of 0.5 and 1 mg/ml, an effect that was not influenced by further removal of the remaining lipid fraction. In summary, the available animal data confirm a predominant role of K/C in GST induction by coffee, but the presence of K/C was not an absolute requirement. Because of lack of sufficient data, no summarizing statement can as yet be made on NAT.

Humans

Another advantage of whole coffee studies may be the possibility to investigate parameters of phase II xenobiotic metabolism in humans. Still, such data are scarce, and, to our knowledge, there are none concerning NAT. As far as GSH-related detoxification is concerned, significant, although rather marginal, increases in GSH of 15% and 8%, respectively, were observed in plasma and colorectal mucosa of a group of consumers of 1 l unfiltered coffee with high K/C content daily (Grubben *et al.*, 2000). The coffee-exposed colorectal samples of the same group of volunteers displayed unmodified GST-CDNB and 4–15% increases in the proteins of GST-α, GST-μ, and GST-π, however, all lacking statistical significance. An almost identical marginal increase in plasma GSH (+16%) was observed in another study that investigated the intake of an over fourfold lower volume of coffees with only intermediate K/C content (mocha and espresso) (Esposito *et al.*, 2003). However, with effects of such a small degree, it is difficult to draw firm conclusions on the contribution of individual components. In contrast, very clear (i.e., threefold–fourfold) increases in GST-

CDNB were observed in the saliva of persons who had consumed between 0.5 and 1.2 l of coffee daily and, in the same study, a tendency toward elevation was also found in plasma (Sreerama *et al.*, 1995). In blood samples of other volunteers who had each consumed 1 l of coffee for 5 consecutive days, GST-CDNB was only marginally enhanced, and GST-α protein remained unmodified (Steinkellner *et al.*, 2005). However, the levels of GST-π protein in this study were elevated threefold, which was not modified by filtering, thereby excluding a role of K/C. Such pronounced effects observed at comparatively low doses may suggest that humans could be more sensitive to the potentially beneficial GST induction by coffee than rats, in analogy to the greater sensitivity of humans that was reported for the unfavorable K/C-related elevation of blood cholesterol (Urgert and Katan, 1997). However, it has to be kept in mind that plasma GSTs may also be used as an indicator of cellular damage (Giannini *et al.*, 2000; Mazur *et al.*, 2003) and that increased GST in body fluids must, therefore, not necessarily be a consequence of enzyme induction. Thus, more data on the origin of the elevated GSTs in blood and saliva must be obtained before final conclusions can be drawn. K/C did not increase GST-CDNB in the salivary gland of rats (Huber *et al.*, 2002b).

Effects of Coffee Polyphenols on NAT and GST

NAT

The general scarcity of available data on NAT modifiers also concerns coffee components other than K/C. However, CGA, CA, and FA were found to suppress the acetylation (i.e., the NAT-related metabolism) of 2-aminofuorene in six strains of human gastrointestinal microflora *in vitro* (Lo and Chung, 1999; Tsou *et al.*, 2000). Some potential influence of this phenomenon on human cancer risk is not unlikely, because gastrointestinal microflora may be involved in the metabolism of certain human carcinogens *in vivo* (Knasmüller *et al.*, 2001).

GST

The effects of polyphenols on GSTs seemed to be far less clear-cut than the severalfold overall increases by K/C. In *in vitro* assays with purified rat liver enzymes, CGA, FA, CA, and TA, were all even identified as GST inhibitors (Das *et al.*, 1984; Zhang and Das, 1994), and moderate decreases of GST-CDNB and/or GST-μ protein were also observed in rat liver *in vivo* upon a single gavage with CA (Ploemen *et al.*, 1993) and in mouse liver upon a single intraperitoneal dose of TA (Krajka-Kuzniak and

Baer-Dubowska, 2003). Further *in vivo* studies investigating CA (Debersac *et al.*, 2001; Manson *et al.*, 1997) and FA (Kawabata *et al.*, 2000) in the rat and CA and CGA in the mouse (Kitts and Wijewickreme, 1994) showed unmodified hepatic GST-CDNB activity. However, a comparatively moderate enhancement of hepatic GST-CDNB by 70% was caused by a single gavage of TA, although TA in this rat study was only investigated in combination with not GST-inducing o-toluidine (Szaefer *et al.*, 2003). In three of the rat studies that had remained negative with respect to modification of hepatic GST-CDNB, there were enhancements of specific other parameters of liver GST. CA led to atypically strong 3.8- and 3-fold increases in the content of rGSTA3/A5 protein (Debersac *et al.*, 2001) and in the metabolism of the GST-μ–specific substrate 1,2-dichloro-4-nitrobenzene (DCNB) (Manson *et al.*, 1997), respectively. The latter was also significantly elevated by FA, albeit in this case only by 27% (Kawabata *et al.*, 2000). More frequently than in liver, enhancements of GST-CDNB by coffee-associated polyphenols were observed in extrahepatic organs, but also these changes were not greater than maximally twofold. For instance, a 60% increase in GST-CDNB in rat kidney was discovered after the combined treatment with TA and o-toluidine described previously (Szaefer *et al.*, 2003), and several other studies revealed elevations of a similar degree in mouse organs (e.g., by 50–80% in intestine with CGA and CA [Kitts and Wijewickreme, 1994], by 35% in kidney with TA [Krajka-Kuzniak and Baer-Dubowska, 2003], by up to 60% in the forestomach with TA [Athar *et al.*, 1989], and twofold in epidermis on topical application of CGA (Szaefer *et al.*, 2004) and TA (Baer-Dubowska *et al.*, 1997). In the colons of rats on a diet containing p-CMA, the amount of mRNA of the GST-μ unit GSTM2 was elevated by about a third, whereas the same treatment did not modify GSTM1, GST-π, and GCS (Guglielmi *et al.*, 2003). Thus, in summary, the available data argue for a weaker GST-inducing capacity of polyphenols than of K/C, although this disadvantage in efficiency may be somewhat diminished by the higher content of total polyphenols in coffee. Also, whereas humans ingested K/C more or less exclusively with coffee, polyphenols may originate from various other relevant sources in the human diet. Notably, the polyphenols in most studies described previously were given at doses (up to 1% in the diet and 250 mg/kg body weight by gavage) that were higher than or at least equal to the K/C doses that had caused ≥threefold enhancements of hepatic and renal GST-CDNB (see Fig. 2 and previously). Therefore, a large part of polyphenol-related chemoprevention seems to be due to other mechanisms than enhanced carcinogen deactivation by GST, a concept that is also supported by the repeatedly observed protection by polyphenols applied after the carcinogen (i.e., later than the tumor initiation phase) (Mori et al., 1986, 2000; Morishita et al., 1997; Shimizu et al., 1999).

Effects of Caffeine on NAT and GST

Caffeine is metabolized by NAT and, therefore, widely used as a test substance in human NAT phenotyping (Butler et al., 1992; Vistisen et al., 1992). However, treatment of female F344 rats with 0.1% caffeine in the drinking water caused no significant differences from control in the mRNAs of NAT1 and NAT2 from colon, mammary gland, and liver, although some tendency toward inhibition may have been present in the colon (Takeshita et al., 2003). Notably, in another context it has been hypothesized that the well-known interference of caffeine with sleep quality and melatonin secretion may involve an indirect inhibition of the NAT-related synthesis of melatonin (Shilo et al., 2002).

A certain contribution of caffeine to the induction of GST-CDNB is suggested by a study investigating the effect of black tea in rat liver in which the already very modest GST-induction was further diminished by decaffeination of the tea (Bu-Abbas et al., 1998). However, according to two more recent studies in the rat, 0.34% caffeine in the chow for 2 weeks and 0.0625% in the drinking water for 4 weeks were apparently not sufficient to increase hepatic GST-CDNB and/or the metabolism of three further GST substrates over control level (Maliakal and Wanwimolruk, 2001; Turesky et al., 2003). Likewise, no differences with respect to hepatic GST-CDNB and GSH were observed in a study that compared the effect of caffeinated and decaffeinated coffees in the mouse (Abraham and Singh, 1999), and, in another study, caffeine was even found to deplete hepatic GSH in the rat by 40% (Kalhorn et al., 1990). Not surprisingly because of this overwhelmingly negative evidence, mechanisms of chemo-protection other than modification of xenobiotic metabolism have been suggested for caffeine, for instance, direct scavenging of hydroxyl radicals (Devasagayam et al., 1996; Shi et al., 1991) or induction of apoptosis (He et al., 2003).

Effects of Melanoidins, N-Methylpyridinium Ion, and Fiber on GSTs

Data on the effect of specifically coffee-derived melanoidins on GSTs are thus far lacking. However, increases in GST-CDNB by up to one third were observed with certain melanoidins from roasted malt (Faist et al., 2002) and bread crust (Borrelli et al., 2003; Lindenmeier et al., 2002), both in Caco-2 cells in vitro, whereas melanoidins from biscuit were found to inhibit GST-CDNB in the same system (Borrelli *et al.*, 2003). An *in vivo* study in the rat investigated two components of heated casein, each being representative of a particular stage of the Maillard reaction, and reported 60–90% increases in GST-CDNB and GSH (Wenzel *et al.*, 2002). However, these changes were highly dependent on individual treatment and investi- gated organ, each component only being effective in either kidney or small intestine but not in liver. The N-methylpyridinium ion, a degradation product of trigonelline (the latter contained at 0.3–1.3% in coffee beans [Illy and Viani, 1995]; for formulas, see Fig. 1) and another potentially chemoprotective component of coffee, caused an elevation of GST-CDNB in Caco-2 cells as well (Somoza *et al.*, 2003). However, the increase was slight (22%), and, moreover, the effect was neither dose-dependent nor could it be reproduced in rat liver and kidney *in vivo*. To our knowledge, the effects of coffee fiber on NAT and GST have not been investigated. However, when rats were exposed to a 4-week diet with a rather high content (20%) of wheat bran, 36% of which consisted of fiber, there were modest (<40%) increases in the GST-α proteins GSTA1 and GSTA2 in the colon and in GSTA1 in the liver, which were, however, accompanied by decreases in hepatic GSTA2 and GST-CDNB (Helsby *et al.*, 2000). Be- cause of such contradictions and the as yet incomplete picture, additional research is required to evaluate the possible relevance of GST modifica- tions by melanoidins, N-methylpyridinium, and coffee fiber. This will aid in the interpretation of the health effects of human coffee consumption and the development of chemopreventive strategies.

Summary and Conclusions

In animal and *in vitro* studies, whole coffee and several of its compo- nents were capable of increasing GST, the most efficient components being the diterpenes K/C, which are specific of unfiltered coffee, and to a some- what lesser extent the polyphenols CGA, CA, and TA, which may be ingested from other sources than coffee as well. More information is at present required to draw conclusions on the role of melanoidins, fiber, and the N-methylpyridinium ion, whereas there is no apparent contribution of caffeine. Both K/C in rat liver and cultured rat hepatocytes and polyphe- nols in intestinal microflora also inhibited NAT that until recently was considered to be unmodifiable. These enzymatic changes might partially explain many antitumorigenic/antimutagenic effects in animals that were observed with K/C, polyphenols, and coffee, particularly the protection of K/C against the cooked food mutagen PhIP that is activated by NAT and detoxified by GST. Thus, K/C, for which further beneficial effects but no signs of genotoxic, prooxidant or protumorigenic activity have as yet been identified, may be promising for future strategies of cancer chemoprevention, although caution has to be exercised because of the hypercholesterolemic capacity of these compounds and because of the detoxification potential of NAT toward other carcinogens such as aromatic amines. The K/C doses that caused NAT inhibition and GST induction in the animal were relatively high compared with the ingestion by regular coffee drinkers. Therefore, further research is required to elucidate whether modifications of GST and NAT by K/C and other coffee components are also causally related to the lower risk of cancers of colon, liver, and possibly other organs observed in coffee drinkers.

References

Abraham, S. K. (1991). Inhibitory effects of coffee on the genotoxicity of carcinogens in mice. *Mutat. Res.* **262,** 109–114.

Abraham, S. K., Singh, S. P., and Kesavan, P. C. (1998). *In vivo* antigenotoxic effects of dietary agents and beverages co-administered with urethane: Assessment of the role of glutathione S-transferase activity. *Mutat. Res.* **413,** 103–110.

Abraham, S. K., and Singh, S. P. (1999). Anti-genotoxicity and glutathione S-transferase activity in mice pretreated with caffeinated and decaffeinated coffee. *Food Chem. Toxicol.* **37,** 733–739.

Abraham, S. K., and Stopper, H. (2004). Anti-genotoxicity of coffee against N-methyl-N-nitro-N-nitrosoguanidine in mouse lymphoma cells. *Mutat. Res.* **561,** 23–33.

Adamson, R. H., Thorgeirsson, U. P., and Sugimura, T. (1996). Extrapolation of heterocyclic amine carcinogenesis data from rodents and nonhuman primates to humans. *Arch. Toxicol. Suppl.* **18,** 303–318.

Aeschbacher, H. U., Wolleb, U., Loliger, J., Spadone, J. C., and Liardon, R. (1989). Contribution of coffee aroma constituents to the mutagenicity of coffee. *Food Chem. Toxicol.* **27,** 227–232.

Anese, M., and Nicoli, M. C. (2003). Antioxidant properties of ready-to-drink coffee brews. *J. Agric. Food Chem.* **51,** 942–946.

Athar, M., Khan, W. A., and Mukhtar, H. (1989). Effect of dietary tannic acid on epidermal, lung, and forestomach polycyclic aromatic hydrocarbon metabolism and tumorigenicity in Sencar mice. *Cancer Res.* **49,** 5784–5788.

Baer-Dubowska, W., Gnojkowski, J., and Fenrych, W. (1997). Effect of tannic acid on benzo [a]pyrene-DNA adduct formation in mouse epidermis: Comparison with synthetic gallic acid esters. *Nutr. Cancer* **29,** 42–47.

Borrelli, R. C., Mennella, C., Barba, F., Russo, M., Russo, G. L., Krome, K., Erbersdobler, H. F., Faist, V., and Fogliano, V. (2003). Characterization of coloured compounds obtained by enzymatic extraction of bakery products. *Food Chem. Toxicol.* **41,** 1367–1374.

Borrelli, R. C., Esposito, F., Napolitano, A., Ritieni, A., and Fogliano, V. (2004). Characterization of a new potential functional ingredient: Coffee silverskin. *J. Agric. Food Chem.* **52,** 1338–1343.

Bruning, T., Lammert, M., Kempkes, M., Thier, R., Golka, K., and Bolt, H. M. (1997). Influence of polymorphisms of GSTM1 and GSTT1 for risk of renal cell cancer in workers with long-term high occupational exposure to trichloroethene. *Arch. Toxicol.* **71,** 596–599.

Bu-Abbas, A., Clifford, M. N., Walker, R., and Ioannides, C. (1998). Contribution of caffeine and flavanols in the induction of hepatic Phase II activities by green tea. *Food Chem. Toxicol.* **36**, 617–621.

Buetler, T. M., Gallagher, E. P., Wang, C., Stahl, D. L., Hayes, J. D., and Eaton, D. L. (1995). Induction of phase I and phase II drug-metabolizing enzyme mRNA, protein, and activity by BHA, ethoxyquin, and oltipraz. *Toxicol. Appl. Pharmacol.* **135**, 45–57.

Butler, M. A., Lang, N. P., Young, J. F., Caporaso, N. E., Vineis, P., Hayes, R. B., Teitel, C. H., Massengill, J. P., Lawsen, M. F., and Kadlubar, F. F. (1992). Determination of CYP1A2 and NAT2 phenotypes in human populations by analysis of caffeine urinary metabolites. *Pharmacogenetics* **2**, 116–127.

Cascorbi, I., Brockmoller, J., Mrozikiewicz, P. M., Bauer, S., Loddenkemper, R., and Roots, I. (1996). Homozygous rapid arylamine N-acetyltransferase (NAT2) genotype as a susceptibility factor for lung cancer. *Cancer Res.* **56**, 3961–3966.

Cavin, C., Holzhauser, D., Constable, A., Huggett, A. C., and Schilter, B. (1998). The coffee-specific diterpenes cafestol and kahweol protect against aflatoxin B1-induced genotoxicity through a dual mechanism. *Carcinogenesis* **19**, 1369–1375.

Cavin, C., Mace, K., Offord, E. A., and Schilter, B. (2001). Protective effects of coffee diterpenes against aflatoxin B1-induced genotoxicity: Mechanisms in rat and human cells. *Food Chem. Toxicol.* **39**, 549–556.

Cavin, C., Bezencon, C., Guignard, G., and Schilter, B. (2003). Coffee diterpenes prevent benzo[a]pyrene genotoxicity in rat and human culture systems. *Biochem. Biophys. Res. Commun.* **306**, 488–495.

Chang, S. H., Chen, G. W., Yeh, C. C., Hung, C. F., Lin, S. S., and Chung, J. G. (2001). Effects of the butylated hydroxyanisole and butylated hydroxytoluene on the DNA adduct formation and arylamines N-acetyltransferase activity in human colon tumor cells. *Anticancer Res.* **21**, 1087–1093.

Chen, G. W., Chung, J. G., Hsieh, C. L., and Lin, J. G. (1998). Effects of the garlic components diallyl sulfide and diallyl disulfide on arylamine N-acetyltransferase activity in human colon tumour cells. *Food Chem. Toxicol.* **36**, 761–770.

Chen, Y. S., Ho, C. C., Cheng, K. C., Tyan, Y. S., Hung, C. F., Tan, T. W., and Chung, J. G. (2003). Curcumin inhibited the arylamines N-acetyltransferase activity, gene expression and DNA adduct formation in human lung cancer cells (A549). *Toxicol. In Vitro* **17**, 323–333.

Chung, J. G. (1999). Effects of butylated hydroxyanisole (BHA) and butylated hydroxytoluene (BHT) on the acetylation of 2-aminofluorene and DNA-2-aminofluorene adducts in the rat. *Toxicol. Sci.* **51**, 202–210.

Chung, J. G., Li, Y. C., Lee, Y. M., Lin, J. P., Cheng, K. C., and Chang, W. C. (2003). Aloe-emodin inhibited N-acetylation and DNA adduct of 2-aminofluorene and arylamine N-acetyltransferase gene expression in mouse leukemia L 1210 cells. *Leukot. Res.* **27**, 831–840.

Chung, K. T., Wong, T. Y., Wei, C. I., Huang, Y. W., and Lin, Y. (1998). Tannins and human health: A review. *Crit. Rev. Food Sci. Nutr.* **38**, 421–464.

Coles, B. F., Morel, F., Rauch, C., Huber, W. W., Yang, M., Teitel, C. H., Green, B., Lang, N. P., and Kadlubar, F. F. (2001). Effect of polymorphism in the human glutathione S-transferase A1 promoter on hepatic GSTA1 and GSTA2 expression. *Pharmacogenetics* **11**, 663–669.

Coles, B. F., and Kadlubar, F. F. (2003). Detoxification of electrophilic compounds by glutathione S-transferase catalysis: Determinants of individual response to chemical carcinogens and chemotherapeutic drugs? *Biofactors* **17**, 115–130.

Conney, A. H. (2003). Enzyme induction and dietary chemicals as approaches to cancer chemoprevention: The Seventh DeWitt S. Goodman Lecture. *Cancer Res.* **63**, 7005–7031.

Daglia, M., Racchi, M., Papetti, A., Lanni, C., Govoni, S., and Gazzani, G. (2004). *In vitro* and *ex vivo* antihydroxyl radical activity of green and roasted coffee. *J. Agric. Food Chem.* **52**, 1700–1704.

Das, M., Bickers, D. R., and Mukhtar, H. (1984). Plant phenols as *in vitro* inhibitors of glutathione S-transferase(s). *Biochem. Biophys. Res. Commun.* **120**, 427–433.

Debersac, P., Vernevaut, M. F., Amiot, M. J., Suschetet, M., and Siess, M. H. (2001). Effects of a water-soluble extract of rosemary and its purified component rosmarinic acid on xenobiotic-metabolizing enzymes in rat liver. *Food Chem. Toxicol.* **39**, 109–117.

Devasagayam, T. P., Kamat, J. P., Mohan, H., and Kesavan, P. C. (1996). Caffeine as an antioxidant: Inhibition of lipid peroxidation induced by reactive oxygen species. *Biochim. Biophys. Acta* **1282**, 63–70.

Di Simplicio, P., Jensson, H., and Mannervik, B. (1989). Effects of inducers of drug metabolism on basic hepatic forms of mouse glutathione transferase. *Biochem. J.* **263**, 679–685.

Esposito, F., Morisco, F., Verde, V., Ritieni, A., Alezio, A., Caporaso, N., and Fogliano, V. (2003). Moderate coffee consumption increases plasma glutathione but not homocysteine in healthy subjects. *Aliment. Pharmacol. Ther.* **17**, 595–601.

Faist, V., Lindenmeier, M., Geisler, C., Erbersdobler, H. F., and Hofmann, T. (2002). Influence of molecular weight fractions isolated from roasted malt on the enzyme activities of NADPH-cytochrome c-reductase and glutathione-S-transferase in Caco-2 cells. *J. Agric. Food Chem.* **50**, 602–606.

Farah, A., de Paulis, T., Trugo, L. C., and Martin, P. R. (2005). Effect of roasting on the formation of chlorogenic acid lactones in coffee. *J. Agric. Food Chem.* **53**, 1505–1513.

Franceschi, S., Barra, S., La Vecchia, C., Bidoli, E., Negri, E., and Talamini, R. (1992). Risk factors for cancer of the tongue and the mouth. A case-control study from northern Italy. *Cancer* **70**, 2227–2233.

Fukutome, K., Watanabe, M., Shiraishi, T., Murata, M., Uemura, H., Kubota, Y., Kawamura, J., Ito, H., and Yatani, R. (1999). N-acetyltransferase 1 genetic polymorphism influences the risk of prostate cancer development. *Cancer Lett.* **136**, 83–87.

Gallus, S., Bertuzzi, M., Tavani, A., Bosetti, C., Negri, E., La Vecchia, C., Lagiou, P., and Trichopoulos, D. (2002a). Does coffee protect against hepatocellular carcinoma? *Br. J. Cancer* **87**, 956–959.

Gallus, S., Tavani, A., Negri, E., and La Vecchia, C. (2002b). Does coffee protect against liver cirrhosis? *Ann. Epidemiol.* **12**, 202–205.

Gershbein, L. L. (1994). Action of dietary trypsin, pressed coffee oil, silymarin and iron salt on 1,2-dimethylhydrazine tumorigenesis by gavage. *Anticancer Res.* **14**, 1113–1116.

Ghadirian, P., Simard, A., and Baillargeon, J. (1993). A population-based case-control study of cancer of the bile ducts and gallbladder in Quebec, Canada. *Rev. Epidemiol. Sante Publique* **41**, 107–112.

Giannini, E., Risso, D., Ceppa, P., Botta, F., Chiarbonello, B., Fasoli, A., Malfatti, F., Romagnoli, P., Lantieri, P. B., and Testa, R. (2000). Utility of alpha-glutathione S-transferase assessment in chronic hepatitis C patients with near normal alanine aminotransferase levels. *Clin. Biochem.* **33**, 297–301.

Giovannucci, E. (1998). Meta-analysis of coffee consumption and risk of colorectal cancer. *Am. J. Epidemiol.* **147**, 1043–1052.

Golka, K., Prior, V., Blaszkewicz, M., and Bolt, H. M. (2002). The enhanced bladder cancer susceptibility of NAT2 slow acetylators towards aromatic amines: A review considering ethnic differences. *Toxicol. Lett.* **128**, 229–241.

Gross, G., Jaccaud, E., and Huggett, A. C. (1997). Analysis of the content of the diterpenes cafestol and kahweol in coffee brews. *Food Chem. Toxicol.* **35**, 547–554.

Gross, M., Kruisselbrink, T., Anderson, K., Lang, N., McGovern, P., Delongchamp, R., and Kadlubar, F. (1999). Distribution and concordance of N-acetyltransferase genotype and phenotype in an American population. *Cancer Epidemiol. Biomarkers Prev.* **8**, 683–692.

Grubben, M. J., Van Den Braak, C. C., Broekhuizen, R., De Jong, R., Van Rijt, L., De Ruijter, E., Peters, W. H., Katan, M. B., and Nagengast, F. M. (2000). The effect of unfiltered coffee on potential biomarkers for colonic cancer risk in healthy volunteers: A randomized trial. *Aliment. Pharmacol. Ther.* **14**, 1181–1190.

Guglielmi, F., Luceri, C., Giovannelli, L., Dolara, P., and Lodovici, M. (2003). Effect of 4-coumaric and 3,4-dihydroxybenzoic acid on oxidative DNA damage in rat colonic mucosa. *Br. J. Nutr.* **89**, 581–587.

Hagiwara, A., Hirose, M., Takahashi, S., Ogawa, K., Shirai, T., and Ito, N. (1991). Forestomach and kidney carcinogenicity of caffeic acid in F344 rats and C57BL/6N x C3H/HeN F1 mice. *Cancer Res.* **51**, 5655–5660.

Hagiwara, A., Kokubo, Y., Takesada, Y., Tanaka, H., Tamano, S., Hirose, M., Shirai, T., and Ito, N. (1996). Inhibitory effects of phenolic compounds on development of naturally occurring preneoplastic hepatocytic foci in long-term feeding studies using male F344 rats. *Teratog. Carcinog. Mutagen.* **16**, 317–325.

Hagiwara, A., Boonyaphiphat, P., Tanaka, H., Kawabe, M., Tamano, S., Kaneko, H., Matsui, M., Hirose, M., Ito, N., and Shirai, T. (1999). Organ-dependent modifying effects of caffeine, and two naturally occurring antioxidants alpha-tocopherol and n-tritriacontane-16,18-dione, on 2-amino-1-methyl-6-phenylimidazo[4,5-b]pyridine (PhIP)-induced mammary and colonic carcinogenesis in female F344 rats. *Jpn. J. Cancer Res.* **90**, 399–405.

Harries, L. W., Stubbins, M. J., Forman, D., Howard, G. C., and Wolf, C. R. (1997). Identification of genetic polymorphisms at the glutathione S-transferase Pi locus and association with susceptibility to bladder, testicular and prostate cancer. *Carcinogenesis* **18**, 641–644.

He, Z., Ma, W. Y., Hashimoto, T., Bode, A. M., Yang, C. S., and Dong, Z. (2003). Induction of apoptosis by caffeine is mediated by the p53, Bax, and caspase 3 pathways. *Cancer Res.* **63**, 4396–4401.

Hein, D. W., Doll, M. A., Fretland, A. J., Leff, M. A., Webb, S. J., Xiao, G. H., Devanaboyina, U. S., Nangju, N. A., and Feng, Y. (2000). Molecular genetics and epidemiology of the NAT1 and NAT2 acetylation polymorphisms. *Cancer Epidemiol. Biomarkers Prev.* **9**, 29–42.

Helsby, N. A., Zhu, S., Pearson, A. E., Tingle, M. D., and Ferguson, L. R. (2000). Antimutagenic effects of wheat bran diet through modification of xenobiotic metabolising enzymes. *Mutat. Res.* **454**, 77–88.

Hirose, M., Mutai, M., Takahashi, S., Yamada, M., Fukushima, S., and Ito, N. (1991). Effects of phenolic antioxidants in low dose combination on forestomach carcinogenesis in rats pretreated with N-methyl-N′-nitro-N-nitrosoguanidine. *Cancer Res.* **51**, 824–827.

Hirose, M., Kawabe, M., Shibata, M., Takahashi, S., Okazaki, S., and Ito, N. (1992). Influence of caffeic acid and other o-dihydroxybenzene derivatives on N-methyl-N′-nitro-N-nitrosoguanidine-initiated rat forestomach carcinogenesis. *Carcinogenesis* **13**, 1825–1828.

Hirose, M., Takahashi, S., Ogawa, K., Futakuchi, M., Shirai, T., Shibutani, M., Uneyama, C., Toyoda, K., and Iwata, H. (1999). Chemoprevention of heterocyclic amine-induced carcinogenesis by phenolic compounds in rats. *Cancer Lett.* **143,** 173–178.

Hirose, M., Nishikawa, A., Shibutani, M., Imai, T., and Shirai, T. (2002). Chemoprevention of heterocyclic amine-induced mammary carcinogenesis in rats. *Environ. Mol. Mutagen.* **39,** 271–278.

Hirvonen, A., Pelin, K., Tammilehto, L., Karjalainen, A., Mattson, K., and Linnainmaa, K. (1995). Inherited GSTM1 and NAT2 defects as concurrent risk modifiers in asbestos-related human malignant mesothelioma. *Cancer Res.* **55,** 2981–2983.

Hirvonen, A., Saarikoski, S. T., Linnainmaa, K., Koskinen, K., Husgafvel-Pursiainen, K., Mattson, K., and Vainio, H. (1996). Glutathione S-transferase and N-acetyltransferase genotypes and asbestos-associated pulmonary disorders. *J. Natl. Cancer Inst.* **88,** 1853–1856.

Hou, S. M., Ryberg, D., Falt, S., Deverill, A., Tefre, T., Borresen, A. L., Haugen, A., and Lambert, B. (2000). GSTM1 and NAT2 polymorphisms in operable and non-operable lung cancer patients. *Carcinogenesis* **21,** 49–54.

Huber, W. W., McDaniel, L. P., Kaderlik, K. R., Teitel, C. H., Lang, N. P., and Kadlubar, F. F. (1997). Chemoprotection against the formation of colon DNA adducts from the food-borne carcinogen 2-amino-1-methyl-6-phenylimidazo[4,5-b]-pyridine (PhIP) in the rat. *Mutat. Res.* **376,** 115–122.

Huber, W. W., Prustomersky, S., Delbanco, E. H., Uhl, M., Scharf, G., Turesky, R. J., Thier, R., and Schulte-Hermann, R. (2002a). Enhancement of the chemoprotective enzymes glucuronosyl transferase and glutathione transferase by the coffee components Kahweol and Cafestol in specific organs of the rat. *Arch. Toxicol.* **76,** 209–217.

Huber, W. W., Scharf, G., Rossmanith, W., Prustomersky, S., Grasl-Kraupp, B., Peter, B., Turesky, R. J., and Schulte-Hermann, R. (2002b). The coffee components Kahweol and Cafestol induce γ-glutamylcysteine synthetase, the rate limiting enzyme of chemoprotective glutathione synthesis, in several organs of the rat. *Arch. Toxicol.* **75,** 685–694.

Huber, W. W., Scharf, G., Nagel, G., Prustomersky, S., Schulte-Hermann, R., and Kaina, B. (2003). Coffee and its chemopreventive components Kahweol and Cafestol increase the activity of O6-methylguanine-DNA methyltransferase in rat liver–comparison with phase II xenobiotic metabolism. *Mutat. Res.* **522,** 57–68.

Huber, W. W., Teitel, C. H., Coles, B. F., King, R. S., Wiese, F. W., Kaderlik, K. R., Casciano, D. A., Shaddock, J. G., Mulder, G. J., Ilett, K. F., and Kadlubar, F. F. (2004). Potential chemoprotective effects of the coffee components kahweol and cafestol palmitates via modification of hepatic N-acetyltransferase and glutathione S-transferase activities. *Environ. Mol. Mutagen.* **44,** 265–276.

Hudson, E. A., Dinh, P. A., Kokubun, T., Simmonds, M. S., and Gescher, A. (2000). Characterization of potentially chemopreventive phenols in extracts of brown rice that inhibit the growth of human breast and colon cancer cells. *Cancer Epidemiol. Biomarkers Prev.* **9,** 1163–1170.

IARC. (1991). "Coffee, Tea, Mate, Methylxanthines and Methylglyoxal." WHO, Lyon, France.

Illy, A., and Viani, R. (1995). "Espresso Coffee, the Chemistry of Quality." Academic Press, London, San Diego.

Inoue, M., Yoshimi, I., Sobue, T., and Tsugane, S. (2005). Influence of coffee drinking on subsequent risk of hepatocellular carcinoma: A prospective study in Japan. *J. Natl. Cancer Inst.* **97,** 293–300.

Isaksson, B., Jonsson, F., Pedersen, N. L., Larsson, J., Feychting, M., and Permert, J. (2002). Lifestyle factors and pancreatic cancer risk: A cohort study from the Swedish Twin Registry. *Int. J. Cancer* **98**, 480–482.

Iwai, K., Kishimoto, N., Kakino, Y., Mochida, K., and Fujita, T. (2004). *In vitro* antioxidative effects and tyrosinase inhibitory activities of seven hydroxycinnamoyl derivatives in green coffee beans. *J. Agric. Food Chem.* **52**, 4893–4898.

Jacobsen, B. K., and Heuch, I. (2000). Coffee, K-ras mutations and pancreatic cancer: A heterogeneous aetiology or an artefact? *J. Epidemiol. Commun. Health* **54**, 654–655.

Johns, L. E., and Houlston, R. S. (2000). Glutathione S-transferase mu1 (GSTM1) status and bladder cancer risk: A meta-analysis. *Mutagenesis* **15**, 399–404.

Jordan, S. J., Purdie, D. M., Green, A. C., and Webb, P. M. (2004). Coffee, tea and caffeine and risk of epithelial ovarian cancer. *Cancer Causes Control* **15**, 359–365.

Kadlubar, F. F., and Badawi, A. F. (1995). Genetic susceptibility and carcinogen-DNA adduct formation in human urinary bladder carcinogenesis. *Toxicol. Lett.* **82–83**, 627–632.

Kalhorn, T. F., Lee, C. A., Slattery, J. T., and Nelson, S. D. (1990). Effect of methylxanthines on acetaminophen hepatotoxicity in various induction states. *J. Pharmacol. Exp. Ther.* **252**, 112–116.

Kato, T., Takahashi, S., and Kikugawa, K. (1991). Loss of heterocyclic amine mutagens by insoluble hemicellulose fiber and high-molecular-weight soluble polyphenolics of coffee. *Mutat. Res.* **246**, 169–178.

Kawabata, K., Yamamoto, T., Hara, A., Shimizu, M., Yamada, Y., Matsunaga, K., Tanaka, T., and Mori, H. (2000). Modifying effects of ferulic acid on azoxymethane-induced colon carcinogenesis in F344 rats. *Cancer Lett.* **157**, 15–21.

Kensler, T. W., Egner, P. A., Trush, M. A., Bueding, E., and Groopman, J. D. (1985). Modification of aflatoxin B1 binding to DNA *in vivo* in rats fed phenolic antioxidants, ethoxyquin and a dithiothione. *Carcinogenesis* **6**, 759–763.

Kensler, T. W. (1997). Chemoprevention by inducers of carcinogen detoxication enzymes. *Environ. Health Perspect.* **105**(Suppl. 4), 965–970.

Khan, N. S., Ahmad, A., and Hadi, S. M. (2000). Anti-oxidant, pro-oxidant properties of tannic acid and its binding to DNA. *Chem. Biol. Interact.* **125**, 177–189.

Kim, J. Y., Jung, K. S., and Jeong, H. G. (2004a). Suppressive effects of the kahweol and cafestol on cyclooxygenase-2 expression in macrophages. *FEBS Lett.* **569**, 321–326.

Kim, J. Y., Jung, K. S., Lee, K. J., Na, H. K., Chun, H. K., Kho, Y. H., and Jeong, H. G. (2004b). The coffee diterpene kahweol suppress the inducible nitric oxide synthase expression in macrophages. *Cancer Lett.* **213**, 147–154.

Kitts, D. D., and Wijewickreme, A. N. (1994). Effect of dietary caffeic and chlorogenic acids on *in vivo* xenobiotic enzyme systems. *Plant Foods Hum. Nutr.* **45**, 287–298.

Klatsky, A. L., and Armstrong, M. A. (1992). Alcohol, smoking, coffee, and cirrhosis. *Am. J. Epidemiol.* **136**, 1248–1257.

Knasmuller, S., Steinkellner, H., Hirschl, A. M., Rabot, S., Nobis, E. C., and Kassie, F. (2001). Impact of bacteria in dairy products and of the intestinal microflora on the genotoxic and carcinogenic effects of heterocyclic aromatic amines. *Mutat. Res.* **480–481**, 129–138.

Knudsen, L. E., Norppa, H., Gamborg, M. O., Nielsen, P. S., Okkels, H., Soll-Johanning, H., Raffn, E., Jarventaus, H., and Autrup, H. (1999). Chromosomal aberrations in humans induced by urban air pollution: Influence of DNA repair and polymorphisms of glutathione S-transferase M1 and N-acetyltransferase 2. *Cancer Epidemiol. Biomarkers Prev.* **8**, 303–310.

Kono, Y., Shibata, H., Kodama, Y., and Sawa, Y. (1995). The suppression of the N-nitrosating reaction by chlorogenic acid. *Biochem. J.* **312**(Pt. 3), 947–953.

Kono, Y., Kobayashi, K., Tagawa, S., Adachi, K., Ueda, A., Sawa, Y., and Shibata, H. (1997). Antioxidant activity of polyphenolics in diets. Rate constants of reactions of chlorogenic acid and caffeic acid with reactive species of oxygen and nitrogen. *Biochim. Biophys. Acta* **1335**, 335–342.

Krajka-Kuzniak, V., and Baer-Dubowska, W. (2003). The effects of tannic acid on cytochrome P450 and phase II enzymes in mouse liver and kidney. *Toxicol. Lett.* **143**, 209–216.

Kubik, A. K., Zatloukal, P., Tomasek, L., Pauk, N., Havel, L., Krepela, E., and Petruzelka, L. (2004). Dietary habits and lung cancer risk among non-smoking women. *Eur. J. Cancer Prev.* **13**, 471–480.

Lam, L. K., Sparnins, V. L., and Wattenberg, L. W. (1982). Isolation and identification of kahweol palmitate and cafestol palmitate as active constituents of green coffee beans that enhance glutathione S-transferase activity in the mouse. *Cancer Res.* **42**, 1193–1198.

Lam, L. K., Sparnins, V. L., and Wattenberg, L. W. (1987). Effects of derivatives of kahweol and cafestol on the activity of glutathione S-transferase in mice. *J. Medicinal Chem.* **30**, 1399–1403.

Lam, L. K., and Zhang, J. (1991). Reduction of aberrant crypt formation in the colon of CF1 mice by potential chemopreventive agents. *Carcinogenesis* **12**, 2311–2315.

Lampe, J. W., King, I. B., Li, S., Grate, M. T., Barale, K. V., Chen, C., Feng, Z., and Potter, J. D. (2000). Brassica vegetables increase and apiaceous vegetables decrease cytochrome P450 1A2 activity in humans: Changes in caffeine metabolite ratios in response to controlled vegetable diets. *Carcinogenesis* **21**, 1157–1162.

Lang, N. P., Butler, M. A., Massengill, J., Lawson, M., Stotts, R. C., Hauer Jensen, M., and Kadlubar, F. F. (1994). Rapid metabolic phenotypes for acetyltransferase and cytochrome P4501A2 and putative exposure to food-borne heterocyclic amines increase the risk for colorectal cancer or polyps. *Cancer Epidemiol. Biomarkers Prev.* **3**, 675–682.

Le Bon, A. M., Roy, C., Dupont, C., and Suschetet, M. (1997). *In vivo* antigenotoxic effects of dietary allyl sulfides in the rat. *Cancer Lett.* **114**, 131–134.

Le Marchand, L., Sivaraman, L., Franke, A. A., Custer, L. J., Wilkens, L. R., Lau, A. F., and Cooney, R. V. (1996). Predictors of N-acetyltransferase activity: Should caffeine phenotyping and NAT2 genotyping be used interchangeably in epidemiological studies? *Cancer Epidemiol. Biomarkers Prev.* **5**, 449–455.

Leung, K. H. (1986). Selective inhibition of leukotriene C4 synthesis in human neutrophils by ethacrynic acid. *Biochem. Biophys. Res. Commun.* **137**, 195–200.

Lin, D., Meyer, D. J., Ketterer, B., Lang, N. P., and Kadlubar, F. F. (1994). Effects of human and rat glutathione S-transferase on the covalent DNA binding of the N-acetoxy derivatives of heterocyclic amine carcinogens *in vitro*: A possible mechanism of organ specificity in their carcinogenesis. *Cancer Res.* **54**, 4920–4926.

Lin, Y., Tamakoshi, A., Kawamura, T., Inaba, Y., Kikuchi, S., Motohashi, Y., Kurosawa, M., and Ohno, Y. (2002). Risk of pancreatic cancer in relation to alcohol drinking, coffee consumption and medical history: Findings from the Japan collaborative cohort study for evaluation of cancer risk. *Int. J. Cancer* **99**, 742–746.

Lindenmeier, M., Faist, V., and Hofmann, T. (2002). Structural and functional characterization of pronyl-lysine, a novel protein modification in bread crust melanoidins showing *in vitro* antioxidative and phase I/II enzyme modulating activity. *J. Agric. Food Chem.* **50**, 6997–7006.

Lo, H. H., and Chung, J. G. (1999). The effects of plant phenolics, caffeic acid, chlorogenic acid and ferulic acid on arylamine N-acetyltransferase activities in human gastrointestinal microflora. *Anticancer Res.* **19**, 133–139.

Lou, Y. R., Lu, Y. P., Xie, J. G., Huang, M. T., and Conney, A. H. (1999). Effects of oral administration of tea, decaffeinated tea, and caffeine on the formation and growth of tumors in high-risk SKH-1 mice previously treated with ultraviolet B light. *Nutr. Cancer* **33**, 146–153.

Lubin, F., Ron, E., Wax, Y., and Modan, B. (1985). Coffee and methylxanthines and breast cancer: A case-control study. *J. Natl. Cancer Inst.* **74**, 569–573.

Lutz, U., Lugli, S., Bitsch, A., Schlatter, J., and Lutz, W. K. (1997). Dose response for the stimulation of cell division by caffeic acid in forestomach and kidney of the male F344 rat. *Fundam. Appl. Toxicol.* **39**, 131–137.

Maggi-Capeyron, M. F., Ceballos, P., Cristol, J. P., Delbosc, S., Le Doucen, C., Pons, M., Leger, C. L., and Descomps, B. (2001). Wine phenolic antioxidants inhibit AP-1 transcriptional activity. *J. Agric. Food Chem.* **49**, 5646–5652.

Majer, B. J., Hofer, E., Cavin, C., Lhoste, E., Uhl, M., Glatt, H. R., Meinl, W., and Knasmuller, S. (2005). Coffee diterpenes prevent the genotoxic effects of 2-amino-1-methyl-6-phenylimidazo[4,5-b]pyridine (PhIP) and N-nitrosodimethylamine in a human derived liver cell line (HepG2). *Food Chem. Toxicol.* **43**, 433–441.

Malats, N., Camus-Radon, A. M., Nyberg, F., Ahrens, W., Constantinescu, V., Mukeria, A., Benhamou, S., Batura-Gabryel, H., Bruske-Hohlfeld, I., Simonato, L., Menezes, A., Lea, S., Lang, M., and Boffetta, P. (2000). Lung cancer risk in nonsmokers and GSTM1 and GSTT1 genetic polymorphism. *Cancer Epidemiol. Biomarkers Prev.* **9**, 827–833.

Maliakal, P. P., and Wanwimolruk, S. (2001). Effect of herbal teas on hepatic drug metabolizing enzymes in rats. *J. Pharm. Pharmacol.* **53**, 1323–1329.

Manson, M. M., Ball, H. W., Barrett, M. C., Clark, H. L., Judah, D. J., Williamson, G., and Neal, G. E. (1997). Mechanism of action of dietary chemoprotective agents in rat liver: Induction of phase I and II drug metabolizing enzymes and aflatoxin B1 metabolism. *Carcinogenesis* **18**, 1729–1738.

Mazur, W., Gonciarz, M., Kajdy, M., Mazurek, U., Jurzak, M., Wilczok, T., and Gonciarz, Z. (2003). Blood serum glutathione alpha s-transferase (alpha GST) activity during antiviral therapy in patients with chronic hepatitis C. *Med. Sci. Monit.* **9**(Suppl. 3), 44–48.

McMahon, M., Itoh, K., Yamamoto, M., Chanas, S. A., Henderson, C. J., McLellan, L. I., Wolf, C. R., Cavin, C., and Hayes, J. D. (2001). The Cap'n'Collar basic leucine zipper transcription factor Nrf2 (NF-E2 p45-related factor 2) controls both constitutive and inducible expression of intestinal detoxification and glutathione biosynthetic enzymes. *Cancer Res.* **61**, 3299–3307.

Mensink, R. P., Lebbink, W. J., Lobbezoo, I. E., Weusten-Van der Wouw, M. P., Zock, P. L., and Katan, M. B. (1995). Diterpene composition of oils from Arabica and Robusta coffee beans and their effects on serum lipids in man. *J. Intern. Med.* **237**, 543–550.

Michaud, D. S., Giovannucci, E., Willett, W. C., Colditz, G. A., and Fuchs, C. S. (2001). Coffee and alcohol consumption and the risk of pancreatic cancer in two prospective United States cohorts. *Cancer Epidemiol. Biomarkers Prev.* **10**, 429–437.

Miller, D. P., Liu, G., De Vivo, I., Lynch, T. J., Wain, J. C., Su, L., and Christiani, D. C. (2002). Combinations of the variant genotypes of GSTP1, GSTM1, and p53 are associated with an increased lung cancer risk. *Cancer Res.* **62**, 2819–2823.

Miller, D. P., De Vivo, I., Neuberg, D., Wain, J. C., Lynch, T. J., Su, L., and Christiani, D. C. (2003). Association between self-reported environmental tobacco smoke exposure and lung cancer: Modification by GSTP1 polymorphism. *Int. J. Cancer* **104**, 758–763.

Miller, E. G., Formby, W. A., Rivera-Hidalgo, F., and Wright, J. M. (1988). Inhibition of hamster buccal pouch carcinogenesis by green coffee beans. *Oral Surg. Oral Med. Oral Pathol.* **65**, 745–749.

Miller, E. G., McWhorter, K., Rivera-Hidalgo, F., Wright, J. M., Hirsbrunner, P., and Sunahara, G. I. (1991). Kahweol and cafestol: Inhibitors of hamster buccal pouch carcinogenesis. *Nutr. Cancer* **15**, 41–46.

Minchin, R. F., Kadlubar, F. F., and Ilett, K. F. (1993). Role of acetylation in colorectal cancer. *Mutat. Res.* **290**, 35–42.

Mitrunen, K., Jourenkova, N., Kataja, V., Eskelinen, M., Kosma, V. M., Benhamou, S., Vainio, H., Uusitupa, M., and Hirvonen, A. (2001). Glutathione S-transferase M1, M3, P1, and T1 genetic polymorphisms and susceptibility to breast cancer. *Cancer Epidemiol. Biomarkers Prev.* **10**, 229–236.

Mori, H., Tanaka, T., Shima, H., Kuniyasu, T., and Takahashi, M. (1986). Inhibitory effect of chlorogenic acid on methylazoxymethanol acetate-induced carcinogenesis in large intestine and liver of hamsters. *Cancer Lett.* **30**, 49–54.

Mori, H., Kawabata, K., Matsunaga, K., Ushida, J., Fujii, K., Hara, A., Tanaka, T., and Murai, H. (2000). Chemopreventive effects of coffee bean and rice constituents on colorectal carcinogenesis. *Biofactors* **12**, 101–105.

Morishita, Y., Yoshimi, N., Kawabata, K., Matsunaga, K., Sugie, S., Tanaka, T., and Mori, H. (1997). Regressive effects of various chemopreventive agents on azoxymethane-induced aberrant crypt foci in the rat colon. *Jpn. J. Cancer Res.* **88**, 815–820.

Mukhtar, H., Das, M., Khan, W. A., Wang, Z. Y., Bik, D. P., and Bickers, D. R. (1988). Exceptional activity of tannic acid among naturally occurring plant phenols in protecting against 7,12-dimethylbenz(a)anthracene-, benzo(a)pyrene-, 3-methylcholanthrene-, and N-methyl-N-nitrosourea-induced skin tumorigenesis in mice. *Cancer Res.* **48**, 2361–2365.

Nagao, M., Fujita, Y., Wakabayashi, K., Nukaya, H., Kosuge, T., and Sugimura, T. (1986). Mutagens in coffee and other beverages. *Environ. Health Perspect.* **67**, 89–91.

Nardini, M., Natella, F., Gentili, V., Di Felice, M., and Scaccini, C. (1997). Effect of caffeic acid dietary supplementation on the antioxidant defense system in rat: An *in vivo* study. *Arch. Biochem. Biophys.* **342**, 157–160.

Nijhoff, W. A., and Peters, W. H. (1994). Quantification of induction of rat oesophageal, gastric and pancreatic glutathione and glutathione S-transferases by dietary anti-carcinogens. *Carcinogenesis* **15**, 1769–1772.

Nomura, T. (1980). Timing of chemically induced neoplasia in mice revealed by the antineoplastic action of caffeine. *Cancer Res.* **40**, 1332–1340.

Nomura, T. (1983). Comparative inhibiting effects of methylxanthines on urethan-induced tumors, malformations, and presumed somatic mutations in mice. *Cancer Res.* **43**, 1342–1346.

Norppa, H. (2001). Genetic polymorphisms and chromosome damage. *Int. J. Hyg. Environ. Health* **204**, 31–38.

Nowell, S., Coles, B., Sinha, R., MacLeod, S., Luke Ratnasinghe, D., Stotts, C., Kadlubar, F. F., Ambrosone, C. B., and Lang, N. P. (2002). Analysis of total meat intake and exposure to individual heterocyclic amines in a case-control study of colorectal cancer: Contribution of metabolic variation to risk. *Mutat. Res.* **506–507**, 175–185.

Pannala, A. S., Razaq, R., Halliwell, B., Singh, S., and Rice-Evans, C. A. (1998). Inhibition of peroxynitrite dependent tyrosine nitration by hydroxycinnamates: Nitration or electron donation? *Free Radic. Biol. Med.* **24**, 594–606.

Petrek, J. A., Sandberg, W. A., Cole, M. N., Silberman, M. S., and Collins, D. C. (1985). The inhibitory effect of caffeine on hormone-induced rat breast cancer. *Cancer* **56**, 1977–1981.

Petridou, E., Koukoulomatis, P., Dessypris, N., Karalis, D., Michalas, S., and Trichopoulos, D. (2002). Why is endometrial cancer less common in Greece than in other European Union countries? *Eur. J. Cancer Prev.* **11**, 427–432.

Pezzuto, J. M., Nanayakkara, N. P., Compadre, C. M., Swanson, S. M., Kinghorn, A. D., Guenthner, T. M., Sparnins, V. L., and Lam, L. K. (1986). Characterization of bacterial mutagenicity mediated by 13-hydroxy-ent-kaurenoic acid (steviol) and several structurally-related derivatives and evaluation of potential to induce glutathione S-transferase in mice. *Mutat. Res.* **169**, 93–103.

Ploemen, J. H., van Ommen, B., de Haan, A., Schefferlie, J. G., and van Bladeren, P. J. (1993). *In vitro* and *in vivo* reversible and irreversible inhibition of rat glutathione S-transferase isoenzymes by caffeic acid and its 2-S-glutathionyl conjugate. *Food Chem. Toxicol.* **31**, 475–482.

Porta, M., Malats, N., Alguacil, J., Ruiz, L., Jariod, M., Carrato, A., Rifa, J., and Guarner, L. (2000). Coffee, pancreatic cancer, and K-ras mutations: Updating the research agenda. *J. Epidemiol. Commun. Health* **54**, 656–659.

Rao, C. V., Chou, D., Simi, B., Ku, H., and Reddy, B. S. (1998). Prevention of colonic aberrant crypt foci and modulation of large bowel microbial activity by dietary coffee fiber, inulin and pectin. *Carcinogenesis* **19**, 1815–1819.

Reddy, B. S., and Maeura, Y. (1984). Dose-response studies of the effect of dietary butylated hydroxyanisole on colon carcinogenesis induced by methylazoxymethanol acetate in female CF1 mice. *J. Natl. Cancer Inst.* **72**, 1181–1187.

Rodriguez, T., Altieri, A., Chatenoud, L., Gallus, S., Bosetti, C., Negri, E., Franceschi, S., Levi, F., Talamini, R., and La Vecchia, C. (2004). Risk factors for oral and pharyngeal cancer in young adults. *Oral Oncol.* **40**, 207–213.

Rothen, J. P., Haefeli, W. E., Meyer, U. A., Todesco, L., and Wenk, M. (1998). Acetaminophen is an inhibitor of hepatic N-acetyltransferase 2 *in vitro* and *in vivo*. *Pharmacogenetics* **8**, 553–559.

Ruiz del Castillo, M. L., Herraiz, M., and Blanch, G. P. (1999). Rapid analysis of cholesterol-elevating compounds in coffee brews by off-line high-performance liquid chromatography/high-resolution gas chromatography. *J. Agric. Food Chem.* **47**, 1525–1529.

Rundlof, T., Olsson, E., Wiernik, A., Back, S., Aune, M., Johansson, L., and Wahlberg, I. (2000). Potential nitrite scavengers as inhibitors of the formation of N-nitrosamines in solution and tobacco matrix systems. *J. Agric. Food Chem.* **48**, 4381–4388.

Saarikoski, S. T., Voho, A., Reinikainen, M., Anttila, S., Karjalainen, A., Malaveille, C., Vainio, H., Husgafvel-Pursiainen, K., and Hirvonen, A. (1998). Combined effect of polymorphic GST genes on individual susceptibility to lung cancer. *Int. J. Cancer* **77**, 516–521.

Saroja, M., Balasenthil, S., Ramachandran, C. R., and Nagini, S. (2001). Coffee enhances the development of 7,12-dimethylbenz[a]anthracene (DMBA)-induced hamster buccal pouch carcinomas. *Oral Oncol.* **37**, 172–176.

Savolainen, H. (1992). Tannin content of tea and coffee. *J. Appl. Toxicol.* **12**, 191–192.

Scarpato, R., Hirvonen, A., Migliore, L., Falck, G., and Norppa, H. (1997). Influence of GSTM1 and GSTT1 polymorphisms on the frequency of chromosome aberrations in lymphocytes of smokers and pesticide-exposed greenhouse workers. *Mutat. Res.* **389**, 227–235.

Scharf, G., Prustomersky, S., Knasmuller, S., Schulte-Hermann, R., and Huber, W. W. (2003). Enhancement of glutathione and γ-glutamylcysteine synthetase, the rate limiting enzyme of glutathione synthesis, by chemoprotective plant-derived food and beverage components in the human hepatoma cell line HepG2. *Nutr. Cancer* **45**, 74–83.

Schilter, B., Perrin, I., Cavin, C., and Huggett, A. C. (1996). Placental glutathione S-transferase (GST-P) induction as a potential mechanism for the anti-carcinogenic effect of the coffee-specific components cafestol and kahweol. *Carcinogenesis* **17**, 2377–2384.

Schröder, K. R., Wiebel, F. A., Reich, S., Dannappel, D., Bolt, H. M., and Hallier, E. (1995). Glutathione-S-transferase (GST) theta polymorphism influences background SCE rate. *Arch. Toxicol.* **69**, 505–507.

Sherratt, P. J., Manson, M. M., Thomson, A. M., Hissink, E. A., Neal, G. E., van Bladeren, P. J., Green, T., and Hayes, J. D. (1998). Increased bioactivation of dihaloalkanes in rat liver due to induction of class theta glutathione S-transferase T1-1. *Biochem. J.* **335**, 619–630.

Shi, X., Dalal, N. S., and Jain, A. C. (1991). Antioxidant behaviour of caffeine: Efficient scavenging of hydroxyl radicals. *Food Chem. Toxicol.* **29**, 1–6.

Shilo, L., Sabbah, H., Hadari, R., Kovatz, S., Weinberg, U., Dolev, S., Dagan, Y., and Shenkman, L. (2002). The effects of coffee consumption on sleep and melatonin secretion. *Sleep Med.* **3**, 271–273.

Shimizu, M., Yoshimi, N., Yamada, Y., Matsunaga, K., Kawabata, K., Hara, A., Moriwaki, H., and Mori, H. (1999). Suppressive effects of chlorogenic acid on N-methyl-N-nitrosourea-induced glandular stomach carcinogenesis in male F344 rats. *J. Toxicol. Sci.* **24**, 433–439.

Silverman, D. T., Swanson, C. A., Gridley, G., Wacholder, S., Greenberg, R. S., Brown, L. M., Hayes, R. B., Swanson, G. M., Schoenberg, J. B., Pottern, L. M., Schwartz, A. G., Fraumeni, J. F., Jr., and Hoover, R. N. (1998). Dietary and nutritional factors and pancreatic cancer: A case-control study based on direct interviews. *J. Natl. Cancer Inst.* **90**, 1710–1719.

Slebos, R. J., Hoppin, J. A., Tolbert, P. E., Holly, E. A., Brock, J. W., Zhang, R. H., Bracci, P. M., Foley, J., Stockton, P., McGregor, L. M., Flake, G. P., and Taylor, J. A. (2000). K-ras and p53 in pancreatic cancer: Association with medical history, histopathology, and environmental exposures in a population-based study. *Cancer Epidemiol. Biomarkers Prev.* **9**, 1223–1232.

Somoza, V., Lindenmeier, M., Wenzel, E., Frank, O., Erbersdobler, H. F., and Hofmann, T. (2003). Activity-guided identification of a chemopreventive compound in coffee beverage using *in vitro* and *in vivo* techniques. *J. Agric. Food Chem.* **51**, 6861–6869.

Sparnins, V. L., Venegas, P. L., and Wattenberg, L. W. (1982). Glutathione S-transferase activity: Enhancement by compounds inhibiting chemical carcinogenesis and by dietary constituents. *J. Natl. Cancer Inst.* **68**, 493–496.

Sreerama, L., Hedge, M. W., and Sladek, N. E. (1995). Identification of a class 3 aldehyde dehydrogenase in human saliva and increased levels of this enzyme, glutathione S-transferases, and DT-diaphorase in the saliva of subjects who continually ingest large quantities of coffee or broccoli. *Clin. Cancer Res.* **1**, 1153–1163.

Srivastava, R. C., Husain, M. M., Hasan, S. K., and Athar, M. (2000). Green tea polyphenols and tannic acid act as potent inhibitors of phorbol ester-induced nitric oxide generation in rat hepatocytes independent of their antioxidant properties. *Cancer Lett.* **153**, 1–5.

Stalder, R., Bexter, A., Wurzner, H. P., and Luginbuhl, H. (1990). A carcinogenicity study of instant coffee in Swiss mice. *Food Chem. Toxicol.* **28**, 829–837.

Steinhoff, C., Franke, K. H., Golka, K., Thier, R., Romer, H. C., Rotzel, C., Ackermann, R., and Schulz, W. A. (2000). Glutathione transferase isozyme genotypes in patients with prostate and bladder carcinoma. *Arch. Toxicol.* **74**, 521–526.

Steinkellner, H., Hoelzl, C., Uhl, M., Cavin, C., Haidinger, G., Gsur, A., Schmid, R., Kundi, M., Bichler, J., and Knasmüller, S. (2005). Coffee consumption induces GST pi and protects against (±)-antibenzo[a]pyrene-7,8-dihydrodiol-9,10-epoxide induced DNA damage in humans. *Mutat. Res.* in press.

Stucker, I., de Waziers, I., Cenee, S., Bignon, J., Depierre, A., Milleron, B., Beaune, P., and Hemon, D. (1999). GSTM1, smoking and lung cancer: A case-control study. *Int. J. Epidemiol.* **28,** 829–835.

Stucker, I., Hirvonen, A., de Waziers, I., Cabelguenne, A., Mitrunen, K., Cenee, S., Koum-Besson, E., Hemon, D., Beaune, P., and Loriot, M. A. (2002). Genetic polymorphisms of glutathione S-transferases as modulators of lung cancer susceptibility. *Carcinogenesis* **23,** 1475–1481.

Sumiyoshi, H., and Wargovich, M. J. (1990). Chemoprevention of 1,2-dimethylhydrazine-induced colon cancer in mice by naturally occurring organosulfur compounds. *Cancer Res.* **50,** 5084–5087.

Sweeney, C., Coles, B. F., Nowell, S., Lang, N. P., and Kadlubar, F. F. (2002). Novel markers of susceptibility to carcinogens in diet: Associations with colorectal cancer. *Toxicology* **181–182,** 83–87.

Szaefer, H., Jodynis-Liebert, J., Cichocki, M., Matuszewska, A., and Baer-Dubowska, W. (2003). Effect of naturally occurring plant phenolics on the induction of drug metabolizing enzymes by o-toluidine. *Toxicology* **186,** 67–77.

Szaefer, H., Cichocki, M., Brauze, D., and Baer-Dubowska, W. (2004). Alteration in phase I and II enzyme activities and polycyclic aromatic hydrocarbons-DNA adduct formation by plant phenolics in mouse epidermis. *Nutr. Cancer* **48,** 70–77.

Takeshita, F., Ogawa, K., Asamoto, M., and Shirai, T. (2003). Mechanistic approach of contrasting modifying effects of caffeine on carcinogenesis in the rat colon and mammary gland induced with 2-amino-1-methyl-6-phenylimidazo[4,5-b]pyridine. *Cancer Lett.* **194,** 25–35.

Takezaki, T., Hirose, K., Inoue, M., Hamajima, N., Kuroishi, T., Nakamura, S., Koshikawa, T., Matsuura, H., and Tajima, K. (1996). Risk factors of thyroid cancer among women in Tokai, Japan. *J. Epidemiol.* **6,** 140–147.

Tanaka, T., Nishikawa, A., Shima, H., Sugie, S., Shinoda, T., Yoshimi, N., Iwata, H., and Mori, H. (1990). Inhibitory effects of chlorogenic acid, reserpine, polyprenoic acid (E-5166), or coffee on hepatocarcinogenesis in rats and hamsters. *Basic Life Sci.* **52,** 429–440.

Tanaka, T., Kojima, T., Kawamori, T., Wang, A., Suzui, M., Okamoto, K., and Mori, H. (1993). Inhibition of 4-nitroquinoline-1-oxide-induced rat tongue carcinogenesis by the naturally occurring plant phenolics caffeic, ellagic, chlorogenic and ferulic acids. *Carcinogenesis* **14,** 1321–1325.

Tavani, A., and La Vecchia, C. (2000). Coffee and cancer: A review of epidemiological studies, 1990–1999. *Eur. J. Cancer Prev.* **9,** 241–256.

Tavani, A., Gallus, S., Dal Maso, L., Franceschi, S., Montella, M., Conti, E., and La Vecchia, C. (2001). Coffee and alcohol intake and risk of ovarian cancer: An Italian case-control study. *Nutr. Cancer* **39,** 29–34.

Tavani, A., Bertuzzi, M., Talamini, R., Gallus, S., Parpinel, M., Franceschi, S., Levi, F., and La Vecchia, C. (2003). Coffee and tea intake and risk of oral, pharyngeal and esophageal cancer. *Oral Oncol.* **39,** 695–700.

Tavani, A., and La Vecchia, C. (2004). Coffee, decaffeinated coffee, tea and cancer of the colon and rectum: A review of epidemiological studies, 1990–2003. *Cancer Causes Control* **15,** 743–757.

Terry, P., Vainio, H., Wolk, A., and Weiderpass, E. (2002). Dietary factors in relation to endometrial cancer: A nationwide case-control study in Sweden. *Nutr. Cancer* **42,** 25–32.

Thier, R., Taylor, J. B., Pemble, S. E., Humphreys, W. G., Persmark, M., Ketterer, B., and Guengerich, F. P. (1993). Expression of mammalian glutathione S-transferase 5-5 in

Salmonella typhimurium TA1535 leads to base-pair mutations upon exposure to dihalomethanes. *Proc. Natl. Acad. Sci. USA* **90**, 8576–8580.

Thier, R., Pemble, S. E., Kramer, H., Taylor, J. B., Guengerich, F. P., and Ketterer, B. (1996). Human glutathione S-transferase T1-1 enhances mutagenicity of 1,2-dibromoethane, dibromomethane and 1,2,3,4-diepoxybutane in Salmonella typhimurium. *Carcinogenesis* **17**, 163–166.

Thier, R., Bruning, T., Roos, P. H., Rihs, H. P., Golka, K., Ko, Y., and Bolt, H. M. (2003). Markers of genetic susceptibility in human environmental hygiene and toxicology: The role of selected CYP, NAT and GST genes. *Int. J. Hyg. Environ. Health* **206**, 149–171.

Tsou, M. F., Hung, C. F., Lu, H. F., Wu, L. T., Chang, S. H., Chang, H. L., Chen, G. W., and Chung, J. G. (2000). Effects of caffeic acid, chlorogenic acid and ferulic acid on growth and arylamine N-acetyltransferase activity in Shigella sonnei (group D). *Microbios.* **101**, 37–46.

Tsuda, H., Sekine, K., Uehara, N., Takasuka, N., Moore, M. A., Konno, Y., Nakashita, K., and Degawa, M. (1999). Heterocyclic amine mixture carcinogenesis and its enhancement by caffeine in F344 rats. *Cancer Lett.* **143**, 229–234.

Turesky, R. J., Richoz, J., Constable, A., Curtis, K. D., Dingley, K. H., and Turteltaub, K. W. (2003). The effects of coffee on enzymes involved in metabolism of the dietary carcinogen 2-amino-1-methyl-6-phenylimidazo[4,5-b]pyridine in rats. *Chem. Biol. Interact.* **145**, 251–265.

Tverdal, A., and Skurtveit, S. (2003). Coffee intake and mortality from liver cirrhosis. *Ann. Epidemiol.* **13**, 419–423.

Urgert, R., van der Weg, G., Kosmeijer-Schuil, T. G., van de Bovenkamp, P., Hovenier, R., and Katan, M. B. (1995). Levels of the cholesterol-elevating diterpenes Cafestol and Kahweol in various coffee brews. *J. Agric. Food Chem.* **43**, 2167–2172.

Urgert, R., and Katan, M. B. (1997). The cholesterol-raising factor from coffee beans. *Annu. Rev. Nutr.* **17**, 305–324.

Vatten, L. J., Solvoll, K., and Loken, E. B. (1990). Coffee consumption and the risk of breast cancer. A prospective study of 14,593 Norwegian women. *Br. J. Cancer* **62**, 267–270.

Veierod, M. B., Thelle, D. S., and Laake, P. (1997). Diet and risk of cutaneous malignant melanoma: A prospective study of 50,757 Norwegian men and women. *Int. J. Cancer* **71**, 600–604.

Villeneuve, P. J., Johnson, K. C., Hanley, A. J., and Mao, Y. (2000). Alcohol, tobacco and coffee consumption and the risk of pancreatic cancer: Results from the Canadian Enhanced Surveillance System case-control project. Canadian Cancer Registries Epidemiology Research Group. *Eur. J. Cancer Prev.* **9**, 49–58.

Vistisen, K., Poulsen, H. E., and Loft, S. (1992). Foreign compound metabolism capacity in man measured from metabolites of dietary caffeine. *Carcinogenesis* **13**, 1561–1568.

Wang, D. Y., Yeh, C. C., Lee, J. H., Hung, C. F., and Chung, J. G. (2002). Berberine inhibited arylamine N-acetyltransferase activity and gene expression and DNA adduct formation in human malignant astrocytoma (G9T/VGH) and brain glioblastoma multiforms (GBM 8401) cells. *Neurochem. Res.* **27**, 883–889.

Wargovich, M. J. (1987). Diallyl sulfide, a flavor component of garlic (Allium sativum), inhibits dimethylhydrazine-induced colon cancer. *Carcinogenesis* **8**, 487–489.

Wargovich, M. J., Jimenez, A., McKee, K., Steele, V. E., Velasco, M., Woods, J., Price, R., Gray, K., and Kelloff, G. J. (2000). Efficacy of potential chemopreventive agents on rat colon aberrant crypt formation and progression. *Carcinogenesis* **21**, 1149–1155.

Wattenberg, L. W., Lam, L. K., and Fladmoe, A. V. (1979). Inhibition of chemical carcinogen-induced neoplasia by coumarins and alpha-angelicalactone. *Cancer Res.* **39**, 1651–1654.

Wattenberg, L. W. (1983). Inhibition of neoplasia by minor dietary constituents. *Cancer Res.* **43,** 2448s–2453s.

Weiderpass, E., Partanen, T., Kaaks, R., Vainio, H., Porta, M., Kauppinen, T., Ojajarvi, A., Boffetta, P., and Malats, N. (1998). Occurrence, trends and environment etiology of pancreatic cancer. *Scand. J. Work Environ. Health* **24,** 165–174.

Welsch, C. W., DeHoog, J. V., and O'Connor, D. H. (1988). Influence of caffeine and/or coffee consumption on the initiation and promotion phases of 7,12-dimethylbenz(a) anthracene-induced rat mammary gland tumorigenesis. *Cancer Res.* **48,** 2068–2073.

Wenzel, E., Tasto, S., Erbersdobler, H. F., and Faist, V. (2002). Effect of heat-treated proteins on selected parameters of the biotransformation system in the rat. *Ann. Nutr. Metab.* **46,** 9–16.

Weusten-Van der Wouw, M. P., Katan, M. B., Viani, R., Huggett, A. C., Liardon, R., Liardon, R., Lund-Larsen, P. G., Thelle, D. S., Ahola, I., Aro, A., Meyboom, S., and Beynen, A. C. (1994). Identity of the cholesterol-raising factor from boiled coffee and its effects on liver function enzymes [published erratum appears in *J. Lipid. Res.* 1994 Aug;35(8):1510]. *J. Lipid Res.* **35,** 721–733.

Woutersen, R. A., van Garderen-Hoetmer, A., Bax, J., and Scherer, E. (1989). Modulation of dietary fat-promoted pancreatic carcinogenesis in rats and hamsters by chronic coffee ingestion. *Carcinogenesis* **10,** 311–316.

Wu, L. T., and Chung, J. G. (1998). Effects of vitamin C on arylamine N-acetyltransferase activity in human liver tumor cells. *Anticancer Res.* **18,** 3481–3486.

Wu, L. T., Chung, J. G., Chen, J. C., and Tsauer, W. (2001). Effect of norcantharidin on N-acetyltransferase activity in HepG2 cells. *Am. J. Chin. Med.* **29,** 161–172.

Yamanaka, N., Oda, O., and Nagao, S. (1997). Prooxidant activity of caffeic acid, dietary non-flavonoid phenolic acid, on Cu2+-induced low density lipoprotein oxidation. *FEBS Lett.* **405,** 186–190.

Yang, C. C., Chen, G. W., Lu, H. F., Wang, D. Y., Chen, Y. S., and Chung, J. G. (2003). Paclitaxel (taxol) inhibits the arylamine N-acetyltransferase activity and gene expression (mRNA NAT1) and 2-aminofluorene-DNA adduct formation in human bladder carcinoma cells (T24 and TSGH 8301). *Pharmacol. Toxicol.* **92,** 287–294.

Zeegers, M. P., Dorant, E., Goldbohm, R. A., and van den Brandt, P. A. (2001a). Are coffee, tea, and total fluid consumption associated with bladder cancer risk? Results from the Netherlands Cohort Study. *Cancer Causes Control* **12,** 231–238.

Zeegers, M. P., Tan, F. E., Goldbohm, R. A., and van den Brandt, P. A. (2001b). Are coffee and tea consumption associated with urinary tract cancer risk? A systematic review and meta-analysis. *Int. J. Epidemiol.* **30,** 353–362.

Zhang, K., and Das, N. P. (1994). Inhibitory effects of plant polyphenols on rat liver glutathione S-transferases. *Biochem. Pharmacol.* **47,** 2063–2068.

Zhong, S., Wyllie, A. H., Barnes, D., Wolf, C. R., and Spurr, N. K. (1993). Relationship between the GSTM1 genetic polymorphism and susceptibility to bladder, breast and colon cancer. *Carcinogenesis* **14,** 1821–1824.

[21] Activation of Alkyl Halides by Glutathione Transferases

By F. PETER GUENGERICH

Abstract

Glutathione (GSH) transferases catalyze the conjugation of the tripeptide GSH with alkyl halides and related compounds. If a second leaving group is present, the substrate is at least a potential bis-electrophile and the initial conjugate may be susceptible to further attack by the sulfur atom. This process can yield potent electrophiles that modify DNA and are genotoxic. Much of the chemistry is understood in the context of the halide order and size of rings generated in reactive sulfonium ions. Similar chemistry has been demonstrated with the active site cysteine residue in the DNA repair protein O^6-alkylguanine DNA-alkyltransferase.

Introduction

Glutathione (GSH) transferases (GST) catalyze the conjugation of GSH with electrophilic substrates (Habig *et al.*, 1974) The GST gene family is reviewed elsewhere in this volume. The substrates are rather diverse and include important endogenous modulators, as well as xenobiotic chemicals and the oxidation products of the transformation of xenobiotics by other enzymes. The roles of GSTs in these latter groups of reactions with xenobiotic chemicals have led to the general view that GSTs are detoxication enzymes.

GSH conjugations of alkyl halides (X designates halide in this chapter) can be detoxication or bioactivation reactions. A general list of the GSH conjugations catalyzed by GSTs is presented in Fig. 1. Reaction 1, with a monohalide, is a detoxication reaction. The other reactions, with ≥2 halides, can activate the substrate because of the instability of the product (Fig. 1), leading to reactions with nucleophiles, particularly DNA and proteins. Mechanisms have been studied most extensively in the case of the *vic*-dihaloethanes (also termed ethylene dihalides) (Guengerich, 1994, 2003; Peterson *et al.*, 1988) With the substrates in reactions 1–4 of Fig. 1 (at least with relatively small alkyl halides), the theta class GST enzymes are the most active (Meyer *et al.*, 1991; Ploemen *et al.*, 1995; Thier *et al.*, 1993), although other mammalian GSTs also have some activities (Cmarik *et al.*, 1990). The zeta class GSTs have been implicated in the conjugation of dihalo acids (Fig. 1, reaction 5) (Tong *et al.*, 1998).

METHODS IN ENZYMOLOGY, VOL. 401   0076-6879/05 $35.00
Copyright 2005, Elsevier Inc. All rights reserved.
DOI: 10.1016/S0076-6879(05)01021-9

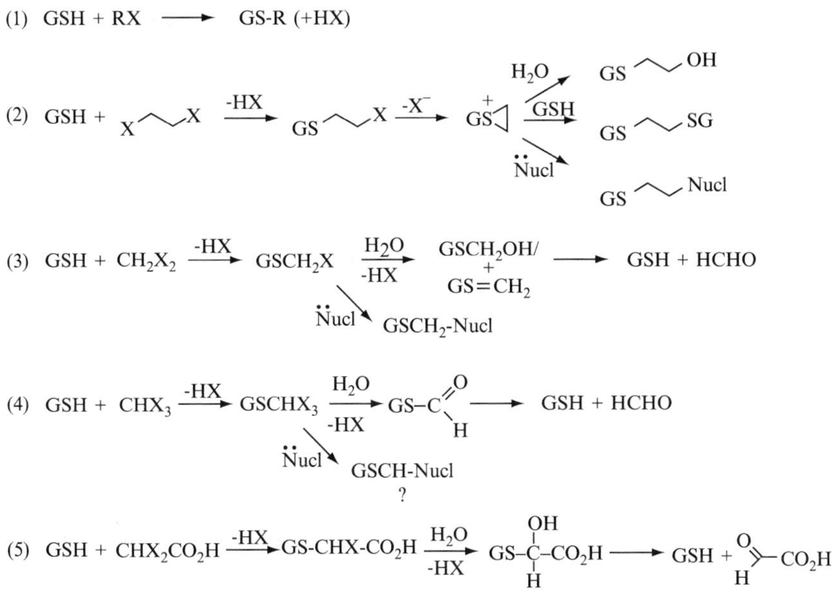

(1) $GSH + RX \longrightarrow GS\text{-}R\ (+HX)$

(2) $GSH +$ X$\sim\sim$X $\xrightarrow{-HX}$ GS$\sim\sim$X $\xrightarrow{-X^-}$ GS$\overset{+}{\triangleleft}$
    $\xrightarrow{H_2O}$ GS$\sim$OH
    $\xrightarrow{GSH}$ GS$\sim$SG
    $\xrightarrow{Nucl}$ GS$\sim$Nucl

(3) $GSH + CH_2X_2 \xrightarrow{-HX} GSCH_2X \xrightarrow[-HX]{H_2O} \begin{matrix} GSCH_2OH/ \\ + \\ GS=CH_2 \end{matrix} \longrightarrow GSH + HCHO$
    $\xrightarrow{Nucl}$ GSCH$_2$-Nucl

(4) $GSH + CHX_3 \xrightarrow{-HX} GSCHX_3 \xrightarrow[-HX]{H_2O} GS\text{-}C\overset{O}{\underset{H}{}} \longrightarrow GSH + HCHO$
    $\xrightarrow{Nucl}$ GSCH-Nucl
    ?

(5) $GSH + CHX_2CO_2H \xrightarrow{-HX} GS\text{-}CHX\text{-}CO_2H \xrightarrow[-HX]{H_2O} GS\text{-}\overset{OH}{\underset{H}{C}}\text{-}CO_2H \longrightarrow GSH + \overset{O}{\underset{H}{}}\text{-}CO_2H$

FIG. 1. Reactions of GSH with alkyl halides. Nucl, nucleophile.

Experimental Details

GST DM11

The plasmid pE1962 coding for *Methylophilus* GST DM11 was obtained from Dr. Stephane Vuilleumier (Strasbourg, France). For purification, expression was done in *Escherichia coli* BL21(DE3)pLys S cells under the control of isopropyl-β-D-thiogalactoside (IPTG) (16 h at 26°) (Wheeler *et al.*, 2001a). The protein was purified to electrophoretic homogeneity using $(NH_4)_2SO_4$ fractionation and DEAE cellulose chromatography (Wheeler *et al.*, 2001a).

For use in the modified Ames test, DM11 was subcloned from pE1962 into the vector pTrc99A (Pharmacia, Piscataway, NJ), with slight modification (Wheeler *et al.*, 2001a) and expressed in *Salmonella typhimurium* TA1535. Quantitation of levels of expression of DM11 can be done using immunoelectrophoresis and a polyclonal rabbit antibody raised against purified DM11 (Wheeler *et al.*, 2001a).

Rat GST 5-5

Expression for protein purification used the plasmid pKK233-2 (Thier *et al.*, 1993) in *E. coli* DHFα F'IQ (Gibco-BRL, Grand Island, NY). Induction with IPTG was done for 18 h at 28° (Wheeler *et al.*, 2001a). The protein was purified to electrophoretic homogeneity using a combination of DEAE-Sepharose and hydroxylapatite chromatography (Wheeler et al., 2001a). The transformation of S. typhimurium TA1535 with the vector pKK233-2 expressing GST 5-5 is described (Thier et al., 1993). Estimation of levels of expression can be made using a polyclonal rabbit antibody raised against purified GST 5-5 (Wheeler et al., 2001a).

Human GST T1-1

GST T1-1 was expressed in *E. coli* HMS174 (DE3) cells using the vector pET24d (Novagen) containing the GST T1 DNA (18 h, 28°, IPTG) (Wheeler et al., 2001a). This construct contains an N-terminal His tag. The protein was purified to electrophoretic homogeneity using Ni^{2+}-nitriloacetic acid chromatography, and fractions were assayed by polyacrylamide gel electrophoresis and for conjugation activity with the substrate 1,2-epoxy-3-(*p*-nitrophenoxy)propane.

Expression of GST T1-1 n *S. typhimurium* TA100 was done with the vector pKK233-2 (Wheeler et al., 2001a). Estimation of the expression of the protein can be done by immunoelectrophoresis with a polyclonal rabbit antibody raised against purified GST T1-1 (Wheeler et al., 2001a).

Assays of Catalytic Activity

Conjugation of Alkyl Halides (Fig. 1, Reaction 1)

Although the disappearance of substrate can be monitored by chromatography (Wheeler et al., 2001b), the most sensitive and relevant assays involve high-performance liquid chromatography (HPLC) determination of GSH conjugates. Radioactive substrates facilitate assays. GSH conjugates can be monitored (without radioactivity) by HPLC/ultraviolet (UV) measurements (A_{214}, relatively insensitive) or by combined HPLC–mass spectrometry (MS). The GSH conjugates can be derivatized (also the residual GSH) to add a visible (Reed et al., 1999) or fluorescent (Wheeler et al., 2001a) chromophore, which considerably enhances the sensitivity. One option is the use of radiolabeled GSH, but the high K_m values have the effect of reducing the sensitivity or raising the cost.

Conjugation of Ethylene Dihalides

The same HPLC approaches mentioned previously for alkyl halides can be used to measure *S*-(2-hydroxyethyl)GSH (the ethylene bridge of the substrate could be substituted, e.g., with an alkyl group). In the assays, the excess GSH reacts with the product (*S*-(2-haloethyl)GSH or episulfonium ion) nonenzymatically to produce a large fraction of the ethylene-*bis*-GSH product, which can also be assayed by HPLC (Cmarik *et al.*, 1990).

DNA conjugates can also be assayed (Humphreys *et al.*, 1990), although the fraction of the episulfonium ion reacting with DNA is relatively low. The assays can be useful but do not provide an estimate of total GSH conjugation.

The mechanism for the activation of ethylene dihalides (1,2-dihaloalkanes) is reasonably well understood and involves anchimeric assistance (Peterson *et al.*, 1988; Webb *et al.*, 1987). The initial conjugation of longer α,ω-dihaloalkanes is faster because of an inductive effect, but the 3-carbon compounds are more stable because of the difficulty of generating a sulfonium ion with a four-membered ring (Inskeep and Guengerich, 1984). 1,4-Dibromobutane can yield a five-membered ring sulfonium compound (thialonium), but these are stable (Knipe, 1981; Thier *et al.*, 1995). GSTs 5-5 and T1-1 expression accentuate the mutagenicity of 1,2,3,4-diepoxybutane (butadiene diepoxide) and several other bifunctional electrophiles, most containing a halogen and an epoxide (Shimada *et al.*, 1996; Thier *et al.*, 1995, 1996) The activation can be rationalized using an episulfonium mechanism (Thier *et al.*, 1995), but more chemical evidence is needed.

gem-Dihaloalkanes

The conjugation of *gem*-dihalomethanes results in the formation of HCHO, which can be measured by several methods. One approach involves colorimetric estimation, which is most commonly done with the Nash assay (Guengerich, 2001; Nash, 1953) but can also be done with 4-amino-5-hydrazino-3-mercapto-1,2,4-triazole (Purpald) (Avigad, 1983; Small and Jones, 1990). These assays do have finite blank interference, which can limit sensitivity. Another approach involves the (*in situ*) conversion of NAD^+ to NADH by *Pseudomonas putida* formaldehyde dehydrogenase, which provides for a continuous assay (Wheeler *et al.*, 2001b). HCHO can also be trapped as the 2,4-dinitrophenylhydrazone (Bell and Guengerich, 1997; Shriner *et al.*, 1965) or as other hydrazone derivatives (Guengerich *et al.*, 1996) and quantified by HPLC/UV or gas chromatography (GC)/mass spectrometry (MS) measurements (Guengerich *et al.*, 1996).

Haloforms (trihalomethanes)

The methods for these assays are less well defined than for the other assays. HCO_2H is a product, but the assays are not trivial, and radiometric assays have merit (Ross and Pegram, 2003). However, the availability of radiolabeled forms of some of the trihalomethanes is an issue.

Dihaloacetic Acids

Assays for these reactions involve measurement of glyoxylic acid (or the 2-carbonyl carboxylic acid corresponding to the substrate, e.g., pyruvic acid from 2,2-dichloropropionic acid) using HPLC methods (e.g., hydrazones) or GC-MS (Tong *et al.*, 1998).

Results and Discussion

Enzyme Selectivity and Kinetic Mechanism

Most of the early work demonstrating the activation of *vic*-diha-loethanes was done with cytosolic fractions or crude preparations of GSTs (Ozawa and Guengerich, 1983; Rannug *et al.*, 1978; van Bladeren *et al.*, 1980). These studies established the role of enzymatic GSH conjugation in DNA binding and genotoxicity (Koga *et al.*, 1986). Several rat and human GSTs can catalyze the reaction (Cmarik *et al.*, 1990; Inskeep and Guengerich, 1984), although the theta class enzymes are usually most active (Ploemen *et al.*, 1995; Thier *et al.*, 1993).

The theta class GSTs show very high K_m values for substrates (non-GSH) in the reactions they catalyze. This kinetic behavior was first observed in some of the assays of CH_2Cl_2 conjugation with cytosolic preparations, and comparisons with the dose-response relationship for tumorigenicity led to the hypothesis that CH_2Cl_2 conjugation, not oxidation, was involved in carcinogenicity (Andersen *et al.*, 1987; Reitz *et al.*, 1989) The limited saturability of the reaction was subsequently confirmed with purified mammalian theta class enzymes for CH_2Cl_2 (Meyer *et al.*, 1991; Wheeler *et al.*, 2001b) and other alkyl halides (Wheeler *et al.*, 2001a).

The K_m values for CH_2Cl_2 by rat and human theta GSTs are very high, and the limited solubility of dihalomethanes makes accurate determination of both k_{cat} and K_m difficult (the ratio k_{cat}/K_m is a better descriptor and is approximated by the slope of plots of v *vs*. S) (Fig. 2). This kinetic behavior is generally seen in the conjugation of haloalkanes and *vic*-dihaloethanes as well (Fig. 2) (Wheeler *et al.*, 2001a), suggesting that the phenomenon is an inherent property of these GSTs and not related to the subsequent reaction of the primary reaction product (Fig. 1, reaction 3).

However, the bacterial GST DM11 catalyzes the conjugation of diha-lomethanes with a relatively low K_m (2–50 μM) (Stourman *et al.*, 2003; Wheeler *et al.*, 2001b). This kinetic behavior extends to monohaloalkanes and *vic*-dihaloethanes (Wheeler *et al.*, 2001a) and indicates an inherent difference in the catalytic behavior of these enzymes. Another phenome-non is the effect of the halide order. With the mammalian theta class

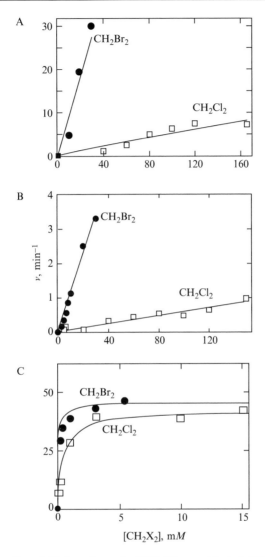

FIG. 2. Conjugation of dihalomethanes by GSTs (Guengerich *et al.*, 2003; Wheeler *et al.*, 2001b). (A) Rat GST 4-4; (B) human GST T1-1; (C) bacterial GST DM11. CH_2Cl_2 (o); CH_2Br_2 (λ).

enzymes, the parameter k_{cat}/K_m increases with decreased strength of the C-X bond (i.e., catalysis of C-Br cleavage being more efficient than of a C–Cl bond). However, with GST DM11 the change of the halide to a less

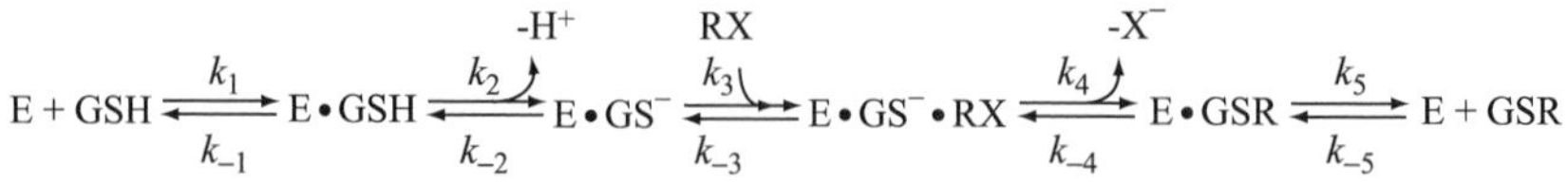

$$E + GSH \underset{k_{-1}}{\overset{k_1}{\rightleftharpoons}} E \bullet GSH \underset{k_{-2}}{\overset{k_2 \,\,\, -H^+}{\rightleftharpoons}} E \bullet GS^- \underset{k_{-3}}{\overset{k_3 \,\,\, RX}{\rightleftharpoons}} E \bullet GS^- \bullet RX \underset{k_{-4}}{\overset{k_4 \,\,\, -X^-}{\rightleftharpoons}} E \bullet GSR \underset{k_{-5}}{\overset{k_5}{\rightleftharpoons}} E + GSR$$

FIG. 3. General scheme for conjugation of alkyl halides by GSTs.

electronegative species results in a decrease in K_m but not k_{cat} (Guengerich et al., 2003; Wheeler et al., 2001a).

An explanation is found in a more detailed consideration of a general kinetic scheme for GSTs (Fig. 3). (This scheme is abbreviated in that a separate step is not included for the dissociation of the halide ion, which may or may not contribute here.) A key study was the demonstration that conjugation shows burst kinetics with GST DM11, and a step following product formation must be rate limiting (Stourman et al., 2003). This kinetic behavior has two effects on the observed K_m: (1) the K_m value is lower than predicted by the K_d value for binding of the alkyl halide (e.g., k_{-3}/k_3 in this scheme), and (2) a decrease in k_4 will raise the K_m (Guengerich et al., 2003). These are general phenomena (Walsh, 1979) and had been observed in our own work on the oxidation of ethanol by P450 2E1 (Bell and Guengerich, 1997; Bell-Parikh and Guengerich, 1999).

The results of experiments with the mammalian theta class GSTs and bacterial DM11 can be fit into the model of Fig. 3 and simulated with a reasonable set of rate constants (Guengerich et al., 2003). This information is of use in predicting the behavior of various substrates. Changing the halide from F to Cl to Br in a series will decrease K_m for a conjugation catalyzed by GST DM11 but not the k_{cat}. With the mammalian theta GSTs, the reactions will not be very saturable, and the enzyme efficiency, k_{cat}/K_m, will increase with the halide order. With a substrate containing multiple halogens (e.g., reactions 2–4 in Fig. 1), the main factor is the weakest C–X bond, because this is involved in the first reaction with GSH. The primary reaction product stability is not so great as to show the reaction, unless the product has an F (e.g., $GSCH_2F$, $GSCH_2CH_2F$). If the degradation products of these products are assayed, the apparent kinetics will be dominated by the product stability (Blocki et al., 1994; Wheeler et al., 2001a). Whether the slow step after product release for GST DM11 is actually k_5 of Fig. 3 or a hidden conformational change (or halide release) is unknown.

These kinetic principles apply to both the detoxication and bioactivation reactions catalyzed by these enzymes (vide infra). One aspect of the latter (bioactivation) with DM11 is that the mutagenic product is sequestered in the enzyme and may undergo some chemical degradation before release into the cellular environment (Stourman et al., 2003).

Genotoxicity of GSH-haloalkane Conjugates

The genotoxicity of the GST-generated GSH-haloalkane conjugates has been evaluated generally as either DNA adduct formation (Humphreys *et al.*, 1990; Inskeep and Guengerich, 1984; Inskeep *et al.*, 1986; Kim and Guengerich, 1990; Koga *et al.*, 1986; Ozawa and Guengerich, 1983) or bacterial mutagenicity, using primarily either the Ames *S. typhimurium* (*his*) or *E. coli* (*lac*) test systems (Cmarik *et al.*, 1992; Humphreys *et al.*, 1990; Wheeler *et al.*, 2001a,b) The cited references provide details of experimental manipulations of the bacterial systems, as well as for the analysis of S-[2-(N^7-guanylethyl)]GSH, the major DNA adduct derived from *vic*-dihaloethanes.

The conjugates derived from monohaloalkanes (Fig. 1, reaction 1) are not genotoxic in the absence of any further activation (e.g., by cysteine conjugate β-lyase [Cooper *et al.*, 2002]). The point should be made that the stability of the conjugates formed in reactions 2–4 of Fig. 1 can be an issue in genotoxicity assays. S-(2-Bromoethyl)GSH is more reactive than S-(2-chloroethyl)GSH, less stable, and less genotoxic. S-(2-Fluoroethyl)GSH is more stable and more genotoxic than either the Cl or Br analog (Wheeler *et al.*, 2001a). Thus, the order contrasts with the rates of activation (and the patterns of genotoxicity). With regard to dihalomethanes, compounds of type $RSCH_2X$ are very unstable (e.g., $t_{1/2}$ of 18 ms for CH_3SCH_2Cl in aqueous buffer at neutral pH) (Stourman *et al.*, 2003). S-(1-Acetoxymethyl)GSH can be synthesized in CH_3OH; its $t_{1/2}$ in neutral aqueous buffer is 12 s (Marsch *et al.*, 2001). Although this conjugate reacts with DNA to give adducts (Marsch *et al.*, 2001, 2004; Thier *et al.*, 1993), it is not genotoxic when added to bacteria (Thier *et al.*, 1993). GST activation of dihalomethanes was ineffective in producing genotoxicity when the activation was done outside of the bacteria, presumably because of instability crossing the membrane (Thier *et al.*, 1993). Thus, expression of GSTs within bacteria is the preferred approach. For comparative studies, however, provision may need to be made for quantifying GST expression (Wheeler *et al.*, 2001a).

Reaction 2 of Fig. 1 is relevant *in vivo*; ethylene dibromide is carcinogenic and produces high levels of S-[2-(N^7-guanyl)]GSH adducts in rodent models (Inskeep *et al.*, 1986). The level of adducts can be attenuated by GSH depletion and enhanced by inhibiting the oxidation of ethylene by inhibiting the oxidation of ethylene dibromide, thus diverting more substrate to the GSTs (Kim and Guengerich, 1990). DNA adducts can be generated with CH_2Cl_2 and GSTs *in vitro* (Marsch *et al.*, 2004), although the adduct instability has thus far precluded demonstration of adducts *in vivo*. With trihalomethanes (reaction 4 of Fig. 1), very limited information about DNA damage and specific adducts is available beyond demonstration of mutagenicity in bacterial strains expressing GSTs (Ross and Pegram, 2004).

The chemistry involved in the activation of haloalkanes is not unique to the GSH conjugation. The DNA repair protein O^6-methylguanine DNA alkyltransferase (AGT) has been demonstrated to activate ethylene dibromide and CH_2Br_2 to cause genotoxicity (Liu *et al.*, 2002, 2004) The mechanisms are similar to those shown in Fig. 1, reactions 2 and 3, except that the highly reactive Cys145 of (human) AGT is the thiolate involved instead of activated GSH (Liu *et al.*, 2004). Other aspects of the phenomenon include crosslinking of AGT to DNA and depurination, although depurination does not account for all of the mutations (Liu *et al.*, 2004).

Activation of Alkyl Halides to Products that Modify Proteins

The topic has not been studied in this laboratory, but some examples are known. Reed's laboratory has studied the modification of thioredoxin by the conjugate *S*-(2-chloroethyl)GSH. The sites of modification are cysteine residues (Erve *et al.*, 1995).

Modification of proteins by di- and trihalomethanes has not been reported.

The conjugation of α-dihaloacids by zeta class GSTs results in inactivation of the GST. Some of the details of the process have been explained using MS (Anderson *et al.*, 2002).

Conclusions

GST-catalyzed conjugation of GSH with alkyl halides is a relatively facile process. When the substrate contains another leaving group, the initial conjugation may have the potential to react with nucleophilic sites on DNA and proteins. These reactions are most commonly observed with, but not restricted to, the theta and zeta family GSTs.

The presence of the entities shown in Scheme 1 is an indicator for caution in the use of these chemicals. The risk to human health can be predicted using knowledge of these pathways and comparisons of the enzymatic reactions in experimental animals and humans (Reitz *et al.*, 1989).

References

Andersen, M. E., Clewell, H. J., III, Gargas, M. L., Smith, F. A., and Reitz, R. H. (1987). Physiologically based pharmacokinetics and the risk assessment process for methylene chloride. *Toxicol. Appl. Pharmacol.* **87**, 185–205.
Anderson, W. B., Liebler, D. C., Board, P. G., and Anders, M. W. (2002). Mass spectral characterization of dichloroacetic acid-modified human glutathione transferase zeta. *Chem. Res. Toxicol.* **15**, 1387–1397.
Avigad, G. (1983). A simple spectrophotometric determination of formaldehyde and other aldehydes: Application to periodate-oxidized glycol systems. *Anal. Biochem.* **134**, 499–504.

Bell-Parikh, L. C., and Guengerich, F. P. (1999). Kinetics of cytochrome P450 2E1-catalyzed oxidation of ethanol to acetic acid via acetaldehyde. *J. Biol. Chem.* **274**, 23833–23840.

Bell, L. C., and Guengerich, F. P. (1997). Oxidation kinetics of ethanol by human cytochrome P450 2E1. Rate-limiting product release accounts for effects of isotopic hydrogen substitution and cytochrome b_5 on steady-state kinetics. *J. Biol. Chem.* **272**, 29643–29651.

Blocki, F. A., Logan, M. S. P., Baoli, C., and Wackett, L. P. (1994). Reaction of rat liver glutathione *S*-transferases and bacterial dichloromethane dehalogenase with dihalomethanes. *J. Biol. Chem.* **269**, 8826–8830.

Cmarik, J. L., Humphreys, W. G., Bruner, K. L., Lloyd, R. S., Tibbetts, C., and Guengerich, F. P. (1992). Mutation spectrum and sequence alkylation selectivity resulting from modification of bacteriophage M13mp18 with *S*-(2-chloroethyl)glutathione. Evidence for a role of *S*-[2-(N^7-guanyl)ethyl]glutathione as a mutagenic lesion formed from ethylene dibromide. *J. Biol. Chem.* **267**, 6672–6679.

Cmarik, J. L., Inskeep, P. B., Meyer, D. J., Meredith, M. J., Ketterer, B., and Guengerich, F. P. (1990). Selectivity of rat and human glutathione S-transferases in activation of ethylene dibromide by glutathione conjugation and DNA binding and induction of unscheduled DNA synthesis in human hepatocytes. *Cancer Res.* **50**, 2747–2752.

Cooper, A. J., Bruschi, S. A., and Anders, M. W. (2002). Toxic, halogenated cysteine *S*-conjugates and targeting of mitochondrial enzymes of energy metabolism. *Biochem. Pharmacol.* **64**, 553–564.

Erve, J. C. L., Barofsky, E., Barofsky, D. F., Deinzer, M. L., and Reed, D. J. (1995). Alkylation of *Escherichia coli* thioredoxin by *S*-(2-chloroethyl)glutathione and identification of the adduct on the active site cysteine-32 by mass spectrometry. *Chem. Res. Toxicol.* **8**, 934–941.

Guengerich, F. P. (1994). Metabolism and genotoxicity of dihaloalkanes. *In* "Conjugation-Dependent Carcinogenicity and Toxicity of Foreign Compounds, Advances in Pharmacology, Vol. 27" (M. W. Anders and W. Dekant, eds.), pp. 211–236. Academic Press, Orlando, FL.

Guengerich, F. P. (2001). Analysis and characterization of enzymes and nucleic acids. *In* "Principles and Methods of Toxicology" (A. W. Hayes, ed.), pp. 1625–1687. Taylor & Francis, Philadelphia.

Guengerich, F. P. (2003). Activation of dihaloalkanes by thiol-dependent mechanisms. *J. Biochem. Mol. Biol.* **36**, 20–27.

Guengerich, F. P., McCormick, W. A., and Wheeler, J. B. (2003). Analysis of the kinetic mechanism of haloalkane conjugation by mammalian theta class glutathione transferases. *Chem. Res. Toxicol.* **16**, 1493–1499.

Guengerich, F. P., Yun, C.-H., and Macdonald, T. L. (1996). Evidence for a one-electron oxidation mechanism in N-dealkylation of *N,N*-dialkylanilines by cytochrome P450 2B1. Kinetic hydrogen isotope effects, linear free energy relationships, comparisons with horseradish peroxidase, and studies with oxygen surrogates. *J. Biol. Chem.* **271**, 27321–27329.

Habig, W. H., Pabst, M. J., and Jakoby, W. B. (1974). Glutathione S-transferases: The first enzymatic step in mercapturic acid formation. *J. Biol. Chem.* **249**, 7130–7139.

Humphreys, W. G., Kim, D.-H., Cmarik, J. L., Shimada, T., and Guengerich, F. P. (1990). Comparison of the DNA alkylating properties and mutagenic responses caused by a series of *S*-(2-haloethyl)-substituted cysteine and glutathione derivatives. *Biochemistry* **29**, 10342–10350.

Inskeep, P. B., and Guengerich, F. P. (1984). Glutathione-mediated binding of dibromoalkanes to DNA: Specificity of rat glutathione *S*-transferases and dibromoalkane structure. *Carcinogenesis* **5**, 805–808.

Inskeep, P. B., Koga, N., Cmarik, J. L., and Guengerich, F. P. (1986). Covalent binding of 1,2-dihaloalkanes to DNA and stability of the major DNA adduct, S-[2-(N^7-guanyl)ethyl] glutathione. *Cancer Res.* **46,** 2839–2844.

Kim, D.-H., and Guengerich, F. P. (1990). Formation of the DNA adduct S-[2-(N^7-guanyl) ethyl]glutathione from ethylene dibromide: Effects of modulation of glutathione and glutathione S-transferase levels and the lack of a role for sulfation. *Carcinogenesis* **11,** 419–424.

Knipe, A. C. (1981). Reactivity of sulphonium salts. *In* "The Chemistry of the Sulphonium Group" (C. J. M. Stirling and S. Patai, eds.), pp. 313–385. Wiley, New York.

Koga, N., Inskeep, P. B., Harris, T. M., and Guengerich, F. P. (1986). S-[2-(N^7-Guanyl)ethyl] glutathione, the major DNA adduct formed from 1,2-dibromoethane. *Biochemistry* **25,** 2192–2198.

Liu, L., Hachey, D. L., Valadez, J. G., Williams, K. M., Guengerich, F. P., and Pegg, A. E. (2004). Characterization of a mutagenic DNA adduct formed from 1,2-dibromoethane by O^6-alkylguanine-DNA alkyltransferase. *J. Biol. Chem.* **279,** 4250–4259.

Liu, L., Pegg, A. E., Williams, K. M., and Guengerich, F. P. (2002). Paradoxical enhancement of the toxicity of 1,2-dibromoethane by O^6-alkylguanine-DNA alkyltransferase. *J. Biol. Chem.* **277,** 37920–37928.

Marsch, G. A., Botta, S., Martin, M. V., McCormick, W. A., and Guengerich, F. P. (2004). Formation and mass spectrometric analysis of DNA and nucleoside adducts by S-(1-acetoxymethyl)glutathione and by glutathione S-transferase-mediated activation of dihalomethanes. *Chem. Res. Toxicol.* **279,** 4250–4259.

Marsch, G. A., Mundkowski, R. G., Morris, B. J., Manier, M. L., Hartman, M. K., and Guengerich, F. P. (2001). Characterization of nucleoside and DNA adducts formed by S-(1-acetoxymethyl)glutathione and implications for dihalomethane-glutathione conjugates. *Chem. Res. Toxicol.* **14,** 600–608.

Meyer, D. J., Coles, B., Pemble, S. E., Gilmore, K. S., Fraser, G. M., and Ketterer, B. (1991). Theta, a new class of glutathione transferases purified from rat and man. *Biochem. J.* **274,** 409–414.

Nash, T. (1953). The colorimetric estimation of formaldehyde by means of the Hantzsch reaction. *Biochem. J.* **55,** 416–421.

Ozawa, N., and Guengerich, F. P. (1983). Evidence for formation of an S-[2-(N^7-guanyl)ethyl] glutathione adduct in glutathione-mediated binding of 1,2-dibromoethane to DNA. *Proc. Natl. Acad. Sci. USA* **80,** 5266–5270.

Peterson, L. A., Harris, T. M., and Guengerich, F. P. (1988). Evidence for an episulfonium ion intermediate in the formation of S-[2-(N^7-guanyl)ethyl]glutathione in DNA. *J. Am. Chem. Soc.* **110,** 3284–3291.

Ploemen, J. H. T. M., Wormhoudt, L. W., van Ommen, B., Commandeur, J. N. M., Vermeulen, N. P. E., and van Bladeren, P. J. (1995). Polymorphism in the glutathione conjugation activity of human erythrocytes towards ethylene dibromide and 1,2-epoxy-3-(*p*-nitrophenoxy)-propane. *Biochim. Biophys. Acta* **1243,** 469–476.

Rannug, U., Sundvall, A., and Ramel, C. (1978). The mutagenic effect of 1,2-dichloroethane on *Salmonella typhimurium*. I. Activation through conjugation with glutathione *in vitro*. *Chem. Biol. Interact.* **20,** 1–16.

Reed, D. J., Babson, J. R., Beatty, P. W., Brodie, A. E., Ellis, W. W., and Potter, D. W. (1999). High-performance liquid chromatography analysis of nanomole levels of glutathione, glutathione disulfide, and related thiols and disulfides. *Anal. Biochem.* **106,** 55–62.

Reitz, R. H., Mendrala, A., and Guengerich, F. P. (1989). *In vitro* metabolism of methylene chloride in human and animal tissues: Use in physiologically-based pharmacokinetic models. *Toxicol. Appl. Pharmacol.* **97**, 230–246.

Ross, M. K., and Pegram, R. A. (2003). Glutathione transferase theta 1-1-dependent metabolism of the water disinfection byproduct bromodichloromethane. *Chem. Res. Toxicol.* **16**, 216–226.

Ross, M. K., and Pegram, R. A. (2004). *In vitro* biotransformation and genotoxicity of the drinking water disinfection byproduct bromodichloromethane: DNA binding mediated by glutathione transferase theta 1-1. *Toxicol. Appl. Pharmacol.* **195**, 166–181.

Shimada, T., Yamazaki, H., Oda, Y., Hiratsuka, A., Watabe, T., and Guengerich, F. P. (1996). Activation and inactivation of carcinogenic dihaloalkanes and other compounds by glutathione S-transferase 5-5 in *Salmonella typhimurium* tester strain NM5004. *Chem. Res. Toxicol.* **9**, 333–340.

Shriner, R. L., Fuson, R. C., and Curtin, D. Y. (1965). "The Systematic Identification of Organic Compounds." Wiley, New York.

Small, W. C., and Jones, M. E. (1990). A quantitative colorimetric assay for semialdehydes. *Anal. Biochem.* **185**, 156–159.

Stourman, N. V., Rose, J. A., Vuilleumier, S., and Armstrong, R. N. (2003). Catalytic mechanism of dichloromethane dehalogenase from *Methylophilus* sp. strain DM11. *Biochemistry* **42**, 11048–11056.

Thier, R., Müller, M., Taylor, J. B., Pemble, S. E., Ketterer, B., and Guengerich, F. P. (1995). Enhancement of bacterial mutagenicity of bifunctional alkylating agents by expression of mammalian glutathione S-transferase. *Chem. Res. Toxicol.* **8**, 465–472.

Thier, R., Pemble, S., Kramer, H., Taylor, J. B., Guengerich, F. P., and Ketterer, B. (1996). Human glutathione S-transferase T1-1 enhances mutagenicity of 1,2-dibromoethane, dibromomethane, and 1,2,3,4-diepoxybutane in *Salmonella typhimurium*. *Carcinogenesis* **17**, 163–166.

Thier, R., Pemble, S. E., Taylor, J. B., Humphreys, W. G., Persmark, M., Ketterer, B., and Guengerich, F. P. (1993). Expression of mammalian glutathione S-transferase 5-5 in *Salmonella typhimurium* TA1535 leads to base-pair mutations upon exposure to dihalomethanes. *Proc. Natl. Acad. Sci. USA* **90**, 8576–8580.

Tong, Z., Board, P. G., and Anders, M. W. (1998). Glutathione transferase zeta-catalyzed biotransformation of dichloroacetic acid and other alpha-haloacids. *Chem. Res. Toxicol.* **11**, 1332–1338.

van Bladeren, P. J., Breimer, D. D., Rotteveel-Smijs, G. M. T., de Jong, R. A. W., Buijs, W., van der Gen, A., and Mohn, G. R. (1980). The role of glutathione conjugation in the mutagenicity of 1,2-dibromoethane. *Biochem. Pharmacol.* **29**, 2975–2982.

Walsh, C. (1979). "Enzymatic Reaction Mechanisms." W. H. Freeman Co., San Francisco.

Webb, W. W., Elfarra, A. A., Webster, K. D., Thom, R. E., and Anders, M. W. (1987). Role for an episulfonium ion in S-(2-chloroethyl)-DL-cysteine-induced cytotoxicity and its reaction with glutathione. *Biochemistry* **26**, 3017–3023.

Wheeler, J. B., Stourman, N. V., Armstrong, R. N., and Guengerich, F. P. (2001a). Conjugation of haloalkanes by bacterial and mammalian glutathione transferases: Mono- and vicinal dehaloethanes. *Chem. Res. Toxicol.* **14**, 1107–1117.

Wheeler, J. B., Stourman, N. V., Thier, R., Dommermuth, A., Vuilleumier, S., Rose, J. A., Armstrong, R. N., and Guengerich, F. P. (2001b). Conjugation of haloalkanes by bacterial and mammalian glutathione transferases: mono- and dihalomethanes. *Chem. Res. Toxicol.* **14**, 1118–1127.

[22] Peptide Phage Display for Probing GST–Protein Interactions

By MARYAM H. EDALAT and BENGT MANNERVIK

Abstract

Phage display is a powerful strategy for identifying protein–peptide interactions. Glutathione transferases (GSTs) play prominent roles in the cellular protection against oxidative stress by catalyzing detoxication reactions. In addition, GSTs seem to act in signaling pathways by means of interaction with other macromolecules such as protein kinases. This chapter describes how the technique of peptide phage display can be used to identify possible partners in GST–protein complexes.

Introduction

Genome sequencing projects have provided the scientific community with an immense amount of data. For example, most of the primary protein structures of many organisms are now known. However, the full spectrum of activities that these proteins share responsibility for still remains to be unraveled. It is not uncommon that a given protein participates in more that one separate task. Numerous biological functions of proteins, if not all, are dependent on protein–protein interactions. In the study of cellular signaling, understanding the interactions among proteins has received particular attention because of its obvious importance. Many signaling proteins interact with each other by molecular recognition of a short linear amino acid sequence in the cognate protein structure (Yaffe and Cantley, 2000). In this chapter, we outline the use of peptide phage display to investigate potential protein–protein interactions as exemplified by peptide sequences with affinity for glutathione transferases (GSTs).

The GSTs were first discovered as detoxication enzymes (Booth et al., 1961; Combes and Stakelum, 1961). Later these proteins were also proposed to act as carrier proteins and were named ligandin (Litwack et al., 1971). Today, despite 40 years of further research, the scope of functions for what is now recognized as a GST superfamily is more complex than ever.

GSTs exist both as cytosolic and membrane-bound proteins. The human cytosolic, or soluble, GSTs are not only present in the cytoplasm but may also be localized in the mitochondria, the nucleus, or other subcellular

METHODS IN ENZYMOLOGY, VOL. 401
Copyright 2005, Elsevier Inc. All rights reserved.
0076-6879/05 $35.00
DOI: 10.1016/S0076-6879(05)01022-0

compartments (Johansson and Mannervik, 2001a). GSTs are primarily known as the cellular catalysts conjugating the nucleophilic sulfhydryl group of glutathione with various electrophilic toxic compounds. However, the catalytic function of GSTs is not restricted to detoxication. Some GSTs isoenzymes seem to have a metabolic role in the biosynthesis of steroids and prostaglandins (Beuckmann et al., 2000; Johansson and Mannervik, 2001b) or the degradation of aromatic amino acids (Blackburn et al., 1998). Lately, research on GSTs has unraveled yet another function, namely a regulatory role in cellular signaling (Adler et al., 1999; Cho et al., 2001). This function has added a new dimension to the importance of this large family of proteins.

To probe possible interactions between a given GST and other proteins, it is possible to investigate their physical association by means of affinity methods for trapping and isolation of GST–protein complexes from diverse cellular fractions. However, this approach is hampered by temporal and spatial differences in the expression of the proteome such that important combinations can be missed. In crude biological samples, the interacting protein may also be present in low concentration or be outcompeted by other components in the sample and, therefore, remain undiscovered.

An alternate method is to use a simpler biochemical system, in which the interacting target species are more accurately defined, such as a library of random peptide sequences from which members with affinity for a GST can be identified. In one format, peptides are displayed on the surface of phage particles (Smith, 1985), which are probed against immobilized GST protein. This embodiment of affinity selection is described in this chapter. As an example, studies of human GSTs are used (Edalat et al., 2002, 2003). The peptide sequences identified may serve as importance leads to new discoveries in functional proteomics. Obviously, any affinity discovered between a peptide and a protein has to be verified as a biologically relevant interaction by complementary functional studies.

Materials and Methods

The procedures described have been used in previous publications, where further details may be found (Edalat et al., 2002, 2003). *Escherichia coli* K91 was available from the laboratory of Dr. Mats, A. A. Persson at the Karolinska Institutet, Stockholm, Sweden. Enzymes for recombinant DNA work were from New England Biolabs Inc. (Beverly, MA). Maxisorp immunoplates were from Nunc (Roskilde, Denmark). Transfection kit FuGENE 6^{TM} was from Roche Diagnostics (Mannheim, Germany). Luciferase assay system was from Promega (Madison, WI). Cell culture medium was purchased from Invitrogen (Paisley, UK). All other chemicals were from Sigma-Aldrich (St. Louis, MO).

Preparation of Culture Media

Luria-Bertani broth (LB): Bacto-yeast extract, 5 g; Bacto-tryptone, 10 g; NaCl, 5 g

Dissolve in 1 liter of distilled or deionized water, adjust to pH 7.0 with 1 M NaOH, and then autoclave. This medium is called 2YT.

LB/amp: 2YT fortified with ampicillin (100 μg/ml)
LB/tet: 2YT fortified with tetracycline (10 μg/ml)
LB agar plates/tet: 15 g Bacto-agar is dissolved in 1 liter LB, adjusted to pH 7.0 with 1 M NaOH, and then autoclaved. The medium is cooled to 50°, tetracycline (10 μg/ml) is added, and the solution is poured into Petri plates.

Solutions

Phosphate-buffered saline (PBS): 137 mM NaCl, 3 mM KCl, 8 mM Na_2HPO_4, 1.5 mM KH_2PO_4. Adjust to pH 7.2 with HCl and autoclave.
Blocking solution: PBS containing 5% (w/v) nonfat dried milk.
Washing buffer: PBS containing 0.05% Tween 20.
Elution buffer: 0.1 M glycine-HCl buffer. Adjust pH to 2.2 and autoclave.
Neutralization buffer: 2 M Tris base solution. Autoclave.

Random Peptide Library Displayed on Phage

The library of random cyclic peptide used in the studies presented here was transfected into *E. coli* K91. The phage particles displayed sequences of 9 or 15 amino acid residues, flanked by Ala and Cys at the N- and C-terminal ends (Fig. 1). This mixture originated from two separate peptide libraries (Lundin *et al.*, 1996). The peptides expressed were fused to the N-terminus of the gene III (gIII) product of filamentous phage: N-terminal-AEDLELCA $(X)_{9/15}$ ACTS-gIII. (The amino acids are designated by the one-letter code; X denotes any amino acid). Approximately 30% of the 9-mer peptides consisted of linear sequences having a PALG-TETS linker replacing ACTS at the C-terminus. Most of the peptides formed cyclic structures owing to the presence of two Cys residues. The phage vector fAST, carrying the tetracycline-resistant gene in addition to

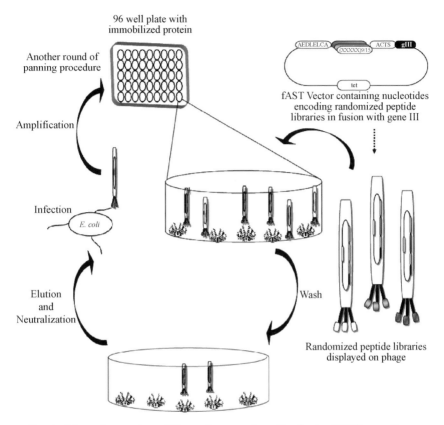

FIG. 1. Schematic overview of biopanning procedure. Randomized DNA encoding pep-tides were fused to gene III and cloned into the fAST vector. The peptide (9/15-mer) library was used to isolate ligands to specific GSTs. After three to five rounds of binding, washing, and elution, characteristic peptide sequences were obtained.

the random peptide sequences, was a modified version of fUSE2 (Lundin *et al.*, 1996).

Peptide libraries displayed on phage are also available commercially from (e.g., New England Biolabs Inc. [Beverly, MA]).

Purified human GST A1-1, GST P1-1, and GST M2-2 were obtained in our laboratory by heterologous expression (Kolm *et al.*, 1995; Johansson *et al.*, 1999; Stenberg *et al.*, 1992). GST A1-1 and GST P1-1 are also available from Invitrogen Corp. (San Diego, CA).

Biopanning Procedures

Figure 1 shows an outline of the selection procedure used in the phage display selection. The target GST protein (100 μg) in 50 μl PBS is immobilized on the plastic surface of individual wells of a 96-well Nunc-Immuno Plate, Maxisorp. The plate is incubated at 4° overnight. The wells are then treated with blocking solution for 1 h at room temperature (RT) before the addition of phage libraries. Before being transferred to GST-coated microtiter wells, 50 × 10⁶ colony-forming units (cfu) from the random-peptide library are added to a preblocked well, treated with block solution only, and incubated for 1 h at RT. This extra step of preincubation of phage libraries in preblocked wells, before addition to the GST-coated well, is introduced to minimize the unspecific binding of phage to the plastic of the wells. The phage is incubated in the protein-coated wells at RT for 2 h. After removal of unbound phage, the wells are rinsed with 300 μl washing solution: twice in the first round, five in the second round, and 10 times in the later rounds of selection. Specifically bound phage is then acid-eluted with 100 μl 0.1 M glycine-HCl buffer, pH 2.2, and the eluate neutralized with 6 μl 2 M Tris base solution.

The eluted phage is used to infect *E. coli* K91 cells growing in LB/tet (OD$_{600}$ = 0.6). Aliquots of 100 and 10 μl are plated on LB agar/tet plates and the remainder of the infected cells propagated at 37° overnight in 50 ml LB/tet liquid medium. The amplified phage is precipitated (see later) and used for a new round of selection.

Precipitation of Phage

Cultures of phage-infected bacteria grown overnight are centrifuged at 3,950g for 15 min to remove the cells. To precipitate the phage, 4% (w/v) polyethylene glycol (PEG) 8000 and 3% (w/v) NaCl are added to the phage-containing supernatant. The mixture is incubated for 30 min on ice before centrifugation at 9,900g for 30 min. The pellet is resuspended in 1 ml PBS and the suspension centrifuged at 16,000g for 5 min to remove cell debris.

Identification of Cloned Peptides by DNA Sequencing

Seven to 10 phage clones selected for binding to different target GSTs were randomly chosen for DNA analysis after the third round of panning. The nucleotide sequences encoding the peptide sequences derived from the random-peptide library were determined.

Three rounds of selection were conducted with GST M2-2 and GST P1-1, and five rounds of biopanning were performed with GST A1-1. For example, the fraction of phage binding to GST M2-2 increased 100-fold after the third round, indicating specific enrichment (Table I). The biopanning procedure was discontinued when the DNA sequencing results showed clear selection of clones with a particular sequence (Table II).

The three GSTs have distinct peptide-binding selectivities (Table II). GST P1-1 and GST M2-2 have affinity for constrained cyclic peptides. Interestingly, GST A1-1 seems to have a preference for the linker sequence PALGTETS. In the selection of peptides with affinity for GST P1-1, six clones were identical of 10 isolated and analyzed (Table II), indicating a clear enrichment of one specific peptide (Pp). All 10 clones contained two Cys residues and were predicted to have a cyclic peptide structure because of the formation of an intracellular disulfide bond.

In the selection using GST A1-1 as a target, two clones of eight were identical (Pa). Surprisingly, five of eight clones from this selection contained the PALGTETS linker (Palink) at the C-terminus, which thus seems to be a GST A1-1 binding sequence.

TABLE I

THE RATIO OF INPUT/OUTPUT PHAGE RECOVERED FROM THE SELECTION ROUNDS

Target protein	First round	Second round	Third round	Fourth round	Fifth round
GST A1-1	3.3×10^{-5}	1.1×10^{-5}	2.7×10^{-7}	1.5×10^{-5}	4.0×10^{-6}
GST M2-2	2.7×10^{-8}	1.6×10^{-8}	1.2×10^{-6}		
GST P1-1	2.7×10^{-7}	2.0×10^{-7}	4.0×10^{-7}		

5×10^7 cfu (input) of each peptide library was mixed together and added to GST-coated microtiter wells, unbound phage was removed, and the wells were washed. Specifically, bound phage was eluted and the cfu were counted (output).

TABLE II

GST-BINDING PEPTIDE SEQUENCES IDENTIFIED BY PHAGE DISPLAY

Phage clone	Variant peptide sequence	Number of identical clones total number of clones sequenced
(A) Pa	AEDLELCAEYRARGWGGWCRPALGTETS	2/8
Palink	PALGTETS	5/8
(B) Pp	AEDLELCAGGGVWAVSGACTS	6/10
(C) Pm	AEDLELCAGVVRGPSRGACTS	5/7

Amino acid sequences of the peptides displayed on phage that were selected with human GST A1–1 (A), GST P1–1 (B), and GST M2–2 (C). The phage clones shown were randomly picked up for sequencing after fourth round (A) or third round (B and C) of selection. The underlined letters indicate segments derived from the random sequences.

TABLE III

DATABASE SEQUENCES WITH SIMILARITY TO THE AMINO ACID SEQUENCES IDENTIFIED
BY PEPTIDE PHAGE DISPLAY

Query	Hits	Recognized amino acid residues	Human protein
PALGTETS	PGLGTET	279–285	C-terminal PDZ domain ligand of neuronal NO-synthase
PALGTETS	PALGSET	185–191	TRAF4 associated factor 1
PALGTETS	PALGTE	228–234	MRGX3
PALGTETS	PALSTET	430–436	Protocadherin gamma A1
PALGTETS	ALGTESS	150–156	Vacuolar protein sorting 18
PALGTETS	PALGPET	950–956	Regulatory factor X1
PALGTETS	PALGPESS	268–275	Dystrobrevin binding protein 1
GVVRGPSRG	GVARGPGSRG	2–11	Phosphatidic-acid phosphatase (type 2, isoform 3)
GVVRGPSRG	VRGPSR	365–370	Vacuolar protein sorting 4B
GVVR-PSRGA	GVVKGQPSPSGA	408–419	JNK3
AGVVRGPSRGA	AGVLRGPGR	427–435	Bcl3
AGVVRGPSRGA	AGGVRGAARGA	23–33	Thioredoxin reductase TR3
AGVVRGPSRGA	VVREPPRGA	237–244	Spinster-like protein
AGVVRGPSRGA	GVVRGHARG	43–51	Mitochondrial solute carrier protein
AGVVRGPSRGA	VHRGPSRG	498–505	GRB2-associated binding protein 2 isoform a
AGVVRGPSRGA	RGVSRG	608–613	Cytoplasmic activation/prolifer ation-associated protein 1
AGVVRGPSRGA	GVSRGGSRGA	556–565	GPI-anchored protein p137
AGVVRGPSRGA	RGPSRG	141–146	Zinc finger protein 342
YRARGWGGW	YCARGWGGW	64–72	Immunoglobulin heavy chain variable region
YRARGWG	YQARGWG	814–820	LYST-interacting protein LIP8
GGGVWAVSGA	WAVSGA	53–58	Tumor necrosis factor (ligand) superfamily, member 9
GGGVWAVSGA	GGGVWA	23–28	Calsyntenin 1
GGGVWAVSGA	VWEVSGA	49–55	T-cell receptor beta chain
GGGVWAVSGA	GIWAVS	167–172	Orexin receptor-1; OX1R; G protein-coupled receptor

Database Analysis

To find protein sequences with similarities to selected peptides, the following site can be used: http://www.ncbi.nlm.nih.gov/BLAST/search.shtml

Table III shows a number of proteins with sequence similarities to one or the other of the selected peptide sequences binding to GST A1-1, GST

M2-2, or GST P1-1. To make the search less complicated, we only searched for human proteins.

Competitive Binding Studies with Phage Displayed Peptide

The binding specificity of Pp to the GST P1-1 was examined by comparing the binding of the phage from the fifth round of selection (Pp5) with that of an unrelated control peptide (Pneg). Pneg is a linear 15-mer peptide with affinity to mouse microglia phage clones. GST P1-1 and GST M2-2 were immobilized on the surface of individual wells (100 μg in each well) in a 96-well Maxisorp plate and incubated overnight at 4°. The wells were then blocked with 5% (w/v) nonfat milk in PBS for 1 h at RT before addition of 10^9 cfu of phage displaying selected peptide for binding GST P1-1 (Pp) or Pneg to each well. After six washes with PBS containing 0.05% (v/v) Tween 20, 50 μg competing GST P1–1 protein was added to two wells containing GST P1-1 + Pp and GST P1-1 + Pneg. Three additional washes were then conducted before acid elution with 100 μl of a 0.1 M glycine/HCl solution, pH 2.2, and neutralization with 6 μl of 2 M Tris base. The eluted phage was quantified by infecting an *E. coli* K91 culture. The phage-displayed Pp5 was separately tested for binding to immobilized GST P1-1. Pp demonstrated substantially higher affinity than Pneg for GST P1-1 (Fig. 2A).

Furthermore, the binding specificity of the peptide Pp5 was tested by conducting the binding experiment with an alternative GST, namely immobilized human GST M2-2. The peptide Pp displayed on phage binds about 50-fold more strongly to GST P1-1 than to GST M2–2 (Fig. 2A). Finally, competitive binding experiments showed a lower recovery of phage displaying Pp when GST P1-1 was added in solution to compete with the immobilized GST P1-1 (Fig. 2A). Similarly, Pm was found to bind approximately 30-fold more tightly to GST M2-2 than does Pneg, and GST M2-2 in solution could compete with the immobilized GST M2-2 in binding Pm (Fig. 2B).

Construction of GST M2-2 and GST A1-1 into pSTC HA Tagged
  Expression Vector

The coding sequences of GST M2-2 (Johansson *et al.*, 1999) and GST A1-1 (Stenberg *et al.*, 1992) were amplified by polymerase chain reaction (PCR), digested with *Bam*HI and *Xba*I, purified on 1% agarose gel followed by 1 h ligation at RT into pSTC-HA3.X556 tagged (a derivative of a vector kindly provided by Dr. Rusconi, Universität Zurich, Switzerland) expression vector that was digested with the same restriction enzymes. The expression of GST M2-2 and GST A1-1 from the vector construct in

A

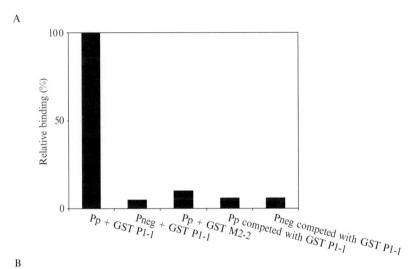

B

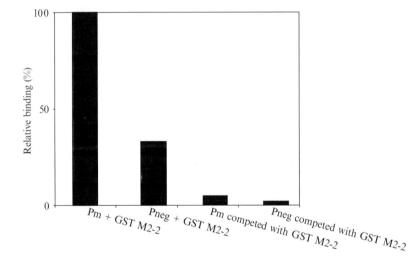

FIG. 2. Examination of the specificity of the binding affinity of the selected peptides to targeted GSTs. Binding studies with (A) 100 μg immobilized GST P1-1 and Pp or Pneg displayed on phage. The binding of Pp was further more outcompeted with 50 μg GST P1-1 in solution. Relative binding values are the ratio of number of "out phage" and "in phage" plotted with the positive binding value as 100%. (B) Similar binding studies were conducted with GST M2-2 and pm (From Edalat et al., 2002, 2003, with permission).

mammalian cells was confirmed by Western blot analysis. Obviously, many alternate expression systems could be used for introduction of the GSTs into targeted cells.

Transfection of Human Cell Lines with GST M2-2 Up-Regulates
Transcriptional Activity of Activating Protein-1 (AP-1)

The human embryonic kidney (HEK) 293 cells were maintained at 37°
in minimal essential medium with L-glutamine (MEM) supplemented with
10% (v/v) fetal calf serum. The human embryonic retinoblast 911 cells
were maintained at 37° in Dulbecco's modified Eagle's medium (DMEM)
supplemented with 10% (v/v) fetal calf serum. The cells were grown in
35-mm dishes to 30–40% confluence and then transfected using FuGENE
6™ (Roche). The cells were stimulated with 100 ng/ml phorbol 12-myr-
istate 13-acetate (PMA) 20–24 h before harvesting. Transfection efficiencies were normalized using β-galactosidase activity. The transcriptional
activity was assayed using the luciferase assay system. The Bcl3 expression vector was kindly provided by Dr. Claus Scheidereit, Max-Delbrück
Center, Berlin, Germany, and the AP-1 reporter construct was a gift from
Dr. Hans Wolf-Watz, Umeå University, Sweden.

Overexpression of GST M2-2 up-regulates AP-1 transcription almost
fourfold (Fig. 3A). The coactivating effect was increased in a dose-
dependent manner with increasing amounts of GST M2-2 DNA used for
transcription. Specificity of the effect of human GST M2-2 on AP-1 transcription was demonstrated by overexpressing a related enzyme, GST
A1-1. With this alternate GST, no significant stimulation of transcriptional
activity was noted (Fig. 3B). All the *in vivo* experiments were repeated
three times independently in HEK 293 cells, as well as in human embryonic
retinoblast 911 cells.

Discussion

Phage display is a powerful strategy for identifying protein–peptide
interactions. The principle behind phage display is to connect phenotype
and genotype. The fusion of DNA encoding a peptide or a protein to the
gene III, which encodes a minor capsid protein of filamentous bacteriophage M13, makes it possible to express foreign polypeptide segments on
the surface of the phage particle (Smith, 1985). Every displayed peptide has
an addressable tag by means of the DNA encoding this same peptide
displayed by the phage. As a consequence, DNA sequence analysis reveals
the primary structure of the peptide at the surface of the phage particle that
by way of its affinity for a target protein has been selected from a peptide
library.

In this chapter, the principles introduced by Smith (1985) were applied
to the identification of peptide sequences with affinity for human GST
A1-1, GST M2-2, and GST P1-1 (Fig. 1). The enrichment of peptide

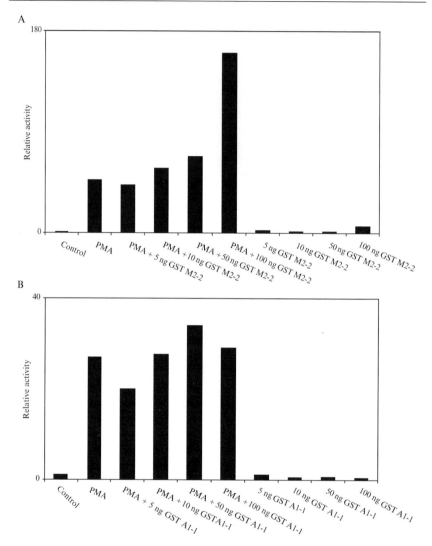

FIG. 3. Up-regulation of AP-1 transcriptional activity by GST M2-2. HEK 293 cells were cotransfected with AP1-luciferase reporter plasmid, pcDNA3, and (A) psctGST M2-2 (B) psctGST A1-1 plasmids. After 24 h, the cells were treated with 100 ng/ml PMA. After another 24 h, the cells were harvested, lysed, and tested for AP-1 transcriptional activity by means of a luciferase reporter assay. In all the experiments, the cells were also transfected with the same amount of pcDNAβ-gal to normalize the transfection efficiency by assaying for β-galactosidase activity expressed from the vector (From Edalat et al., 2002, 2003, with permission).

sequences that differ among the GST targets indicates specific binding (Table II), further confirmed by competitive binding studies (Fig. 2). These results together with a database search suggest a large number of proteins that potentially could interact with GSTs (Table III). As evidence for the regulatory role of GST M2-2 by binding to JNK3,

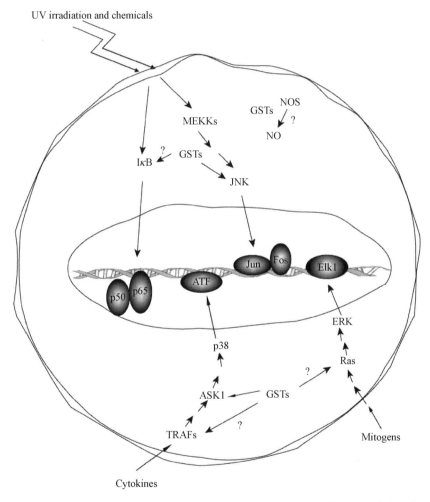

FIG. 4. Possible contributions of GSTs in the regulation of different cell-signaling pathways responsive to oxidative stress. Some of the components in the mitogen-activated protein kinase signal transduction pathway (Dunn *et al.*, 2002) are indicated as well as the nitric oxide synthase (NOS).

we have shown that overexpression of GST M2-2 coactivates AP-1 transcription in two different human cell lines (Fig. 3). The findings presented in this chapter provide important clues to signal transduction pathways in which GSTs may be involved. Figure 4 shows some of the pathways for which indications of GST interactions are provided by peptide phage display.

Acknowledgments

Financial support from the Swedish Cancer Society and the Swedish Research Council is gratefully acknowledged.

References

Adler, V., Yin, Z., Fuchs, S. Y., Benezra, M., Rosario, L., Tew, K. D., Pincus, M. R., Sardana, M., Henderson, C. J., Wolf, C. R., Davis, R. J., and Ronai, Z. (1999). Regulation of JNK signaling by GSTp. *EMBO J.* **18**, 1321–1334.

Beuckmann, C. T., Fujimori, K., Urade, Y., and Hayaishi, O. (2000). Identification of mu-class glutathione transferases M2-2 and M3-3 as prostaglandin E synthases in the human brain. *Neurochem. Res.* **25**, 733–738.

Blackburn, A. C., Woollatt, E., Sutherland, G. R., and Board, P. G. (1998). Characterization and chromosome location of the gene GSTZ1 encoding the human Zeta class glutathione transferase and maleylacetoacetate isomerase. *Cytogenet. Cell Genet.* **83**, 109–114.

Booth, J., Boyland, E., and Sims, P. (1961). An enzyme from rat liver catalyzing conjugations with glutathione. *Biochem. J.* **79**, 516–524.

Cho, S. G., Lee, Y. H., Park, H. S., Ryoo, K., Kang, K. W., Park, J., Eom, S. J., Kim, M. J., Chang, T. S., Choi, S. Y., Shim, J., Kim, Y., Dong, M. S., Lee, M. J., Kim, S. G., Ichijo, H., and Choi, E. J. (2001). Glutathione S-transferase mu modulates the stress-activated signals by suppressing apoptosis signal-regulating kinase 1. *J. Biol. Chem.* **276**, 12749–12755.

Combes, B., and Stakelum, G. S. (1961). A liver enzyme that conjugates sulfobromophthalein sodium with glutathione. *J. Clin. Invest.* **40**, 981–988.

Dunn, C., Wiltshire, C., Mac Laren, A., and Gillespie, D. A. F. (2002). Molecular mechanism and biological functions of c-Jun N-terminal kinase signaling via the c Jun transcription factor. *Cell. Signalling* **14**, 585–593.

Edalat, M., Pettersson, S., Persson, M. A. A., and Mannervik, B. (2002). Probing biomolecular interactions of glutathione transferase M2-2 by using peptide phage display. *ChemBioChem.* **3**, 823–828.

Edalat, M., Persson, M. A. A., and Mannervik, B. (2003). Selective recognition of peptide sequences by glutathione transferases: A possible mechanism for modulation of cellular stress-induced signaling pathways,. *Biol. Chem.* **384**, 645–651.

Johansson, A.-S., Bolton-Grob, R., and Mannervik, B. (1999). Use of silent mutations in cDNA encoding human glutathione transferase M2-2 for optimized expression in *Escherichia coli. Protein Expression Purif.* **17**, 105–112.

Johansson, A.-S., and Mannervik, B. (2001a). Interindividual variability of glutathione transferase expression. *In* "Interindividual Variability in Human Drug Metabolism" (G. M. Pacifici and O. Pelkonen, eds.), pp. 460–519. Taylor & Francis, London.

Johansson, A.-S., and Mannervik, B. (2001b). Human glutathione transferase A3-3, a highly efficient catalyst of double-bond isomerization in the biosynthetic pathway of steroid hormones. *J. Biol. Chem.* **276**, 33061–33065.

Kolm, R. H., Stenberg, G., Widersten, M., and Mannervik, B. (1995). High-level bacterial expression of human glutathione transferase P1-1 encoded by semisynthetic DNA. *Protein Expression Purif.* **6**, 265–271.

Litwack, G., Ketterer, B., and Arias, I. M. (1971). Ligandin: A hepatic protein which binds steroids, bilirubin, carcinogens and a number of exogenous organic anions. *Nature* **234**, 466–467.

Lundin, K., Samuelsson, A., Jansson, M., Hinkula, J., Wahren, B., Wigzell, H., and Persson, M. A. A. (1996). Peptides isolated from random peptide libraries on phage elicit a neutralizing anti-HIV-1 response: Analysis of immunological mimicry. *Immunology* **89**, 579–586.

Smith, G. P. (1985). Filamentous fusion phage: Novel expression vectors that display cloned antigens on the virion surface. *Science* **228**, 1315–1317.

Stenberg, G., Björnestedt, R., and Mannervik, B. (1992). Heterologous expression of recombinant human glutathione transferase A1-1 from a hepatoma cell line. *Protein Expression Purif.* **3**, 80–84.

Yaffe, M. B., and Cantley, L. C. (2000). Mapping specificity determinants for protein–protein association using protein fusions and random peptide libraries. *Methods Enzymol.* **328**, 157–170.

[23] Fosfomycin Resistance Proteins: A Nexus of Glutathione Transferases and Epoxide Hydrolases in a Metalloenzyme Superfamily

By Rachel E. Rigsby, Kerry L. Fillgrove, Lauren A. Beihoffer, and Richard N. Armstrong

Abstract

Three similar but mechanistically distinct fosfomycin resistance proteins that catalyze the opening of the oxirane ring of the antibiotic are known. FosA is a Mn(II) and K^+-dependent glutathione transferase. FosB is a Mg^{2+}-dependent L-cysteine thiol transferase. FosX is a Mn(II)-dependent fosfomycin-specific epoxide hydrolase. The expression, purification, kinetic, and physical characteristics of six fosfomycin resistance proteins including the FosA proteins from transposon TN2921 and *Pseudomonas aeruginosa*, the FosB proteins from *Bacillus subtilis* and *Staphylococcus aureus*, and the FosX proteins from *Mesorhizobium loti* and *Listeria monocytogenes* are reported.

METHODS IN ENZYMOLOGY, VOL. 401   0076-6879/05 $35.00
Copyright 2005, Elsevier Inc. All rights reserved.   DOI: 10.1016/S0076-6879(05)01023-2

Introduction

Fosfomycin, (*1R, 2S*)-epoxypropylphosphonic acid, is a broad-spectrum antibiotic produced by certain strains of *Streptomyces* (Hendlin, 1969). The antibiotic functions by covalent modification of the enzyme (UDP-N-acetylglucosamine-3-enolpyruvyl transferase or MurA) that catalyzes the first committed step in peptidoglycan biosynthesis (Kahan, 1974). Shortly after its introduction into the clinic, plasmid-mediated resistance to the drug was observed (Llaneza *et al.*, 1985; Mendoza *et al.*, 1980; Villar *et al.*, 1986). The resistance gene was subsequently established to encode a 16-kDa polypeptide (FosA) that catalyzed the addition of glutathione (GSH) to the antibiotic rendering it inactive (Fig. 1) (Arca *et al.*, 1988, 1990). A detailed biochemical analysis revealed that FosA was a Mn(II)-dependent metalloenzyme with a requirement for K$^+$ for optimum catalytic activity (Bernat and Armstrong, 2001; Bernat *et al.*, 1997, 1999) Sequence analysis and the three-dimensional structure of FosA places the protein in the vicinal oxygen chelate (VOC) superfamily of proteins, which includes extradiol dioxygenases, glyoxalase I, and methylmalonyl-CoA epimerase (Armstrong, 2000; Pakhomova *et al.*, 2004; Rife *et al.*, 2002).

Queries of microbial genome sequences indicated the existence of several homologs of FosA. For example, *Pseudomonas aeruginosa* harbors a gene encoding a FosA homologue (FosA^{PA}) that shares a high degree (60%) of sequence identity with the plasmid-encoded protein described previously. Structural and mechanistic analysis of this protein established that it catalyzes the Mn(II) and K$^+$-dependent addition of GSH

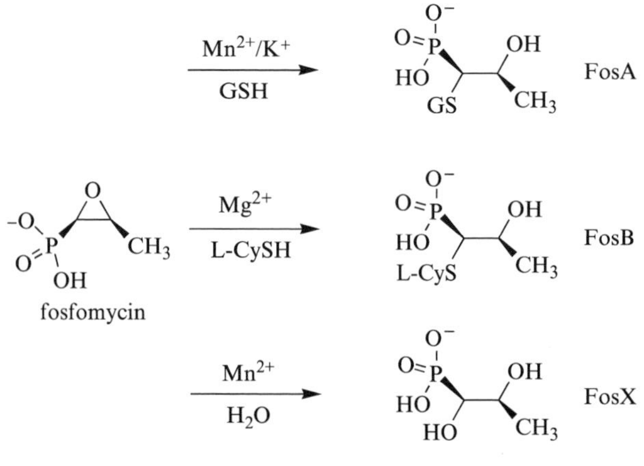

FIG. 1. Reactions catalyzed by the fosfomycin resistance proteins FosA, FosB, and FosX.

to fosfomycin and confers resistance to fosfomycin in the biological context of *Escherichia coli* (Rife *et al.*, 2002; Rigsby *et al.*, 2004).

Several other more distantly related FosA homologs that exhibit 30–40% sequence identity led to the discovery of two additional mechanistically distinct classes of fosfomycin resistance proteins. The genomes of *Bacillus subtilis* and *Staphylococcus aureus* and related microorganisms encode proteins that catalyze the addition of L-cysteine to fosfomycin (Fig. 1) (Cao *et al.*, 2001; Rigsby and Armstrong, unpublished results). These enzymes, called FosB, require Mg^{2+} for optimum activity and do not show a dependence on K^+ or any other monovalent cation. The evolution of this distinct activity was driven in part by the fact that these microorganisms do not make GSH. The FosB enzymes characterized to date seem to be genuine fosfomycin resistance proteins in that they confer measurable resistance to the antibiotic in *Escherichia coli* even though the concentration of L-cysteine in this organism is quite low (Rigsby and Armstrong, unpublished results).

A third class of resistance proteins, FosX, has been shown to have fosfomycin hydrolase activity (Fig. 1). Fosfomycin hydrolases encoded in the genomes of *Mesorhizobium loti* and the pathogen *Listeria monocytogenes* have been characterized with respect to their structure and mechanism (Fillgrove *et al.*, 2003). The regiochemistry of the addition of water is the same as that observed for FosA and FosB. Like FosA, these two proteins require Mn(II) for catalysis. However, neither requires a monovalent cation for optimal activity. The enzyme from *M. loti* ($FosX^{ML}$) is not a very good resistance protein because of its low catalytic activity. However, $FosX^{ML}$ is unique in that it is catalytically promiscuous having both GSH transferase and fosfomycin hydrolase activities, although it does neither reaction very well. The gene encoding $FosX^{ML}$ is part of a putative phn operon in this organism that is presumably involved in phosphonate metabolism. Although the biological role of $FosX^{ML}$ is probably not resistance to the antibiotic, it has been proposed to be a possible progenitor of the fosfomycin resistance proteins (Fillgrove *et al.*, 2003). In contrast, $FosX^{LM}$ from *Listeria monocytogenes* is a very efficient fosfomycin hydrolase that confers robust resistance to the antibiotic in *E. coli*.

This chapter reports the procedures for expression and purification of FosA, FosB, and FosX proteins from six microbial sources. The quantification of enzyme activities by a fluorescence-detected high-performance liquid chromatography (HPLC) assay (FosA and FosB) and a gas chromatography-mass spectrometry (GC-MS) (FosX) assay are described. In addition, a rapid semiquantitative assay of FosA and FosX activities by ^{31}P-NMR is reported. The chapter concludes with the kinetic constants and biological activity for the six resistance proteins.

Expression and Purification of Fosfomycin Resistance Proteins

FosA, a Fosfomycin-Specific Glutathione Transferase

Plasmid-Encoded FosA from Transposon TN2921 (FosA^{TN}) (Bernat et al., 1997). The TN2921 coding sequence flanked by *Nde*I and *Eco*RI restriction sites was synthesized from multiple oligonucleotides by successive rounds of polymerase chain reaction (PCR) and confirmed by sequencing as described previously (Bernat *et al.*, 1997). The final PCR product was ligated into pET20b(+) expression vector. Protein was overexpressed in *E. coli* BL21(DE3) cells transformed with this plasmid. Bacteria were grown in LB media containing 100 μg/ml ampicillin at 37° to an $OD_{600} = 1.0$ before induction of transcription by the addition of IPTG (0.4 mM final concentration). After 3 h, bacteria were harvested by centrifugation. All buffers used for dialysis and purification contained 1 mM DTT. Cells were resuspended in buffer A (20 mM Tris/HCl [pH 8.0]) and lysed by sonication. After centrifugation at 4° for 35 min at 30,000g, the supernatant was dialyzed overnight against buffer A. Protein was then loaded onto a 2.5 × 16 cm DEAE Sepharose column equilibrated with buffer A and washed with 5 column volumes of buffer. Protein was eluted with a linear gradient of NaCl (0–400 mM). Fractions containing FosA were identified by activity, pooled, and dialyzed against buffer B (20 mM sodium phosphate [pH 7.0]). Protein was subsequently passed through a 2.5 × 16 cm hydroxylapatite column (BioRad, Hercules, CA) equilibrated with buffer B and eluted isocratically with the same buffer. Fractions containing enzyme were concentrated, and any bound metal was removed by dialysis against buffer C (20 mM MES [pH 6.0]) containing 5 mM EDTA followed by dialysis in buffer C. Purified protein was then passed through a 1 × 25 cm Chelex 100 (H$^+$ form, 100–200 mesh) column equilibrated with buffer C. Glassware used for the removal of metal was washed with 3 M HCl before use. Protein was concentrated using a nitrogen pressure cell fitted with an Amicon 10K molecular weight cutoff membrane and stored at −80°.

Genomically Encoded FosA from Pseudomonas aeruginosa *(FosA^{PA}) (Rife et al., 2002).* The gene encoding the FosA from *Pseudomonas aeruginosa*, PA1129, was amplified by PCR from cosmid PMO013326 (Pseudomonas Genetic Stock Center, East Carolina University School of Medicine, Greenville, NC). Primers were designed to incorporate *Nde*I and *Bam*HI restriction sites at the 5' and 3' ends, respectively, of the coding sequence. The PCR product was purified using the QiaQuick PCR purification kit (Qiagen Inc., Valencia, CA). The purified PCR product and pET20b(+) DNA were digested with the indicated restriction enzymes, purified by gel electrophoresis, and ligated to form the PA1129 expression plasmid. This plasmid was used to transform *E. coli* BL21(DE3) cells. Protein was over-expressed in *E. coli* as described previously, except 2 l of cells were grown to an $OD_{600} = 0.8$, and protein was allowed to express for 4 h. Bacteria were harvested by centrifugation and stored at $-20°$.

Bacteria were resuspended in buffer D (20 m*M* Tris/HCl [pH 7.5]) and lysed by sonication. Cellular debris was removed by centrifugation ($30,000g$ for 30 min), and the supernatant was dialyzed against buffer D. All dialysis steps were conducted at $4°$, and all dialysis buffers contained 1 m*M* DTT. Crude lysate was passed through a 2.5×17 cm Fast Flow DEAE cellulose column equilibrated with buffer D, washed with buffer D, and eluted with a linear gradient (20–500 m*M*) of NaCl over 8 column volumes. Fractions containing FosA as determined by the absorbance at 280 nm and sodium dodecyl sulfate-polyacrylamide gel electrophoresis (SDS-PAGE) were pooled and treated to remove any bound metal by dialysis for at least 8 h against buffer D containing 5 m*M* EDTA and 3 g Chelex 100 resin (Na^+ form). This was followed by dialysis against buffer D. Protein was then passed through a 1×16 cm HiPrep Q column (Pharmacia, Piscataway, NJ) using an FPLC system. The column was equilibrated with buffer D before loading protein and was developed with a linear gradient of 0–220 m*M* NaCl in buffer D. Fractions containing FosA were pooled and dialyzed against buffer E (20 m*M* potassium phosphate [pH 6.8]) overnight. Protein was then passed through a 2×7.5 cm hydroxylapatite column equilibrated with buffer E and eluted isocratically with the same buffer. Fractions containing FosA were pooled and treated to remove any bound metal by dialysis for at least 8 h against buffer F (20 m*M* TMAHEPES [pH 8.0]) containing 5 m*M* EDTA and 3 g Chelex 100 resin (Na^+ form) followed by dialysis in buffer F. Protein was concentrated using a nitrogen pressure cell fitted with an Amicon 10K molecular weight cut-off membrane and stored at $-80°$.

FosB, A Fosfomycin-Specific L-Cysteine Thiol Transferase

Genomically Encoded FosB from Bacillus subtilis *(FosB[BS]) (Cao et al., 2001)*. An expression plasmid (pET 17b containing the coding sequence) for the FosB protein from *Bacillus subtilis* was provided by the Helmann laboratory (Department of Microbiology, Cornell University). This plasmid was used to transform *E. coli* BL-21(DE3) pLysS cells. Bacteria were grown and protein expressed as described previously for FosA[PA]. Harvested bacteria were suspended in buffer D, lysed, and dialyzed as described for FosA[PA]. All dialysis steps were conducted at $4°$, and all dialysis buffers contained 1 m*M* DTT. Crude lysate was passed through a

2.5 × 17 cm Fast Flow DEAE cellulose column equilibrated with buffer D, which was then washed with buffer D, and eluted with a linear 20–250 mM NaCl gradient over 8 column volumes. Fractions containing FosB as determined by the absorbance at 280 nm and SDS-PAGE were pooled and dialyzed for at least 8 h against buffer E. Protein was loaded onto a 2 × 7.5 cm hydroxylapatite column equilibrated with buffer D, then washed with D, and eluted with a phosphate gradient (50–350 mM) over 6 column volumes. Fractions containing FosB were pooled and dialyzed against buffer D containing 100 mM NaCl and 0.1 mM EDTA before concentration using a nitrogen pressure cell fitted with an Amicon 10K molecular weight cutoff membrane to approximately 10 mg/ml. Protein was then loaded onto a 1.6 × 60 cm Sephacryl S-100 column equilibrated with the same buffer and eluted isocratically at a flow rate of 1 ml/min. Protein was dialyzed in buffer D containing 5 mM EDTA and 3 g Chelex 100 (Na$^+$ form, 100–200 mesh) to remove protein-bound metal. The protein was dialyzed against 20 mM HEPES (pH 8.0) and concentrated for storage as described for FosA^{PA}.

Genomically Encoded FosB from Staphylococcus aureus *(FosB^{SA}).* The gene encoding the FosB protein from *Staphylococcus aureus* was amplified from genomic DNA by PCR. *Nde*I and *Eco*RI sites were added to the 5′ and 3′ ends, respectively. A silent mutation (CA**T** to CA**G**) was incorporated in codon 134 to remove an *Nde*I site from the coding sequence. An expression vector was prepared as described for FosA^{PA}. Protein was expressed as described except after induction with IPTG (0.2 mM) cells were grown at 30° for 5 h. Bacteria harvested from 2 l of culture were suspended in 100 ml buffer E, lysed, and the resulting suspension centrifuged as described previously. The supernatant was treated with Benzonase Nuclease (Novagen, Madison, WI) according to the manufacturer's directions, thereby reducing or eliminating undesired interactions of FosB with nucleic acids. Benzonase treatment was followed by dialysis against buffer E. All dialysis was performed at 4°, and all buffers contained 1 mM DTT. Protein was passed through a 1.6 × 60 cm SP-Sepharose column equilibrated with buffer E, the column was then washed with buffer E and the protein was eluted with a 10–500 mM NaCl gradient (in buffer E) over 8 column volumes. Fractions containing FosB were identified by the absorbance at 280 nm and SDS-PAGE. These fractions were pooled and dialyzed against buffer E. Protein was then loaded onto a 2 × 7.5 cm hydroxylapatite column equilibrated with buffer E and eluted with a linear phosphate gradient (20–500 mM) in buffer E. Fractions containing FosB^{SA} were pooled and dialyzed against buffer G (20 mM MOPS [pH 7.0]) containing 5 mM EDTA and 3 g Chelex 100 (Na$^+$ form, 100–200 mesh) to remove protein-bound metal, followed by dialysis against buffer G. Protein was then concentrated and stored as described for Fo-sA^{PA}.

FosX, A Fosfomycin-Specific Hydrolase

Genomically Encoded FosX from Mesorhizobium loti *(FosX^{ML}) (Fillgrove et al., 2003).* An expression plasmid for the *mlr3345* gene from *Mesorhizobium loti* was prepared from genomic DNA as described for FosA^{PA} gene, except *Eco*RI was used as the 3′ restriction site. Overexpression of the FosX protein in *E. coli* was accomplished as described for FosA^{PA}. Bacteria were harvested from 1 l of culture and were suspended in 20 ml buffer H (25 mM Tris, 50 mM NaCl [pH 7.5]) containing 0.1 mg/ml lysozyme. The lysozyme incubation was conducted for 15 min at room temperature followed by 15 min on ice. After this incubation, 50 ml buffer H was added. Bacteria were then lysed and centrifuged as described. The supernatant was treated with 1% (w/v) streptomycin sulfate followed by centrifugation at 30,000g for 30 min before dialysis at 4° against buffer H containing 0.25 mM EDTA. All dialysis buffers also contained 1 mM DTT. Protein was loaded onto a 2.5 × 17 cm DEAE-cellulose column (Whatman, Maidstone, UK) and eluted with a linear gradient (50–500 mM) of NaCl. Fractions containing the FosX protein were pooled and dialyzed at least 8 h against buffer H containing 5 mM EDTA and 3 g Chelex 100 (Na^{+} form, 100–200 mesh) to remove protein bound metal. After additional dialysis against buffer H, the FosX was further purified using a 6-ml Resource Q column (Pharmacia, Piscataway, NJ) developed with a linear NaCl gradient (50–350 mM). Protein was dialyzed against buffer I (20 mM Tris/HCl [pH 7.3]) overnight before loading onto a 2.5 × 12 cm hydroxyapatite column. Protein was eluted with a linear phosphate gradient (0–300 mM) in buffer I. Fractions containing FosX^{ML} were pooled and dialyzed against 25 mM HEPES (pH 7.5) before concentration and storage at −80°.

Genomically Encoded FosX from Listeria monocytogenes *(FosX^{LM}) (Fillgrove et al., 2003).* An expression plasmid for *lmo*1702, the gene encoding the FosX from *Listeria monocytogenes*, was prepared as described for FosX^{ML}. Plasmid DNA was used to transform *E. coli* Rosetta (DE3) cells (Novagen, Madison, WI). These cells were used because they are engineered to express tRNAs for codons of rare use in *E. coli*, and 13 of 134 codons in the *lmo*1702 gene are considered rare. Bacteria were grown and protein expressed as described for the FosA^{PA} protein, except chloramphenicol (12 μg/ml) was added to the media, and overexpression was induced with 0.2 mM IPTG. Bacterial lysate was treated as described for FosX^{ML} protein and was loaded onto a 2.5 × 12 cm Fast Flow DEAE-sepharose column (Amersham Biosciences, Uppsala, Sweden).

The column was washed, and protein was eluted with a 75–400 mM gradient of NaCl. Fractions containing FosX were pooled and dialyzed against buffer J (30 mM KH$_2$PO$_4$ [pH 7.0]) overnight before loading onto a 2.5 × 7 cm hydroxyapatite column equilibrated with buffer J. Protein was eluted with a linear 30–250 mM phosphate gradient. Fractions containing FosX were pooled and dialyzed against buffer K (20 mM HEPES [pH 7.5]) containing 100 mM NaCl, 5 mM EDTA and 3 g Chelex100 (Na$^+$ form, 100–200 mesh). Protein was then concentrated to approximately 10 mg/ml and passed through a 1.6 × 60 cm Sephacryl S-100 column equilibrated with buffer K containing 100 mM NaCl and 0.25 mM EDTA. Purified protein was dialyzed against buffer K containing 3 g Chelex 100 resin before concentration and storage at −80°.

Assay of FosA, FosB, and FosX Activities

Assay of FosA and FosB by Fluorescence-Detected HPLC (Bernat et al., 1997, 1998; Cao et al., 2001)

Activity assays for FosA and FosB rely on the detection of the free amino group of the peptide or amino acid adduct of the antibiotic after derivatization with the fluorescent reagent 6-aminoquinolyl-N-hydroxysuccinimidyl carbamate (AQC) and separation of the derivatives by reversed-phase HPLC. For FosA activity assays, typically 50–150 nM enzyme was preincubated with MnCl$_2$ and KCl in 20 mM TMA/HEPES, pH 8.0, before mixing with glutathione (GSH). Reactions were equilibrated to 25° and were initiated by the addition of fosfomycin (tetramethylammonium salt, TMAfos). Typical reactions contained 50 μM Mn(II), 100 mM K$^+$, 1–100 mM GSH, and 0.1–5 mM TMAfos in a final volume of 100 μl. After 2–5 min, reactions were quenched by the addition of 200 μl 5% (w/v) trichloroacetic acid followed by vortexing and the addition of 100 μl 0.8 M NaOH and 100 μl 1 mM valine (internal standard). A 5-μl aliquot was then added to 80 μl borate buffer and mixed with 15 μl of the AQC reagent (Waters, Milford, ME) solubilized according to manufacturer's directions. Reactions were heated for 10 min at 55° and diluted with 400 μl 50 mM NaOAc, pH 5.0, before injection onto a 0.046 × 25 cm Ultrasphere C-18 column (Beckman, Fullerton, CA). Samples were eluted with a linear 0–35% MeCN gradient with a flow rate of 1 ml/min over 15 min. GSH-fos, GSH, and valine had retention times of 9.3, 11.2, and 14.7 min, respectively, as illustrated in Fig. 2. The ratio of product to valine peak areas was used to calculate product concentrations from a calibration curve constructed from area ratios of product/internal standard versus substrate concentration.

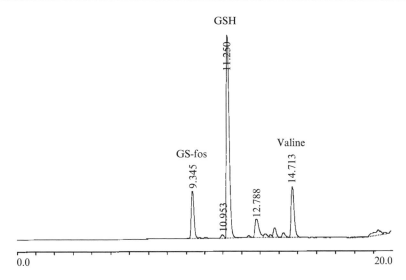

GSH

GS-fos

Valine

9.345

11.250

10.953

12.788

14.713

0.0   20.0

FIG. 2. High-performance liquid chromatogram of a typical FosA reaction mixture.

FosB reactions were done similarly, except 300–600 nM enzyme was incubated with 1 mM MgCl$_2$ in 20 mM HEPES, pH 8.0, before the addition of 1–15 mM L-cysteine followed by addition of 0.1–4 mM fosfomycin. Reactions were quenched and derivatized as described before HPLC analysis. Cysteine and Cys-fos had retention times of 10.0 and 13.6 min, respectively.

Assay of FosX by Gas Chromatography/Mass Spectrometry

Because no convenient fluorescent or UV absorbance derivatizing agents have been identified for the FosX reaction product for HPLC analysis, a GC-MS method was developed to assay FosX activity. Enzyme was preincubated with MnCl$_2$ in 10 mM HEPES or KH$_2$PO$_4$ (pH 7.5). Reactions were initiated by adding enzyme to varying concentrations of fosfomycin in a total volume of 750 μl. Reactions were incubated at 25° for 4 min and quenched by adding CHCl$_3$, vortexing, and immediately freezing on dry ice. After a minimum of 30 min, reactions were thawed, and *n*-butylphosphonic acid (7.5 μl of a 24.8 mM solution) was added as an internal standard. After centrifugation (13,000g for 5 min), 5 μl of the aqueous layer was removed and taken to dryness in silanized glass vials. Product was derivatized by the addition of 50 μl of a 1:1 mixture of CH$_2$Cl$_2$:

bis-trimethylsilylacetamide (Pierce, Rockford, IL). Vials were sealed and left at room temperature for 8 h. Samples were then injected onto a 0.25 mm ID × 20 m RTX-1701 column (Restek, Bellafonte, PA) attached to a ThermoFinnigan DSQ GC-MS system. A linear temperature gradient (80°–210°, 1°/min) at a constant carrier gas (He) flow rate of 1 ml/min was used to elute the samples. A constant MS source temperature (250°) was maintained throughout runs. TMS derivatives of n-butylphosphonic acid, fosfomycin, and 1,2-dihydroxypropylphosphonic acid were detected by selected ion monitoring in EI mode and had retention times of 9.27, 9.71, and 12.35 min, respectively. A calibration plot of the area ratio of analyte/internal standard (5–7 points) versus analyte concentration was prepared. The concentrations of product in the reactions were determined using this calibration plot.

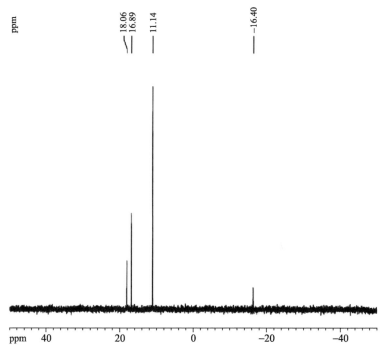

FIG. 3. 121 mHz ^{31}P spectrum of a reaction of FosX^{ML} in the presence of GSH and fosfomycin (11.14 ppm) showing production of GS-fos (18.06 ppm) and 1,2-dihydroxypropyl-phosphonic acid (16.89 ppm). The internal standard triphenylphosphate appears at −16.40 ppm.

Assay of FosA, FosB, and FosX by ^{31}P-NMR Spectroscopy

Although this assay is not appropriate for obtaining detailed kinetic constants for these enzyme-catalyzed reactions, it does provide a relatively quick semiquantitative assay for all three activities, simultaneously if desired. Because the substrate and all products can be detected by ^{31}P NMR spectroscopy, this was used as a quick method for obtaining preliminary kinetic data for the fosfomycin resistance proteins, particularly for the FosX class of resistance protein. Typically, enzyme was equilibrated with Mn(II) in 20 mM HEPES (pH 7.5) before the addition of 25–150 mM GSH. Reactions were initiated by the addition of 25 mM fosfomycin and allowed to proceed at room temperature for several minutes to several hours. Reactions were quenched by vortexing with CHCl$_3$ followed by freezing on dry ice. After thawing and centrifugation, the aqueous layer was removed to a fresh tube. The reaction was incubated with frequent mixing with a small amount of Chelex 100 resin (Na$^+$ form) for at least 3 h to remove Mn(II), which causes a paramagnetic line broadening of the ^{31}P signal. Reactions were centrifuged to pellet the Chelex resin followed by removal of the supernatant to a fresh tube. The Chelex was washed with distilled, deionized H$_2$O, centrifuged, and the supernatant removed and combined with the first supernatant. After addition of D$_2$O, the ^{31}P spectra of the reactions were collected at 121 mHz. The promiscuous FosX and FosA activities of the FosX^{ML} enzyme are illustrated in Fig. 3, where GS-fos, 1,2-dihydroxypropylphosphonic acid, fosfomycin, and triphenyl phosphate (standard) had ^{31}P

TABLE I

CATALYTIC, BIOLOGICAL, AND PHYSICAL CHARACTERISTICS OF THE FOSFOMYCIN
RESISTANCE PROTEINS

Protein	$k_{cat}{}^a$ (s^{-1})	$k_{cat}/K_M{}^{fos}$ (M^{-1} s^{-1})	MIC^b mg/ml	MW^c	$\varepsilon_{280}{}^d$ M^{-1}cm^{-1}
FosA^{TN}	660 ± 10	$(1.4 \pm 0.1) \times 10^7$	>20	15889	31300
FosA^{PA}	175 ± 6	$(9.0 \pm 1.4) \times 10^5$	>20	15114	25670
FosB^{BS}	4.8 ± 0.3	$(4.0 \pm 0.5) \times 10^3$	0.1	17173	14390
FosB^{SA}	0.99 ± 0.02	$(9.2 \pm 0.1) \times 10^3$	0.4	16637	21280
FosX^{ML}	0.15 ± 0.02	$(5.0 \pm 0.6) \times 10^2$	0.025	16181	12090
FosX^{LM}	34 ± 2	$(9 \pm 2) \times 10^4$	>20	15655	13370
None			≤0.025		

^{a} The FosA and FosB proteins were assayed with fixed concentrations of GSH (mM) and L-cysteine (mM), respectively.
^{b} Minimum inhibitory concentration of fosfomycin toward *Escherichia coli* transformed with expression plasmids of the resistance protein.
^{c} Calculated from amino acid sequence.
^{d} Calculated from the phenylalanine, tyrosine, and tryptophan content.

chemical shifts of 18.1, 16.9, 11.1, and −16.4 ppm, respectively. Following the law of conservation of mass, product and substrate peak heights are measured and amount of product formed calculated from the ratio between peak heights and total substrate used in the reaction.

Physical and Catalytic Characteristics

The physical and catalytic characteristics of the fosfomycin resistance proteins described in this article are listed in Table I. All of the proteins are dimers in solution. There is a correlation between the $k_{cat}/K_M{}^{fos}$ and the biological efficacy of the enzyme in conferring resistance to fosfomycin in *E. coli*.

References

Arca, P., Hardisson, C., and Suarez, J. E. (1990). Purification of a glutathione S-transferase that mediates fosfomycin resistance in bacteria. *Antimicrob. Agents Chemother.* **34**, 844–848.

Arca, P., Rico, M., Brana, A. F., Villar, C. J., Hardisson, C., and Suarez, J. E. (1988). Formation of an adduct between fosfomycin and glutathione: A new mechanism of antibiotic resistance in bacteria. *Antimicrob. Agents Chemother.* **32**, 1552–1556.

Armstrong, R. N. (2000). Mechanistic diversity in a metalloenzyme superfamily. *Biochemistry* **39**, 13625–13632.

Bernat, B. A., and Armstrong, R. N. (2001). Elementary steps in the aquisition of Mn^{2+} by the fosfomycin resistance protein (Fos A). *Biochemistry* **40**, 12712–12718.

Bernat, B. L., Laughlin, T., and Armstrong, R. N. (1997). Fosfomycin resistance protein (FosA) is a manganese metalloglutathione transferase related to glyoxalase I and the extradiol dioxygenases. *Biochemistry* **36**, 3051–3055.

Bernat, B. L., Laughlin, T., and Armstrong, R. N. (1998). Regiochemical and stereochemical course of the reaction catalyzed by the fosfomycin resistance protein, FosA. *J. Organic Chem.* **63**, 3778–3780.

Bernat, B. L., Laughlin, T., and Armstrong, R. N. (1999). Elucidation of a monovalent cation dependence and characterization of the divalent cation binding site of the fosfomycin resistance proteins. *Biochemistry* **38**, 7462–7469.

Cao, M., Bernat, B. A., Wang, Z., Armstrong, R. N., and Helmann, J. D. (2001). FosB, a cysteine-dependent fosfomycin resistance protein under the control of s^W, an extra-cytoplasmic-function σ factor in *Bacillus subtilis*. *J. Bacteriol.* **183**, 2380–2383.

Fillgrove, K. L., Pakhomova, S., Newcomer, M. N., and Armstrong, R. N. (2003). Mechanistic diversity of fosfomycin resistance in pathogenic microorganisms. *J. Am. Chem. Soc.* **125**, 15730–15731.

Kahan, F. M., Kahan, J. S., and Cassidy, P. J. (1974). The mechanism of action of fosfomycin. *Ann. NY Acad. Sci.* **235**, 364–386.

Llaneza, J., Villar, C. J., Salas, J. A., Suarez, J. E., Mendoza, M. C., and Hardisson, C. (1985). Plasmid-mediated fosfomycin resistance is due to enzymatic modification of the antibiotic. *Antimicrob. Agents Chemother.* **28**, 163–164.

Mendoza, M. C., Garcia, J. M., Llaneza, J., Mendez, J. F., Hardisson, C., and Ortiz, J. M. (1980). *Antimicrob. Agents Chemother.* **18**, 215–219.

Pakhomova, S., Rife, C. L., Armstrong, R. N., and Newcomer, M. E. (2004). Structure of fosfomycin resistance protein A from transposon TN2921. *Protein Sci.* **13**, 1260–1265.

Rife, C. L., Pharris, R. E., Newcomer, M. E., and Armstrong, R. N. (2002). Crystal structure of a genomically encoded fosfomycin resistance protein (FosA) at 1.19 A resolution by MAD phasing off the L-III edge of Tl^+. *J. Am. Chem. Soc.* **124**, 11001–11003.

Rigsby, R. E., Rife, C. L., Fillgrove, K. L., Newcomer, M. E., and Armstrong, R. N. (2004). Phosphonoformate: A minimal transition state analogue inhibitor of the fosfomycin resistance protein, FosA. *Biochemistry* **43**, 13666–13673.

Villar, C. J., Hardisson, C., and Suarez, J. E. (1986). Cloning and molecular epidemiology of plasmid-determined fosfomycin resistance. *Antimicrob. Agents Chemother.* **29**, 309–314.

[24] Regulation of 4-Hydroxynonenal Mediated Signaling By Glutathione S-Transferases

By YOGESH C. AWASTHI, G. A. S. ANSARI, and SANJAY AWASTHI

Abstract

4-Hydroxy-trans-2-nonenal (HNE) was initially considered to be merely a toxic end product of lipid peroxidation that contributed to oxidative stress–related pathogenesis. However, in recent years its physiological role as an important "signaling molecule" has been established. HNE can modulate various signaling pathways in a concentration-dependent manner. Glutathione S-transferases (GSTs) are major determinants of the intracellular concentration of HNE, because these enzymes account for the metabolism of most cellular HNE through its conjugation to glutathione. Evidence is emerging that GSTs are involved in the regulation of the HNE-mediated signaling processes. Against the backdrop of our current understanding on the formation, metabolism, and role of HNE in signaling processes, the physiological role of GSTs in regulation of HNE-mediated signaling processes is critically evaluated in this chapter. Available evidence strongly suggests that besides their well-established pharmacological role of detoxifying xenobiotics, GSTs also play an important physiological role in the regulation of cellular signaling processes.

Introduction

Glutathione S-transferases (EC.2.5.18) represent a family of multifunctional enzymes whose role in detoxification of electrophilic xenobiotics or electrophilic metabolites is well established. In general, GSTs catalyze the conjugation of a wide variety of structurally dissimilar compounds containing electrophilic carbon, nitrogen, or sulfur atom to glutathione (GSH). After the initial report of its existence in rat liver (Booth *et al.*, 1961), much of the early studies with GST isozymes focused on their ability to inactivate environmental toxicants and carcinogens such as the electrophilic metabolites of benz(a)pyrene and other polycyclic aromatic hydrocarbons by catalyzing their conjugation to GSH (Chasseuad, 1979). The

METHODS IN ENZYMOLOGY, VOL. 401   0076-6879/05 $35.00
Copyright 2005, Elsevier Inc. All rights reserved.   DOI: 10.1016/S0076-6879(05)01024-4

chemopreventive effect of various micronutrients has been shown to correlate with their ability to induce GSTs (Awasthi et al., 1996; Coles and Kadlubar, 2003, reviews). Although GSTs have been found in various subcellular fractions, the cytosolic GSTs have been extensively studied. At least seven gene families of cytosolic GSTs exist. Details about these gene families and their role in drug metabolism and detoxification of xenobiotics, including chemical carcinogens, are elegantly covered in numerous reviews, and for brevity only few are cited here (Hayes and Pulford, 1995; Hayes et al., 2004; Jakoby, 1978; Mannervik and Danielson, 1989; Salinas and Wong, 1999; Sheehan et al., 2001). Also, aspects of GST structures and functions are covered elsewhere in this volume.

Although the pharmacological role of GSTs in metabolism and detoxification of xenobiotics is well established, the significance of GST-catalyzed conjugation of endogenous compounds is not well understood. All endogenous compounds having the α, β-unsaturated carbons (Michael acceptor group) are potential substrates of GSTs. Some of these endogenous compounds, particularly those formed during the oxidative degradation of cellular components, have drawn a great deal of attention because of their ability to attack cellular nucleophiles that can lead to toxicity. GSTs can catalyze the conjugation of these compounds including 4-hydroxy-2-trans-nonenal (HNE) and other α, β-unsaturated alkenyls, malonaldialdehyde, acrolein, and propenal bases such as cytosine propenal, thymine propenal, and uracil propenal (Alin et al., 1985; Berhane et al., 1994) with GSH and provide protection against these potential toxicants. Because of their electrophilic nature, these compounds can also nonenzymatically react with GSH, but this reaction is enhanced by an estimated 500–600 fold in the presence of GSTs.

Among the previously listed endogenous substrates of GSTs, HNE and its homologous α, β-unsaturated aldehydes, which are generated as stable end-products of lipid peroxidation, have drawn maximum attention in recent years. HNE first discovered by Estabauer and his group as the major end-product of the oxidative degradation of n-6 polyunsaturated fatty acids (PUFA) (Esterbauer et al., 1991, review) was generally perceived mainly as a toxic end-product contributing to the deleterious effect of oxidative stress-induced lipid peroxidation and a causative factor in the etiology of age-related degenerative disorders such as Alzheimer's disease (Sayre et al., on's disease (Yoritaka et al., 1996), atherosclerosis (Negre-Salvayre et al., 2003; Yang et al., 2004), cataractogenesis (Awasthi et al., 1996) and cancer (Hu et al., 2002). Recently, HNE seems to have attained the status of an important signaling molecule that is involved in the signaling pathways for differentiation, proliferation, cell cycle control, and the regulation of expression of various stress response genes (Awasthi et al., 2003, 2004/

reviews; Barrera *et al.*, 2004; Dianzani, 2003). It has been suggested that HNE may interact with various cellular targets and differentially affect signaling pathways, depending on the cell type and its activation stage (Nakashima *et al.*, 2003; Sharma *et al.*, 2004). More importantly, HNE seems to affect some of these processes in a concentration-dependent manner. For example, it is well known that at high concentrations, HNE causes necrosis, differentiation, and apoptosis, but the cells in which constitutive HNE is depleted, proliferate at a higher growth rate (Cheng *et al.*, 1999, 2001a). Likewise, the expression of protein kinase C (PKC) isozymes is differentially affected by HNE and is dependent on its concentration (Marinari *et al.*, 2003). This has led to the suggestion that the intracellular concentration of HNE (and perhaps of other homologous alkenyls) could be a determinant for the signaling for apoptosis, differentiation, or proliferation (Awasthi, *et al.*, 2003, 2004).

In view of the concentration-dependent effect of HNE on signaling pathways, the factors that regulate HNE concentration within the cells are likely to be involved in modulation of signaling processes. Because HNE is primarily formed through the oxidative degradation of PUFA initiated by reactive oxygen species (ROS) during oxidative stress, it is logical to assume that the intracellular concentration of HNE should be determined primarily through its metabolism and exclusion from the cells. Most HNE is excluded from cells through its GST-catalyzed conjugation to GSH and subsequent transport of the GSH conjugate (GS–HNE). Therefore, GSTs and the transporter(s) catalyzing the efflux of GS–HNE are likely to play a major role in regulating HNE homeostasis in cells. Consequently, these proteins should be relevant to the mechanisms that regulate HNE-mediated signaling for apoptosis, differentiation, proliferation, etc. Therefore, GST-catalyzed conjugation of HNE to GSH and its possible implications in the cellular signaling processes is the focus of this chapter. In the following sections of this chapter, the formation, metabolism, regulation of intracellular levels, and biological activities of HNE are briefly covered. The major thrust of the article is on the role of GSTs in the regulation of the intracellular concentrations of HNE and in the modulation of oxidative stress and HNE-mediated signaling processes.

Formation, Metabolism, and Disposition of HNE

Formation

HNE is a stable end product of the oxidative degradation of *n*-6 PUFA and is formed by the decomposition of the hydroperoxides of these acids (Schaur *et al.*, 2003). It has been shown that 9(S) hydroperoxyoctadecadienoic acid C–C and O–O bonds split results in the formation of nonenal and 9-oxo-nonaic acid (Schneider *et al.*, 2001). Nonenal can undergo hydroperoxidation at position 4 either nonenzymatically or through enzymatic reaction catalyzed by alkenyl oxygenase to form 4-hydroperoxynonenal, which can be reduced to HNE. Similarly, decomposition of 13(S) hydroperoxyoctadecadienoic acid also leads to formation of HNE (Schaur, 2003).

The relative contribution of the nonenzymatic and the enzymatic processes in the formation of HNE is not clear. However, increased formation of 4-HNE under even very mild stress conditions (Cheng *et al.*, 2001b; Yang *et al.*, 2003) suggests that in cells HNE formation may occur primarily through nonenzymatic processes. Because of the ever-present ROS in aerobic organisms, the constitutive "physiological" levels of HNE are expected to be present in cells/tissues, but it is difficult to exactly assess these levels because of its uncontrolled formation and rapid metabolism. During the conditions of oxidative stress, HNE concentrations in plasma may be significantly elevated. Therefore, HNE levels in different tissues should depend on the presence of ROS together with the level of defense enzymes such as glutathione peroxidase (GPx) and GSTs, which prevent propagation of lipid peroxidation. It is, therefore, not surprising that there is a sizable variation in the reported concentrations of HNE in tissues. The reported HNE concentrations in various mammalian tissues and in plasma vary in the range of approximately $0.5~\mu M$–$6~\mu M$ (Grune *et al.*, 1997; Siems *et al.*, 1995, 2002). In diseases in which patients are under continuous oxidative stress, a threefold to 10-fold increase in HNE levels over the physiological levels is reported (Siems and Grune, 2003). On the other hand, a threefold decrease in HNE concentration has been reported in cells overexpressing GSTA1-1 or GSTA2-2, which suppress lipid peroxidation and limit HNE formation (Yang *et al.*, 2001, 2002). These findings suggest that among the defense enzymes, GSTs may be the major determinants of the constitutive HNE levels in cells, because these enzymes can regulate HNE concentration not only through its metabolism but also by limiting its formation. It is important to consider that even though a constitutive level of HNE is likely to be present in most of the tissues and in plasma, it is difficult to assess its steady-state levels because of variations in factors leading to HNE formation, enzymes which limit lipid peroxidation (e.g., GST_s and GP_x), and the enzymes such as GSTs, aldoketoreductase (AKRs), and aldehyde dehydrogenase (ALDHs), which metabolize HNE. A brief account of our current understanding of HNE metabolism is given in the following.

Metabolism of HNE

The fact that within 48 h of an intravenous administration of [³H] HNE to rats, more than two thirds of the radioactivity is excreted in urine strongly suggests that an efficient detoxification machinery exists within cells to metabolize and exclude this potentially toxic product from the system (Alary *et al.*, 2003). The mercapturic acids of 4-hydroxynonenoic acid (HNA), its lactone, ω-hydroxy HNA-lactone, and dihydroxynonane constitute a major portion of HNE metabolites in urine (Alary *et al.*, 1995, 2003), suggesting that GSTs, which catalyze the first conjugation step in the mercapturic acid pathway, are heavily involved in the metabolism and detoxification of HNE. This has been confirmed by studies showing that in perfused rat heart, about two thirds of HNE is conjugated to GSH, presumably through a GST-catalyzed reaction (Srivastava *et al.*, 1998). The aldehyde group of HNE can be reduced to 1,4-dihydroxynonenol (DHN) by AKRs including aldolase reductase (AR) (Alary *et al.*, 2003; Srivastava *et al.*, 1995, 1998). Alternately, it can be oxidized to HNA by ALDHs (Reichard *et al.*, 2002). In humans, ALDH3A1 seems to have specificity toward HNE and other aliphatic aldehydes (Townsend *et al.*, 2001). Mercapturic acids of HNA and its lactone have been characterized in urine, suggesting that these or their precursors are substrates of GSTs. Although GSTs, ALDHs, and AKRs are among the major enzymes responsible for the metabolism of HNE, several other enzymes are also involved in pathways leading to elimination of HNE. For example, HNA or its lactone can be hydroxylated at the ω-position of the alkyl chain by Cyp_{450} 4A family of enzymes from rodent liver (Gueraud *et al.*, 1999). Surprisingly, the mercapturic acids of ω-hydroxylated HNE metabolite constitute a significant portion of HNE-derived polar mercapturic acids isolated from rat urine. This may have some physiological significance, because HNE metabolites, including DHN and its GSH-conjugate (GS-DHN), are implicated in modulation of specific signaling processes (Ramana *et al.*, 2002, 2003), and it may not be surprising if the polar ω-hydroxylated derivatives of HNE and GS-HNE are also involved in signaling. Other minor enzymes involved in the metabolisms of HNE and its derivatives include the NADPH-dependent alkenyl/one oxidoreductase at position 2 to give the saturated aldehyde, 4-hydroxynonanal (Dick *et al.*, 2001). In Table I, the enzymes involved in the metabolism of HNE are listed, and the major metabolic pathways for HNE disposition are outlined in Fig 1.

Stereospecificity may also contribute to the physiological role, because HNE has a chiral center at carbon 4, and biologically formed HNE exists as a racemic mixture of S- and R-enantiomers. On conjugation with GSH and

TABLE I
MAJOR ENZYMES INVOLVED IN THE METABOLISM OF HNE

Enzymes	Substrates	Product(s)
Glutathione	HNE ω-hydroxy	GS-HNE (hemiacetal)
S-transferases[a](GSTs)	HNA-lactone	GS-conjugate
Aldoketoreductases (AKRs)[b]	HNE	DHN
Aldose reductase	GS-HNE	GS-DHN
Aldehyde dehydrogenases	HNE	HNA
	GS-HNE	GS-HNA lactone
Cyp450[c]	HNA-lactone	ω-hydroxy HNA-lactone
β-oxidation enzymes	HNA	CO_2, H_2O
Alkenal oxidoreductase	HNE	4-hydroxynonanal

[a] All human GSTs have measurable activity toward HNE, but the α-class GSTA4-4 and the human GST 5.8 have substrate preference for HNE. These enzymes show about two orders of magnitude higher activity toward HNE than other GST isozymes. References are cited in the text.

[b] Among the AKRs, aldose reductase has higher substrate preference for GS-HNE than HNE.

[c] Cyp4504A family catalyzes this reaction.

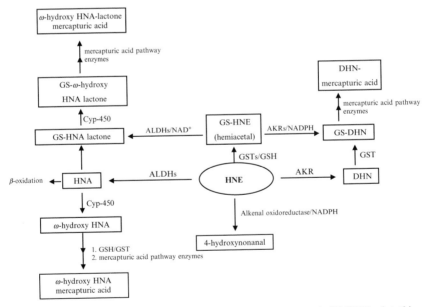

FIG. 1. Metabolic pathways for 4-HNE. HNE, 4-hydroxynonenal; GS-HNE, glutathione conjugate of HNE; HNA, 4-hydroxynonenoic acid; GS-HNA lactone, glutathione conjugate of the HNA lactone; DHN, 1,4,dihydroxynonenol; GS-DHN, glutathione conjugate of DHN; ALDH, aldehyde dehydrogenase; AKR, aldoketoreductases including aldose reductase.

subsequent hemiacetal formation, the conjugated GS–HNE can theoretically exist in eight diastereoisomeric forms. It has been reported that mainly the GS–HNE diastereoisomers derived from R-HNE enantiomer are reduced to GS–DHN, suggesting stereoselectivity of enzymes reducing GS–HNE to GS–DHN. This may have significant physiological relevance because GS–DHN seems to have an important role in NfκB-mediated signaling (Ramana *et al.*, 2003). It may be important to investigate the putative physiological significance of the chirality of HNE on its metabolism and the role of GST, AKR, and ALDH gene families.

Disposition of GSH Conjugates

The GSH conjugates of HNE and its metabolites (e.g., GS–HNA, GS–DHN) are converted to corresponding mercapturic acids (Fig. 2). The major site of mercapturic acid formation is kidney, and some of the cells in humans do not contain a complete battery of mercapturic acid pathway enzymes. Thus, the GSH conjugates need to be transported from cells because their accumulation can impair the function of GST and other detoxification enzymes (Awasthi *et al.*, 1993), resulting in toxicity. GSH conjugates are transported through ATP-dependent primary active

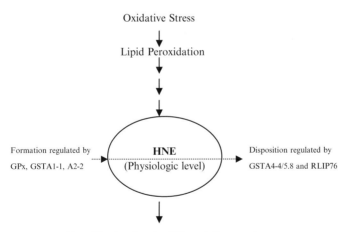

Multispecific signaling by differentially targeting various cellular nucleophiles depending on the intracellular concentration of HNE. Available evidence suggests that the rise in HNE levels promote signaling for apoptosis, while subphysiologic levels promote proliferation.

FIG. 2. Role of HNE and GSTs in the signaling process.

transport system (LaBelle *et al.*, 1986, 1988). It was proposed that multi-drug resistance–associated protein (MRP1) represents the major GSH conjugate (GS-X) efflux pump in cells (Ishikawa, 1992). Recent studies have, however, established that in human tissues and derived cell lines, the Ral-binding protein 1 (Ral BP-1), also designated as RLIP76, is the major transporter of GS-X, and it accounts for about two thirds of the activity mediating the ATP-dependent efflux of GSH conjugates of HNE and xenobiotics (Awasthi *et al.*, 2000, 2002, 2003a,b; Sharma *et al.*, 2001). Studies in our laboratories have clearly demonstrated that up to 70% of GS–HNE and DNP–SG transport from erythrocytes and various cancer cell lines is inhibited by anti-RLIP76, whereas only less than 30% transport is inhibited by anti-MRP1 antibodies (Awasthi *et al.*, 2003a,b; Sharma *et al.*, 2001; Singhal *et al.*, 2003).

RLIP76-mediated transport of GS–HNE and other GSH conjugates from cells, as well as by proteoliposomes reconstituted with homogenous RLIP76 has been demonstrated (Awasthi *et al.*, 2000; Sharma *et al.*, 2002; Singhal *et al.*, 2001). ATP and temperature dependence of this transport have been established, and it has been demonstrated that RLIP76 can mediate transport in liposomes against a concentration gradient. More importantly, it has been shown that the K_m of GS–HNE for RLIP76 is within the estimated physiological concentration of GS–HNE in cells. The physiological significance of the disposition of HNE by coordinated actions of GSTs and RLIP76 is indicated by studies showing that cells overexpressing RLIP76 and GST acquire resistance to HNE and H_2O_2-mediated cytotoxicity and apoptosis (Awasthi *et al.*, 2000; Cheng *et al.*, 2001a,b). RLIP76-induction is one of the early protective responses to oxidative stress and UVA irradiation, and the cells with induced RLIP76 transport GS–HNE at a higher efflux rate acquire resistance to H_2O_2, and UVA cytotoxicity (Yang *et al.*, 2003). Thus, it seems that whereas both MRP1 and RLIP76 mediate ATP-dependent transport of GS–HNE from cells, RLIP76 is the major GS–X efflux pump in most human tissues and coordinated actions of GSTs and RLIP76 seems to be the major mechanism to eliminate HNE from cells (Awasthi *et al.*, 2002).

Measurement of GS–HNE Transport and HNE Levels

Experiments to measure the transport of GS–HNE from cells, membrane vesicles, and proteoliposomes are designed in our laboratory adopting the classical approach used for the functional reconstitution of ion pumps in artificial liposomes. A brief protocol for measuring GS–HNE transport in RLIP76 proteoliposomes is described here.

Reconstitution of RLIP76 into Proteoliposomes

Reconstitution of purified RLIP76 into proteoliposomes is performed by the method of Awasthi *et al.* (2000). The purified RLIP76 is dialyzed against reconstitution buffer (10 mM Tris/HCl, pH 7.4, 100 mM KCl, 1 mM EGTA, 2 mM MgCl$_2$, 40 mM sucrose, 2.8 mM β-mercaptoethanol, and 0.025% (by vol) polidocanol. A mixture of soybean lipids (Asolectin; 40 mg/ml) and cholesterol (10 mg/ml) is prepared in the reconstitution buffer by sonication, and a 100-μl aliquot of this mixture is added to 0.9 ml of dialyzed purified RLIP76 containing 10 μg of protein. The mixture is sonicated for 15 s at 50 W after which 100 mg of SM-2 bio-beads pre-equilibrated with reconstitution buffer (without polidocanol) are added to initiate vesiculation by removal of detergent. After incubation for 4 h at 4°, SM-2 beads are removed by centrifugation. The suitability of the proteo-liposomes for transport measurement is assessed by electron microscopy. In our laboratories, this procedure results in heterogeneous population of vesicles with an average radius of 249 ± 99 nm and an average intravesicular volume of 6.5×10^{-11} μl (Awasthi *et al.*, 1998), and these proteolipo-somes are suitable for transport studies (Awasthi *et al.*, 2000, 2001, 2003a,b; Sharma *et al.*, 2002).

Transport Measurement

ATP-dependent transport of [^{3}H]GS-HNE by RLIP76 proteoliposomes can be measured by the rapid filtration technique described by Awasthi *et al.* (1994). Proteoliposomes are equilibrated at 37° for 5 min in 120 μl of reconstitution buffer containing 10 μM [^{3}H]GS-HNE (100 cpm/pmol). The transport reaction is started by the addition of ATP (final concentration 4 mM), and the reaction mixture is incubated for an appropriate time at 37° with gentle shaking. Control proteoliposomes are incubated with 4 mM 5′ AMP in place of ATP. The reactions are stopped by placing the tubes on ice. Aliquots (30 μl) of the reaction mixture are filtered through nitrocellu-lose membrane (0.45 μm) in a 96-well filtration plate under a uniform suction. The plate is then air-dried, and the filters are cut with a Millipore punch hole assembly. The filters are individually placed in glass scintilla-tion vials containing 10 ml of scintillation fluid, and radioactivity is deter-mined in a scintillation counter (Beckman LS 6800). Measurements are performed in triplicate, and the mean ± SD is determined. Net ATP-dependent transport of GS–HNE is calculated by subtracting vesicle-associated radioactivity in the presence of AMP from that determined in the presence of ATP. The dependence of GS-HNE transport on RLIP76 protein is demonstrated by comparing uptake by protein-containing lipo-somes and protein-free liposomes in the absence and presence of ATP.

Kinetic constants for the ATP-dependent uptake of GS–HNE can be determined by varying the concentration of [^{3}H]-GSH-HNE from 0.5 μM–20 μM, and that of ATP from 0.5–8 mM.

Methods for HNE Measurement

Variations in ROS-induced formation and rapid elimination of HNE from cells make it rather difficult to assess its steady-state concentrations in tissue, and it is, therefore, not surprising that reproducible measurements of HNE in tissues and cell lines are usually difficult. Earlier methods for HNE determination involved use of separation and estimation of its derivatives (e.g., 2,4-dinitrophenyl hydrazone derivative), but in recent years reliable and reproducible methods for separation and quantitation of HNE and its metabolites by GC, HPLC have been developed (Grune et al., 1997; Siems and Grune, 2003). These techniques are highly useful to study the metabolism and elimination of HNE. For example, detailed analysis of HNE-derived compounds in urine demonstrates that most administered HNE in rats is excreted as mercapturic acids derived from GS–HNE, GS–DHN, and GS–HNA or its lactone, which reaffirms a central role for GSTs in HNE-metabolism in vivo (Alary et al., 2003). Likewise, characterization of the polar mercapturic acids in urine provides convincing evidence that HNE and/or HNA are substrates for Cyp-450, and ω-hydroxylation of HNE-metabolites is involved in its pathways for elimination. More importantly, changes in HNE homeostasis during stress conditions or during the pathological conditions associated with oxidative stress can be accurately assessed by comparing HNE levels. For example, with these methods, it has been shown (Grune et al., 1997; Siems et al., 2002) that under chronic oxidative stress conditions such as those exist in various diseases, including rheumatoid arthritis, systemic sclerosis or lupus erythematous, chronic lymphedema, or chronic renal failure, a 3–10-fold change in plasma HNE levels is observed (Siems and Grune, 2003). These chromatographic methods are very useful and have immensely contributed to the understanding of the physiological/pathophysiological significance of HNE. A brief description of these methods is given in the following.

Chromatographic Analysis of HNE and Its Metabolites

Both gas chromatography (GC) and high-performance liquid chromatography (HPLC) could be used for quantitative analysis of HNE and its metabolites. For GC analysis because of limitation of volatility, the samples of HNE, DHN, and HNA need to be derivatized, whereas their GSH conjugates and mercapturic acid are not suitable to GC analysis because of their low volatility. For HPLC, there is no such limitation, and all these metabolites could be analyzed including R- and S-isomers of HNE. The derivatization, HNE, DHN, and HNA for GC analysis could be achieved by using N-methyl-N-(t-butyldimethylsilyl)-trifluoacetamide in diethyl ether and then can be resolved on capillary DB-5–fused silica column using the conditions described by Srivastava et al. (1998). The peaks can be analyzed either by flame ionization detector on the basis of their retention time or by mass spectrometer as has been used by Srivastava et al. (1998). The advantage of mass analysis is that it provides the molecular weight and fragmentation pattern by which the chemical structures could be confirmed. HNE and its metabolites could be separated by HPLC using a reversed-phase C18 column and a solvent gradient of acetonitrile in 0.1% trifluoroacetic acid (Cheng et al., 2001b; Srivastava et al., 1998). The effluent can be monitored at 224 and 254 nm. Individual peaks can be identified by retention time compared with standards. Effluent corresponding to peak area can be collected and analyzed by mass spectrophotometry (Cheng et al., 2001b; Srivastava et al., 1998). R- and S-isomers of HNE could also be resolved on a chiral column. A racemic mixture of HNE needs to be derivatized to oxime using methyl oxime in pyridine and then analyzed at chiralpak AD column and hexane/ethanol mixture as the eluting solvent with detection at 235nm (Schneider et al., 2004).

Spectrophotometric Methods for HNE Determination

In many laboratories focusing on the biological activities of HNE, the sophisticated instruments such as MS-ES, GC, and HPLC may not be available for routine use. To address this problem, in recent years several spectrophotometric assays involving chromophores that react with HNE and other α, β-unsaturated aldehydes to form adducts having suitable absorption wave lengths and extinction coefficients have been developed and are available commercially as kits. Among these kits, the Bioxytech® LPO-586™ for lipid peroxidation products is widely used. In our laboratory, this kit is routinely used for comparing HNE and MDA levels in tissues/cells, particularly in experiments in which the effect of heat, oxidative stress, and UVA on HNE homeostasis and its effect on signaling mechanisms are examined. The principle of the assay is based on reactivity of N-methyl-2-phenylindole, which reacts with MDA or HNE to yield a stable chromophore with absorption maxima at 586 nm. The details of standard curves, for this assay, detection limits, and potential pitfalls of the assay are provided in the manufacturers' protocol along with ready-to-use stock solutions of the reagents and standards of HNE and MDA. In our laboratory, this method seems to be highly reproducible and offers significant advantage over the classical method used to determine TBARS (Yagi et al.,

1998) as an index of lipid peroxidation. It is to be pointed out that the methods for spectroscopic determination for HNE are being continually developed, refined, and updated by various commercial companies, and interested investigators should keep abreast of the newer commercial kits suitable to meet their needs. A detailed procedure for determining HNE/MDA using Bioxytech® LPO-586™ as followed routinely in our laboratory is described in the following.

In a typical experiment, wild-type, vector transfected, and *hGSTA1*-transfected K562 cells were collected by centrifugation at 500g for 10 min (1 × 10^7 cells) and after washing twice with PBS were resuspended in 0.2 ml of 20 mM Tris-HCl, pH 7.4, containing 5 mM BHT (buffer A). To each sample, 650 μl of N-methyl-2-phenylindole and 150 μl of either 12 N HCl (for MDA determination) in 15.4 M methanesulfonic acid (for 4-HNE plus MDA determination) were added. The reaction mixture was vortexed and incubated for 60 min at 45°. The samples were then centrifuged at 15,000g for 10 min, and the absorbance of the supernatants was determined at 586 nm. The extinction coefficients for MDA and 4-HNE prepared by the hydrolysis of 1,1,3,3-tetramethoxypropane in HCl and of HNE-diethylacetal in methanesulfonic acid, respectively (1.1 × 10^5 M^{-1}cm^{-1} and 1.3 × 10^5 M^{-1}cm^{-1}, respectively) determined from the standard curves, and the values were expressed as pmol of 4-HNE/mg of protein (Yang *et al.*, 2001). Cell samples, after collection and processing, can be stored at –70° for determinations at convenient times. This method can also be used for tissue samples. An equivalent amount of protein present in approximately 10^7–10^8 cells is suitable for obtaining reproducible results. It is critical that the estimated amount of HNE/MDA falls within the linear range of HNA + MDA and MDA concentrations in the standard curves. The needed methods for establishing standard curves and standard samples are provided with the kit supplied by the manufacturers.

HNE-Mediated Signaling Processes

Historically, lipid peroxidation has been associated with toxicity. For example, the toxicity of various xenobiotics and oxidants (e.g., CCl$_4$, doxorubicin, paraquat, and various quinonoids) has been attributed to lipid peroxidation initiated and propagated by these agents. Even though it has been long speculated that lipid peroxidation products may be physiologically relevant, little or no concrete information existed on the physiological significance of lipid peroxidation before Esterbauer and his colleagues identified HNE and showed its role in the cellular signaling mechanism (Esterbauer *et al.*, 1991, review). In recent years, there has been a tremendous interest in the role of HNE as a second signaling molecule, and the number of publications on HNE has grown exponentially. The chemostatic activity on neutrophils (Curzio *et al.*, 1982) was perhaps the first reported physiological effect of HNE. Since then, numerous biological effects of HNE showing its involvement in the regulation of cell cycle events, proliferation, differentiation, apoptosis, and expression of important cell cycle regulatory genes such as P53, c-myc and other oncogenes, PKC, phospholipase C, tyrosine kinase receptor (TKR), retinoblastoma protein (pRb), MAPK, JNK, heat shock proteins, cyclins, and integrin have been demonstrated (Awasthi *et al.*, 2003, 2004; Barrera *et al.*, 2004; Dianzani *et al.*, 2003, reviews). There is ample evidence that HNE is an important second messenger molecule in stress-mediated signaling. This idea is consistent with our recent studies showing that HNE is a common denominator in heat, UVA, oxidative stress, and xenobiotic-induced apoptosis of various cell types in culture (Cheng *et al.*, 2001b, Yang *et al.*, 2001, 2003).

Being a potent alkylating agent, HNE can react with a variety of nucleophilic sites of DNA, proteins, and lipids to form corresponding HNE adducts. Results from different laboratories using a variety of cell lines convincingly demonstrate that inclusion of HNE in cell cultures can activate SAPK/JNK, a member of the MAPK family that is involved in apoptosis, perhaps through alternate mechanisms in different cells. In hepatic stellate cells, HNE seems to activate JNK by direct binding and not through phosphorylation (Parola *et al.*, 1998) but it seems to activate JNK through the redox-sensitive MAPK kinase cascade in other cell lines (Cheng *et al.*, 2001a,b; Uchida *et al.*, 1999; Yang *et al.*, 2001, 2002, 2003). HNE-induced activation of JNK promotes its translocation in the nucleus where JNK-dependent phosphorylation of c-Jun and the transcription factor activator protein (AP-1) binding take place, leading to the transcription of a number of genes having AP-1 consensus sequences in their promoter regions (Camandola *et al.*, 2000). Activation of JNK by 4-HNE is accompanied by the activation of caspase 3 and eventual apoptosis (Camandola *et al.*, 2000; Cheng *et al.*, 2001a; Yang *et al.*, 2004). HNE-induced activation of p38 MAPK has been suggested to occur because of its ability to induce COX-2 through stabilization of its mRNA (Uchida *et al.*, 2003). HNE can modulate expression of various genes, including PKC βII (Chiarpotto *et al.*, 1999), c-myc (Fazio *et al.*, 1992), procollagen type I (Parola *et al.*, 1993), aldose reductase (Spycher *et al.*, 1996), c-myb (Barrera *et al.*, 1996), and transforming growth factor β1 (Fazio *et al.*, 1992). In HL-60 cells, physiological concentrations of 4-HNE inhibit proliferation and induce a granulocyte-like differentiation (Barrera *et al.*, 1996). As discussed before, the "physiological" concentrations of HNE vary from cell to cell, depending on the extent of lipid peroxidation and levels of antioxidant enzymes including GSTs, Thus, the preceding reported effects of HNE are likely to vary from cell to cell. For example, unpublished studies in our laboratory have shown that whereas K562 cells rapidly undergo differentiation when exposed to 10 μM 4-HNE, under similar conditions other cells such as HLE B-3 and RPE remain unaffected.

Concentration-Dependent Effect of 4-HNE

Interestingly, HNE seems to affect the signaling processes in a concentration-dependent manner. Although numerous studies have shown that 4-HNE is proapoptotic, it has been shown to stimulate proliferation of some cell types (Barrera et al., 1991; Ruef et al., 1998) at relatively lower concentrations. In K562 cells, HNE induces differentiation at moderate concentrations and apoptosis at relatively high concentrations (Cheng et al., 1999, 2001a). HNE inhibits PKCα activity of rat hepatocytes in culture, but it activates PKCβ isoforms at lower concentrations, whereas at higher concentrations, it has an inhibitory effect (Marinari et al., 2003). HNE also has a biphasic effect on tyrosine kinase receptor (RTK) signaling. A transient exposure to low levels of HNE leads to autophosphorylation of EGFR and PDGFR, resulting in an activation of downstream signaling, which may regulate proliferation in smooth muscle cells. Exposure to a higher concentration of HNE for longer periods causes inhibition of RTK-mediated signaling and other pathways (Negre-Salvary et al., 2003). The concentration-dependent effect of HNE on cell cycle events is further suggested by recent studies demonstrating that overexpression of GSTA4-4 in attached cells, HLE B-3 and CCL-75, leads to their transformation and immortalization, and this activity is dependent on the depletion of intracellular 4-HNE by GSTA4-4 (Sharma et al., 2004). Lowering of HNE levels in these cells seems to profoundly affect the expression of key cell cycle genes. Our unpublished studies with mGSTA4-null mice show that cell cycle regulatory gene p53 is up-regulated because of higher 4-HNE levels in tissues of these mice. These studies are consistent with a significant physiological role for HNE and suggest that GSTs are modulators of HNE-mediated signaling processes and may regulate cell cycle events by affecting the homeostasis of 4-HNE in cells.

It has been suggested that the differential response of HNE to signaling pathways may be due to its interaction with multiple targets among the components of various signaling cascades, which may ultimately determine the final outcome in the form of suppression, activation, or arrest of specific cellular functions (Nakashima et al., 2003). HNE can conjugate with proteins through cysteine, lysine, and histidine and affect a wide number of biological processes as discussed previously. Differential interactions with target molecules in a concentration-dependent manner may explain apparently opposite effects of HNE, depending on its intracellular concentration. If this conjecture is true, then differential activity of the three reactive groups of HNE (i.e., the Michael acceptor group, the carbonyl group, and the hydroxyl group) toward the target groups in proteins, DNA, and membrane lipids may provide the mechanistic basis for multiple and often confounding effects of HNE on cell cycle arrest, proliferation, and programmed cell death in different cell lines (Awasthi *et al.*, 2004).

GSTs as Modulators of HNE-Induced Signaling

As discussed previously, HNE seems to affect cell cycle events and various signaling cascades in a concentration-dependent manner. This has led to the suggestion that the intracellular concentration of HNE may be a major determinant of whether the cell would undergo proliferation, programmed cell death, differentiation, or growth arrest. Therefore, the factors governing HNE homeostasis in cells may modulate signaling pathways by regulating the generation and disposition of HNE in cells. By and large, formation of HNE occurs by uncontrolled processes (e.g., unforeseen exposure to oxidants and stress), and its intracellular concentrations must be regulated by its metabolism. Because GST-mediated conjugation of HNE is the major pathway for HNE metabolism, GSH and GSTs should be the main determinants of HNE concentration in cells. Because GS–HNE needs to be transported from cells because its accumulation would inhibit GSTs, the enzymes catalyzing the transport of GS–HNE should also play a role in regulating HNE homeostasis. Studies in our laboratories during the past few years have focused on the possible role of GSTs in the regulation of HNE-mediated signaling for apoptosis, proliferation, differentiation, and gene expression.

Catalytic Activity of Various GSTs for HNE

Because of the presence of a Michael acceptor group, HNE can react with GSH to form the GS–HNE conjugate. However, HNE is a substrate for GSTs (Alin *et al.*, 1985), and, perhaps, most HNE in cells is conjugated to GSH through GST-catalyzed reaction. If GSTs can, indeed, affect HNE signaling, the isozymes with high activity toward HNE may play an important regulatory role. Most of the GST isozymes investigated for their activity toward HNE show variable activity toward this substrate (Table II). A subgroup of the α-class GSTs identified in mammals shows about 2 orders of magnitude higher activity toward HNE and other

TABLE II
ACTIVITIES OF DIFFERENT GLUTATHIONE S-TRANSFERASES TOWARD 4-HYDROXYNONENAL

Source	Isoenzyme	Specific Activity (μmoles/mg/min)	Reference
Human	hGST 5.8	168–200	Singhal et al., 1994a,b
	hGSTA4-4	189	Hubastch et al., 1998
	Mixture of μ class of GSTs	3.2	Singhal et al., 1994b
	Mixture of hGSTA1-1, A2-2, A3-3	0.56	Singhal et al., 1994b
	hGST P1-1	0.56–1.6	Singhal et al., 1994b
Mouse	mGSTA4-4	55–65	Zimniak et al., 1994
Rat	rGSTA4-4	170	Sternberg et al., 1992
			Hubastch et al., 1998
Drosophila	DmGSTD1-1	32	Sawicki et al., 2003

endogenously generated α, β-unsaturated carbonyls compared with other GST isozymes (Zimniak et al., 1992). For example, rat GSTA4-4 (Sterberg et al., 1992), mouse GSTA4-4 (Zimniak et al., 1994), and human GSTA4-4 (Hubatsch et al., 1998) have high activity toward HNE, whereas the human μ- and π-class GSTs show only minimal activity (Singhal et al., 1994b). A GST isozyme with substrate preference for HNE has also been characterized from the fruit fly (Drosophila melanogaster) by Sawicki et al. (2003). It seems that in aerobic organisms, this group of HNE-metabolizing isozymes provides defense against HNE and other α, β-unsaturated carbonyls. In humans, two immunologically distinct GST isozymes, hGSTA4-4 and hGST5.8, with remarkably high catalytic efficiency ($K_{cat}/K_m > 2000$ S^{-1} mM^{-1}) are present (Cheng et al., 2001c; Hubatsch et al., 1998; Singhal et al., 1994a). hGSTA4-4, cDNA has been cloned, but the complete primary structure of hGST5.8 is still unknown and is provisionally designated as hGST5.8 because of its isoelectric focusing point being close to 5.8. Kinetic studies with tissue-purified hGST5.8 show its high substrate preference for HNE. The enzyme is recognized by antibodies raised against mouse and rat GSTA4-4 but not by hGSTA4-4 antibodies. Even under mild stress conditions, hGST5.8 is induced (Cheng et al., 2001b), indicating an important role of this and other HNE-metabolizing GST isozymes in the regulation of HNE homeostasis.

Other isozymes besides GSTA4-4 also contribute to HNE metabolism in mammals. For example, the isozymes GSTA1-1 and GSTA2-2 not only mitigate HNE formation by reducing hydroperoxides, these isozymes can also catalyze conjugation of HNE to GSH, even though this activity is up to two magnitudes lower than GSTA4-4. Likewise, mammalian GSTs belonging to the μ and π classes also have significant measurable activity toward

HNE. Although the specific activities of these enzymes are much less than GSTA4-4 or hGST5.8, it is important to note that GSTA1-1, A2-2 and the π and μ class GSTs constitute the bulk of GST protein in mammalian tissues. Therefore, these isozymes also contribute substantially to the metabolism of HNE. This is indicated by recent studies showing that in the tissues of *mGSTA4* null mice, approximately 30% of the activity for conjugating HNE to GSH is retained (Engles *et al.*, 2004). Considering numerous recent studies showing that an increase in HNE concentrations leads to apoptosis in various cell lines, it was expected that *mGSTA* null mice might have severe pathogenesis. However, these mice show an almost normal phenotype, indicating that other GST isozymes compensate for the loss of *mGSTA4* (Engles *et al.*, 2004). It is interesting to note that all the major enzymes (e.g., GSTs, ALDHs, AKRs) involved in HNE metabolism are members of multigene families and have varying degrees of activity toward HNE and other endogenous α, β-unsaturated carbonyls. The redundancy and overlapping activity of these enzymes perhaps provides backup mechanisms for regulating HNE homeostasis and underscores the role of HNE in physiological/pathophysiological processes.

Overexpression of α-Class GSTs Protects Against Stress-Induced Apoptosis

Overexpression of GST isozymes with substrate preference of HNE results in lower HNE levels in cells (Hé *et al.*, 1996). Likewise, overexpression of GSTA1-1 or GSTA2-2, which attenuate lipid peroxidation, leads to lower HNE levels in cells (Yang *et al.*, 2001). Therefore, we have overexpressed GSTA1-1, GSTA2-2, or GSTA4-4 and compared stress-induced signaling events in these cells and mock transfected cells. The conjugates of GSH with electrophiles including GS-HNE are transported from cells by RLIP76 and MRP in an ATP-dependent manner, and it has been shown that in human cells investigated so far, RLIP76 accounts for more than two thirds of the total GS-HNE transport (Sharma *et al.*, 2001). Therefore, we also examined whether HNE-induced signaling for apoptosis can be accentuated by blocking GS–HNE transport by anti-RLIP76 antibodies. These studies summarized in the following strongly suggest that HNE and perhaps other lipid peroxidation products act as second messengers for stress-mediated signaling that can be regulated by GSTs.

Our earlier studies demonstrated that when mGSTA4-4 was overexpressed in Chinese hamster ovary cells, it provided protection against doxorubicin-induced lipid peroxidation and oxidant cytotoxicity (Hé *et al.*, 1996; Zimniak *et al.*, 1997). This was consistent with the role of GSTs in detoxification of HNE, the toxic end product of lipid peroxidation. In mGSTA4-transfected K562 cells, up to 90% depletion of cellular HNE was observed (Cheng et al., 1999). These cells with lower HNE levels grew at an accelerated pace as suggested by their doubling time, which was approximately two thirds of the wild-type or vector-transfected cells. These studies suggested that HNE could be involved in cell cycle events, which was consistent with earlier studies showing proliferation of vascular smooth muscle cells by low concentrations of HNE (Ruef et al., 1998). Both K562 and HL60 cells treated with 10–20 μm HNE undergo a substantial degree of apoptosis within 2–8 h. K562 cells treated with HNE also show erythroid differentiation as indicated by expression of hemoglobin (Cheng et al., 1999, 2001a). These effects of HNE can be blocked by GSTA4-4 overexpression. Other GST isozymes with only minimal activity toward HNE do not have such an effect, suggesting HNE to be the causative factor. GSTA4-4 overexpression in these cells does not affect the expression of the antioxidant enzymes including SOD, catalase, and GPx. Blockage of H_2O_2-induced apoptosis and differentiation in GSTA4-4 overexpressing cells suggests that H_2O_2 (ROS)-induced signaling for apoptosis is transduced through HNE, at least in part.

Oxidative Stress-Induced Apoptosis Can Be Blocked by Limiting HNE Formation

The role of lipid peroxidation products in general, and of HNE in particular, in the signaling pathways is also suggested by our studies in which the overexpression of GSTs has been used to suppress lipid peroxidation in the cultured cells. The α-class GSTs A1-1 and A2-2 have relatively high glutathione peroxidase (GPx) activity and are capable of GSH-dependent reduction (Zhao et al., 1999) of fatty acid (FA-OOH) and phospholipid hydroperoxides (PL-OOH). We have shown that transfection of HLE-B3, K562, and other cells with GST A1-1 or GST A2-2 protect these cells from lipid peroxidation and oxidant toxicity (Yang et al., 2001, 2002). In cells transfected with hGSTA1 or hGSTA2, oxidative stress–induced lipid peroxidation is prevented, and the MDA and HNE levels remain unaffected in the transfected cells even after treatment with H_2O_2 and Fe^{2+}, suggesting a major role of GSTs A1-1 and A2-2 in preventing ROS-induced lipid peroxidation. GSTA1-1 or GSTA2-2 overexpressing K562 and HLE-B3 cells are resistant to H_2O_2-induced apoptosis. Likewise, these cells are resistant to PL-OOH– and FA-OOH–induced apoptosis. On the other hand, HNE (10– 20 μm) causes apoptosis in both the wild-type and GSTA1-1 or GSTA2-2 overexpressing cells. These findings clearly indicate a role of lipid peroxidation products in H_2O_2-induced apoptosis in these cells, because neither GSTA1-1 nor GSTA2-2 can reduce H_2O_2 and use only organic hydroperoxides (e.g., PL-OOH, FA-OOH) as substrates (Zhao et al., 1999). The findings that both PL-OOH and FA-OOH cause apoptosis in these cells and that it can be blocked by GSTA1-1/2-2 overexpression further suggests a role of lipid hydroperoxides or their downstream products in the cascade of lipid peroxidation reaction in the mechanism of H_2O_2 (ROS)-induced signaling for programmed cell death. HNE is formed from the cleavage of PL-OOH. Therefore, it is not surprising that GSTA1-1/2-2 overexpression does not protect cells from HNE-induced apoptosis, because these isozymes have only minimal activity to conjugate HNE to GSH.

Oxidative Stress-Induced Activation of JNK and Caspase Can Be Blocked by GSTs

A sustained activation of c-Jun N-terminal kinase leading to activation of caspase 3 and eventual apoptosis is one of the more understood pathways among various apoptotic-signaling pathways (Cheng et al., 2001a; Uchida et al., 1999). During H_2O_2-induced apoptosis in the wild-type or vector-transfected K562 cells, a sustained activation of JNK and caspase 3 is observed, but in GSTA1-1/2-2 overexpressing cells, H_2O_2 fails to activate JNK or caspase 3, suggesting that prevention of lipid peroxidation by these GST isozymes blocks H_2O_2-induced apoptosis by inhibiting the activation of JNK and caspase 3 (Yang et al., 2001, 2003). Oxidants, such as naphthalene and doxorubicin, generate ROS by redox cycling from quinonoid intermediates and cause lipid peroxidation in cells. These agents along with other oxidants can induce apoptosis, presumably through ROS. We have shown that GSTA1-1/A2-2–overexpressing HLE-B3 cells are protected against naphthalene-induced JNK activation, caspase 3 activation, and apoptosis (Yang et al., 2002), whereas the wild-type or empty vector–transfected cells undergo apoptosis through activation of JNK and caspase 3.

HNE, a Common Denominator in Stress-Mediated Signaling for Apoptosis

Induction of GST and RLP76, an Early Response to Stress

In a series of experiments (Cheng et al., 2001b; Yang et al., 2003), K562 cells were exposed to mild stress conditions of heat (42°, 30 min), UVA (357nm, 5 min), or H_2O_2 (50 μM, 20 min), and HNE levels were measured immediately after exposure. Stress caused a 40–50% increase in the HNE levels. Reasoning that a certain constitutive HNE level is present in cells and the fact that higher than the physiological levels of HNE causes toxicity, we studied whether the enzymes responsible for HNE disposition were induced as one of the early responses to stress. After allowing the cells to rest under normal serum conditions to have time for expression of relevant genes, cell extracts were subjected to Western blot analysis for the antioxidant enzymes including catalase, superoxide dismutase, Se-dependent glutathione peroxidase 1, glutathione reductase, glucose-6-phosphate dehydrogenase, GSTs, and the heat shock protein hsp70. None of these proteins showed any significant induction, except that a specific GST isozyme hGST 5.8 was induced about 10-fold in the stress-conditioned cells. This isozyme has remarkably high catalytic efficiency for conjugating HNE to GSH (Singhal et al., 1994a). Gamma-glutamate cysteine ligase (GCL), which is the key enzyme in the GSH biosynthetic pathway, was also induced severalfold, suggesting that in response to mild stress, cells acquire enhanced capability to conjugate HNE without a limitation of GSH availability. GCL is known to be induced by HNE (Dickinson et al., 2002). In the stressed cells, RLIP76, which catalyzes the transport of GS–HNE, was also induced by threefold to fivefold. On loading the cells with [^{3}H] HNE, the efflux of GS-[^{3}H]HNE from the stress-preconditioned cells was approximately threefold higher than the control cells (without any stress preconditioning), indicating that the induction of hGST5.8 and RLIP76 leads to accelerated disposition of HNE from cells. More importantly, these findings suggest that one of the initial responses of the cells to stress is to acquire enhanced capability to exclude HNE from the cellular environment and that this adaptive response occurs even before the induction of prototypical stress response proteins such as HSP70 or antioxidant enzymes. On sustained stress conditions (e.g., heat shock for 3 h) induction of hGST5.8 and RLIP76 was ablated, and severalfold induction of HSP70 was observed, indicating that whereas the initial response of the cells is to limit the increase of HNE levels, cells acquire other defense mechanisms to combat stress before being eventually overwhelmed by its adverse effects. Thus, it seems in response to stress-induced overproduction of HNE, the first response of the cells is to acquire the capacity to more efficiently dispose of HNE. When these studies were extended to several other human cell lines in culture, similar results were obtained. RPE (retinal pigment epithelium) cells and HLE-B3 (human lens epithelial) cells, which grow as attached cells, showed severalfold induction of hGST5.8 and RLIP76 on mild stress preconditioning. Likewise, two lung cancer cell lines that grow in suspension also showed remarkable induction of hGST5.8 and RLIP76 in response to mild transient stress. All these stress-preconditioned cells transported GS-HNE at an accelerated pace compared with the untreated control, suggesting this stress response to be common among these cells.

Stress Preconditioning Confers Resistance to Apoptosis by UVA and Oxidative Stress

If the mechanisms to maintain HNE homeostasis are geared up to protect cells from apoptosis caused by rising HNE levels, the stress preconditioned cells should acquire resistance to apoptosis caused not only by

HNE but also by the apoptotic agents that cause lipid peroxidation and HNE formation. This prediction was confirmed, because our studies (Cheng *et al.*, 2001b; Yang *et al.*, 2003) showed that stress-preconditioned cells were relatively much more resistant to apoptosis caused by HNE as well as by H_2O_2, superoxide anion and UVA, all of which cause lipid peroxidation and HNE formation. More importantly, the cells preconditioned with mild UVA showed resistance to apoptosis caused by H_2O_2, superoxide anion as well as by HNE, which strongly suggests that HNE is a common denominator in the signaling pathway for apoptosis by these agents. This is further supported by studies showing that the activation of the JNK and caspase 3 by these stress agents is also blocked in apoptosis-resistant preconditioned cells but not in the control cells that undergo apoptosis (Yang *et al.*, 2003). These results are consistent with the idea that HNE plays a key role in stress signaling pathways.

Inhibition of GS–HNE Transport Sensitizes Stress-Preconditioned Cells to Stress-Induced Apoptosis

If an accelerated elimination of HNE in stress-preconditioned cells because of an early induction of hGST5.8 and RLIP76 is the cause of the acquisition of resistance against stress-induced apoptosis, then this resistance should be compromised by blocking the elimination of HNE from cells. GS–HNE formation is catalyzed by GST, but it has been shown that GSH conjugates are potent inhibitors of GSTs (Awasthi *et al.*, 1993). To sustain detoxification of HNE, GS–HNE must be transported from cells. The efflux of GS–HNE can be blocked by coating the cells with polyclonal antibodies against RLIP76 (Sharma *et al.*, 2001). The resistance of preconditioned cells to HNE and stress-induced apoptosis was compromised when these cells were coated with anti RLIP76 IgG to inhibit GS–HNE efflux from the cells (Cheng *et al.*, 2001b; Yang *et al.*, 2003), which further indicated an important role of HNE in signaling pathways for apoptosis and that its accumulation in cells caused apoptosis. HNE is continually formed in cells because of the presence of ROS generated in the metabolic processes and also because of exogenous stimuli (e.g., exposure to xenobiotics). Its disposition, therefore, is crucial, and a block in this process could lead to apoptosis as indicated by our studies showing that anti RLIP76 IgG, *per se*, can cause apoptosis without treatment with other apoptotic agents (Awasthi *et al.*, 2003b). Regulation of HNE homeostasis, therefore, seems to be an important factor in stress-mediated signaling, and the intracellular concentration of HNE (of which GSTs are primary regulators) may be important in determining whether cells undergo apoptosis, differentiation, or proliferation.

Stress caused by heat, oxidants, irradiations, infection, or any other factors can induce lipid peroxidation and the formation of degradation products. HNE is perhaps the most widely studied product of lipid peroxidation. Its role in signaling mechanisms is supported by voluminous studies, only a portion of which are cited in this article. However, the question of whether HNE occupies a central position in the pathway(s) of stress-mediated signaling is still open, because even though HNE may be a common denominator in signaling for apoptosis by heat, UV, superoxide anions, and chemicals, it can be argued that by depletion of GSH, HNE can cause oxidative stress leading to ROS, which may cause apoptosis. Consistent with this idea, HNE-induced generation of oxides in cells has been demonstrated (Uchida *et al.*, 1999). A counter argument can, however, be made that the generation of ROS may be the consequence of HNE-induced apoptosis and that the stress-induced apoptosis signaling is in fact transduced through HNE (Yang *et al.*, 2003).

Depletion of HNE by GSTA4-4 in HLE-B3 and CCL-75 Cells Leads to Transformation

Whether HNE is a central player in stress-induced apoptosis or not, studies discussed previously clearly show that HNE does, indeed, modulate numerous signaling cascades. Our recent studies (Sharma *et al.*, 2004) have demonstrated that alterations in the intracellular concentrations of HNE can introduce irreversible phenotypic changes in some cells. HLE-B3 cells, which grow as attached cells with a limited life span (Andley *et al.*, 1994), are transferred into round smaller cells that grow indefinitely in suspension on transfection with *hGSTA4* cDNA. Even though the overexpression of hGSTA4-4 in these cells leads to HNE depletion only by approximately 50–60%, within 8 weeks, the clones overexpressing the enzyme are transformed into rounded, smaller cells, with a phenotype distinct from the parent cells and acquires immortality. This dramatic effect of hGSTA4-4 overexpression depends on the depletion of HNE, because cells transfected with mutant hGSTA4 (Y212F) that has no activity toward HNE maintain their original phenotype. The requirement of HNE depletion for the transformation of the cells is also indicated by experiments in which cytosolic microinjection of active hGSTA4-4 causes transformation within 72 h, but microinjection of inactive mutant GSTA4-4 (Y212F) does not cause such transformation. Microinjection of the GST isozyme of *Drosophila* (Dm GSTD1-1) having high activity toward HNE (Sawicki *et al.*, 2003) also transformed the cells, but human GSTs, hGSTA1-1, and hGSTP1-1, which are known to have minimal activity toward HNE, did not cause such a transformation. Similar transformation of lung fibroblast cells (CCL-75) on hGSTA4-4 overexpression (Sharma *et al.*, 2004), further underscores the role of HNE metabolizing GSTs in the signaling processes. These studies indicate that disturbances in HNE homeostasis can profoundly affect the signaling processes and suggest that lower than the basal constitutive levels of HNE favor proliferation, whereas higher levels tend to favor programmed cell death.

Depletion of HNE by GSTA4-4 in HLE-B3 Causes Profound Changes in the Expression of Genes

Our (unpublished) studies show that in the transformed HLE-B3 cells, the expression of TGFβ, ERK, PKCβ, CDK-2, and c-myc is up-regulated, whereas the expression of p53, Bad, Fas, connexin 43, fibronectin, laminin-γ-1, and integrin α-6 is almost completely suppressed. In general, this alteration in the expression of genes seems to be consistent with the phenotype of the transformed cells that grow as suspended cells and acquire immortality. It is important to note that all these changes in gene expression and altered phenotype stringently correlated with the depletion of cellular HNE, because hGSTA4-4 mutant (Y212F), which has no activity toward HNE (but maintains its activity toward 1-chloro-2,4-dinitrobenzene), does not affect either the phenotype or gene expression in HLE-B3 cells. These studies also suggest that alteration in HNE homeostasis profoundly affects signaling and propagates extensive cross-talks between various signaling cascades as indicated by the dramatic phenotypic changes in HLE-B3 and CCL-75 observed by changing only a single parameter (i.e., change in the expression of a HNE metabolizing GST isozyme). Implications of these findings could perhaps be enormous and should be explored to unravel the mystery of the concentration-dependent effects of HNE on the cellular processes.

The mechanisms through which 4-HNE affects signaling processes in a concentration-dependent manner seem to be complex. 4-HNE is a strong electrophile that can react with nucleophilic groups of proteins (Uchida et al., 1993; Yoritaka et al., 1996), nucleic acids (Chung et al., 2000; Hu et al., 2002), and lipids (Guichardant et al., 1998). It can interact with thiols and also with nucleophilic nitrogen atoms in proteins and phospholipids with varying affinity. It may be postulated that, at low concentrations, 4-HNE may selectively affect pathways favoring proliferation by interacting with nucleophilic targets that have a high affinity for this compound, and such interactions favor proliferation. On the other hand, at higher concentrations of 4-HNE, the effects of these interactions may be overwhelmed by reactions with low-affinity groups of cellular nucleophiles to trigger the pathways favoring apoptosis. Further studies on the chemical interaction of 4-HNE with cellular nucleophiles, including proteins, nucleic acids, and lipids, and the possible correlation between these interactions and signaling cascades may provide clues to the mechanisms by which 4-HNE affects signaling events in cells.

Acknowledgment

Supported in part by NIH Grants EY04396, ES012171 to Y. C. A., CA77495 to S. A., and ES11584 to G. A. S. A. The help of Dr. Shaheen Dhanani, MD, in the preparation of the manuscript is gratefully acknowledged.

References

Alary, J., Geuraud, F., and Cravedi, J. P. (2003). Fate of 4-hydroxynonenal *in vivo*: Disposition and metabolic pathways. *Mol. Aspects Med.* **24,** 177–187.

Alary, J., Bravis, F., Cravedi, J. P., Debrauwer, L., Rao, D., and Bories, G. (1995). Mercapturic acids as urinary end-metabolites of lipid peroxidation product 4-hydroxynonenal in rat. *Chem. Res. Toxicol.* **8,** 34–39.

Alin, P., Danielson, U. H., and Mannervik, B. (1985). 4-Hydroxyalk-2-enals are substrates for glutathione transferase. *FEBS Lett.* **179,** 267–270.

Andley, U. P., Rhim, J. S., Chylack, L. T., Jr., and Fleming, T. P. (1994). Propagation and immortalization of human lens epithelial cells in culture. *Invest. Ophthalmol. Vis. Sci.* **35,** 3094–3102.

Awasthi, S., Singhal, S. S., Cheng, J. Z., Zimniak, P., and Awasthi, Y. C. (2003a). Role of RLIP76 in lung cancer doxorubicin resistance II. *Int. J. Oncol.* **22,** 713–720.

Awasthi, S., Singhal, S. S., Singhal, J., Yang, Y., Zimniak, P., and Awasthi, Y. C. (2003b). Anti-RLIP76 antibodies trigger apoptosis in lung cancer cells and synergistically increase doxorubicin cytotoxicity III. *Int. J. Oncol.* **22,** 721–732.

Awasthi, S., Sharma, R., Singhal, S. S., Zimniak, P., and Awasthi, Y. C. (2002). RLIP76, a novel transporter catalyzing ATP-dependent efflux of xenobiotics. *Drug Metab. Disposition* **30,** 1300–1310.

Awasthi, S., Cheng, J. Z., Singhal, S. S., Pandya, U., Sharma, R., Singh, S. V., Zimniak, P., and Awasthi, Y. C. (2001). Functional reassembly of ATP-dependent xenobiotic transport by the N- and C-terminal domains of RLIP76 and identification of ATP binding sequences. *Biochemistry* **40,** 4159–4168.

Awasthi, S., Cheng, J. Z., Singhal, S. S., Saini, M. K., Pandya, U., Pikula, S., Bandorowicz-Pikula, J., Singh, S. V., Zimniak, P., and Awasthi, Y. C. (2000). A novel function of human RLIP76: ATP-dependent transport of glutathione conjugates and doxorubicin. *Biochemistry* **39,** 9327–9334.

Awasthi, S., Singhal, S. S., Pikula, S., Piper, J. T., Srivastava, S.K, Torman, R. T., Bandorowicz-Pikula, J., Zimniak, P., and Awasthi, Y. C. (1998). ATP–dependent human erythrocyte glutathione-conjugate transporter; functional reconstitution of transport activity. *Biochemistry* **37,** 5239–5248.

Awasthi, S., Srivastava, S. K., Piper, J. T., Singhal, S. S., Chaubey, M., and Awasthi, Y. C. (1996). Curcumin protects against 4-hydroxynonenal induced cataract formation in rat lenses. *Am. J. Clin. Nutr.* **64,** 761–766.

Awasthi, S., Singhal, S. S., Srivastava, Sanjay, K., Zimniak, P., Bajpai, K. K., Saxena, M., Sharma, R., Ziller, S. A., Frankel, E., Singh, S. V., Hé, N. G., and Awasthi, Y. C. (1994). ATP-dependent transport of doxorubicin, daunomycin, and vinblastine in human tissues by a mechanism distinct from P-glycoprotein. *J. Clin. Invest.* **93,** 958–965.

Awasthi, S., Srivastava, S. K., Ahmad, F., Ahmad, H., and Ansari, G. A. (1993). Interaction of glutathione S-transferase-pi with ethacrynic acid and its glutathione conjugate. *Biochim. Biophys. Acta* **1164,** 173–178.

Awasthi, Y. C., Yang, Y., Tiwari, N. K., Patrick, B., Sharma, A., Li, J., and Awasthi, S. (2004). Regulation of 4-hydroxynonenal mediated signaling by glutathione S-transferases. *Free Radic. Biol. Med.* **37,** 607–619.

Awasthi, Y. C., Sharma, R., Cheng, J. Z., Yang, Y., Sharma, A., Singhal, S. S., and Awasthi, S. (2003). Role of 4-hydroxynonenal in stress-mediated apoptosis signaling. *Mol. Aspects Med.* **24,** 219–230.

Awasthi, Y. C., Singhal, S. S., and Awasthi, S. (1996). Mechanisms of anticarcinogenic effects of antioxidant nutrients. *In* "Nutrition and Cancer Prevention" (R. R. Watson and S. I. Mufti, eds.), pp. 139–172. CRC Press Inc., Boca Raton.

Barrera, G., Pizzimenti, S., and Dianzani, M. U. (2004). 4-Hydroxynonenal and regulation of cell cycle: Effects on pRb/E2F pathway. *Free Radic. Biol. Med.* **37,** 597–606.

Barrera, G., Pizzimenti, S., Serra, A., Ferretti, C., Fazio, V. M., Saglio, G., and Dianzani, M. U. (1996). 4-Hydroxynonenal specifically inhibits c-myb but does not affect c-fos expressions in HL-60 cells. *Biochem. Biophys. Res. Commun.* **227,** 589–593.

Barrera, G., Brossa, O., Fazio, V. M., Farace, M. G., Paradisi, L., Gravela, E., and Dianzani, M. U. (1991). Effects of 4-hydroxynonenal, a product of lipid peroxidation, on cell proliferation and ornithine decarboxylase activity. *Free Radic. Res. Commun.* **14,** 81–89.

Berhane, K., Widersten, M., Angstrom, A., Kozarich, J. W., and Mannervik, B. (1994). Detoxification of base propenals and other α, β- unsaturated aldehyde products of radical reactions and lipid peroxidation by glutathione S-transferases. *Proc. Natl. Acad. Sci. USA* **91,** 1480–1484.

Booth, J., Bayland, E., and Sims, P. (1961). An enzyme from rat liver catalyzing conjugation with glutathione. *Biochem. J.* **79,** 516–521.

Camandola, S., Poli, G., and Mattson, M. P. (2000). The lipid peroxidation product 4-hydroxy-2,3-nonenal increases AP-1 binding activity through caspase activation in neurons. *J. Neurochemistry* **74,** 159–168.

Cheng, J. Z., Singhal, S. S., Sharma, A., Saini, M., Yang, Y., Awasthi, S., Zimniak, P., and Awasthi, Y. C. (2001a). Transfection of mGSTA4 in HL-60 cells protects against 4-hydroxynonenal–induced apoptosis by inhibiting JNK-mediated signaling. *Arch. Biochem. Biophys.* **392,** 197–207.

Cheng, J. Z., Sharma, R., Yang, Y., Singhal, S. S., Sharma, A., Saini, M. K., Singh, S. V., Zimniak, P., Awasthi, S., and Awasthi, Y. C. (2001b). Accelerated metabolism and exclusion of 4-hydroxynonenal through induction of RLIP76 and hGST5.8 is an early adaptive response of cells to heat and oxidative stress. *J. Biol. Chem.* **276,** 41213–41223.

Cheng, J. Z., Yang, Y., Singh, S. P., Singhal, S. S., Awasthi, S., Pan, S. S., Singh, S. V., Zimniak, P, and Awasthi, Y. C. (2001c). Two distinct 4-hydroxynonenal metabolizing glutathione S-transferase isozymes are differentially expressed in human tissues. *Biochem. Biophys. Res. Commun.* **282,** 1268–1274.

Cheng, J. Z., Singhal, S. S., Saini, M. K., Singhal, J., Piper, J. T., Van Kuijk, F. J., Zimniak, P., Awasthi, Y. C., and Awasthi, S. (1999). Effects of mGST A4 transfection on 4-hydroxynonenal–mediated apoptosis and differentiation of K562 human erythroleukemia cells. *Arch. Biochem. Biophys.* **372,** 29–36.

Chiarpotto, E., Domenicotti, C., Paola, D., Vitali, A., Nitti, M., Pronzato, M. A., Biasi, F., Cottalasso, D., Marinari, U. M., Dragonetti, A., Cesaro, P., Isodoro, C., and Poli, G. (1999). Regulation of rat hepatocyte protein kinase C beta isoenzymes by the lipid peroxidation product 4-hydroxy-2,3-nonenal: A signaling pathway to modulate vesicular transport of glycoproteins. *Hepatology* **29,** 1565–1572.

Chung, F. L., Nath, R. G., Ocando, J., Nishikawa, A., and Zhang, L. (2000). Deoxyguanosine adducts of t-4-hydroxy-2-nonenal are endogenous DNA lesions in rodents and humans: Detection and potential sources. *Cancer Res.* **60,** 1507–1511.

Coles, B. F., and Kadlubar, F. F. (2003). Detoxification of electrophilic compounds by glutathione S-transferase catalysis. Determinants of individual response to chemical carcinogens and chemotherapeutic drugs. *Biofactors* **17,** 115–130.

Curzio, M., Torrielli, M. V., Geroud, J. P., Esterbauer, H., and Dianzani, M. U. (1982). Neutrophil chemotactic response to aldehydes. *Res. Commun. Chem. Pathol. Pharmacol.* **36**, 463–472.

Dianzani, M. U. (2003). 4-Hydroxynonenal from pathology to physiology. *Mol. Aspects Med.* **24**, 263–272.

Dick, R. A., Kwak, M. K., Sutter, T. R., and Kensler, T. W. (2001). Antioxidative functions and substrate specificity of NAD (P) H-dependent alkenal/one oxidoreductase a new role for leukotriene B-4 12-hydroxydehydrogenase/15-oxoprostaglandin 13-reducatse. *J. Biol. Chem.* **276**, 40803–40810.

Dickinson, D. A., Iles, K. E., Watanabe, N., Iwamoto, T., Zhang, H., Karzywanski, D. M., and Foreman, H. J. (2002). 4-Hydroxynonenal induces glutamate cysteine ligase through JNK in HBE1 cells. *Free Radic. Biol. Med.* **33**, 974–987.

Engles, M. R., Singh, S. P., Czernik, P. J., Gaddy, D., Montague, D., Ceci, J. P., Yang, Y., Awasthi, S., Awasthi, Y. C., and Zimniak, P. (2004). Physiological role of mGSTA4-4, a glutathione S-transferase metabolizing 4-hydroxynonenal: Generation and analysis of mGsta4 null mouse. *Toxicol. Appl. Pharmacol.* **194**, 296–308.

Fazio, V. M., Barrera, G., Martinotti, S., Farace, M. G., Giglioni, B., Frati, L., Manzari, V., and Dianzani, M. U. (1992). 4-Hydroxynonenal, a product of cellular lipid peroxidation, which modulates c-myc and globin gene expression in K5622 erythroleukemic cells. *Cancer Res.* **52**, 4866–4871.

Gueraud, F., Alary, J., Costet, P., Debrauwer, L., Dolo, L., Pineau, T., and Paris, A. (1999). *In vivo* involvement of cytochrome P450 4A family in the oxidative metabolism of the lipid peroxidation product trans-4 hydroxy-2-nonenal, using PPAR α-deficit mice. *J. Lipid Res.* **40**, 152–159.

Grune, T., Michel, P., Sitte, N., Eggert, W., Albert-Nebe, H., Esterbaur, H., and Siems, W. G. (1997). Increased levels of 4-hydroxynonenal modified proteins in plasma of children with autoimmune disease. *Free Radic. Biol. Med.* **23**, 357–360.

Guichardant, M., Taibi-Tronche, P., Fay, L. B., and Lagarde, M. (1998). Covalent modifications of aminophospholipids by 4-hydroxynonenal. *Free Radic. Biol. Med.* **25**, 1049–1056.

Hayes, J. D., Flangan, J. U., and Jorsey, I. R. (2004). Glutathione transferases. *Annu. Rev. Pharmacol. Toxicol.* **45**, 51–88.

Hayes, J. D., and Pulford, D. J. (1995). The glutathione S-transferase super gene family: Regulation of GST and contribution of the isozymes to cancer chemoprevention and drug resistance. *Crit. Rev. Biochem. Mol. Biol.* **30**, 445–600.

Hé, N. G., Singhal, S. S., Chaubey, M., Awasthi, S., Zimniak, P., Patridge, C. A., and Awasthi, Y. C. (1996). Purification and characterization of a 4-hydroxynonenal metabolizing glutathione S-transferase isozyme from bovine pulmonary microvessel endothelial cells. *Biochim. Biophys. Acta* **1291**, 182–188.

Hu, W., Feng, Z., Eveleigh, J., Iyer, G., Pan, J., Amin, S., Chung, F. L., and Tang, M. S. (2002). The major lipid peroxidation product, trans-4-hydroxy-2-nonenal, preferentially forms DNA adducts at codon 249 of human p53 gene, a unique mutational hotspot in hepatocellular carcinoma. *Carcinogenesis* **11**, 1781–1789.

Hubatsch, I., Ridderstrom, M., and Mannervik, B. (1998). Human glutathione S-transferase A4-4; An alpha class enzyme with high specificity for 4-hydroxynonenal and other genotoxic products of lipid peroxidation. *Biochem. J.* **330**, 175–179.

Ishikawa, T. (1992). The ATP-dependent glutathione S-conjugate pump. *Trends Biochem. Sci.* **17**, 463–468.

Jakoby, W. B. (1978). The isozymes of glutathione transferase. *Adv. Enzymol. Rel. Areas Mol. Biol.* **57**, 383–414.

LaBelle, E. F., Singh, S. V., Ahmed, H., Wronski, L., Srivastava, S. K., and Awasthi, Y. C. (1988). A novel S-dinitrophenylglutathione stimulated ATPase is present in human erythrocyte membrane. *FEBS Lett.* **228**, 53–56.

LaBelle, E. F., Singh, S. V., Srivastava, S. K., and Awasthi, Y. C. (1986). Dinitrophenyl glutathione efflux from human erythrocytes is primarily active ATP-dependent transport. *Biochem. J.* **238**, 443–449.

Mannervik, B., and Danielson, U. H. (1989). Glutathione S-transferases—structure and catalytic activity. *CRC Crit. Rev. Biochem. Mol. Biol.* **23**, 283–337.

Marinari, U. M., Nitti, M., Pronzato, M. A., and Domenicotti, C. (2003). Role of PKC-dependent pathways in HNE-induced cell protein transport and secretion. *Mol. Aspects Med.* **24**, 205–211.

Nakashima, I., Liu, W., Akhand, A. A., Takeda, K., Kawamoto, Y., Kato, M., and Suzuki, H. (2003). 4-Hydroxynonenal triggers multistep signal transduction cascades for suppression of cellular functions. *Mol. Aspects Med.* **24**, 231–238.

Negre-Salvayre, A., Vieira, O., Escargueil-Blane, I., and Salvayre, R. (2003). Oxidized LDL and 4-hydroxynonenal modulate tyrosine kinase receptor activity. *Mol. Aspects Med.* **24**, 251–261.

Parola, M., Robino, G., Marra, F., Pinzano, M., Bellomo, G., Leonarduzzi, G., Chiarugi, P., Camandola, S., Poli, G., Waeg, G., Gentilini, P., and Dianzani, M. U. (1998). HNE interacts directly with JNK isoforms in human hepatic stellate cells. *J. Clin. Invest.* **102**, 1942–1950.

Parola, M., Pinzani, M., Casini, A., Albano, E., Poli, G., Gentilini, A., Gentilini, P., and Dianzani, M. U. (1993). Stimulation of lipid peroxidation or 4-hydroxynonenal treatment increases procollagen alpha 1 (I) gene expression in human liver fat-storing cells. *Biochem. Biophys. Res. Commun.* **194**, 1044–1050.

Ramana, K. V., Friedrich, B., Bhatnagar, A., and Srivastava, S. K. (2003). Aldose reductase mediates cytotoxic signals of hyperglycemia and TNF-alpha in human lens epithelial cells. *FASEB J.* **17**, 315–319.

Ramana, K. V., Chandra, D., Srivastava, S., Bhatnagar, A., Aggarwal, B. B., and Srivastava, S. K. (2002). Aldose reductase mediates mitogenic signaling in vascular smooth muscle. *J. Biol. Chem.* **277**, 32063–32070.

Reichard, J. F., Vasiliou, V., and Petersen, D. R. (2002). Characterization of 4-hydroxy-2-metabolism in stellate cell lines derived from normal and cirrhotic rat liver. *Biochim. Biophys. Acta* **1487**, 222–232.

Ruef, J., Rao, G. N., Li, F., Bode, C., Patterson, C., Bhatnagar, A., and Runge, M. S. (1998). Induction of rat aortic smooth muscle cell growth by the lipid peroxidation product 4-hydroxy-2-nonenal. *Circulation* **97**, 1071–1078.

Salinas, A. E., and Wang, M. G. (1999). Glutathione S-transferases—A review. *Curr. Med. Chem.* **6**, 279–309.

Sawicki, R., Singh, S. P., Mondal, A. K., Benes, H., and Zimniak, P. (2003). Cloning, expression and biochemical characterization of one Epsilon-class (GST-3) and ten delta class (GST-1) glutathione S-transferases from *Drosophila melanogaster* and identification of additional nine members of the Epsilon class. *Biochem. J.* **370**, 661–669.

Sayre, L. M., Zealsko, D. A., Harris, P. L. A., Perry, G., Saloman, R. G., and Smith, M. A. (1997). 4- Hydroxynonenal-derived advanced lipid peroxidation products are increased in Alzheimer's disease. *J. Neurosci.* **35**, 1335–1344.

Schaur, R. J. (2003). Basic aspects of the biochemical reactivity of 4-hydroxynonenal. *Mol. Aspects Med.* **24**, 149–159.

Schneider, C., Porter, N. A., and Brash, A. R. (2004). Autooxidative transformation of chiral $\omega6$ hydroxy linoleic and arachidonic acids to chiral 4-hydroxy-2E-nonenal. *Chem. Res. Toxicol.* **17**, 937–941.

406   PHASE II: GLUTATHIONE TRANSFERASES   [24]

Schneider, C., Tallman, K. A., Porter, N. A., and Brash, A. R. (2001). Two distinct pathways
    of formation of 4-hydroxynonenal and of the 9- and 13-hydroperoxides of linoleic acid to
    4-hydroxyalkenals. *J. Biol. Chem.* **276,** 20831–20838.
Sharma, R., Brown, D., Awasthi, S., Yang, Y., Sharma, A., Patrick, B., Saini, M. K., Singh,
    S. P., Zimniak, P., Singh, S. V., and Awasthi, Y. C. (2004). Transfection with
    4-hydroxynonenal-metabolizing glutathione S-transferase isozymes leads to phenotypic
    transformation and immortalization of adherent cells. *Eur. J. Biochem.* **271,** 1690–1701.
Sharma, R., Sharma, A., Yang, Y., Singhal, S. S., Awasthi, S., Zimniak, P., and Awasthi, Y. C.
    (2002). Functional reconstitution of RLIP76 in proteoliposomes catalyzing transport of
    glutathione-conjugate of 4-hydroxynonenal. *Acta Biochimica. Polonica* **49,** 693–701.
Sharma, R., Singhal, S. S., Cheng, J. Z., Yang, Y., Sharma, A., Zimniak, P., Awasthi, S., and
    Awasthi, Y. C. (2001). RLIP76 is the major transporter of glutathione-conjugate and
    doxorubicin in human erythrocytes. *Arch. Biochem. Biophys.* **391,** 71–79.
Sheehan, D., Meade, G., Foley, V. M., and Dowd, C. A. (2001). Structure function and
    evolution of glutathione S-transferases: Implications for classification of non-mammalian
    members of an ancient enzyme super family. *Biochem. J.* **360,** 1–16.
Siems, W., and Grune, T. (2003). Intracellular metabolism of 4-hydroxynonenal. *Mol. Aspects
    Med.* **24,** 167–175.
Siems, W., Carluccio, F., Grune, T., Jackstad, M., Quast, S., Hampl, H., and Sommerberg, O.
    (2002). Elevated serum concentration of cardiotoxic lipid peroxidation products in chronic
    renal failure in reaction of degree of renal anemia. *Clin. Nephrol.* **58**(suppl. 1), 20–25.
Siems, W., Grune, T., and Esterbaur, E. (1995). 4-Hydroxynonenal formation during ischemia
    and reperfusion of rat small intestine. *Life Sci.* **57,** 785–789.
Singhal, S. S., Singhal, J., Sharma, R., Singh, S. V., Zimniak, P., Awasthi, Y. C., and Awasthi, S.
    (2003). The APTase activity of RLIP76 correlates with doxorubicin and 4-hydroxynonenal
    resistance in lung cancer cells I. *Int. J. Oncol.* **22,** 365–375.
Singhal, S. S., Singhal, J., Cheng, J., Pikula, S., Sharma, R., Zimniak, P., Awasthi, Y. C., and
    Awasthi, S. (2001). Purification and functional reconstitution of intact rat-binding GTPase
    activating protein RLIP76 in artificial liposomes. *Acta Biochim. Polonica* **48,** 551–562.
Singhal, S. S., Zimniak, P., Sharma, R., Srivastava, S. K., Awasthi, S., and Awasthi, Y. C.
    (1994a). A novel glutathione S-transferase isozyme similar to GST 8-8 of rat is selectively
    expressed in human tissues. *Biochim. Biophys. Acta* **1204,** 279–286.
Singhal, S. S., Zimniak, P., Awasthi, S., Piper, J. J., Hé, N. G., Teng, J. I., Petersen, D. R., and
    Awasthi, Y. C. (1994b). Several closely related Glutathione S-transferase isozymes
    catalyzing conjugation of 4-hydroxynonenal are differentially expressed in human tissues.
    Arch. Biochem. Biophys. **311,** 242–250.
Spycher, S., Tabataba-Vakili, S., O'Donnell, V. B., Palomba, L., and Azzi, A. (1996).
    4-Hydroxy-2,3-trans-nonenal induces transcription and expression of aldose reductase.
    Biochem. Biophys. Res. Commun. **226,** 512–516.
Srivastava, S., Chandra, A., Wang, L. F., Seifert, W. E., Jr., Dague, B. B., Ansari, N. H.,
    Srivastava, S. K., and Bhatnagar, A. (1998). Metabolism of the lipid peroxidation product
    4-hydroxy-trans-2 nonenal, in isolated perfused rat heart. *J. Biol. Chem.* **273,** 197–205.
Srivastava, S., Chandra, A., Bhatnagar, A., Srivastava, S. K., and Ansari, N. H. (1995).
    Lipid peroxidation product 4-hydroxynonenal and its conjugate with GSH are excell-
    ent substrates of bovine lens aldose reductase. *Biochem. Biophys. Res. Commun.* **217,** 741–746.
Sternberg, G., Ridderström, M., Engström, Å., Pemlble, S. E., and Mannervik, B. (1992).
    Cloning and heterologous expression of cDNA encoding class Alpha rat glutathione
    transferase 8-8, an enzyme with high catalytic activity towards genotoxic α, β-unsaturated
    carbonyl compounds. *Biochem. J.* **284,** 313–319.

Townsend, A. T., Leonekabler, S., Haynes, R. L., Wu, Y. H., Szweda, L., and Bunting, K. D. (2001). Selective protection by stable transfected human Aldh3A1(but not human Aldh1A1) against toxicity of aliphatic aldehydes in V79 cells. *Chemico-Biol. Interact.* **130**, 261–273.

Uchida, K., and Kumagai, T. (2003). 4-Hydroxy-2-nonenal as a COX-2 inducer. *Mol. Aspects Med.* **24**, 213–218.

Uchida, K., Shiraishi, M., Naito, Y., Torii, Y., Nakamura, Y., and Osawa, T. (1999). Activation of stress signaling pathways by the end product of lipid peroxidation. 4-Hydroxy-2-nonenal is a potential inducer of intracellular peroxide production. *J. Biol. Chem.* **274**, 2234–2242.

Uchida, K., Szweda, L. I., Chae, H. Z., and Stadtman, E. R. (1993). Immunochemical detection of 4-hydroxynonenal protein adducts in oxidized hepatocytes. *Proc. Natl. Acad. Sci. USA* **90**, 8742–8746.

Yagi, K. (1998). Simple procedure for specific assay of lipid hydroperoxides in serum or plasma. *Free Radic. Antioxidant Protocols* **108**, 101–106.

Yang, Y., Yang, Y., Trent, M. B., Hé, N. G., Lick, S. D., Zimniak, P., Awasthi, Y. C., and Boor, P. J. (2004). Glutathione-S-transferase A4-4 modulates oxidative stress in endothelium: Possible role in human atherosclerosis. *Atherosclerosis* **173**, 211–221.

Yang, Y., Sharma, A., Patrick, B., Singhal, S. S., Zimniak, P., Awasthi, S., and Awasthi, Y. C. (2003). Cells preconditioned with mild UVA irradiation acquire resistance to oxidative stress and UVA-induced apoptosis. *J. Biol. Chem.* **278**, 41380–41388.

Yang, Y., Sharma, R., Cheng, J. Z., Saini, M. K., Ansari, N. H., Andley, U., and Awasthi, Y. C. (2002). Protection against hydrogen peroxide and naphthalene induced lipid peroxidation and apoptosis by transfection with hGSTA1 and hGSTA2. *Invest. Ophthalmol. Vis. Sci.* **43**, 434–445.

Yang, Y., Cheng, J. Z., Singhal, S. S., Saini, M., Pandya, U., Awasthi, S., and Awasthi, Y. C. (2001). Role of glutathione S-transferases in protection against lipid peroxidation. Over expression of hGSTA2-2 in K562 cells protects against hydrogen-peroxide induced apoptosis and inhibits JNK and caspase 3 activation. *J. Biol. Chem.* **276**, 19220–19230.

Yoritaka, A., Hattori, N., Uchida, K., Tanaka, M., Stadtman, E. R., and Mizuno, Y. (1996). Immunohistochemical detection of 4-hydroxynonenal protein adducts in Parkinson disease. *Proc. Natl. Acad. Sci. USA* **93**, 2696–2701.

Zhao, T. T., Singhal, S. S., Piper, J. T., Cheng, J. Z., Pandya, U., Clark-Woronski, J., Awasthi, S., and Awasthi, Y. C. (1999). The role of human glutathione S-transferases hGSTA1-1 and hGSTA2-2 in protection against oxidative stress. *Arch. Biochem. Biophys.* **367**, 216–224.

Zimniak, L., Awasthi, S., Srivastava, S. K., and Zimniak, P. (1997). Increased resistance to oxidative stress in transfected cultured cells overexpressing glutathione S-transferase mGSTA4-4. *Toxicol. Appl. Pharmacol.* **143**, 221–229.

Zimniak, P., Singhal, S. S., Srivastava, S. K., Awasthi, S., Sharma, R., Hayden, J. B., and Awasthi, Y. C. (1994). Estimation of genomic complexity and enzymatic characterization of mouse glutathione S-transferase mGSTA4-4 (GST5.7). *J. Biol. Chem.* **269**, 992–1000.

Zimniak, P., Eckles, M. A., Saxena, M., and Awasthi, Y. C. (1992). A subgroup of alpha class glutathione S-transferases. Cloning of cDNA for mouse lung glutathione S-transferase GST 5.7. *FEBS Lett.* **313**, 173–176.

Further Reading

Chasseaud, L. F. (1979). The role of glutathione and glutathione S-transferases in metabolism of chemical carcinogens and other electrophilic agents. *Adv. Cancer Res.* **29**, 175–274.

Esterbaur, H., Schaur, R. J., and Zollner, H. (1991). Chemistry and biochemistry of 4-hydroxynonenal, malonaldehyde and related aldehydes. *Free Radic. Biol. Med.* **23**, 81–128.

[25]   Gene Expression of γ-Glutamyltranspeptidase

By YOSHITAKA IKEDA and NAOYUKI TANIGUCHI

Abstract

γ-Glutamyltranspeptidase is a heterodimeric glycoprotein that cata-lyzes the transpeptidation and hydrolysis of the γ-glutamyl group of gluta-thione and related compounds. It is known that the enzyme plays a role in the metabolism of glutathione and in salvaging constituents of glutathione. In the adult animal, high levels of γ-glutamyltranspeptidase are constitu-tively expressed in the kidney, intestine, and epididymis. On the other hand, although γ-glutamyltranspeptidase is up-regulated in the liver during the perinatal stage, its expression is nearly undetectable in the adult. In addition, it has long been observed that the intake of certain xenobiotics, including carcinogens and drugs, induces the hepatic expression of the enzyme. This induction seems to be associated with both transcriptional regulation and the growth of certain types of cells in the injured liver. A number of studies have been carried out to explain the mechanism by which γ-glutamyltranspeptidase expression is regulated. 5'-Untranslated regions of mRNAs of the enzyme differ in a tissue-specific manner but share a common protein coding region, and the tissue-specific and devel-opmental stage–specific expression, as well as hepatic induction, are con-ferred by different promoters. As suggested by the capability of enzymatic activity–independent induction of osteoclasts, the expression of γ-gluta-myltranspeptidase may also be involved in various biological processes that are not directly associated with glutathione metabolism. This chapter brief-ly summarizes studies to date concerning the tissue-specific expression and induction of γ-glutamyltranspeptidase and transcriptional regulation by the multiple promoter system is discussed.

γ-Glutamyltranspeptidase and Glutathione Metabolism

Mammalian γ-glutamyltranspeptidase (EC 2.3.2.2) is a membrane-bound glycoprotein that consists of two polypeptide chains composed of a small and a large subunit. The protein is anchored to the cell surface by means of an N-terminal hydrophobic transmembrane domain. The enzyme catalyzes the hydrolysis of a γ-glutamyl group of glutathione, a unique tripeptide that is involved in a variety of biological events such as intracel-lular redox regulation, drug metabolism, and the detoxification and supply

METHODS IN ENZYMOLOGY, VOL. 401
Copyright 2005, Elsevier Inc. All rights reserved.
0076-6879/05 $35.00
DOI: 10.1016/S0076-6879(05)01025-6

of cysteine (Hughey and Curthoys, 1976; McIntyre and Curthoys, 1979; Meister, 1992; Meister and Larsson, 1995; Meister and Tate, 1976; Meister *et al.*, 1981; Taniguchi and Ikeda, 1998; Tate and Meister, 1985). The enzyme is capable of hydrolyzing glutathione-related compounds such as leukotriene C, prostaglandins, and several γ-glutamyl amino acids and also catalyzes a transpeptidation reaction in which a γ-glutamyl group is transferred from glutathione to dipeptides and amino acids (Anderson *et al.*, 1982; Cagen *et al.*, 1976).

The biosynthesis of glutathione is intracellularly initiated by the action of γ-glutamylcysteine synthetase, which conjugates glutamate and cysteine in an adenosine triphosphate (ATP)-dependent manner (Huang *et al.*, 1993a; 1993b; Orlowski and Meister, 1971), and the resulting dipeptide is further ligated with glycine by glutathione synthetase (Gali and Board, 1996; Huang *et al.*, 1995; Oppenheimer *et al.*, 1979). In contrast to intracellular synthesis, glutathione is metabolized to the constituent amino acids in the extracellular space by γ-glutamyltranspeptidase, the amino acids of which are incorporated into the cells and become available for the synthesis of glutathione (Meister, 1973; Meister and Larsson, 1995). The primary role of this enzyme seems to be the degradation of glutathione into glutamate and cysteinyl glycine, the latter of which is further hydrolyzed into cysteine and glycine by dipeptidase and is generally believed to be involved in the cellular recovery of cysteine and the other two amino acids, as indicated by studies using knock-out mice (Lieberman *et al.*, 1996). It was found that the lack of γ-glutamyltranspeptidase gives rise to the nutritional deficiency of cysteine, and dysfunction as the result of the deficiency is, at least in part, recovered by providing cystine. Because the chemistry and biological consequences of this are described in other chapters, this chapter will not deal with these topics.

Biosynthesis of γ-Glutamyltranspeptidase

This enzyme is translated in the form of a single precursor that is catalytically inactive (Tate, 1986) and then processed into the active heterodimeric form (Nash and Tate, 1982, 1984). This cleavage has recently been proposed to be an autocatalytic process, as has been suggested for the bacterial enzyme (Hiratake *et al.*, 2002; Inoue *et al.*, 2000; Suzuki and Kumagai, 2002), although it was previously thought that the precursor protein of GGT is processed by a putative processing protease. In addition to the processing of the bacterial enzyme, the autocatalytic processing of the single-chain precursor to a heterodimeric form seems reasonable on the basis of the structural analogy with the N-terminal hydrolase class of enzymes (Brannigan *et al.*, 1995) such as proteasome and lysosomal glycosyl asparaginase, which is also known as aspartyl glucosaminidase. Although it has been reported that HepG2 cells express an active single polypeptide form of γ-glutamyltranspeptidase (Tate and Galbraith, 1987, 1988), there may be different mechanisms for conferring proteolytic processing in mammalian enzymes, which remains to be examined in more detail.

Tissue-Specific Expression of γ-Glutamyltranspeptidase

γ-Glutamyltranspeptidase activity is relatively higher in the kidney, intestine, and epididymis among tissues, and the enzyme is constitutively expressed in these tissues (Meister et al., 1981). The enzyme is the most abundant in the proximal tubule of the kidney. The expression of γ-glutamyltranspeptidase is regulated in a tissue-specific manner, and it seems that the difference in the expression of the enzyme is closely associated with differences in glutathione metabolism.

In general, the activity of the enzyme is relatively high in the luminal surface of ductal tissues. The distribution profile seems to be consistent with the suggestion that the enzyme is frequently expressed in tissues that are associated with the transport of various biological compounds. As shown by studies using γ-glutamyltranspeptidase–deficient mice (Lieberman et al., 1996), in the kidney γ-glutamyltranspeptidase plays a critical role in the recovery of cysteine from urinary glutathione, and the lack of γ-glutamyltranspeptidase leads to the excretion of large amounts of cysteine in the urine. Because cysteine content, as glutathione, is larger than the free form, the availability of this amino acid for protein synthesis and other metabolic pathways is greatly dependent on glutathione as the source rather than free cysteine and cystine (Higashi et al., 1977, 1983; Tateishi et al., 1977, 1980). Thus, the nutrient roles of glutathione have often been emphasized, and the reabsorption of the amino acid from urine occurs by the action of kidney γ-glutamyltranspeptidase (Hanigan, 1995; Hanigan and Ricketts, 1993; Lieberman et al., 1996). Such a critical role of the enzyme has not been intensively investigated in other tissues and organs and, thus, is not fully understood. Similarly, as indicated by the evidence that most cells are not able to uptake intact forms of both reduced and oxidized glutathione (Anderson and Meister, 1989; Puri and Meister, 1983), it seems more likely that γ-glutamyltranspeptidase would enable a fraction of cyst(e)ine, which is contained in extracellular glutathione, to be available to cells. This possible increase in availability seems to be associated with drug resistance in some cancer cells, in which γ-glutamyltranspeptidase, as well as γ-glutamylcysteine synthetase, is up-regulated (Godwin et al., 1992).

In contrast to tissues such as the kidney and intestine, it is well known that the activity of γ-glutamyltranspeptidase is low and nearly undetectable in the adult liver (Alberts *et al.*, 1970; Hanigan and Pitot, 1985; Iannaccone and Koizumi, 1983; Tateishi *et al.*, 1976). It is known that glutathione is synthesized by almost all mammalian cells examined to date, and, among the tissues examined, the liver produces glutathione in the largest quantities. This tissue is relatively unique in terms of the absence of activity with respect to degrading glutathione. When the activity of γ-glutamyltranspeptidase was carefully examined in the rodent liver, a very low activity was detected. It is likely that such a detectable, but low, activity in the liver would be displayed by intrahepatic biliary cells but not hepatocytes (Iannaccone and Koizumi, 1983; Shiozawa *et al.*, 1990; Suzuki *et al.*, 1987).

Although γ-glutamyltranspeptidase activity is very low in the adult liver, a transient increase in activity is observed in the liver at the perinatal stage during development (Iannaccone and Koizumi, 1983; Tateishi *et al.*, 1976; Tsuchida *et al.*, 1979). On the other hand, the activity of the enzyme in the kidney is initially low in the neonate. The enzyme level increases as the neonate grows and reaches a maximal level at maturity (Tateishi *et al.*, 1976; Tsuchida *et al.*, 1979). In both rats and mice, although fetal liver hepatocytes express the enzyme at significant levels, its expression is suppressed after birth. In conjunction with gene suppression, the activity becomes lower and finally decreases to undetectable levels. Although the profiles for the alteration in activity during embryonic development seem to vary with the species and strain, it has generally been thought that the glucocorticoid hormone plays a role in occurrence of the transient increase (Cotariu *et al.*, 1988; Lambiotte *et al.*, 1973; Tongiani *et al.*, 1984). It has been observed that corticosterone levels in fetal serum at the latest stage of embryo are significantly increased, and the fetal liver is able to respond to exogenously administered hydrocortisone in terms of the increase in activity of γ-glutamyltranspeptidase activity. In the fetal liver, the activity of the enzyme is the highest in the periportal area (zone 1) and modest in the midzone (zone 2), whereas the activity is absent in the pericentral area (zone 3) (Tongiani and Paolicchi, 1989; Tongiani *et al.*, 1984). The increase in activity seems to be closely associated with biliary ductular morphogenesis, as suggested by the report showing that the addition of glucocorticoid is required for bile duct formation in *in vitro* cultures of hepatocytes isolated from fetal liver (Lambiotte *et al.*, 1973).

Bilateral adrenalectomy in the adult female Wister rat led to a slight decrease, to almost 50%, in the hepatic activity of γ-glutamyltranspeptidase, and this decrease was recovered by treating the rat with hydrocortisone, suggesting that the expression of γ-glutamyltranspeptidase in the liver might be regulated by glucocorticoids (Billon *et al.*, 1980). The difference in responsiveness between adult and fetal animals could be due to the numbers of cells in which the expression of γ-glutamyltranspeptidase is up-regulated in response to the steroid hormone. In 12-day-old and 21-day-old rats, *in vivo* treatment with hydrocortisone induced an increase in the activity of γ-glutamyltranspeptidase, up to almost 30-fold, in the liver, whereas no increase was observed in the cases of the 28-day-old and mature rats (Tongiani *et al.*, 1984).

Because the transient elevation in the activity of γ-glutamyltranspeptidase coincides with the active growth of cells in the fetal liver, it was proposed that the growth rate of the cells is associated with the levels of γ-glutamyltranspeptidase activity. However, such a relationship does not necessarily seem to be consistent in various types of cells, as indicated, for example, by studies using a series of Morris hepatomas (Fiala *et al.*, 1976; Richards *et al.*, 1982).

Induction of Hepatic γ-Glutamyltranspeptidase

Even in the liver of adult animals, however, the enzyme activity is elevated in conjunction with an increase in mRNA levels in response to the administration or intake of xenobiotics, carcinogens, and alcohol. Nearly 30 years ago, it was found that γ-glutamyltranspeptidase activity in the rat liver is elevated on hepatocarcinogenesis by the long-term administration of 3'-methyl-4-dimethyl-amino azobenzene despite being undetectable in normal liver (Fiala and Fiala, 1969). Further studies have found that the activity of γ-glutamyltranspeptidase in the liver is also increased by a variety of other carcinogens: 4-dimethyl-aminoazobenzene (Taniguchi *et al*, 1974), diethylnitrosamine (Solt *et al.*, 1977), 2-N-fluorenylacetamide (Laishes *et al.*, 1978), aflatoxin B1 (Kalengayi *et al.*, 1975), N-methyl-N-nitrosourea (Solt *et al.*, 1980), and 7,12-dimethylbenz[a]anthracene (Pereira *et al.*, 1983). These inducing effects were investigated on the basis of an increase in enzyme activity and protein level as determined by histochemistry and enzyme immunoassay (Suzuki *et al.*, 1987; Taniguchi *et al.*, 1983). As reported later for some of the compounds, it seemed likely that such effects are mostly due to increased mRNA levels (Deguchi and Pitot, 1995; Power *et al.*, 1987; Yan *et al.*, 1998). A histochemical study showed that the increase in hepatic γ-glutamyltranspeptidase is due to both an increase of biliary cell numbers, which intrinsically express the enzyme, and the emergence of unique γ-glutamyltranspeptidase-positive cells (Daoust, 1982).

As shown in earlier studies, the induction of γ-glutamyltranspeptidase in the rat liver by various chemicals exhibits temporally a biphasic alteration (Manson, 1983; Taniguchi *et al.*, 1974). When rats are fed a carcinogen such as 3′-methyl-4-dimethylaminoazobenzene, the activity is initially increased after about 7 weeks and is reduced thereafter. After further 3–4 weeks, however, the reduced activity is re-increased, being associated with hepatocarcinogenesis. When administration of the chemical is stopped before the later peak in the elevation, the activity can be restored to the normal level. Thus, the initial elevation in activity is reversible. On the other hand, the later peak of activity is irreversible, and discontinuation of the chemical does not cancel the increase in the level of γ-glutamyltranspeptidase. Such an alteration in γ-glutamyltranspeptidase level seems to be consistent with Solt-Farber's initiation/promotion model of chemical carcinogenesis (Farber, 1984a; Solt *et al.*, 1977). The multistep process of liver carcinogenesis encompasses three different steps: initiation, promotion, and progression, and the proceeding reversible elevation of γ-glutamyltranspeptidase activity corresponds to the initiation step. In the liver at this state, γ-glutamyltranspeptidase–positive foci and hyperplastic nodules are observed, and it is thought that such histological alterations can be extinguished by re-differentiation of the cells and tissue remodeling (Pereira *et al.*, 1984; Russell *et al.*, 1987). The late irreversible increase in enzyme activity seems to be an event associated with carcinogenesis.

In addition to a number of chemical carcinogens, it was found that chemical compounds that do not solely cause cancer also are able to induce the hepatic expression of γ-glutamyltranspeptidase. Many promoters in the initiation/promotion model elevate the levels of γ-glutamyltranspeptidase in the liver, and phenobarbital, polychlorobiphenyls, and a choline-deficient/low-methionine diet serve as promoters that are directed to the liver (Farber, 1984b). Some of these are also known agents that induce drug-metabolizing enzymes, and γ-glutamyltranspeptidase, at least in part, also behaves as such an enzyme class in terms of expression pattern.

As described previously, γ-glutamyltranspeptidase is considered to be a useful marker enzyme for the formation of preneoplastic nodules and hepatocarcinogenesis, as well as other tumor marker enzymes such as glutathione S-transferase isozyme (Satoh *et al.*, 1985). However, the usefulness of γ-glutamyltranspeptidase would be a clinical application for the diagnosis of alcoholism. In clinical examinations of human serum, an assay of γ-glutamyltranspeptidase activity is frequently used to detect alcoholism and obstructive jaundice. In the case of jaundice, elevated levels in the serum are most probably caused in general by "solubilization" from the biliary tract, with or without the induction of expression of γ-glutamyltranspeptidase. On the other hand, an increase in the serum level accompanying alcoholism involves the elevated expression of γ-glutamyltranspeptidase in the liver

(Teschke and Petrides, 1982), but the mechanism of the release to the serum has not been sufficiently investigated. The induction of γ-glutamyltranspeptidase by ethanol and propanol, the latter of which is more effective, was also confirmed *in vitro* by use of a hepatocellular carcinoma cell line (Barouki *et al.*, 1983).

γ-Glutamyltranspeptidase Gene Structure

The gene structure and the mechanism of the tissue-specific expression and induction of γ-glutamyltranspeptidase have long been of interest. Despite many efforts to explore the mechanisms underlying the tissue-specific expression and induction of γ-glutamyltranspeptidase, the molecular basis of its expression was not known with certainty until its gene structure and promoter were analyzed. Later, several research groups succeeded in the molecular cloning of cDNAs for mammalian γ-glutamyltranspeptidases, leading to the determination of their primary structures (Coloma and Pitot, 1986; Goodspeed *et al.*, 1989; Laperche *et al.*, 1986; Papandrikopoulou *et al.*, 1989; Rajpert-De Meyts *et al.*, 1988; Sakamuro *et al.*, 1988; Shi *et al.*, 1995). These studies found that a large and a small subunit are produced by limited proteolysis of single precursor polypeptide, consistent with a previous finding involving pulse-chase experiments (Nash and Tate, 1982).

Characterization of the structure and organization of γ-glutamyltranspeptidase gene in the rat and mouse has been carried out by several research groups. The γ-glutamyltranspeptidase gene was found to be a single copy gene in both rodents , whereas it seems that the human gene is a multicopy (Chobert *et al.*, 1990; Figlewicz *et al.*, 1993; Pawlak *et al.*, 1988; Rajagopalan *et al.*, 1993). The rodent γ-glutamyltranspeptidase genes contained 12 coding exons, interrupted by 11 introns, and a range of 12 kb (Rajagopalan *et al.*, 1990; Shi *et al.*, 1995). The organization of the γ-glutamyltranspeptidase gene is well conserved in rat and mouse, and multiforms of mRNA arise from the single copy gene in a tissue-specific manner. Multipromoter and splicing variations allow transcriptions of several species of mRNA for γ-glutamyltranspeptidase (Brouillet *et al.*, 1994; Carter *et al.*, 1994, 1995; Chobert *et al.*, 1990; Darbouy *et al.*, 1991; Habib *et al.*, 1995; Kurauchi *et al.*, 1991; Lahuna *et al.*, 1992; Lieberman *et al.*, 1995; Okamoto *et al.*, 1994; Rajagopalan *et al.*, 1990, 1993; Sepulveda *et al.*, 1994). In the rat, five different promoters, promoter-I to -V (also referred to as P1–P5) have been identified, and, at least six transcripts, including a splicing variant, mRNAs-I, II, III, IV-1, IV-2, and V, have been characterized. It should be noted that the numbering of the types of transcripts and responsible promoters is different between the mouse and rat, but it seems probable that the mechanisms underlying the tissue-specific and developmental stage–specific expression of γ-glutamyltranspeptidase are essentially the same. Although these various mRNAs differ in the structure of 5′ untranslated region, all of the transcripts encode an identical protein sequence, with a common leader sequence immediately upstream of the coding region. Promoters-I, -II, and -IV are active in the rat kidney, small intestine, and epididymis, in which the highest enzyme activity is expressed among the organs, and seem to serve for the constitutive expression of γ-glutamyltranspeptidase. The promoter-II contains a TATA-like sequence at the position expected from the transcription start site. In the case of the epididymis, promoter-III can also be functional (Palladino and Hinton, 1994; Palladino et al., 1994). Promoter-I, -II, and -IV are not active in other rat tissues, for example, in the liver, neoplastic lesions, mammary glands, and hepatoma cells, all of which have been examined in terms of γ-glutamyltranspeptidase expression. These findings and other studies revealed that the tissue-specific expression of γ-glutamyltranspeptidase is conferred by multiple promoters, even though all the different transcripts share an identical coding sequence.

Multiple Promoters and Hepatic Expression of γ-Glutamyltranspeptidase

As described previously, an alteration in the activity of γ-glutamyltranspeptidase in the liver is observed during development from the embryo to the neonate and in chemical carcinogenesis. In the development of the liver, the activity of the enzyme is transiently increased at the perinatal stage, whereas the activity is restricted to biliary cells in the adult liver. Thus, in the normal liver through fetal development and the adult stage, γ-glutamyltranspeptidase is expressed essentially in biliary cells and immature perinatal hepatocytes (Iannaccone and Koizumi, 1983). However, the issue of whether transcriptional regulation in constitutively expressing organs is also functional in such varied expression in the liver was unknown. Studies involving molecular biology have revealed that the expression of γ-glutamyltranspeptidase in liver and other organs such as the kidney is distinctly regulated. Although the transcript from the most proximal promoter is the major mRNA species in the kidney, the gene is transcribed by different promoters, promoter-IV (P4) in biliary cells and promoter-III (P3) in immature hepatocytes, respectively (Brouillet et al., 1998).

In liver development, immature hepatic cells are produced and differentiate into hepatocytic and biliary lineages. These bipotential precursor cells, which are generally referred to as hepatoblasts, alter the expression pattern of mRNA and the use of promoters in conjunction with differentiation. The transcription of the γ-glutamyltranspeptidase gene is initiated by promoters P3, P4, and P5 in rat hepatic precursor cells, and differentiation into the hepatocytic pathway leads to the extinction of promoter P4 and P5 activities. On the other hand, when the hepatoblasts differentiate into the biliary lineage, promoters P3 and P5 become inactive, thus resulting in the predominant expression of transcript, mRNA-IV, by promoter P4 (Holic et al., 2000).

Another type of the cells that express γ-glutamyltranspeptidase can also be found in a liver that is injured during the process of chemical carcinogenesis or by treatment with noncarcinogenic agents such as galactosamine and tetrachloromethane. The cells, known as oval cells, are also a bipotential precursor in the liver and capable of differentiating into both hepatocytes and biliary cells. A liver injury such as necrosis or inflammation results in the rapid proliferation of oval cells, which is subsequently involved in the liver regeneration (Dabeva and Shafritz, 1993; Lemire et al., 1991). When oval cells proliferate, the γ-glutamyltranspeptidase gene is transcribed from promoters P3, P4, and P5, thus leading to the accumulation of mRNAs-III, -IV, and -V (Holic et al., 2000). The subsequent differentiation of oval cells into immature hepatocytes is accompanied by the loss of expression of mRNA-IV and -V from promoters P4 and P5, and, as a result, mRNA-III becomes a major transcript species. As the cells further differentiate into mature hepatocytes, transcription from promoter P3 is also extinguished. On the other hand, differentiation into a biliary lineage leads to the extinction of expression from promoters P3 and P5, and the cells consequently express γ-glutamyltranspeptidase dependent on transcription by means of promoter P4. Expression pattern of mRNAs during oval cell differentiation associated with regeneration is very similar to that for the differentiation of hepatoblasts during development.

Transcription Factors Involved in
γ-Glutamyltranspeptidase Expression

mRNAs-IV and -III are expressed as the major transcripts in biliary cells and immature hepatocytes, respectively, whereas mRNA-V from promoter P5 is also expressed in the undifferentiated state (Holic et al., 2000). Consistent with this finding, undifferentiated hepatoma cells uniquely express transcripts from promoter P5, yielding two species of mRNA-V1 and -V2 by alternative splicing (Nomura et al., 1997). Because these transcripts are not expressed in more highly differentiated hepatoma cells, the transcriptional activity of promoter P5 seems to be associated with the extent of differentiation in liver cells. This distal promoter contains a cis-element that is involved in the expression of mRNA-V species, and it has been suggested that the binding of both hepatocyte nuclear factor-3 and activator protein-1 to the composite site in the element is absolutely required for transcription from promoter P5.

In the rat epididymis, mRNA-IV is a major transcript of γ-glutamyltranspeptidase, and expression from promoter P4 would play a role in maintaining the high enzyme activity of γ-glutamyltranspeptidase in the initial segment (Palladino and Hinton, 1994; Palladino et al., 1994). Polyomavirus enhancer activator 3 (Xin et al., 1992), a member of Ets transcriptional factor family, is expressed at high levels in the initial segment of the epididymis. Promoter P4 contains several binding sites for this transcription factor, and it has been suggested that some sites that are located proximal from the transcription start site are involved in the transcriptional activation of promoter P4 on the γ-glutamyltranspeptidase gene (Lan et al., 1999). However, the transfection of polyomavirus enhancer activator 3 into other types of cells failed to enhance the expression of γ-glutamyltranspeptidase. It seems that transcription from promoter P4, which yields mRNA-IV, is regulated by this Ets family transcription factor and the inhibitory protein, the ETS2 repressor factor.

Human γ-Glutamyltranspeptidase Gene and Its Expression

Although the γ-glutamyltranspeptidase gene is present as a single copy in mice and rats, the human gene is a multigene family that encompasses at least seven different genes (Courtay et al., 1994). Most members of the gene family are localized on chromosome 22 (Figlewicz et al., 1993). Human γ-glutamyltranspeptidase is also transcribed under the control of multipromoters, and several mRNA species have been identified, as has been found in mice and rats. Alternative splicing and transcription from multiple genes also seems to contribute to the diversity of the transcripts (Visvikis et al., 2001). Some of the transcripts do not code for an entire protein sequence but encode truncated forms of the protein, in which the open reading frame comprises sequences that largely consist of either large or small subunits (Leh et al., 1996; 1998; Pawlak et al., 1990; Wetmore et al., 1993).

Translation has been found to be regulated by the 5' untranslated region of an mRNA for γ-glutamyltranspeptidase from a human hepatocellular carcinoma cell line, HepG2 cells, and, thus, the region seems to serve as a tissue-specific active translational enhancer (Diederich et al., 1993). Such a regulating mechanism for translational efficiency would confer more flexible control of the expression of γ-glutamyltranspeptidase, as well as transcriptional regulation by the multipromoter system.

γ-Glutamyltranspeptidase-Related Gene Products

On the basis of homology of the primary structure, an enzyme that is structurally related to but distinct from γ-glutamyltranspeptidase has been identified in humans (Heisterkamp et al., 1991). This enzyme was referred to as the γ-glutamyltranspeptidase–related enzyme. It catalyzes the hydrolysis of glutathione and a glutathione-conjugate, leukotriene C4. Leukotriene D4 is produced by means of the hydrolysis by this related enzyme, as catalyzed by γ-glutamyltranspeptidase. However, a synthetic substrate, γ-glutamyl-p-nitroanilide that is widely used in the assay of γ-glutamyltranspeptidase does not serve as a substrate for the related enzyme. Years later, using γ-glutamyltranspeptidase knock out mice, it was found that the enzyme activity for converting leukotriene C4 into D4 remains even in the absence of γ-glutamyltranspeptidase (Carter et al., 1997). The enzyme responsible for the conversion of the leukotriene was cloned and is now referred to as γ-glutamylleukotrienase (Carter et al., 1998; Shi et al., 2001). As shown by a comparison of primary structures, γ-glutamylleukotrienase from mouse seems to be identical to the γ-glutamyltranspeptidase–related enzyme from humans. However, the enzymatic properties of these γ-glutamyltranspeptidase–like activities seem to be different in terms of substrate specificity (e.g., the mouse enzyme is not capable of hydrolyzing glutathione). γ-Glutamylleukotrienase is abundant in the spleen and is involved in the inflammatory response. Although it had previously been thought that γ-glutamyltranspeptidase is a unique enzyme involved in the metabolism of glutathione and related molecules, the discovery of the structurally and functionally related enzymes led to the suggestion that the metabolism is more complicated than previously thought to appropriately respond to various biological events. It has recently been reported that γ-glutamyltranspeptidase possesses biological activity that is not associated with its enzymatic activity toward glutathione and that it induces osteoclast formation in bone marrow cultures (Niida et al., 2004). Because the novel activity of γ-glutamyltranspeptidase is not impaired by a specific inhibitor, acivicin, the enzyme molecule may behave as a ligand for an unknown receptor or receptor-like molecule. It is possible that a structurally related protein with or without enzymatic activity could also serve a similar function. Such proteins related to γ-glutamyltranspeptidase may play a distinctly different role from that in the metabolism of glutathione and cysteine.

References

Albert, Z., Rzucidlo, Z., and Starzyk, H. (1970). Comparative biochemical and histochemical studies on the activity of γ-glutamyltranspeptidase in the organs of fetuses, newborns and adult rats. *Acta Histochem.* **37,** 34–39.

Anderson, M. E., and Meister, A. (1989). Glutathione monoesters. *Anal. Biochem.* **183,** 16–20.

Anderson, M. E., Allison, R. D., and Meister, A. (1982). Interconversion of leukotrienes catalyzed by purified γ-glutamyltranspeptidase: Concomitant formation of leukotriene D4 and γ-glutamylamino acids. *Proc. Natl. Acad. Sci. USA* **79,** 1088–1091.

Barouki, R., Chobert, M. N., Finidori, J., Aggerbeck, M., Nalpas, B., and Hanoune, J. (1983). Ethanol effects in a rat hepatoma cell line: Induction of γ-glutamyltransferase. *Hepatology* **3,** 323–329.

Billon, M. C., Dupre, G., and Hanoune, J. (1980). *In vivo* modulation of rat hepatic γ-glutamyltransferase activity by glucocorticoids. *Mol. Cell. Endocrinol.* **18,** 99–108.

Brannigan, J. A., Dodson, G., Duggleby, H. J., Moody, P. C., Smith, J. L., Tomchick, D. R., and Murzin, A. G. (1995). A protein catalytic framework with an N-terminal nucleophile is capable of self-activation. *Nature* **378,** 416–419.

Brouillet, A., Darbouy, M., Okamoto, T., Chobert, M. N., Lahuna, O., Garlatti, M., Goodspeed, D., and Laperche, Y. (1994). Functional characterization of the rat γ-glutamyltranspeptidase promoter that is expressed and regulated in the liver and hepatoma cells. *J. Biol. Chem.* **269,** 14878–14884.

Brouillet, A., Holic, N., Chobert, M. N., and Laperche, Y. (1998). The γ-glutamyltranspeptidase gene is transcribed from a different promoter in rat hepatocytes and biliary cells. *Am. J. Pathol.* **152,** 1039–1048.

Cagen, L. M., Fales, H. M., and Pisano, J. J. (1976). Formation of glutathione conjugates of prostaglandin A1 in human red blood cells. *J. Biol. Chem.* **251,** 6550–6554.

Carter, B. Z., Habib, G. M., Sepulveda, A. R., Barrios, R., Wan, D. F., Lebovitz, R. M., and Lieberman, M. W. (1994). Type VI RNA is the major γ-glutamyltranspeptidase RNA in the mouse small intestine. *J. Biol. Chem.* **269,** 24581–24585.

Carter, B. Z., Habib, G. M., Shi, Z. Z., Sepulveda, A. R., Lebovitz, R. M., and Lieberman, M. W. (1995). Rat small intestine expresses primarily type VI γ-glutamyltranspeptidase RNA. *Biochem. Mol. Biol. Int.* **35,** 1323–1330.

Carter, B. Z., Shi, Z. Z., Barrios, R., and Lieberman, M. W. (1998). γ-Glutamylleukotrienase, a γ-glutamyltranspeptidase gene family member, is expressed primarily in spleen. *J. Biol. Chem.* **273,** 28277–28285.

Carter, B. Z., Wiseman, A. L., Orkiszewski, R., Ballard, K. D., Ou, C. N., and Lieberman, M. W. (1997). Metabolism of leukotriene C4 in γ-glutamyltranspeptidase-deficient mice. *J. Biol. Chem.* **272,** 12305–12310.

Chobert, M. N., Lahuna, O., Lebargy, F., Kurauchi, O., Darbouy, M., Bernaudin, J. F., Guellaen, G., Barouki, R., and Laperche, Y. (1990). Tissue-specific expression of two γ-glutamyltranspeptidase mRNAs with alternative 5' ends encoded by a single copy gene in the rat. *J. Biol. Chem.* **265,** 2352–2357.

Coloma, J., and Pitot, H. C. (1986). Characterization and sequence of a cDNA clone of γ-glutamyltranspeptidase. *Nucleic Acids Res.* **14,** 1393–1403.

Cotariu, D., Barr-Nea, L., Papo, N., and Zaidman, J. L. (1988). Induction of γ-glutamyltransferase by dexamethasone in cultured fetal rat hepatocytes. *Enzyme* **40,** 212–216.

Courtay, C., Heisterkamp, N., Siest, G., and Groffen, J. (1994). Expression of multiple γ-glutamyltransferase genes in man. *Biochem. J.* **297,** 503–508.

Dabeva, M. D., and Shafritz, D. A. (1993). Activation, proliferation, and differentiation of progenitor cells into hepatocytes in the D-galactosamine model of liver regeneration. *Am. J. Pathol.* **143**, 1606–1620.

Daoust, R. (1982). The histochemical demonstration of γ-glutamyl transpeptidase activity in different populations of rat liver during azo dye carcinogenesis. *J. Histochem. Cytochem.* **30**, 312–316.

Darbouy, M., Chobert, M. N., Lahuna, O., Okamoto, T., Bonvalet, J. P., Farman, N., and Laperche, Y. (1991). Tissue-specific expression of multiple γ-glutamyltranspeptidase mRNAs in rat epithelia. *Am. J. Physiol.* **261**, C1130–C1137.

Deguchi, T., and Pitot, H. C. (1995). Expression of c-myc in altered hepatic foci induced in rats by various single doses of diethylnitrosamine and promotion by 0.05% phenobarbital. *Mol. Carcinog.* **14**, 152–159.

Diederich, M., Wellman, M., Visvikis, A., Puga, A., and Siest, G. (1993). The 5′ untranslated region of the human γ-glutamyltransferase mRNA contains a tissue-specific active translational enhancer. *FEBS Lett.* **332**, 88–92.

Farber, E. (1984a). Cellular biochemistry of the stepwise development of cancer with chemicals. *Cancer Res.* **44**, 5463–5474.

Farber, E. (1984b). The multistep nature of cancer development. *Cancer Res.* **44**, 4217–4223.

Fiala, S., and Fiala, A. E. (1969). Activation of glutathionase in rat liver during carcinogenesis. *Naturwissenschaften* **56**, 565.

Fiala, S., Mohindru, A., Kettering, W. G., Fiala, A. E., and Morris, H. P. (1976). Glutathione and γ glutamyl transpeptidase in rat liver during chemical carcinogenesis. *J. Natl. Cancer Inst.* **57**, 591–598.

Figlewicz, D. A., Delattre, O., Guellaen, G., Krizus, A., Thomas, G., Zucman, J., and Rouleau, G. A. (1993). Mapping of human γ-glutamyltranspeptidase genes on chromosome 22 and other human autosomes. *Genomics* **17**, 299–305.

Godwin, A. K., Meister, A., O'Dwyer, P. J., Huang, C. S., Hamilton, T. C., and Anderson, M. E. (1992). High resistance to cisplatin in human ovarian cancer cell lines is associated with marked increase of glutathione synthesis. *Proc. Natl. Acad. Sci. USA* **89**, 3070–3074.

Goodspeed, D. C., Dunn, T. J., Miller, C. D., and Pitot, H. C. (1989). Human γ-glutamyltranspeptidase cDNA: Comparison of hepatoma and kidney mRNA in the human and rat. *Gene* **76**, 1–9.

Habib, G. M., Carter, B. Z., Sepulveda, A. R., Shi, Z. Z., Wan, D. F., Lebovitz, R. M., and Lieberman, M. W. (1995). Identification of a sixth promoter that directs the transcription of γ-glutamyltranspeptidase type III RNA in mouse. *J. Biol. Chem.* **270**, 13711–13715.

Hanigan, M. H. (1995). Expression of γ-glutamyl transpeptidase provides tumor cells with a selective growth advantage at physiologic concentrations of cyst(e)ine. *Carcinogenesis* **16**, 181–185.

Hanigan, M. H., and Pitot, H. C. (1985). γ-Glutamyltranspeptidase—its role in hepatocarcinogenesis. *Carcinogenesis* **6**, 165–172.

Hanigan, M. H., and Ricketts, W. A. (1993). Extracellular glutathione is a source of cysteine for cells that express γ-glutamyltranspeptidase. *Biochemistry* **32**, 6302–6306.

Heisterkamp, N., Rajpert-De Meyts, E., Uribe, L., Forman, H. J., and Groffen, J. (1991). Identification of a human γ-glutamylcleaving enzyme related to, but distinct from, γ-glutamyl transpeptidase. *Proc. Natl. Acad. Sci. USA* **88**, 6303–6307.

Higashi, T., Tateishi, N., and Sakamoto, Y. (1983). Liver glutathione as a reservoir of L-cysteine. *Prog. Clin. Biol. Res.* **125**, 419–434.

Higashi, T., Tateishi, N., Naruse, A., and Sakamoto, Y. (1977). A novel physiological role of liver glutathione as a reservoir of L-cysteine. *J. Biochem. (Tokyo)* **82**, 117–124.

Hiratake, J., Inoue, M., and Sakata, K. (2002). γ-Glutamyltranspeptidase and γ-glutamylpeptide ligases: Fluorophosphonate and phosphonodifluoromethyl ketone analogs as probes of tetrahedral transition state and γ-glutamyl-phosphate intermediate. *Methods Enzymol.* **354**, 272–295.

Holic, N., Suzuki, T., Corlu, A., Couchie, D., Chobert, M. N., Guguen-Guillouzo, C., and Laperche, Y. (2000). Differential expression of the rat γ-glutamyltranspeptidase gene promoters along with differentiation of hepatoblasts into biliary or hepatocytic lineage. *Am. J. Pathol.* **157**, 537–548.

Huang, C. S., Anderson, M. E., and Meister, A. (1993b). Amino acid sequence and function of the light subunit of rat kidney γ-glutamylcysteine synthetase. *J. Biol. Chem.* **268**, 20578–20583.

Huang, C. S., Chang, L. S., Anderson, M. E., and Meister, A. (1993a). Catalytic and regulatory properties of the heavy subunit of rat kidney γ-glutamylcysteine synthetase. *J. Biol. Chem.* **268**, 19675–19680.

Huang, C. S., He, W., Meister, A., and Anderson, M. E. (1995). Amino acid sequence of rat kidney glutathione synthetase. *Proc. Natl. Acad. Sci. USA* **92**, 1232–1236.

Hughey, R. P., and Curthoys, N. (1976). Comparison of the size and physical properties of γ-glutamyltranspeptidase purified from rat kidney following solubilization with papain or with Triton. *J. Biol. Chem.* **251**, 8763–8770.

Iannaccone, P. M., and Koizumi, J. (1983). Pattern and rate of disappearance of γ-glutamyltranspeptidase activity in fetal and neonatal rat liver. *J. Histochem. Cytochem.* **31**, 1312–1316.

Inoue, M., Hiratake, J., Suzuki, H., Kumagai, H., and Sakata, K. (2000). Identification of catalytic nucleophile of *Escherichia coli* γ-glutamyltranspeptidase by γ-monofluorophosphono derivative of glutamic acid: N-terminal thr-391 in small subunit is the nucleophile. *Biochemistry* **39**, 7764–7771.

Kalengayi, M. M., Ronchi, G., and Desmet, V. J. (1975). Histochemistry of γ-glutamyltranspeptidase in rat liver during aflatoxin B1-induced carcinogenesis. *J. Natl. Cancer Inst.* **55**, 579–588.

Kurauchi, O., Lahuna, O., Darbouy, M., Aggerbeck, M., Chobert, M. N., and Laperche, Y. (1991). Organization of the 5' end of the rat γ-glutamyltranspeptidase gene: Structure of a promoter active in the kidney. *Biochemistry* **30**, 1618–1623.

Lahuna, O., Brouillet, A., Chobert, M. N., Darbouy, M., Okamoto, T., and Laperche, Y. (1992). Identification of a second promoter which drives the expression of γ-glutamyltranspeptidase in rat kidney and epididymis. *Biochemistry* **31**, 9190–9196.

Laishes, B. A., Ogawa, K., Roberts, E., and Farber, E. (1978). γ-Glutamyltranspeptidase: A positive marker for cultured rat liver cells derived from putative premalignant and malignant lesions. *J. Natl. Cancer Inst.* **60**, 1009–1016.

Lambiotte, M., Vorbrodt, A., and Benedetti, E. L. (1973). Expression of differentiation of rat foetal hepatocytes in cellular culture under the action of glucocorticoids: Appearance of bile canaliculi. *Cell Differ.* **2**, 43–53.

Lan, Z. J., Lye, R. J., Holic, N., Labus, J. C., and Hinton, B. T. (1999). Involvement of polyomavirus enhancer activator 3 in the regulation of expression of γ-glutamyltranspeptidase messenger ribonucleic acid-IV in the rat epididymis. *Biol. Reprod.* **60**, 664–673.

Laperche, Y., Bulle, F., Aissani, T., Chobert, M. N., Aggerbeck, M., Hanoune, J., and Guellaen, G. (1986). Molecular cloning and nucleotide sequence of rat kidney γ-glutamyltranspeptidase cDNA. *Proc. Natl. Acad. Sci. USA* **83**, 937–941.

Leh, H., Chikhi, N., Ichino, K., Guellaen, G., Wellman, M., Siest, G., and Visvikis, A. (1998). An intronic promoter controls the expression of truncated human γ-glutamyltransferase mRNAs. *FEBS Lett.* **434,** 51–56.

Leh, H., Courtay, C, Gerardin, P., Wellman, M., Siest, G., and Visvikis, A. (1996). Cloning and expression of a novel type (III) of human γ-glutamyltransferase truncated mRNA. *FEBS Lett.* **394,** 258–262.

Lemire, J. M., Shiojiri, N., and Fausto, N. (1991). Oval cell proliferation and the origin of small hepatocytes in liver injury induced by D-galactosamine. *Am. J. Pathol.* **139,** 535–552.

Lieberman, M. W., Barrios, R., Carter, B. Z., Habib, G. M., Lebovitz, R. M., Rajagopalan, S., Sepulveda, A. R., Shi, Z. Z., and Wan, D. F. (1995). γ-Glutamyltranspeptidase. What does the organization and expression of a multipromoter gene tell us about its functions? *Am. J. Pathol.* **147,** 1175–1185.

Lieberman, M. W., Wiseman, A. L., Shi, Z. Z., Carter, B. Z., Barrios, R., Ou, C. N., Chevez-Barrios, P., Wang, Y., Habib, G. M., Goodman, J. C., Huang, S. L., Lebovitz, R. M., and Matzuk, M. M. (1996). Growth retardation and cysteine deficiency in γ-glutamyltrans-peptidase-deficient mice. *Proc. Natl. Acad. Sci. USA* **93,** 7923–7926.

Manson, M. M. (1983). Biphasic early changes in rat liver γ-glutamyl transpeptidase in response to aflatoxin B1. *Carcinogenesis* **4,** 467–472.

McIntyre, T. M., and Curthoys, N. P. (1979). Comparison of the hydrolytic and transfer activities of rat renal γ-glutamyltransferase. *J. Biol. Chem.* **254,** 6499–6504.

Meister, A., and Tate, S. S. (1976). Glutathione and related γ-glutamyl compounds: Biosynthesis and utilization. *Annu. Rev. Biochem.* **45,** 559–604.

Meister, A. (1973). On the enzymology of amino acid transport. *Science* **180,** 33–39.

Meister, A. (1992). On the antioxidant effects of ascorbic acid and glutathione. *Biochem. Pharmacol.* **44,** 1905–1915.

Meister, A., and Larsson, A. (1995). *In* "The Metabolic and Molecular Bases of Inherited Disease, 7th ed, Vol I" (C. R. Scriver A. L. Beaudet W. S. Sly and D. Valle, eds.), pp. 1461–1477. McGraw-Hill, New York.

Meister, A., Tate, S. S., and Griffith, O. W. (1981). γ-Glutamyltranspeptidase. *Methods Enzymol.* **77,** 237–253.

Nash, B., and Tate, S. S. (1982). Biosynthesis of rat renal γ-glutamyltranspeptidase. Evidence for a common precursor of the two subunits. *J. Biol. Chem.* **257,** 585–588.

Nash, B., and Tate, S. S. (1984). *In vitro* translation and processing of rat kidney γ-glutamyltranspeptidase. *J. Biol. Chem.* **259,** 678–685.

Niida, S., Kawahara, M., Ishizuka, Y., Ikeda, Y., Kondo, T., Hibi, T., Suzuki, Y., Ikeda, K., and Taniguchi, N. (2004). γ-Glutamyltranspeptidase stimulates receptor activator of nuclear factor-kappaB ligand expression independent of its enzymatic activity and serves as a pathological bone-resorbing factor. *J. Biol. Chem.* **279,** 5752–5756.

Nomura, S., Lahuna, O., Suzuki, T., Brouillet, A., Chobert, M. N., and Laperche, Y. (1997). A specific distal promoter controls γ-glutamyl transpeptidase gene expression in undifferentiated rat transformed liver cells. *Biochem. J.* **326,** 311–320.

Okamoto, T., Darbouy, M., Brouillet, A., Lahuna, O., Chobert, M. N., and Laperche, Y. (1994). Expression of the rat γ-glutamyltranspeptidase gene from a specific promoter in the small intestine and in hepatoma cells. *Biochemistry* **33,** 11536–115343.

Oppenheimer, L., Wellner, V. P., Griffith, O. W., and Meister, A. (1979). Glutathione synthetase. Purification from rat kidney and mapping of the substrate binding sites. *J. Biol. Chem.* **254,** 5184–5190.

Orlowski, M., and Meister, A. (1971). Partial reactions catalyzed by γ-glutamylcysteine synthetase and evidence for an activated glutamate intermediate. *J. Biol. Chem.* **246,** 7095–7105.

Palladino, M. A., and Hinton, B. T. (1994). Expression of multiple γ-glutamyltranspeptidase messenger ribonucleic acid transcripts in the adult rat epididymis is differentially regulated by androgens and testicular factors in a region-specific manner. *Endocrinology* **135,** 1146–1156.

Palladino, M. A., Laperche, Y., and Hinton, B. T. (1994). Multiple forms of γ-glutamyltranspeptidase messenger ribonucleic acid are expressed in the adult rat testis and epididymis. *Biol. Reprod.* **50,** 320–328.

Papandrikopoulou, A., Frey, A., and Gassen, H. G. (1989). Cloning and expression of γ-glutamyltranspeptidase from isolated porcine brain capillaries. *Eur. J. Biochem.* **183,** 693–698.

Pawlak, A., Cohen, E. H., Octave, J. N., Schweickhardt, R., Wu, S. J., Bulle, F., Chikhi, N., Baik, J. H., Siegrist, S., and Guellaen, G. (1990). An alternatively processed mRNA specific for γ-glutamyltranspeptidase in human tissues. *J. Biol. Chem.* **265,** 3256–3262.

Pawlak, A., Lahuna, O., Bulle, F., Suzuki, A., Ferry, N., Siegrist, S., Chikhi, N., Chobert, M. N., Guellaen, G., and Laperche, Y. (1988). γ-Glutamyltranspeptidase: A single copy gene in the rat and a multigene family in the human genome. *J. Biol. Chem.* **263,** 9913–9916.

Pereira, M. A., Herren-Freund, S. L., Britt, A. L., and Khoury, M. M. (1983). Effects of strain, sex, route of administration and partial hepatectomy on the induction by chemical carcinogens of γ-glutamyltranspeptidase foci in rat liver. *Cancer Lett.* **20,** 207–214.

Pereira, M. A., Herren-Freund, S. L., Britt, A. L., and Khoury, M. M. (1984). Effect of coadministration of phenobarbital sodium on N-nitrosodiethylamine-induced γ-glutamyl-transferase-positive foci and hepatocellular carcinoma in rats. *J. Natl. Cancer Inst.* **72,** 741–744.

Power, C. A., Griffiths, S. A., Simpson, J. L., Laperche, Y., Guellaen, G., and Manson, M. M. (1987). Induction of γ-glutamyl transpeptidase mRNA by aflatoxin B1 and ethoxyquin in rat liver. *Carcinogenesis* **8,** 737–740.

Puri, R. N., and Meister, A. (1983). Transport of glutathione, as γ-glutamylcysteinylglycyl ester, into liver and kidney. *Proc. Natl. Acad. Sci. USA* **80,** 5258–5260.

Rajagopalan, S., Park, J. H., Patel, P. D., Lebovitz, R. M., and Lieberman, M. W. (1990). Cloning and analysis of the rat γ-glutamyltransferase gene. *J. Biol. Chem.* **265,** 11721–11725.

Rajagopalan, S., Wan, D. F., Habib, G. M., Sepulveda, A. R., McLeod, M. R., Lebovitz, R. M., and Lieberman, M. W. (1993). Six mRNAs with different 5′ ends are encoded by a single γ-glutamyltransferase gene in mouse. *Proc. Natl. Acad. Sci. USA* **90,** 6179–6183.

Rajpert-De Meyts, E., Heisterkamp, N., and Groffen, J. (1988). Cloning and nucleotide sequence of human γ-glutamyltranspeptidase. *Proc. Natl. Acad. Sci. USA* **85,** 8840–8844.

Richards, W. L., Tsukada, Y., and Potter, V. R. (1982). Phenotypic diversity of γ-glutamyltranspeptidase activity and protein secretion in hepatoma cell lines. *Cancer Res.* **42,** 1374–1383.

Russell, J. J., Staffeldt, E. F., Wright, B. J., Prapuolenis, A., Carnes, B. A., and Peraino, C. (1987). Effects of rat strain, diet composition, and phenobarbital on hepatic γ-glutamyltranspeptidase histochemistry and on the induction of altered hepatocyte foci and hepatic tumors by diethylnitrosamine. *Cancer Res.* **47,** 1130–1134.

Sakamuro, D., Yamazoe, M., Matsuda, Y., Kangawa, K., Taniguchi, N., Matsuo, H., Yoshikawa, H., and Ogasawara, N. (1988). The primary structure of human γ-glutamyltranspeptidase. *Gene* **73,** 1–9.

Satoh, K., Kitahara, A., Soma, Y., Inaba, Y., Hatayama, I., and Sato, K. (1985). Purification, induction, and distribution of placental glutathione transferase: A new marker enzyme for preneoplastic cells in the rat chemical hepatocarcinogenesis. *Proc. Natl. Acad. Sci. USA* **82,** 3964–3968.

Sepulveda, A. R., Carter, B. Z., Habib, G. M., Lebovitz, R. M., and Lieberman, M. W. (1994). The mouse γ-glutamyltranspeptidase gene is transcribed from at least five separate promoters. *J. Biol. Chem.* **269,** 10699–10705.

Shi, Z. Z., Habib, G. M., Lebovitz, R. M., and Lieberman, M. W. (1995). Cloning of cDNA and genomic structure of the mouse γ-glutamyltranspeptidase-encoding gene. *Gene* **167,** 233–237.

Shi, Z. Z., Han, B., Habib, G. M., Matzuk, M. M., and Lieberman, M. W. (2001). Disruption of γ-glutamylleukotrienase results in disruption of leukotriene D(4) synthesis *in vivo* and attenuation of the acute inflammatory response. *Mol. Cell. Biol.* **21,** 5389–5395.

Shiozawa, M., Hiraoka, Y., Yasuda, K., Imamura, T., Sakamuro, D, Taniguchi, N., Yamazoe, M., and Yoshikawa, H. (1990). Synthesis of human γ-glutamyl transpeptidase (GGT) during the fetal development of liver. *Gene* **87,** 299–303.

Solt, D. B., Cayama, E., Sarma, D. S., and Farber, E. (1980). Persistence of resistant putative preneoplastic hepatocytes induced by N-nitrosodiethylamine or N-methyl-N-nitrosourea. *Cancer Res.* **40,** 1112–1118.

Solt, D. B., Medline, A., and Farber, E. (1977). Rapid emergence of carcinogen-induced hyperplastic lesions in a new model for the sequential analysis of liver carcinogenesis. *Am. J. Pathol.* **88,** 595–618.

Suzuki, H., and Kumagai, H. (2002). Autocatalytic processing of γ-glutamyltranspeptidase. *J. Biol. Chem.* **277,** 43536–43543.

Suzuki, Y., Ishizuka, H., Kaneda, H., and Taniguchi, N. (1987). γ-Glutamyltranspeptidase in rat liver during 3'-Me-DAB hepatocarcinogenesis: Immunohistochemical and enzyme histochemical study. *J. Histochem. Cytochem.* **35,** 3–7.

Taniguchi, N., and Ikeda, Y. (1998). γ-Glutamyltranspeptidase: Catalytic mechanism and gene expression. *Adv. Enzymol. Relat. Areas Mol. Biol.* **72,** 239–278.

Taniguchi, N., Tsukada, Y., Mukuo, K., and Hirai, H. (1974). Effect of hepatocarcinogenic azo dyes on glutathione and related enzymes in rat liver. *Gann.* **65,** 381–387.

Taniguchi, N., Yososawa, N., Iizuka, S., Sako, F., Tsukada, Y., Satoh, M., and Dempo, K. (1983). γ-Glutamyltranspeptidase of rat liver and hepatoma tissues: An enzyme immunoassay and immunostaining studies. *Ann. NY Acad. Sci.* **417,** 203–212.

Tate, S. S. (1986). Single-chain precursor of renal γ-glutamyltranspeptidase. *FEBS Lett.* **194,** 33–38.

Tate, S. S., and Galbraith, R. A. (1987). A human hepatoma cell line expresses a single-chain form of γ-glutamyltranspeptidase. *J. Biol. Chem.* **262,** 11403–11406.

Tate, S. S., and Galbraith, R. A. (1988). *In vitro* translation and processing of human hepatoma cell (Hep G2) γ-glutamyltranspeptidase. *Biochem. Biophys. Res. Commun.* **154,** 1167–1173.

Tate, S. S., and Meister, A. (1985). γ-Glutamyltranspeptidase from kidney. *Methods Enzymol.* **113,** 400–419.

Tateishi, N., Higashi, T., Nakashima, K., and Sakamoto, Y. (1980). Nutritional significance of increase in γ-glutamyltransferase in mouse liver before birth. *J. Nutr.* **110,** 409–415.

Tateishi, N., Higashi, T., Naruse, A., Nakashima, K., and Shiozaki, H. (1977). Rat liver glutathione: Possible role as a reservoir of cysteine. *J. Nutr.* **107**, 51–60.

Tateishi, N., Higashi, T., Nomura, T., Naruse, A., and Nakashima, K. (1976). Higher transpeptidation activity and broad acceptor specificity of γ-glutamyltransferases of tumors. *Gann* **67**, 215–222.

Teschke, R., and Petrides, A. S. (1982). Hepatic γ-glutamyltransferase activity: Its increase following chronic alcohol consumption and the role of carbohydrates. *Biochem. Pharmacol.* **31**, 3751–3756.

Tongiani, R., and Paolicchi, A. (1989). γ-Glutamyltransferase induction by glucocorticoids in rat liver: Age-dependence, time-dependence, dose-dependence, and intralobular distribution. *Acta Histochem.* **86**, 51–61.

Tongiani, R., Paolicchi, A., and Chieli, E. (1984). γ-Glutamyltranspeptidase induction by cortisol in liver parenchyma of unweaned rats. *Biosci. Rep.* **4**, 203–211.

Tsuchida, S., Hoshino, K., Sato, T., Ito, N., and Sato, K. (1979). Purification of γ-glutamyltransferases from rat hepatomas and hyperplastic hepatic nodules, and comparison with the enzyme from rat kidney. *Cancer Res.* **39**, 4200–4205.

Visvikis, A., Pawlak, A., Accaoui, M. J., Ichino, K., Leh, H., Guellaen, G., and Wellman, M. (2001). Structure of the 5′ sequences of the human γ-glutamyltransferase gene. *Eur. J. Biochem.* **268**, 317–325.

Wetmore, L. A., Gerard, C., and Drazen, J. M. (1993). Human lung expresses unique γ-glutamyltranspeptidase transcripts. *Proc. Natl. Acad. Sci. USA* **90**, 7461–7465.

Xin, J. H., Cowie, A., Lachance, P., and Hassell, J. A. (1992). Molecular cloning and characterization of PEA3, a new member of the Ets oncogene family that is differentially expressed in mouse embryonic cells. *Genes Dev.* **6**, 481–496.

Yan, Y., Higashi, K., Yamamura, K., Fukamachi, Y., Abe, T., Gotoh, S., Sugiura, T., Hirano, T., Higashi, T., and Ichiba, M. (1998). Different responses other than the formation of DNA-adducts between the livers of carcinogen-resistant rats (DRH) and carcinogen-sensitive rats (Donryu) to 3′-methyl-4-dimethylaminoazobenzene administration. *Jpn. J. Cancer Res.* **89**, 806–813.

Further Reading

Gali, R. R., and Board, P. G. (1995). Sequencing and expression of a cDNA for human glutathione synthetase. *Biochem. J.* **310**, 353–358.

[26] γ-Glutamyltranspeptidase: Disulfide Bridges, Propeptide Cleavage, and Activation in the Endoplasmic Reticulum

By CAROL L. KINLOUGH, PAUL A. POLAND, JAMES B. BRUNS, and REBECCA P. HUGHEY

Abstract

γ-Glutamyltranspeptidase (γGT) is found primarily on the apical surface of epithelial and endothelial cells, where it degrades reduced and oxidized glutathione (γ-GluCysGly) by hydrolysis of the unique γ-glutamyl bond. Glutathione plays a key role in disulfide rearrangement in the endoplasmic reticulum (ER) and acts as a redox buffer. Previous work has shown that overexpression of γGT or an inactive splice variant γGTΔ7 mediates a redox stress response in the endoplasmic reticulum (ER) characterized by increased levels of BiP and induction of CHOP-10. To determine whether a CX₃C motif might be the common feature of γGT and γGTΔ7 that mediates this response, we characterized disulfide bridges in γGT that might form between the six highly conserved Cys residues. Using site-directed mutagenesis of γGT, expression in Chinese Hamster Ovary (CHO) cells, metabolic labeling, and immunoblotting, our data predict disulfide formation between Cys49 and Cys73 and between Cys191 and Cys195 (the CX₃C motif). Potential functions for this CX₃C motif are discussed. In the course of defining the disulfides, we also noted that propeptide cleavage correlated with enzymatic activity. Because recent reports indicate that the homologous *Escherichia coli* γGT is a member of the N-terminal nucleophile (Ntn) hydrolase family, where the amino acid at the new N-terminus functions as the nucleophile for both autocatalytic cleavage and enzymatic activity, the rat γGT was similarly characterized. As predicted, mutations at the propeptide cleavage site coincidentally inhibit both heterodimer formation and γGT enzymatic activity. Analysis of early cleavage events using cell extraction into SDS indicates that propeptide cleavage occurs while γGT is still within the ER. Because activation and cleavage are coincident events, this raises the new question of whether an active glutathionase is present within the ER and what role γGT plays in modulating ER glutathione levels that are so critical for proper redox balance and disulfide formation in this compartment.

METHODS IN ENZYMOLOGY, VOL. 401
Copyright 2005, Elsevier Inc. All rights reserved.
0076-6879/05 $35.00
DOI: 10.1016/S0076-6879(05)01026-8

Key Features of γ-Glutamyltranspeptidase (γGT)

γGT Synthesis and Processing

γGT initiates the degradation of both oxidized and reduced glutathione (γ-GluCysGly) at the cell surface by cleaving the unique γ-glutamyl bond. The subsequent hydrolysis of oxidized or reduced CysGly by aminopeptidase or dipeptidase releases Gly, and cysteine/cystine is recovered for intracellular synthesis of reduced glutathione. Recent mouse models of γGT deficiency have confirmed that γGT plays a major physiological role in providing cysteine to cells for glutathione synthesis and protein synthesis, thereby playing a major role in antioxidant defense and normal growth (Barrios *et al.*, 2001; Chevez-Barrios *et al.*, 2000; Harding *et al.*, 1997; Jean *et al.*, 1998, 2002; Kumar *et al.*, 2000; Levasseur *et al.*, 2003; Lipton *et al.*, 2001; Pardo *et al.*, 2003). However, recent work indicates that γGT may have alternative functions under defined conditions, such as mediating a stress response within the endoplasmic reticulum (ER) (Joyce-Brady *et al.*, 2001), stimulation of bone resorption by osteoclasts (Niida *et al.*, 2004), or generating pro-oxidative damage during glutathione hydrolysis (for review, see [Pompella *et al.*, 2003] and Chapter 29).

γGT is a type II transmembrane protein with an uncleaved signal anchor acting as the transmembrane domain and a cytoplasmic domain consisting of only four polar amino acids (see Fig. 1). The rat γGT is synthesized as a propeptide of 568 amino acids and is cleaved to yield a

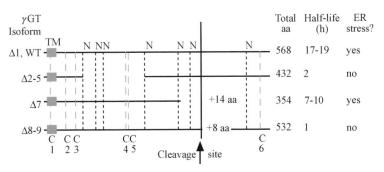

FIG. 1. Isoforms of γ-glutamyltranspeptidase (γGT). The predicted linear structures of splice variants (Δ2-5, Δ7, and Δ8-9) of full-length mouse γGT (Δ1, WT) are drawn to scale to emphasize changes. The position of the propeptide cleavage site (↑), transmembrane domain (TM), N-linked oligosaccharides consensus sites (N), and all six Cys residues (C) are indicated. The predicted half-life of the isoforms and their ability to induce an ER stress response are based on previously published data (Joyce-Brady *et al.*, 2001). See text for details.

stable heterodimer with a large amphipathic subunit (379 amino acids) and a smaller hydrophilic subunit (189 amino acids) (Altman and Hughey, 1986; Capraro and Hughey, 1983; Nash and Tate, 1982, 1984). Amino acids identified within the enzyme's active site are from both subunits (for review, see Taniguchi, 1998 and Chapter 25). N-linked and O-linked glycans are present on both subunits, and the inconsistencies in mobility and appearance of the propeptide and subunits from different sources on SDS-gels are consistent with cell-type–specific heterogeneity in the number and maturation of the glycans (Altman and Hughey, 1986; Blochberger et al., 1989; Capraro and Hughey, 1983). The basis for the lack of cleavage and surface expression of some fraction of the mammalian γGT propeptide is unknown. Interestingly, characterization of propeptide cleavage for the E. coli γGT indicates that the cleavage reaction is autocatalytic and that the amino acid subsequently found at the new N-terminus functions as the nucleophile for both the cleavage reaction and for the enzyme active site (Inoue et al., 2000; Suzuki and Kumagai, 2002). These properties also indicate that all γGTs might belong to the family of N-terminal nucleophile (Ntn) hydrolases as first proposed by Brannigan et al. (1995). Our data presented here support the possibility that the rat γGT is an Ntn hydrolase.

γGT Isoforms and Their Potential Function

Studies by several groups have shown that γGT is derived from a highly regulated gene in several species (for review, see Joyce-Brady et al., 1996; Chikhi et al., 1999; and Chapter 25). The regulation involves alternative splicing coupled with alternative promoter use that generates unique non-coding 5′ ends for multiple γGT cDNAs. In contrast, splicing within coding exons seemed to be constitutive, such that only a single protein product was encoded. However, Jean et al. (1998) found alternative splicing events within coding exons of the γGT gene while characterizing the mutation that blocked γGT expression in the GGT[enu1] mouse (Harding et al., 1997). In more detailed studies, Joyce-Brady et al. (2001) found that alternative splicing events in mouse γGT alter coding exons to produce at least three γGT-related protein isoforms (see Fig. 1). Alternative splicing within exons 2 and 5 deletes residues 96–230 from the large subunit domain of the propeptide (γGTΔ2-5); alternative splicing within intron 7 introduces a premature stop codon in the large subunit domain after addition of 14 unique residues (γGTΔ7); and alternative splicing between exons 8 and 9 deletes residues 401–444 and introduces eight new amino acids in the small subunit domain (γGTΔ8-9). Most importantly, the γGTΔ7 and Δ8-9 splicing events are developmentally regulated in heart, lung, and kidney, whereas the γGTΔ7 splicing event also occurs in human γGT

(Joyce-Brady *et al.*, 2001). Therefore, it seems that these alternative splicing events are physiologically relevant and that the γGT protein isoforms serve unique roles in glutathione metabolism.

Site-directed mutagenesis of the mouse γGT(WT) cDNA (kindly provided by Dr. Michael W. Lieberman, Baylor College of Medicine, Houston, TX) was used to prepare cDNAs with coding sequences corresponding to the γGTΔ1, γGTΔ2-5, γGTΔ7, and γGTΔ8-9 transcripts, and these were expressed in Chinese Hamster Ovary (CHO) cells (Joyce-Brady *et al.*, 2001). γGTΔ1 is identical to γGT(WT) within the coding sequence but exhibits a CAG insert upstream of the start site because of alternative splicing at a CAGCAG motif. This motif is shared between mouse and human γGT genes, and its presence causes a modest reduction in translational efficiency. Interestingly, the occurrence of alternative splicing at such motifs seems to be widespread and to contribute to proteome plasticity (Hiller *et al.*, 2004). Characterization of cells expressing these isoforms showed that only γGTΔ1 produced an enzymatically active protein that was normally processed and expressed on the cell surface. Cells expressing γGTΔ2-5, γGTΔ7, and γGTΔ8-9 produced proteins of 47 kDa, 44 kDa, and 52 kDa, respectively, that were retained in the ER as judged from studies using metabolic labeling, Endo H sensitivity, cell surface biotinylation, and immunofluorescence microscopy. The very short half-life that was observed for γGTΔ2-5 and γGTΔ8-9 was consistent with rapid degradation of improperly folded or slowly folded proteins by ER-associated degradation pathways (ERAD, reviewed by Helenius and Aebi, 2004) (see Fig. 1). However, γGTΔ7 exhibited a considerably longer half-life than γGTΔ2-5 and γGTΔ8-9, suggesting that the ER localization of γGTΔ7 likely reflects a new, previously undefined function for this γGT isoform. This possibility was supported by finding that stable cell lines expressing either the full-length γGTΔ1 or γGTΔ7, but not cell lines expressing either γGTΔ2-5, γGTΔ8-9, or control glycoproteins, exhibited an ER stress response when the cell culture media was switched from DMEM/Ham's F12 (containing cysteine and cystine) to DMEM (containing only cystine) (Joyce-Brady *et al.*, 2001). Northern blot analysis of RNA from these cells showed a dramatic increase in the level of BiP expression and a clear induction of CHOP-10 only in cells expressing γGTΔ1 or γGTΔ7. Induction of BiP and CHOP-10 is a hallmark of the unfolded protein response (UPR, reviewed by Zhang and Kaufman, 2004).

Our present studies are focused on identifying the feature(s) of γGTΔ1 and γGTΔ7 that could signal or facilitate an ER redox stress response. When the sequences of γGTΔ1 and γGTΔ7 were aligned, the most notable feature that was present in both isoforms was an unusual CX_3C motif (L$\underline{C}$EVF$\underline{C}$R) formed by the fourth (C4) and fifth (C5) Cys in the sequence (numbering from the N-terminus, see Fig. 1). This motif is perfectly conserved in yeast, mouse, rat, pig, and human γGT. Closely spaced protein dithiols are present in a variety of proteins and exhibit diverse functions. Thioredoxin reductase has a conserved C̲VNVGC̲ sequence (CX_4C) in its redox catalytic site, whereas thioredoxin (WC̲GPC̲K) and protein disulfide isomerase (WC̲GHC̲K) have conserved CX_2C motifs in their catalytic sites (Edman et al., 1985; Mustacich and Powis, 2000; Powis and Montfort, 2001). Finally, CX_3C sequences are a signature motif for one class of chemokines (Bazan et al., 1997; Wells and Peitsch, 1997), an essential coordinate for copper binding in several yeast proteins (Balatri et al., 2003; Nittis et al., 2001) and an essential motif for the functioning of the redox protein A2.5L of vaccinia virus (Senkevich et al., 2002). Therefore, experiments were carried out to identify disulfide bridges in γGT as a prelude to determining whether the CX_3C motif has a unique function in redox balance of the cell.

Characterization of γGT Maturation

Identification of Disulfide Bridges in γGT

The presence of disulfide bridges in γGT has not been previously addressed, although all six Cys and surrounding residues are highly conserved in eukaryotic γGT. The two subunits formed by γGT propeptide cleavage are not linked by disulfides, because the subunits readily separate on SDS-PAGE in the absence of reducing agents. Because there is only one Cys (C6) in the small subunit, this means that C6 does not form a disulfide bridge. One Cys (C1) is also found within the γGT transmembrane domain. This domain is surrounded by the lipid bilayer, and no evidence exists for dimerization of γGT propeptide or heterodimer, so C1 is also unlikely to form a disulfide bridge. However, intramolecular disulfide bridges are likely formed between the other four Cys residues (C2, C3, C4, and C5). Mutagenesis of Cys residues that normally form disulfide bridges can cause retention in the ER, so we used this characteristic to identify any disulfide bridges present in γGT (Isidoro et al., 1996; Segal et al., 1992).

Each Cys residue in γGT (WT, rat full-length cDNA) was mutated individually to Ala, and the mutants were transiently expressed in CHO cells. The expression of γGT and mutants was assessed by immunoblotting, metabolic labeling with [^{35}S]Met/Cys, and assaying for γGT enzymatic activity in cell extracts. As shown in Fig. 2A, immunoblotting the γGT immunoprecipitates from transfected cells revealed the propeptide (P) and the heterodimer large (L) and small (S) subunits for WT γGT. The

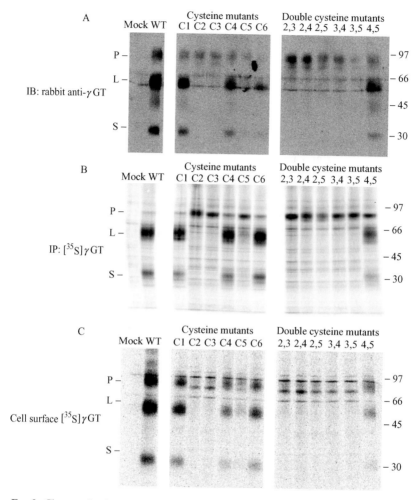

FIG. 2. Characterization of wild-type γGT and Cys-mutants in CHO cells. Either wild-type rat γGT (WT) or mutants with one or two Cys changed to Ala were transiently expressed in CHO cells. Numbering of the six mutated Cys (C1–C6) is based on their order from the amino-terminus as noted in Fig. 1. Steady-state levels of expression were assessed by either (A) immunoblotting with rabbit anti-γGT antibodies or by (B) radiolabeling overnight with [35S]Met/Cys, before immunoprecipitation with the same antibodies for SDS-PAGE. The surface expression of [35S]γGT was determined by treatment of transfected cells on ice with the membrane-impermeant sulfo-NHS-SS-biotin after a 30-min pulse with [35S]Met/Cys and a 2-h chase. Biotinylated [35S]γGT was recovered with avidin-conjugated beads from the immunoprecipitate for SDS-PAGE (C). The pattern of [35S]γGT was obtained with a *BioRad Personal Molecular Imager FX* and *Quantity One* Software. The mobility of the γGT propeptide (P), large subunit (L), or small subunit (S) is indicated at the left of the panels and the mobility of the Amersham rainbow molecular weight markers (kDa) are indicated at the right.

heterodimer (L + S) represented the predominant form of the protein, although all three bands (P, L, and S) were broad and diffuse, consistent with the presence of N-glycans processed to complex type. Processing of N-glycans on both the propeptide and heterodimer was described previously (Capraro and Hughey, 1983). [^{35}S]γGT immunoprecipitated from cells after metabolic labeling overnight with [^{35}S]Met/Cys revealed a similar pattern on SDS-gels, although the propeptide was barely visible (Fig. 2B). However, [^{35}S]propeptide was clearly found on the cell surface along with the [^{35}S]heterodimer after a 30-min pulse with [^{35}S]Met/Cys and a 2-h chase (Fig. 2C). The diffuse bands representing P, L, and S at the cell surface are also consistent with N-glycan processing to complex type during transit through the Golgi complex and *trans*-Golgi network.

As predicted, γGTs with Cys mutations at C1 and C6 produced patterns of [^{35}S]-labeling and immunoblotting nearly identical to γGT(WT), indicating that they are not involved in formation of disulfide bonds (Fig. 2). γGTs with Cys mutations at C2 and C3 were found only as the propeptide and were virtually absent from the cell surface most consistent with their retention in the ER because of the presence of a disulfide bridge between these two Cys residues. Interestingly, γGT with Cys mutated at C4 showed an expression pattern almost identical to the WT γGT, whereas γGT with Cys mutated at C5 was expressed primarily as the propeptide with low levels of the heterodimer subunits and expression at the cell surface. These mixed results with Cys mutants C4 and C5 made any conclusions about disulfide bonds tenuous.

Because mutation of both Cys residues from the same disulfide bridge can sometimes rescue mutant protein expression (Darling *et al.*, 2000; Segal *et al.*, 1992), γGT double mutants were prepared with all combinations between Cys C2, C3, C4, and C5 for further analysis. These γGT mutants were also analyzed by immunoblotting, overnight metabolic labeling, and surface biotinylation of [^{35}S]-labeled cells. Normal processing was found for the double Cys mutant at C4 and C5 (named C4,5) whereas all double mutants involving Cys mutations at C2 and C3 lacked both cleavage and surface expression (see Fig. 2). Therefore, the cumulative data support the likelihood that a disulfide bridge is present between Cys at C4 and C5 within the CX$_3$C motif.

Close inspection of the residues around Cys C4 and C5 (CEVFC) reveals a glutamate residue between the two residues. Reddy *et al.* (1996) found that secretion of free λ light chains with an unpaired Cys^{214} (normally in a disulfide bridge with an IgG heavy chain) was dependent on an adjacent negative residue (Asp^{213}) that blocked ER retention by reducing thiolate conversion. Therefore, it is likely that the glutamate residue adjacent to the closely spaced C4 and C5 stimulated expression of the single Cys mutants of γGT by enhancing their exit from the ER. Interestingly, the glutamate in the CEVFC sequence was more efficient in rescue of the C4 mutant (with C5 unpaired) than the C5 mutant (with C4 unpaired), indicating that the negative charge is most effective when preceding an unpaired Cys as described for free λ light chains (Reddy *et al.*, 1996).

Because the absence of a disulfide bridge could also enhance denaturation and turnover of γGT, we tested the stability of γGT and the mutants both *in vivo* and *in vitro*. The *in vivo* stability of the γGT mutants was addressed by estimating the protein half-life for WT γGT and those mutants that showed sufficient levels of expression by labeling with [^{35}S]Met/Cys. Transiently transfected cells expressing WT γGT or Cys mutants C1, C4, C5, C6, or double Cys mutant C4,5 were pulse labeled for 30 min and chased for either 3 h or 24 h. Immunoprecipitated [^{35}S]γGT was quantified after SDS-PAGE using a BioRad Personal Molecular Imager FX, and the half-life that was calculated from the decrease in [^{35}S]γGT levels over 21 h was no different between WT γGT and all four mutants (data not shown). The *in vitro* stability of WT γGT and mutants was assessed by measuring the thermal stability of γGT enzymatic activity in Triton X-100 extracts of the transfected cells. Cell extracts exhibiting γGT activity above control levels from mock-transfected CHO cells included those expressing WT γGT, γGTs with Cys mutated at C1, C4, C5, C6, and the double Cys mutant C4,5. Preliminary experiments indicated that incubation of cell extracts containing WT γGT for 15 min at 60° reduced the enzymatic activity by one-half when measured at room temperature using γ-glutamyl-*p*-nitroanilide and glycylglycine as substrates (Hughey and Curthoys, 1976). Similar treatment of extracts from transfected cells expressing γGT with Cys mutated at C1 or C6 showed thermal denaturation profiles no different than WT γGT (t$_{1/2}$ ~10 min) (see Fig. 3). However, γGT activity in extracts from cells expressing γGT with Cys mutated at C4 or C5, or the double mutant C4,5, was significantly more sensitive to thermal denaturation (t$_{1/2}$ ~5–7 min) consistent with the presence of a stabilizing disulfide bridge in the rat γGT between C4 and C5. The data also support our earlier conclusions that Cys at C1 and C6 are not involved in the formation of a disulfide bridge.

γGT Enzymatic Activity Correlates with Propeptide Cleavage

In the course of characterizing the expression of γGT with Cys mutations, we noted that the mutants that remained uncleaved (C2, C3, and all double mutants except C4,5) also did not express γGT enzymatic activity above control levels in extracts of transfected CHO cells. This is consistent with the likelihood that rat γGT is a member of the Ntn hydrolase family,

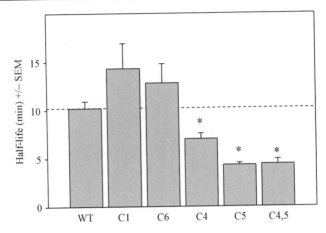

FIG. 3. A disulfide bridge between C4 and C5 enhances the thermal stability of γGT. CHO cells transiently expressing either wild-type rat γGT (WT) or mutants C1, C6, C4, C5, or C4,5 were extracted with octyl-glucoside and assayed for enzymatic activity before and after heating for 15 min at 60°. The half-life was calculated from the percent activity remaining after heating in three separate experiments (mean ±SEM), and the asterisk (*) indicates a statistically significant difference ($p \leq 0.01$) from the half-life of the WT γGT. Cells expressing γGT with C2 and/or C3 mutations did not exhibit sufficient γGT enzymatic activity for the analysis.

where enzymatic activity is dependent on autocatalytic cleavage of the propeptide (Brannigan et al., 1995). As shown in Fig. 4, when γGT activity in cell extracts from transfected cells was compared with the steady-state levels of [^{35}S]γGT propeptide and heterodimer, it was clear that γGT activity correlated best with levels of the [^{35}S]heterodimer (L + S) rather than total levels of [^{35}S]γGT (P + L + S), indicating that the propeptide is inactive. To directly determine whether cleavage of γGT activates the enzyme, we prepared a mutant of rat γGT by changing Thr^{380} to Asn (T380N) within the highly conserved cleavage site ($\downarrow$) to block generation of the heterodimer (DDGG$\downarrow$T^{380}AHL for rat γGT, see Table I). As predicted, transient expression of our T380N mutant in CHO cells produced only the propeptide form of γGT by metabolic labeling with [^{35}S]Met/Cys (30-min pulse and 24-h chase) as predicted, and cell extracts did not exhibit γGT activity above control levels of mock-transfected CHO cells (see Fig. 5). Hashimoto et al. (1995) previously showed that mutation of the corresponding Thr residue to Ala in the E. coli γGT (T391A) blocked its cleavage and recovery of enzymatic activity within the bacterial periplasmic fraction, whereas mutation of Thr^{391} to Ser or Cys partially blocked cleavage and recovery of γGT activity. As shown in Fig. 5, mutation of

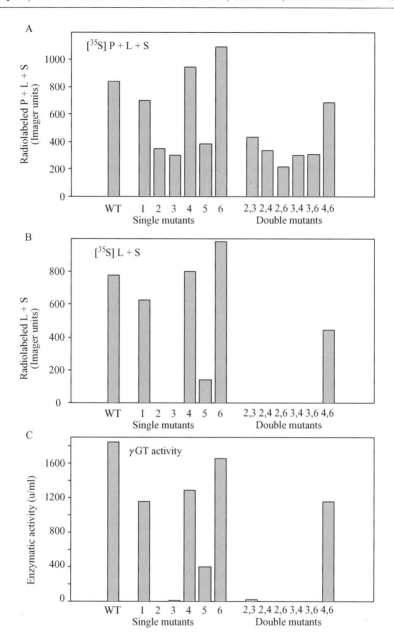

FIG. 4. γGT enzymatic activity correlates with propeptide cleavage. CHO cells transiently expressing either wild-type rat γGT (WT) or Cys mutants were either pulse labeled with [³⁵S] Met/Cys for 30 min and chased for 2 h before immunoprecipitation (A and B) or assayed for enzymatic activity (C). Radioactivity in the propeptide (P), large subunit (L), or small subunit (S) was determined with a molecular imager.

TABLE I
RESIDUES AT THE γGT CLEAVAGE SITE[a]

Rat	DDGG * TAHL
Pig	DDAG * TAHL
Mouse	DDGG * TAHL
Human	DDGG * TAHL
Yeast	NPHG * TAHF
E. coli	ESNQ * TTHY
Pseudomonas	EGSN * TTHY
Mouse GGL[b]	NRMG * TSHV

[a] Site of autocatalytic cleavage is marked with an asterisk.
[b] γ-glutamyl leukotrienase.

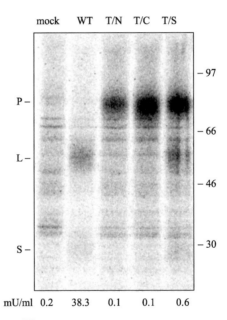

FIG. 5. Mutation of Thr[380] inhibits both γGT propeptide cleavage and enzymatic activity. CHO cells transiently expressing either rat γGT (WT) or γGT with Thr[380] mutated to either Asn (T/N), Cys (T/C) or Ser (T/S) were pulse labeled for 30 min with [35S]Met/Cys and chased for 24 h before enzymatic assay of cell extracts or immunoprecipitation and analysis of [35S]γGT on SDS-gels using a molecular imager. The mobility of the γGT propeptide (P), large subunit (L), or small subunit (S) is indicated at the left of the panel, and the mobility of the Amersham rainbow molecular weight markers (kDa) is indicated at the right. The enzymatic activity for each transfection in milliunits per ml (mU/ml) is indicated below the SDS-gel.

Thr^{380} to Cys (T/C) in rat γGT blocked propeptide cleavage completely, whereas 20% of the propeptide was cleaved with a more conservative mutation of Thr^{380} to Ser (T/S). Because the overall level of T380S mutant expression was significantly higher than for WT γGT in this experiment, the levels of [^{35}S]heterodimer from the mutant and WT γGT were similar. However, there was significantly less enzyme activity (∼2%) found in the T380S cell extract compared with cell extracts containing WT γGT, indicating that substitution of Ser for Thr as the nucleophile in the γGT active site had differential effects on the autocatalytic and enzymatic activity.

Because we had previously observed that the uncleaved propeptide of rat γGT is found at the cell surface along with the heterodimer (Altman and Hughey, 1986), pulse-chase studies were carried out to determine whether the enzymatically inactive rat γGT T380N mutant was also normally processed and expressed at the cell surface. CHO cells expressing T380N were pulse labeled for 30 min with [^{35}S]Met/Cys and chased up to 135 min before biotinylation of the cell surface. Analysis of the [^{35}S]γGT bands with a molecular imager indicates that γGT(T380N) is delivered to the cell surface with kinetics similar to the WT γGT (see Fig. 6). The half-life of [^{35}S]γGT(T380N) was also not significantly different than WT γGT

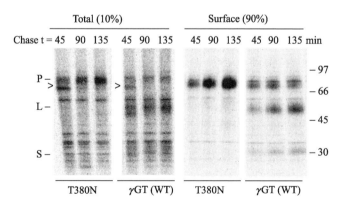

FIG. 6. γGT propeptide cleavage is not required for cell surface expression. CHO cells transiently expressing either rat γGT (WT) or the T380N mutant were pulse labeled for 30 min with [^{35}S]Met/Cys and chased for varying times as indicated. At each time point, the cells were biotinylated on ice. Cell surface γGT (WT) or mutant T380N was recovered from the immunoprecipitates using avidin-conjugated beads before SDS-PAGE and analysis with a molecular imager. The mobility of the γGT propeptide (P), large subunit (L), or small subunit (S) is indicated at the left of the panels, and the mobility of the Amersham rainbow molecular weight markers (kDa) is indicated at the right.

(data not shown). Thus, the absence of propeptide cleavage does not alter the stability or trafficking of the γGT propeptide.

γGT Propeptide Cleavage Occurs within the ER

If γGT is a member of the Ntn hydrolase family, then autocatalytic cleavage and enzymatic activation of γGT can theoretically occur as soon as the protein is properly folded within the ER. To test this possibility, γGT cleavage was studied after its modification with an ER retention signal or after inhibition of vesicular traffic using brefeldin A.

To retain γGT within the ER, a chimera was prepared with the ecto-domain of the rat γGT and the transmembrane and cytoplasmic domain of CD74 (Iip33), the MHC invariant chain isoform that is retained in the ER (Schutze et al., 1994). CD74 is a type-2 transmembrane protein (N^{in}/C^{out}) with two different initiator Met residues; differential use of the two start codons results in cytoplasmic domains of two different lengths for Iip31 and Iip33. Iip33 has 16 more residues than Iip31 including an amino-terminal Arg-rich sequence; and at least two Arg are required for ER retention of Iip33 (Schutze et al., 1994). To determine whether the CD74-γGT chimera is actually retained in the ER, transiently transfected cells were pulse labeled for 30 min and chased for 4 h, a time frame that allows full processing and delivery of WT γGT to the cell surface (data not shown). As shown in Fig. 7, γGT and the non-cleaved mutant T380N exhibited diffuse bands on SDS-PAGE that were resistant to treatment with endoglycosidase H (Endo H), an enzyme that cleaves N-glycans from proteins that are minimally processed before transit through the medial compartment of the Golgi complex. In contrast, two sharp bands representing the heterodimer subunits were observed for the chimera of CD74-γGT. Both were sensitive to treatment with Endo H, indicating that they had not reached the medial Golgi compartment. Approximately 10% of the chimera subunits were found as more diffuse bands that were Endo H resistant and likely represent a small fraction of the total that was aberrantly released from the ER. These conclusions were confirmed by biotinylation of the cell surface after the pulse and chase periods (data not shown). Most notably, assay of the cell extracts also revealed that the CD74-γGT chimera was enzymatically active; and the ratio of *enzymatic activity to [^{35}S]chimera subunit expression* was comparable to the ratio observed for *enzymatic activity to [^{35}S]γGT(WT) subunit expression* in the same experiment (data not shown). Our finding that the CD74-γGT chimera was cleaved despite its apparent retention in the early biosynthetic pathway suggests that γGT is likely to be cleaved as soon as it is properly folded within the ER.

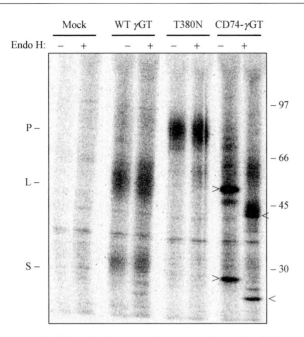

FIG. 7. A mutant of γGT retained in the endoplasmic reticulum is still cleaved. CHO cells transiently expressing either rat γGT (WT), γGT mutant T380N, or a chimera of γGT with the cytoplasmic tail of an ER resident protein CD74 (CD74-γGT) were pulse labeled for 30 min with [³⁵S]Met/Cys and chased for 4 h before immunoprecipitation and treatment with (+) or without (−) Endo H before SDS-PAGE and analysis with a molecular imager. The mobility of the γGT propeptide (P), large subunit (L), or small subunit (S) is indicated at the left of the panels, and the mobility of the Amersham rainbow molecular weight markers (kDa) is indicated at the right. The mobility of the sharp bands representing the CD74-γGT subunits is indicated before (>) and after (<) treatment with Endo H.

Because it was possible that γGT cleavage and activation were actually blocked within the ER but disrupted during extraction of the cells with detergent, the time course of γGT and chimera cleavage was followed after a shorter pulse with [³⁵S]Met/Cys (15 min) and extraction of cells with SDS to block any post-lysis autocatalytic cleavage. Immunoprecipitation of γGT was carried out after dilution of the SDS with an excess of nonionic detergents. As shown in Fig. 8, cleavage of both WT γGT and the CD74-γGT chimera to subunits apparently proceeded in a linear fashion from 0–45 min of chase when cells were extracted with nonionic detergents. However, when cells were directly extracted into SDS, no cleavage was observed after the 15-min pulse (chase t = 0), although a strong propeptide signal was found for both WT γGT and the CD74-γGT chimera. The

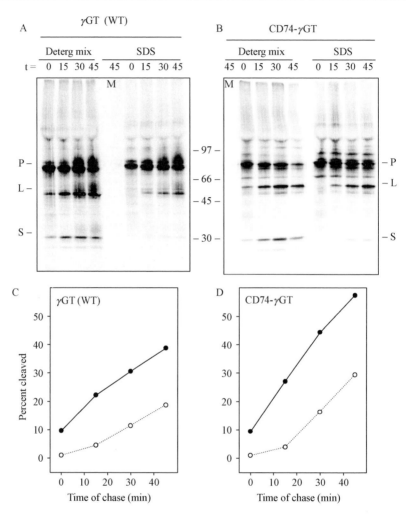

FIG. 8. Cell extraction with SDS reveals a latent autocatalytically active γGT. CHO cells transiently expressing either rat γGT (WT) (A and C) or the CD74-γGT chimera (B and D) were pulse labeled for 15 min with [^{35}S]Met/Cys and chased for the indicated times before extraction with either a mixture of nonionic detergents (closed circles) or SDS (open circles). Mock-transfected cells (no DNA) were chased for 45 min and extracted as indicated (M). Immunoprecipitates were analyzed with a molecular imager after SDS-PAGE (A and B), and the percent cleaved was plotted (C and D) based on radioactivity in (L + S)/(P + L + S). Note that no L or S is present in the t = 0 point of the SDS extraction, although the nascent chains form a slight background blur in the lane. The mobility of the γGT propeptide (P), large subunit (L), or small subunit (S) is indicated at the outside edges of the panels, and the mobility of the Amersham rainbow molecular weight markers (kDa) is indicated between the panels.

apparent *percent cleaved* at the subsequent chase times was also notably reduced by SDS extraction, suggesting that some fraction of the propeptide is competent for post-lysis autocatalytic cleavage if its environment is altered by extraction into nonionic detergents.

γGT was also retained in the ER by inclusion of brefeldin A (BFA) in the culture media during the starvation, pulse, and chase periods of the experiment to block exit of all proteins from the ER. BFA is a fungal metabolite that binds to the guanine nucleotide-exchange protein for the small GTPase ARF1, and thereby blocks vesicular transport to the Golgi complex (Donaldson *et al.*, 1992; Helms and Rothman, 1992). As shown in Fig. 9, treatment of cells with BFA noticeably altered the processing of the glycans on γGT confirming that the protein did not transit the Golgi complex. However, the profile of propeptide cleavage was not altered by retention of the γGT propeptide in the ER. No cleavage was found after the 15-min pulse and extraction into SDS (t = 0), and the percent cleaved at all times of chase were the same with and without BFA treatment,

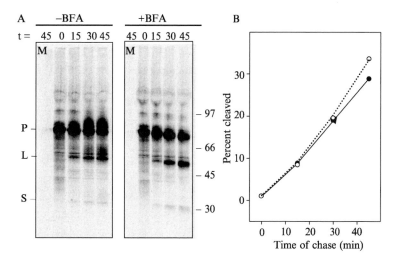

FIG. 9. γGT is cleaved before reaching the Golgi complex. CHO cells transiently expressing rat γGT (WT) were pulse labeled for 15 min with [^{35}S]Met/Cys and chased for the indicated times before extraction with SDS. Brefeldin A (BFA, 10 n*M*) was included in the starvation, pulse, and chase periods as indicated. Immunoprecipitates were analyzed with a molecular imager after SDS-PAGE (A), and the percent cleaved was plotted (B) based on radioactivity in (L + S)/(P + L + S) (closed circles, –BFA and open circles, +BFA). Mock-transfected cells (no DNA) were chased for 45 min as indicated (M). The mobility of the γGT propeptide (P), large subunit (L), or small subunit (S) is indicated at the left of the panel, and the mobility of the Amersham rainbow molecular weight markers (kDa) is indicated at the right.

indicating that γGT propeptide cleavage occurs before its arrival in the Golgi complex.

Discussion

Disulfide Bridges in γ*GT and Functional Implications*

Experiments designed to identify disulfide bridges in mammalian γGT were prompted by our interest in a unique CX_3C motif in both γGT$\Delta1$ (WT) and the naturally occurring inactive isoform γGT$\Delta7$ that specifically mediate an ER redox stress response (Joyce-Brady *et al.*, 2001). The results of our experiments indicate that C1 (Cys23) and C6 (Cys453) in the transmembrane and small subunit domains, respectively, do not form disulfide bridges, because mutation of these residues does not alter the synthesis, processing, surface expression, enzymatic activity, or stability of the enzyme. However, our results do suggest that Cys at C2 (Cys49), C3 (Cys73), C4 (Cys191), and C5 (Cys195) are most likely to form bridges between C2 and C3 and between C4 and C5. Mutation of C2 or C3 blocked propeptide cleavage, N-glycan processing, delivery to the cell surface, and enzymatic activity, whereas mutation of C4 or C5 gave mixed results. Mutation of C4 had almost no effect on γGT processing, whereas mutation of C5 reduced γGT expression dramatically, but mutation of either C4 or C5 reduced the thermal stability of the enzyme equally. Because coincident mutation of C4 and C5 returned γGT expression to normal, whereas the thermal stability of the γGT was still reduced, a disulfide is most likely present between C4 and C5. Although coincident mutation of C2 and C3 did not rescue γGT expression, the similar behavior of the individual mutants is most consistent with a disulfide bridge between these two residues.

There are provocative reports that γGT has biological activities independent of its enzymatic activity. Joyce-Brady *et al.* (2001) reported that either γGT or the inactive γGT$\Delta7$ could mediate an ER stress response, whereas Niida *et al.* (2004) reported that γGT or a form inactivated with acivicin could stimulate bone resorption by osteoclasts. Therefore, it is possible that the unique CX_3C motif has some biological function beyond stabilizing the structure of the γGT. CX_3C sequences have been identified as a signature motif for one class of chemokines (Bazan *et al.*, 1997; Wells and Peitsch, 1997), an essential coordinate for copper binding in several proteins (Balatri *et al.*, 2003; Nittis *et al.*, 2001) and an essential motif in the redox protein A2.5L of vaccinia virus (Senkevich *et al.*, 2002).

Fractalkine/neurotactin is the only member of the CX_3C family of chemokines (for review, see Stievano *et al.*, 2004). However, the crystal structure of the fractalkine chemokine domain reveals that it is similar to other chemokine monomers, particularly MCP-1 and RANTES, where the Cys in the CX_3C motif are involved in two different disulfide bonds (C8–C34 and C12–C50 in fractalkine) (Hoover et al., 2000). Therefore, the CX_3C chemokine is novel, because it is a larger transmembrane mucin-like protein with a chemokine domain rather than because of the novel spacing of the CX_3C Cys residues.

In contrast, the recent solution structure of the *Bacillus subtilis* Sco1, required for proper assembly of mitochondrial cytochrome c oxidase, was solved by NMR and indicates that a highly conserved disulfide within the CX_3C motif is required for Sco1 copper (I) binding (Balatri et al., 2003). Sco1 is a 30-kDa protein with a single transmembrane domain localized to the inner mitochondrial membrane facing the intermembrane space. Characterization of the recombinant anchor minus Sco1 revealed a thioredoxin-type fold characterized by a central β sheet flanked by three helices; this structure is most closely related to a family of antioxidant enzymes known as peroxiredoxins that also show peroxidase activity (Balatri et al., 2003). The CX_3C motif and a highly conserved His residue, implicated in copper (II) binding, are found on two different but structurally close loops; and both loops are found within a negatively charged patch on the surface of Sco1 that is likely to interact directly with a positively charged patch on subunit II of cytochrome c oxidase (Balatri et al., 2003). Interestingly, residues surrounding the CX_3C motif of γGT are also charged (PVLCEVFCRQGK), with the PVLCEVF sequence forming a two-turn helix (personal communication, Carlos Camacho, University of Pittsburgh). This observation indicates that the CX_3C in γGT is probably a solvent-accessible site, raising the possibility that it might also be involved in copper binding. Interestingly, there have been reports linking lipid peroxidation to γGT activity and chelated copper and iron (Glass and Stark, 1997; Paolicchi et al., 1997).

The small α-helical A2.5L protein of poxviruses with a CX_3C motif is one of three thiol oxidoreductases that make up a complete pathway for cytoplasmic formation of disulfides in viral membrane proteins (Senkevich et al., 2002). A2.5L forms a stable disulfide-linked heterodimer with the viral E10R and a transient disulfide-linked complex with the viral G4L. Because E10R is a member of the ERV1/ALR family of FAD-containing sulfhydryl oxidases, oxygen is likely to be the terminal electron acceptor in this pathway. Interestingly, whereas A2.5L is required for the oxidation of E10R, its CX_3C Cys residues are not, and single mutations of Cys to Ser severely reduce, but do not entirely block, oxidation of the downstream viral substrates, suggesting that the A2.5L-E10R heterodimer functions as a complex (Senkevich et al., 2002). Considering the key role that γGT plays in redox balance in the cell, future studies should address the possibility that γGT might exhibit a second function by interacting with other redox-related proteins either within the ER or at the cell surface.

γGT and the Ntn Family of Hydrolases

The results of this present characterization of the rat γGT propeptide cleavage indicate that all mammalian γGTs are likely to be members of the N-terminal nucleophile (Ntn) hydrolase family. Our finding that the T380N mutation in the rat γGT blocked, whereas the more conservative T380S mutation reduced, both propeptide cleavage and recovery of enzymatic activity is consistent with the recent proposal that the bacterial γGT is a member of the Ntn hydrolase family (Inoue et al., 2000; Suzuki and Kumagai, 2002). This structural superfamily, first described by Brannigan et al. (1995), exhibits a four-layered catalytically active αββα-core formed by two antiparallel β-sheets packed against each other and overlaid with a layer of antiparallel α-helices on one side. All of the Ntn-hydrolases exhibit autocatalytic cleavage where the new N-terminus (Thr, Ser, or Cys) becomes the catalytic nucleophile for the enzyme. Based on the crystal structures and known propeptide cleavage events, members of this class include penicillin G acylases; β-subunits 1, 2, and 5 of the proteasome; glutamine 5-phosphoribosyl-1-pyrophosphate aminotransferases (GAT); aspartylglucosaminidase; and L-aminopeptidase-D-Ala-esterase/amidase (for review, see Brannigan et al., 1995; Oinonen and Rouvinen, 2000). Brannigan et al. (1995) were the first to suggest that γGT might be part of this Ntn-hydrolase family because of (1) the known propeptide cleavage event, (2) finding Thr as the new N-terminus, and (3) the ability of γGT to use Gln as a substrate like another family member, GAT. Although only a preliminary crystal structure of bacterial γGT is available (Sakai et al., 1996), Inoue et al. have now shown that a novel mechanism-based affinity reagent, the γ-phosphonic acid monofluoride derivative of glutamic acid, can label the small subunit N-terminal Thr^{391} (Inoue et al., 2000). Because Thr at the N-terminus of the small subunit is the only conserved Cys, Ser, or Thr residue within all known γGTs, the cumulative data support the possibility that all γGTs are members of the Ntn-hydrolase superfamily. Even the novel γ-glutamyl leukotrienase exhibits a conserved sequence consistent with autocatalytic cleavage (see Table I) (Han et al., 2002).

Compared with the mammalian γGT sequences, only the distal side of the propeptide cleavage site is conserved in bacterial γGT ($\downarrow T^{391}TH$, see Table I), and mutation of either Thr^{391} to Ala or mutation of His^{393} to Gly in E. coli γGT completely blocks propeptide cleavage (Hashimoto et al., 1995). However, mutation of Thr^{391} to Ser or Cys, or mutation of the Thr^{392}

to Ala or Ser, only partially blocks the cleavage (Hashimoto et al., 1995; Suzuki and Kumagai, 2002). Most importantly, Hashimoto et al. (1995) showed that there was a clear correlation between the efficiency of γGT propeptide cleavage and levels of enzymatic activity recovered in the periplasmic fraction of each transformed bacterial strain. Interestingly, we found that mutation of Thr[380] to Cys in the rat γGT (corresponding to E. coli Thr[391]) completely blocked, whereas the T380S mutation only partially blocked, γGT cleavage and any recovery of enzymatic activity. Because we have now shown that the rat γGT has six Cys forming only two disulfide bridges, it is possible that the T380C mutation blocks cleavage by interfering with proper pairing of disulfides. E. coli γGT has only one Cys (within the cleaved signal sequence), so the E. coli γGT T391C mutation is unlikely to interfere with its proper folding. Interestingly, the T380S mutation in the rat γGT had a greater effect on enzymatic activity than it did on autocatalytic cleavage, whereas the E. coli γGT T391S mutation showed comparable reduction in processing and enzymatic activity (Hashimoto et al., 1995). Although the E. coli and rat γGTs exhibit 33% identity in amino acids sequence, the enzymes do exhibit notable differences in substrate specificities and enzyme kinetics (Suzuki et al., 1986). Therefore, the reduced enzymatic activity of the rat γGT T380S mutant compared with its autocatalytic activity could simply reflect the unique alignment of the active site for an enzyme that can efficiently catalyze both hydrolysis and transfer reactions with γ-glutamyl substrates.

γGT Cleavage and Activation in the ER

We have also presented data showing that γGT is cleaved before it reaches the Golgi complex. If γGT is activated while still in the ER, this raises the question of how the cell maintains its critical level of glutathione, and what role this localization would have for enzymatically active γGT. The possibility still remains that γGT in the ER is enzymatically inactive because of inhibition by a specific protein chaperone or by small metabolites. Enzymatic assay of detergent extracts from cells expressing the ER-localized CD74-γGT chimera revealed levels of activity that correlated well with the corresponding level of heterodimer subunits regardless of the level of N-glycan processing. Enzymatic activity was also assayed on live cells by replacing the culture media with phophate-buffered saline (PBS) containing γ-glutamyl-*p*-nitroanilide and glycylglycine for 10 min, with and without 0.025% saponin, before measuring absorbance at 410 nm. Saponin at 0.25% will permeabilize cell membranes including the ER but leave the cells intact (Lin et al., 2003). When 0.025% saponin was included in the PBS mix, we observed an increase in γGT activity for cells expressing both the WTγGT (10–20%) and CD74-γGT (370–420%), reflecting their predominant subcellular localization at either the cell surface or within the ER, respectively (data not shown). These results are most consistent with the presence of enzymatically active γGT within the ER. However, either protein or metabolite inhibitors of γGT could be sufficiently diluted by extraction with nonionic detergents (or incubation of cells in saponin) to levels that would cause their dissociation from γGT, and thereby revealing latent activity. In fact, we did observe latent autocatalytic activity of γGT by extraction in nonionic detergents that was blocked by extraction directly into SDS. The results of these experiments showed that γGT propeptide is present in the ER for at least 15 min after its synthesis in a state competent for autocatalytic cleavage. Now, it remains to be determined whether γGT catalytic activity is also latent until γGT exits the ER or reaches its final localization on the plasma membrane, and whether the uncleaved propeptide has a function independent of the catalytically active enzyme.

Acknowledgments

We would like to thank Dr. Ora Weisz and Dr. Martin Joyce-Brady for critical reading of the manuscript, Dr. Carlos Camacha for analysis on p. 442, and Rick Stremple for excellent technical assistance. This work was funded by grants to RPH from The American Lung Association of Pennsylvania and NIH (DK54787 and DK26102).

References

Altman, R. A., and Hughey, R. P. (1986). The identification of two subcellular sites for cleavage of γ-glutamyltranspeptidase propeptide. *Biochem. Int.* **13**, 1009–1017.

Balatri, E., Banci, L., Bertini, I., Cantini, F., and Ciofi-Baffoni, S. (2003). Solution structure of Sco1: A thioredoxin-like protein involved in cytochrome c oxidase assembly. *Structure (Camb.)* **11**, 1431–1443.

Barrios, R., Shi, Z. Z., Kala, S. V., Wiseman, A. L., Welty, S. E., Kala, G., Bahler, A. A., Ou, C. N., and Lieberman, M. W. (2001). Oxygen-induced pulmonary injury in gamma-glutamyl transpeptidase-deficient mice. *Lung* **179**, 319–330.

Bazan, J. F., Bacon, D. B., Hardiman, G., Wang, W., Soo, K., Rossi, D., Greaves, D. R., Zlotnik, S., and Schall, T. J. (1997). A new class of membrane-bound chemokine with a CX₃C motif. *Nature* **385**, 640–644.

Blochberger, T. C., Sabatine, J. M., Lee, Y. C., and Hughey, R. P. (1989). O-linked glycosylation of rat renal gamma-glutamyltranspeptidase adjacent to its membrane anchor domain. *J. Biol. Chem.* **264**, 20718–20722.

Brannigan, J. S., Dodson, G., Duggleby, H. J., Moody, P. C. E., Smith, J. L., Tomchick, D. R., and Murzin, A. G. (1995). A protein catalytic framework with an N-terminal nucleophile is capable of self-activation. *Nature* **378**, 416–419.

Capraro, M. A., and Hughey, R. P. (1983). Processing of the propeptide form of rat renal gamma-glutamyltranspeptidase. *FEBS Lett.* **157**, 139–143.

Chevez-Barrios, P., Wiseman, A. L., Rojas, E., Ou, C. N., and Lieberman, M. W. (2000). Cataract development in gamma-glutamyl transpeptidase-deficient mice. *Exp. Eye Res.* **71**, 575–582.

Chikhi, N., Holic, N., Guellaen, G., and Laperche, Y. (1999). Gamma-glutamyl transpeptidase gene organization and expression: A comparative analysis in rat, mouse, pig and human species. *Comp. Biochem. Physiol.* **122**, 367–380.

Darling, R. J., Ruddon, R. W., Perini, F., and Bedows, E. (2000). Cystine knot mutations affect the folding of the glycoprotein hormone α-subunit. *J. Biol. Chem.* **275**, 15413–15421.

Donaldson, J. G., Finazzi, D., and Klausner, R. D. (1992). Brefeldin A inhibits Golgi membrane-catalysed exchange of guanine nucleotide onto ARF protein. *Nature* **360**, 350–352.

Edman, J. C., Ellis, L., Blacher, R. W., Roth, R. A., and Rutter, W. J. (1985). Sequence of protein disulphide isomerase and implication of its relationship to thioredoxin. *Nature* **317**, 267–270.

Glass, G. A., and Stark, A. A. (1997). Promotion of glutathione-gamma-glutamyl transpeptidase-dependent lipid peroxidation by copper and ceruloplasmin: The requirement for iron and the effects of antioxidants and antioxidant enzymes. *Environ. Mol. Mutagen.* **29**, 73–80.

Han, B., Luo, G., Shi, Z. Z., Barrios, R., Atwood, D., Liu, W., Habib, G. M., Sifers, R. N., Corry, D. B., and Lieberman, M. W. (2002). Gamma-glutamyl leukotrienase, a novel endothelial membrane protein, is specifically responsible for leukotriene D(4) formation *in vivo*. *Am. J. Pathol.* **161**, 481–490.

Harding, C. O., Williams, P., Wagner, E., Chang, D. S., Wild, K., Colwell, R. E., and Wolff, J. A. (1997). Mice with genetic gamma-glutamyl transpeptidase deficiency exhibit glutathionuria, severe growth failure, reduced life spans, and infertility. *J. Biol. Chem.* **272**, 12560–12567.

Hashimoto, W., Suzuki, H., Yamamoto, K., and Kumagai, H. (1995). Effect of site-directed mutations on processing and activity of γ-glutamyltranspeptidase of *Escherichia coli* K-12. *J. Biochem.* **118**, 75–80.

Helenius, A., and Aebi, M. (2004). Roles of N-linked glycans in the endoplasmic reticulum. *Annu. Rev. Biochem.* **73**, 1019–1049.

Helms, J. B., and Rothman, J. E. (1992). Inhibition by brefeldin A of a Golgi membrane enzyme that catalyses exchange of guanine nucleotide bound to ARF. *Nature* **360**, 352–354.

Hiller, M., Huse, K., Szafranski, K., Jahn, N., Hampe, J., Schreiber, S., Backofen, R., and Platzer, M. (2004). Widespread occurrence of alternative splicing at NAGNAG acceptors contributes to proteome plasticity. *Nat. Genet.* **36**, 1255–1257.

Hoover, D. M., Mizoue, L. S., Handel, T. M., and Lubkowski, J. (2000). The crystal structure of the chemokine domain of fractalkine shows a novel quaternary arrangement. *J. Biol. Chem.* **275**, 23187–23193.

Hughey, R. P., and Curthoys, N. P. (1976). Comparison of the size and physical properties of γ-glutamyltranspeptidase purified from rat kidney after solubilization with papain or with Triton X-100. *J. Biol. Chem.* **251**, 7863–7870.

Inoue, M., Hiratake, J., Suzuki, H., Kumagai, H., and Sakata, K. (2000). Identification of catalytic nucleophile of *Escherichia coli* γ-glutamyltranspeptidase by γ-monofluorophosphono derivative of glutamic acid: N-terminal Thr-391 in small subunit is the nucleophile. *Biochemistry* **39**, 7764–7771.

Isidoro, C., Maggioni, C., Demoz, M., Pizzagalli, A., Fra, A. M., and Sitia, R. (1996). Exposed thiols confer localization in the endoplasmic reticulum by retention rather than retrieval. *J. Biol. Chem.* **271**, 26138–26142.

Jean, J. C., Harding, C. O., Oakes, S. M., Yu, Q., Held, P. K., and Joyce-Brady, M. (1998). Gamma-glutamyl transferase (GGT) deficiency in the GGT^{enu1} mouse results from a single point mutation that leads to a stop codon in the first coding exon of GGT mRNA. *Mutagenesis* **13,** 101–106.

Jean, J. C., Liu, Y., Brown, L. A., Marc, R. E., Klings, E., and Joyce-Brady, M. (2002). Gamma-glutamyl transferase deficiency results in lung oxidant stress in normoxia. *Am. J. Physiol Lung Cell Mol. Physiol* **283,** L766–L776.

Joyce-Brady, M., Jean, J.-C., and Hughey, R. P. (2001). Gamma-glutamyltransferase and its isoform mediate an endoplasmic reticulum stress response. *J. Biol. Chem.* **276,** 9468–9477.

Joyce-Brady, M., Oakes, S. M., Wuthrich, D., and Laperche, Y. (1996). Three alternative promoters of the rat gamma-glutamyl transferase gene are active in developing lung and are differentially regulated by oxygen after birth. *J. Clin. Invest* **97,** 1774–1779.

Kumar, T. R., Wiseman, A. L., Kala, G., Kala, S. V., Matzuk, M. M., and Lieberman, M. W. (2000). Reproductive defects in gamma-glutamyl transpeptidase-deficient mice. *Endocrinology* **141,** 4270–4277.

Levasseur, R., Barrios, R., Elefteriou, F., Glass, D. A., Lieberman, M. W., and Karsenty, G. (2003). Reversible skeletal abnormalities in gamma-glutamyl transpeptidase-deficient mice. *Endocrinology* **144,** 2761–2764.

Lin, S., Lu, X., Chang, C. C., and Chang, T. Y. (2003). Human acyl-coenzyme A: Cholesterol acyltransferase expressed in chinese hamster ovary cells: Membrane topology and active site location. *Mol. Biol. Cell* **14,** 2447–2460.

Lipton, A. J., Johnson, M. A., Macdonald, T., Lieberman, M. W., Gozal, D., and Gaston, B. (2001). S-nitrosothiols signal the ventilatory response to hypoxia. *Nature* **413,** 171–174.

Mustacich, D., and Powis, G. (2000). Thioredoxin reductase. *Biochem. J.* **346,** 1–8.

Nash, B., and Tate, S. S. (1982). Biosynthesis of rat renal gamma-glutamyl transpeptidase. Evidence for a common precursor of the two subunits. *J. Biol. Chem.* **257,** 585–588.

Nash, B., and Tate, S. S. (1984). *In vitro* translation and processing of rat kidney gamma-glutamyl transpeptidase. *J. Biol. Chem.* **259,** 678–685.

Niida, S., Kawahara, M., Ishizuka, Y., Ikeda, Y., Kondo, T., Hibi, T., Suzuki, Y., Ikeda, K., and Taniguchi, N. (2004). Gamma-glutamyltranspeptidase stimulates receptor activator of nuclear factor-kappaB ligand expression independent of its enzymatic activity and serves as a pathological bone-resorbing factor. *J. Biol. Chem.* **279,** 5752–5756.

Nittis, T., George, G. N., and Winge, D. R. (2001). Yeast Sco1, a protein essential for cytochrome c oxidase function is a Cu(I)-binding protein. *J. Biol. Chem.* **276,** 42520–42526.

Oinonen, C., and Rouvinen, J. (2000). Structural comparison of Ntn-hydrolases. *Protein Sci.* **9,** 2329–2337.

Paolicchi, A., Tongiani, R., Tonarelli, P., Comporti, M., and Pompella, A. (1997). Gamma-glutamyl transpeptidase-dependent lipid peroxidation in isolated hepatocytes and HepG2 hepatoma cells. *Free Rad. Biol. Med.* **22,** 853–860.

Pardo, A., Ruiz, V., Arreola, J. L., Ramirez, R., Cisneros-Lira, J., Gaxiola, M., Barrios, R., Kala, S. V., Lieberman, M. W., and Selman, M. (2003). Bleomycin-induced pulmonary fibrosis is attenuated in gamma-glutamyl transpeptidase-deficient mice. *Am. J. Respir. Crit Care Med.* **167,** 925–932.

Pompella, A., Visvikis, A., Paolicchi, A., De, T., and Casini, A. F. (2003). The changing faces of glutathione, a cellular protagonist. *Biochem. Pharmacol.* **66,** 1499–1503.

Powis, G., and Montfort, W. R. (2001). Properties and biological activities of thioredoxins. *Annu. Rev. Biophys. Biomol. Struct.* **30,** 412–455.

Reddy, P., Sparvoli, A., Fagioli, C., Fassina, G., and Sitia, R. (1996). Formation of reversible disulfide bonds with the protein matrix of the endoplasmic reticulum correlates with the retention of unassembled Ig light chains. *EMBO J.* **15,** 2077–2085.

Sakai, H., Sakabe, N., Sasaki, K., Hashimoto, W., Suzuki, H., Tachi, H., Kumagai, H., and Sakabe, K. (1996). A preliminary description of the crystal structure of γglutamyltranspeptidase from *E. coli* K-12. *J. Biochem.* **120**, 26–28.

Schutze, M. P., Peterson, P. A., and Jackson, M. R. (1994). An N-terminal double-arginine motif maintains type II membrane proteins in the endoplasmic reticulum. *EMBO J.* **13**, 1696–1705.

Segal, M. S., Bye, J. M., Sambrook, J. F., and Gething, M.-J. H. (1992). Disulfide bond formation during the folding of influenza virus hemagglutinin. *J. Cell Biol.* **118**, 227–244.

Senkevich, T. G., White, C. L., Koonin, E. V., and Moss, B. (2002). Complete pathway for protein disulfide bond formation encoded by poxviruses. *Proc. Natl. Acad. Sci. USA* **99**, 6667–6672.

Stievano, L., Piovan, E., and Amadori, A. (2004). C and CX3C chemokines: Cell sources and physiopathological implications. *Crit Rev. Immunol.* **24**, 205–228.

Suzuki, H., and Kumagai, H. (2002). Autocatalytic processing of gamma-glutamyl transpeptidase. *J. Biol. Chem.* **277**, 43536–43543.

Suzuki, H., Kumagai, H., and Tochikura, T. (1986). Gamma-glutamyltranspeptidase from *Escherichia coli* K-12: Purification and properties. *J. Bacteriol.* **168**, 1325–1331.

Taniguchi, N. (1998). γ-Glutamyl transpeptidase catalytic mechanism and gene expression. *Adv. Enzymol. Rel. Mol. Biol.* **72**, 239–278.

Wells, T. N., and Peitsch, M. C. (1997). The chemokine information source: Identification and characterization of novel chemokines using the World Wide Web and expressed sequence tag databases. *J. Leukoc. Biol.* **61**, 545–550.

Zhang, K., and Kaufman, R. J. (2004). Signaling the unfolded protein response from the endoplasmic reticulum. *J. Biol. Chem.* **279**, 25935–25938.

[27] Gamma-Glutamyl Transpeptidase Substrate Specificity and Catalytic Mechanism

By JEFFREY W. KEILLOR, ROSELYNE CASTONGUAY,* and
CHRISTIAN LHERBET*

Abstract

The enzyme γ-glutamyltranspeptidase (GGT) is critical to cellular detoxification and leukotriene biosynthesis processes, as well as amino acid transport in kidneys. GGT has also been implicated in many important physiological disorders, including Parkinson's disease and inhibition of apoptosis. It binds glutathione as donor substrate and initially forms a γ-glutamyl enzyme that can then react with a water molecule or an acceptor substrate (usually an amino acid or a dipeptide) to form glutamate or a product containing a new γ-glutamyl isopeptide bond, respectively, thus regenerating the free enzyme. Given the importance of GGT in human

* These authors contributed equally to this work.

METHODS IN ENZYMOLOGY, VOL. 401   0076-6879/05 $35.00
Copyright 2005, Elsevier Inc. All rights reserved.
                                                          DOI: 10.1016/S0076-6879(05)01027-X

physiology, we have undertaken studies of its substrate specificity and catalytic mechanism. In the course of these studies, we have developed methods for the indirect evaluation of donor substrate affinity and stereospecificity and applied others for the measurement of steady state and pre-steady state kinetics and linear free-energy relationships. These methods and the pertinent results obtained with them are presented herein.

Introduction

γ-Glutamyl transpeptidase (GGT; EC 2.3.2.2) is a highly glycosylated heterodimeric enzyme found in mammals and plants (Kasai and Larsen, 1980; Taniguchi and Ikeda, 1998). In mammals, it is found in brain, liver, pancreas, and especially in kidneys (Hanigan and Pitot, 1985; Sian et al., 1994; Tate and Meister, 1985). It plays a role in cellular detoxification through formation of mercapturic acids and confers resistance against anti-tumor drugs (Godwin et al., 1992). GGT is involved in the biosynthesis of leukotriene D, which mediates bronchoconstriction in asthma (Bernström et al., 1982; Örning et al., 1980) and in amino acid transport in kidneys (Meister, 1973). It has also been implicated in many physiological disorders, such as Parkinson's disease (Sian et al., 1994), diabetes (Lee et al., 2003), and inhibition of apoptosis (Del Bello et al., 1999; Graber and Losa, 1995). GGT uses glutathione (GSH) as an acyl donor substrate and transfers its γ-glutamyl moiety to acceptor substrates, such as amino acids or dipeptides, to form a product containing a new isopeptide bond. This transamidation role is very important for amino acid transport in the kidney (Griffith et al., 1978; Meister, 1973). The reaction catalyzed by GGT is known to proceed through a modified ping-pong mechanism, as shown in Scheme 1 (Allison, 1985; Taniguchi and Ikeda, 1998). The enzyme binds its donor substrate, is transiently acylated by its γ-glutamyl moiety, and releases the first reaction product (cysteinylglycine, in the case of GSH) during the acylation step. The resulting γ-glutamyl acyl-enzyme can then react with either water (hydrolysis) or an acceptor substrate (typically an amino acid or a dipeptide) in a deacylation step to form either glutamate or a transpeptidated product, respectively.

GGT Substrate Specificity

Donor and Acceptor Substrate Affinities and Potential Amino Acid Interactions

Rat kidney GGT catalyzes the cleavage of the γ-glutamyl bond of GSH and other γ-glutamyl amides. The γ-glutamyl moiety of donor substrates is critical for recognition by GGT, whereas the primary amine-leaving group

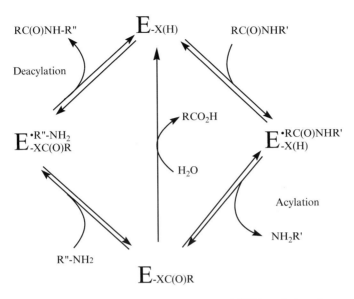

SCHEME 1. Modified ping-pong mechanism of GGT catalytic cycle.

liberated as the first product in the ping-pong mechanism can vary from Cys-Gly, in the case of GSH, to p-substituted anilines (Ménard *et al.*, 2001). The stereoselectivity for donor substrates having the L-configuration at their α-carbon is less elevated than that shown for acceptor substrates (Thompson and Meister, 1976, 1977). In human GGT, arginine 107 (Arg-111 in rat GGT) in the heavy subunit and aspartate 423 (Asp-422 in rat GGT) in the light subunit have been implicated in the interaction with the α-ammonium group and the α-carboxylate group of the γ-glutamyl donor substrate (Heisterkamp *et al.*, 1991; Ikeda *et al.*, 1995a).

The binding site of the acceptor substrate has been proposed to be the site where the cysteine and glycine residues of the donor substrate GSH are bound (Taniguchi and Ikeda, 1998; Thompson and Meister, 1977). GGT exhibits strict stereospecificity toward its acceptor substrates; only L-amino acids and dipeptides are recognized (Allison, 1985). Among the amino acids tested, those showing the greatest affinity for GGT include L-methionine, L-cystine, L-glutamine, and L-alanine. Sterically hindered amino acids such as L-proline do not act as acceptor substrates. Primary amines other than α-amino acids are generally not recognized either (Castonguay *et al.*, 2003). On the other hand, dipeptides are typically even better acceptor substrates than simple amino acids, glycylglycine, cysteinylglycine, and methionylglycine being among the best (Allison, 1985; Thompson and Meister, 1977). Alkyl amide derivatives of L-methionine have also been synthesized as dipeptide mimics and used as acceptor substrates (Castonguay et al., 2003). However, these amides, lacking a terminal carboxylate group, are not recognized as well as natural dipeptides, which may reflect the importance of the putative interaction of this functional group with the ε-ammonium group of a lysine (Lys-99 in rat GGT) in the acceptor substrate–binding site (Stole and Meister, 1991).

Donor Substrate Binding Site Mapping

We have developed an indirect method to permit the mapping of the γ-glutamyl donor substrate–binding site of rat kidney GGT. In this method, derivatives of glutamine were tested as substrates in competition with the chromogenic substrate used in the typical activity assay (Tate and Meister, 1985), L-γ-glutamyl-p-nitroanilide (L-GPNA), in the presence of a saturating concentration of glycylglycine as acceptor substrate (see *Experimental Section*). Inhibition of the colorimetric reaction of L-GPNA was interpreted as being related to the high affinity of the corresponding amide for GGT. In this way, the relative importance of the unmodified α-carboxylate and α-amino groups of various donor substrates was determined (Castonguay et al., 2002). Our results and others (Cook et al., 1987; Griffith and Meister, 1977) have shown the importance of the α-ammonium group for GGT donor substrate recognition and the intolerance of any substitution on the nitrogen of this α-amino group. The α-carboxylate was also shown to be important; however, uncharged derivatives that were either isosteric or only slightly larger (e.g., methyl ester) were still recognized.

Structure-Based GGT Inhibitors

The observation that the substitution of a methyl ester for the α-carboxylate yielded compounds that were recognized by GGT permitted the simplification of synthetic schemes providing access to highly functionalized GGT inhibitors. Derivatives of L-GPNA, as well as its methyl ester, were synthesized containing different heteroatoms at the γ-position of the side chain (Lherbet et al., 2003). Interestingly, attenuation of the γ-glutamyl carbonyl electrophilicity prevented these compounds from serving as donor substrates. This lack of reactivity as γ-glutamyl donors, coupled with the lower affinity for the donor substrate binding site on the loss of the negatively charged carboxylate group, resulted in the direction of these compounds to the acceptor substrate binding site, where they served as reasonably good acceptor substrates.

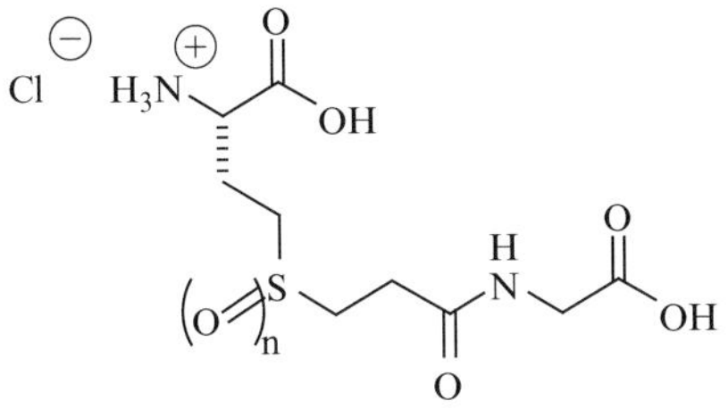

n = 0, L-TEPG
1, L-SPG
2, L-SnPG

FIG. 1. High-affinity sulfoxide inhibitors.

Further modification of the glutamyl side chain led to the discovery of a new class of micromolar inhibitors of GGT. Initially, commercial L-methionine sulfoxide was tested as a mixture of diastereoisomers at the stereogenic sulfur atom. Surprisingly, it was found to be a better competitive inhibitor of the reaction of L-GPNA than many other analogs of L-glutamine. The synthesis and testing of more highly developed GSH analogs, some of which are shown in Fig. 1, confirmed that the inhibition constant of the sulfoxide derivative L-SPG ($K_i = 53$ μM) is 10-fold lower than that of the corresponding thioether (L-TEPG) and 70-fold lower than that of the corresponding sulfone (L-SnPG) (Lherbet and Keillor, 2004; Lherbet et al., 2004). This increased affinity suggests that the distribution of partial charges in the sulfoxide functional group must be important for forming favorable interactions in the active site pocket.

Kinetic Analysis of Diastereomeric Inhibitor Specificity

Our attempts to separate and purify both diastereomers of L-methionine sulfoxide according to methods described in the literature (Holland and Brown, 1998; Holland et al., 1999) were unsuccessful. Therefore, we developed a method for the kinetic analysis of mixtures of diastereomers, where the mole fraction of each diastereomer is known (see *Experimental Section*). Application of this method to inhibition by a series of mixtures of diastereomers of different composition (Fig. 2) allowed the measurement of the competitive inhibition constant for each diastereomer. Remarkably, the sulfoxide having the *S* configuration at the stereogenic sulfur was solely responsible for the observed inhibition. This marked stereospecificity of GGT for the sulfoxide functionality at the δ-position of the donor substrate side chain is highly suggestive of the configuration of

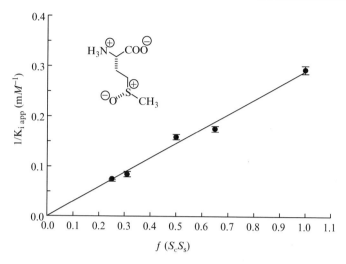

FIG. 2. Kinetic analysis of a mixture of diastereomeric inhibitors.

the tetrahedral intermediate formed in the catalytic reaction with GSH and is indirectly indicative of the configuration of the active site pocket of GGT. Ikeda and co-workers have suggested that two adjacent serines, conserved in all sequences of GGT (namely Ser-450 and Ser-451 in rat GGT), could stabilize the rate-limiting transition state of the catalytic reaction (Ikeda et al., 1995b) in a manner analogous to the "oxyanion hole" present in several hydrolases (Robertus et al., 1972). These putative interactions are represented with others discussed previously in Fig. 3, summarizing many speculations regarding specific interactions that affect GGT specificity.

Catalytic Mechanism of GGT

Acyl-Enzyme Intermediate

As mentioned in the Introduction, GGT catalysis is thought to proceed through a modified ping-pong mechanism (Scheme 1), as evidenced by parallel Lineweaver-Burk plots (Elce and Broxmeyer, 1976; Tate and Meister, 1974). However, it has been suggested that inhibition caused by the binding of acceptor substrates into donor substrate binding sites at high concentrations may also provide parallel plots for a sequential mechanism (Allison, 1985; Gololobov and Bateman, 1994). The isolation of the

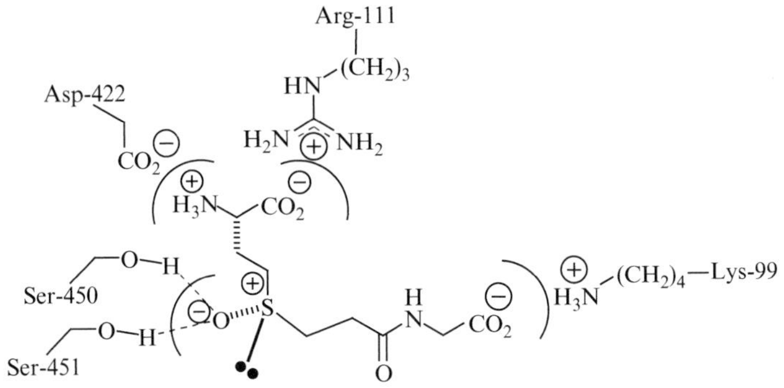

FIG. 3. Speculative interactions contributing to GGT specificity. Residues numbered according to rat kidney GGT sequence. Ligand shown is L-SPG, a transition state analog of GSH.

acyl-enzyme intermediate typical for acyl-transfer reactions would be in direct support of the ping-pong mechanism. Previous experiments designed to isolate the γ-glutamyl enzyme intermediate, formed either in the absence (Elce and Broxmeyer, 1976) or in the presence (Elce, 1980; Smith and Meister, 1995) of inhibitors, were unable to unambiguously identify and characterize the intermediate. Recently, we have used stopped-flow techniques to study the pre-steady state kinetics of the hydrolysis of D-γ-glutamyl-p-nitroanilide (D-GPNA), providing evidence for the accumulation of an acyl-enzyme intermediate using unmodified GGT (Keillor et al., 2004; see Experimental Section). Use of a donor substrate having the D-configuration was crucial for avoiding the autotranspeptidation reaction that is potentially problematic with donor substrates having the L-configuration, thus simplifying analysis of the results. The results obtained for pre-steady state kinetics of the hydrolysis reaction are shown in Fig. 4 (circles). The same experiment was also performed for the aminolysis reaction by repeating the procedure in the presence of an acceptor substrate. L-Methionine was used as an acceptor substrate at a concentration at which it does not inhibit binding at the donor substrate binding site (Allison, 1985) and that was well below its K_M value to allow visualization of the accumulation of the acyl-enzyme intermediate (squares, Fig. 4). These results clearly show the biphasic kinetics characteristic of the rapid release of p-nitroaniline product during the initial accumulation of the γ-glutamyl-enzyme intermediate, followed by the slow turnover of this intermediate, under hydrolytic and aminolytic conditions.

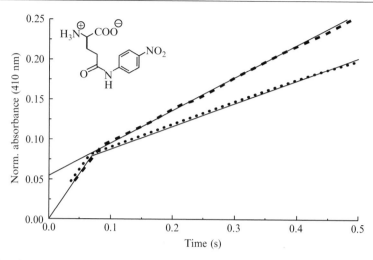

Fig. 4. Pre-steady-state kinetics of GGT acylation by D-GPNA. Circles, hydrolysis reaction; squares, transpeptidation reaction in the presence of 0.3 mM L-Met as acceptor substrate.

Acylation Step of Transpeptidation Reaction

The acylation step of the transpeptidation reaction resembles the reaction catalyzed by proteases and has been studied extensively by many groups (Cook and Peters, 1985; Ménard et al., 2001; Taniguchi and Ikeda, 1998; Tate and Meister, 1974). Our approach to characterizing the transition state of the rate-limiting step of the acylation process involved the use of a series of para-substituted γ-glutamyl anilides as donor substrates (Fig. 5). The concentration of para-substituted aniline released during the acylation step catalyzed by GGT was followed discontinuously using a diazo dye derivation procedure (see Experimental Section). The steady state kinetic data obtained by this method were then correlated to the electronic nature of the para-substituent of the γ-glutamyl anilide. Studies of such substituent effects provide information on the buildup of partial charges at the rate-limiting transition state and represent a powerful tool in the explanation of enzyme mechanisms. For the acylation step in question, a Hammett plot of log (k_{cat}^X/k_{cat}^H) as a function of the σ^- Hammett substituent parameter was obtained (Ménard et al., 2001) and is presented in Fig. 5.

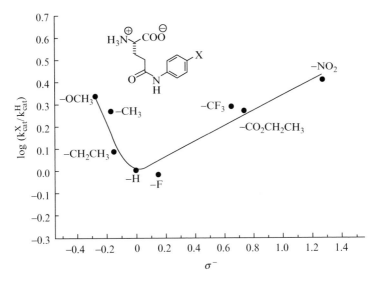

FIG. 5. Hammett plot of GGT acylation by *para*-substituted L-γ-glutamyl anilides.

Typically, Hammett plots display straight lines with positive, negative, or zero slopes. The value of the slope is negatively proportional to the partial charge developed on the aryl-substituted atom at the transition state (Jencks, 1971). A zero slope means that no significant charge is built up on the substituted atom at the transition state of the rate-limiting step, making the formation or rupture of bonds to that atom unlikely. A positive slope indicates the development of partial negative charge at the transition, and a negative slope signifies the development of substantial positive charge on the aryl-substituted atom. Hammett plots showing downward curvature are suggestive of mechanisms where the rate-determining step changes, depending on the substituent effect. However, in our studies of GGT acylation, an upward curve was obtained. Curves of this type have been explained as being indicative of a concerted reaction step where the substituted atom is involved in both bond-making and bond-breaking interactions that show opposite electronic dependency on the substituent (Funderburk *et al.*, 1978). For the acylation reaction in question, the general acid-catalyzed breakdown of the tetrahedral intermediate, formed on attack by the active site nucleophile on the γ-glutamyl carbonyl, would involve the concomitant cleavage of the C–N bond and protonation of the aniline nitrogen. Other experiments including isotope effects, temperature effects, and pH rate profiles provided evidence in support of this acylation mechanism (Ménard et al., 2001).

Deacylation Step of the Transpeptidation Reaction

Although the acylation step of the transpeptidation reaction catalyzed by GGT has been studied extensively, the aminolysis of the acyl-enzyme intermediate by amino acids and dipeptides has been the object of far fewer investigations. This trend does not reflect the lesser importance of the aminolysis reaction, because the transpeptidation reaction has been proposed to be more physiologically relevant than the hydrolytic reaction (Allison and Meister, 1981), but rather the logistical difficulty of studying a reaction step that is typically faster than the acylation step, and in competition with the hydrolytic shunt of the modified ping-pong mechanism. The hydrolytic shunt is of negligible relative importance when high concentrations of dipeptide acceptor substrates such as glycylglycine are used (Tate and Meister, 1985), but when using simple amino acids as acceptor substrates, the hydrolysis reaction is important to consider, even with reasonably good substrates such as L-cystine, L-methionine, and L-alanine (Tate and Meister, 1974). The work of Allison (1985) has provided the equations necessary for the detailed kinetic analysis of the modified ping-pong mechanism. Using these equations, we studied the effect of acceptor substrate nucleophilicity for a series of synthetic L-methionine amide derivatives (Castonguay et al., 2003). Following Allison's approach, we determined, for each acceptor substrate, the partition coefficient (K_{iab}), corresponding to the concentration of acceptor substrate required to achieve similar rates of hydrolysis and aminolysis (Schellenberger et al., 1990), and the specificity constant (V_b/K_b) for aminolysis alone, having corrected for the competing hydrolysis reaction. The latter parameters were then correlated to the pK_a values of the acceptor substrates' primary ammonium groups as a measure of their nucleophilicity (see Experimental Section). The resulting Brønsted plot of the $\log(k_{cat}/K_b)$ against the pK_a of each acceptor substrate conjugate acid shows a positive slope of 0.84 ± 0.13, as shown in Fig. 6.

The sign and value of the slope of a Brønsted plot are directly proportional to the partial charge built up on the substituted atom at the transition state of the rate-limiting step (Jencks, 1971). The positive Brønsted slope (β_{nuc}) we obtained for the aminolytic deacylation of GGT denotes the development of significant positive charge on the nucleophilic nitrogen atom, indicating the advanced nucleophilic attack of the acceptor substrate on the acyl enzyme at the transition state. Conversely, a negative slope

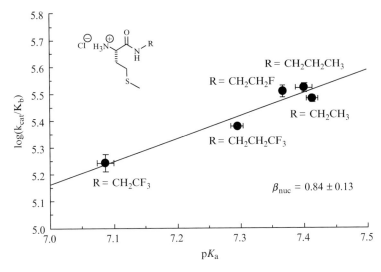

FIG. 6. Brønsted plot of deacylation of γ-glutamyl-GGT by L-Met alkyl amides.

would have denoted the formation of a negative partial charge at the transition state, consistent with the advanced deprotonation of the attacking nitrogen by an active site general base. The value of the slope is also important to consider. Whereas a value of 1.0 would have represented the development of full positive charge at the transition state, the intermediate value of 0.84 that we obtained is suggestive of the concerted deprotonation of the nitrogen during the nucleophilic attack, consistent with general base catalysis. One final level of interpretation of the LFER diagram is to consider whether k_{cat} values or values of k_{cat}/K_M (as in Fig. 6) give a better correlation with substrate basicity. Typically, the better correlation of k_{cat}/K_M is interpreted as representing the reaction of free (acyl-)enzyme with acceptor substrate during the rate-limiting step, whereas the rate-limiting reaction of an enzyme-substrate covalent intermediate would manifest itself as a better correlation of k_{cat} with substrate basicity (Jencks, 1971). The general base catalysis mechanism implied by our Brønsted plot was confirmed through subsequent pH rate profiles and kinetic isotope effects (Castonguay *et al.*, 2003).

Unified Mechanism of GGT-Mediated Transpeptidation

The information obtained through detailed studies of the acylation and (aminolytic) deacylation steps of the GGT catalytic cycle can be summarized by the energy diagram shown in Fig. 7. The rate-limiting transition states for the acylation and deacylation steps are drawn in detail, based on the results of various kinetic techniques, including analysis of linear free-energy relationships (LFERs). These mechanistic studies have permitted a greater comprehension of the potential physiological role of GGT with respect to the relative importance of the hydrolytic and transpeptidation pathways (Castonguay *et al.*, 2003).

Experimental Section

Binding Site Mapping through Competition Experiments

Using 0.1 *M* MOPS pH 7.0 buffer as a solvent, stock solutions of 25–500 m*M* in amide donor substrate analogues were prepared. Subsequent kinetic studies were carried out using different concentrations of L-γ-glutamyl-*p*-nitroanilide (between 50 and 1680 μM), 20 m*M* glycylglycine, and different concentrations of amide substrate (usually from 0 up to 100 m*M* when possible) in 0.1 *M* MOPS pH 7.0 buffer at a final volume of 1 mL. Reactions were initiated by adding 3.38 mU of GGT. The liberation of *p*-nitroaniline was followed spectrophotometrically at 410 nm. An extinction coefficient value of 9200 $M^{-1}cm^{-1}$ at pH 7.0 was used for calculating initial rates (Castonguay *et al.*, 2002). The K_i values for the apparent inhibition of the colorimetric enzymatic reaction of L-γ-glutamyl-*p*-nitroanilide, because of the non-colorimetric enzymatic reaction of each amide substrate, were determined by Lineweaver-Burk and/or Dixon plots.

Kinetic Analysis of Diastereomeric Inhibitors

In Scheme 2 is shown the kinetic scheme representing the competition of two inhibitors for the same enzyme binding site, as one would expect in the case of two structurally similar inhibitors such as diastereomers (Lherbet and Keillor, 2004). From this scheme the following double reciprocal equation (1) can be obtained:

$$\frac{1}{v} = \frac{K_M\left(1 + [I]/K_i + [I']/K_i'\right)}{V_{max}[S]} + \frac{1}{V_{max}} \tag{1}$$

where v is the reaction rate, K_M is the Michaelis constant of the donor substrate, [S] is the concentration of the donor substrate, V_{max} is the maximal rate of the enzymatic reaction, and K_i and K_i' represent the inhibition constants for inhibitors I and I', respectively. The apparent Michaelis constant $K_{M\ app}$ is the Michaelis constant observed in the presence of inhibitors I and I'. The relation of the apparent inhibition constant,

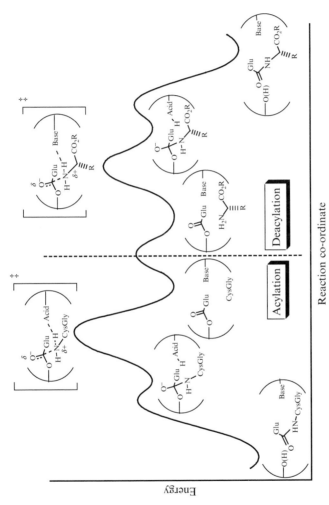

FIG. 7. Energy diagram for GGT-mediated transpeptidation.

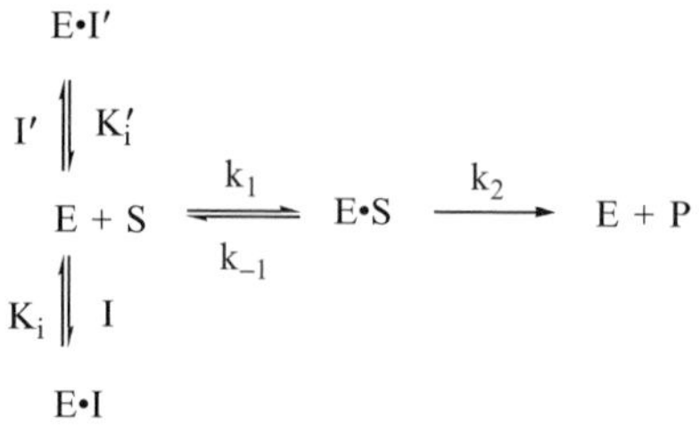

$$E \cdot I'$$

$$I' \Big\| K_i'$$

$$E + S \xrightleftharpoons[k_{-1}]{k_1} E \cdot S \xrightarrow{k_2} E + P$$

$$K_i \Big\| I$$

$$E \cdot I$$

SCHEME 2. Kinetic scheme for inhibition by a mixture of competitive inhibitors.

$K_{i\ app}$, to the total inhibitor concentration, $[I]_{tot}$, can be expressed in terms of individual inhibitor concentrations and inhibition constants as follows:

$$[I]_{tot}/K_{i\ app} = ([I]/K_i + [I']/K_i') \tag{2}$$

From equation (2), the following reciprocal equation can be written:

$$\frac{1}{K_{i\ app}} = f \cdot \left(\frac{1}{K_i} - \frac{1}{K_i'}\right) + \frac{1}{K_i'} \tag{3}$$

where f represents the fraction of the total inhibitor concentration corresponding to inhibitor I, $f = \frac{[I]}{[I]+[I']}$. From Eq. (3), a plot of the reciprocal of the (observed) apparent competitive inhibition constant as a function of the mole fraction of inhibitor I (in our case, one sulfoxide diastereomer – see Fig. 2) in the mixture of inhibitors will give a slope of $\left(\frac{1}{K_i} - \frac{1}{K_i'}\right)$ and an intercept of $\frac{1}{K_i'}$, allowing the apparent inhibition constants for both inhibitors to be determined.

Pre-Steady State Kinetics

GGT was dissolved in pH 8.0 0.1 M Tris-HCl buffer to give an initial concentration of around 8.8 μM and was placed in a 1-ml syringe in an Applied Photophysics SX.18MV stopped-flow apparatus thermostated at 37°. The donor substrate, D-γ-glutamyl-p-nitroanilide, was dissolved in the same buffer to a concentration of 6.64 mM and was loaded into a 0.1-ml syringe. A mixing ratio of 10:1 was thus obtained in the stopped-flow cell, giving final concentrations of 8.0 μM and 600 μM, respectively. Time traces were obtained in triplicate. A blank experiment was performed by replacing alternatively the enzyme or the donor substrate. The absorbance at 410 nm over a 1-cm path length was recorded for 10 s on a logarithmic scale. Slopes of absorbance vs time were transformed into reaction rates using an extinction coefficient of 8,800 $M^{-1}cm^{-1}$ for p-nitroaniline (Tate and Meister, 1985). The same procedure was followed to study the accumulation of the acyl-enzyme intermediate during the transpeptidation reaction by adding L-methionine, an acceptor substrate known to be a poor inhibitor of the donor substrate–binding site (Allison, 1985), at a final concentration of 0.3 mM, well below its K_M value.

Diazotization Assay

The colorimetric determination of the concentration of released anilines, through derivation and quantitation of their corresponding diazo dyes, was based on the protocol of Goldbarg *et al.* (1963) and modified for application to enzyme kinetic studies (Ménard *et al.*, 2001). A test tube was charged with 25–500 μl of a 5 mM stock solution of the *para*-substituted donor substrate L-γ-glutamylanilide to be studied and 1500 μl of 0.1 M Tris-HCl buffer, pH 8.0. To this was added 500 μl of 0.1 M glycylglycine (pH 8.0) and the volume was completed to 2.5 ml with the addition of 0 – 475 μl of 0.1 M Tris-HCl buffer, pH 8.0. Anilides were added as solids to achieve the highest concentrations needed. The reaction mixture was mixed and incubated at 37° for 10 min, before initiation of the enzymatic reaction by the addition of 5–20 μl of a 0.26-μg/ml solution of GGT. Aliquots (375 μl) of the reaction mixture were taken every 5–10 min and added to 125 μl of 40% TCA. At the end of the enzymatic reaction, a volume of 125 μl of a freshly prepared sodium nitrite solution (4 mg/ml in water) was added to each quenched reaction aliquot, and the solution was allowed to react for 3 min. After this time, 125 μl of an aqueous ammonium sulfamate solution (20 mg/ml) was added, and the solution was allowed to stand for 2 min. Finally, 250 μl of a solution of *N*-(1-naphthyl)ethylenediamine dihydrochloride (1.5 mg/ml in ethanol) was added, and the color of the solution was allowed to develop, the time and the temperature of this development depending on the aniline being derivatized (see Table I). Absorbance values were measured at specific wavelengths and divided by the corresponding apparent extinction coefficient as shown in Table I.

Linear Free-Energy Relationship (LFER) Analysis—Hammett Plot

Kinetic parameters obtained from steady-state kinetics of GGT acylation by a series of *para*-substituted γ-glutamyl anilides (using the diazotization assay) were plotted against the σ^- value of the corresponding *para* substituent. The specific type of Hammett substituent parameter to use (σ, σ^+, σ^-, σ^*, etc.) depends on the type of aryl group to which they are attached and the type of reaction taking place (Jencks, 1971). For aniline leaving groups, the σ^- parameter is most appropriate. Plots should be constructed using both log (k_{cat}) and log (k_{cat}/K_M) parameters to determine which (if either) gives the best correlation. In our investigation of GGT

TABLE I

EXPERIMENTAL CONDITIONS USED FOR THE DERIVATION OF PARA-SUBSTITUTED ANILINES AND
THE WAVELENGTH AND APPARENT EXTINCTION COEFFICIENTS USED FOR THE COLORIMETRIC
DETERMINATION OF THE DIAZO DYE COMPOUNDS THUS OBTAINED

Aniline	Temp (°)	Time (min)	λ (nm)	ε^{app} ($mM^{-1}cm^{-1}$)
p-Methoxyaniline	37	30	583	15.3 ± 0.3
p-Methylaniline	25	15	567	24.7 ± 0.3
p-Ethylaniline	25	15	572	33.1 ± 0.4
Aniline	25	15	560	45.1 ± 0.3
p-Fluoroaniline	25	10	558	86.0 ± 1.6
p-Trifluoromethylaniline	25	10	546	67.1 ± 0.3
p-Ethoxycarbonylaniline	25	10	551	114 ± 1
p-Nitroaniline	25	10	560	57.3 ± 0.4

acylation (Ménard et al., 2001), the Hammett plot of log (k_{cat}) vs σ^- was
found to give the best correlation, albeit non-linear (Fig. 5).

Linear Free-Energy Relationship (LFER) Analysis—Brønsted Plot

The kinetic reactions of GGT with 5–150 μM concentrations of donor
substrate (D-GPNA) and 0–20 mM concentrations of L-Met alkyl amides in
0.1 M Tris-HCl buffer (pH 8.0) were initiated by the addition of around 39
mU of rat kidney GGT. The release of p-nitroaniline was followed spectrophotometrically at 410 nm. Lineweaver–Burk plots of the initial rates were
obtained, and the ordinate intercepts were used to make a replot, according to the method previously published by Allison (1985). The replot graph
of the reciprocal of the difference between the ordinate intercept of the
hydrolysis reaction and that of the transpeptidation reaction, plotted
against the reciprocal of the concentration of the acceptor substrate, leads
to the determination of the partition coefficient (K_{iab}) for the acceptor
substrate, from which the value of k_{cat}/K_M for the aminolytic deacylation
step can be determined accurately, correcting for the competitive hydrolytic shunt. In our investigation of the deacylation step of GGT-mediated
transpeptidation reaction (Castonguay et al., 2003), we used a series of
synthetic L-methionine alkyl amides, bearing varying numbers of electron-
withdrawing fluorine atoms, modifying the pK_a of the α-ammonium compound. Over such a narrow interval of pK_a values, it is critical to measure
the pK_a experimentally in the reaction conditions of the kinetic experiments to be able to interpret the effect of acceptor substrate basicity (as a
measure of nucleophilicity) with accuracy. Plots should be constructed
using both log (k_{cat}) and log (k_{cat}/K_M) parameters to determine which (if

either) gives the best correlation. In our investigation of GGT deacylation (Castonguay *et al.*, 2003), the Brønsted plot of log (k_{cat}/K_M) vs ammonium pK_a was found to give the best correlation with a slope (β_{nuc}) consistent with rate-limiting general base-assisted nucleophilic attack (Fig. 6).

Acknowledgments

We thank the Natural Sciences and Engineering Research Council (NSERC) of Canada for financial support. We thank also the NSERC for a postgraduate scholarship (RC) and the Université de Montréal for a Bourse d'excellence (CL).

References

Allison, R. D. (1985). γ-Glutamyl transpeptidase: Kinetics and mechanism. *Methods Enzymol.* **113**, 419–437.

Allison, R. D., and Meister, A. (1981). Evidence that transpeptidation is a significant function of γ-glutamyl transpeptidase. *J. Biol. Chem* **256**, 2988–2992.

Bernström, K., Örning, L., and Hammarström, S. (1982). γ-Glutamyl transpeptidase, a leukotriene metabolizing enzyme. *Methods Enzymol.* **86**, 38–45.

Castonguay, R., Lherbet, C., and Keillor, J. W. (2002). Mapping of the active site of γ-glutamyl transpeptidase using activated esters and their amide derivatives. *Bioorg. Med. Chem.* **10**, 4185–4191.

Castonguay, R., Lherbet, C., and Keillor, J. W. (2003). Kinetic studies of rat kidney γ-glutamyl transpeptidase deacylation reveal a general-base-catalyzed mechanism. *Biochemistry* **42**, 11504–11513.

Cook, N. D., and Peters, T. J. (1985). The effect of pH on the transpeptidation and hydrolytic reactions of rat kidney γ-glutamyltransferase. *Biochim. Biophys. Acta* **832**, 205–212.

Cook, N. D., Upperton, K. P., Challis, B. C., and Peters, T. J. (1987). The donor specificity and kinetics of the hydrolysis reaction of γ-glutamyltransferase. *Biochim. Biophys. Acta* **914**, 240–245.

Del Bello, B., Paolicchi, A., Comporti, M., Pompella, A., and Maellaro, E. (1999). Hydrogen peroxide produced during γ-glutamyl transpeptidase activity is involved in prevention of apoptosis and maintenance of proliferation in U937 cells. *FASEB J.* **13**, 69–79.

Elce, J. S. (1980). Active-site amino acid residues in γ-glutamyltransferase and the nature of the γ-glutamyl-enzyme bond. *Biochem. J.* **185**, 473–481.

Elce, J. S., and Broxmeyer, B. (1976). γ-Glutamyltransferase of rat kidney. simultaneous assay of the hydrolysis and transfer reactions with (glutamate-^{14}C) glutathione. *Biochem. J.* **153**, 223–232.

Funderburk, L. H., Aldwin, L., and Jencks, W. P. (1978). Mechanisms of general acid and base catalysis of the reactions of water and alcohols with formaldehyde. *J. Am. Chem. Soc.* **100**, 5444–5459.

Godwin, A. K., Meister, A., O'Dwyer, P. J., Huang, C. S., Hamilton, T. C., and Anderson, M. E. (1992). High resistance to cisplatin in human ovarian cancer cell lines is associated with marked increase of glutathione synthesis. *Proc. Natl. Acad. Sci. USA* **89**, 3070–3074.

Goldbarg, J. A., Pineda, E. P., Smith, E. E., Friedman, O. M., and Rutenburg, A. M. (1963). A method for the colorimetric determination of *gamma*-glutamyl transpeptidase in human serum; enzymatic activity in health and disease. *Gastroenterology* **44**, 127–133.

Gololobov, M. Y., and Bateman, R. C., Jr. (1994). γ-Glutamyltranspeptidase-catalysed acyl-transfer to the added acceptor does not proceed via the ping-pong mechanism. *Biochem. J.* **304,** 869–876.

Graber, R., and Losa, G. A. (1995). Apoptosis in human lymphoblastoid cells induced by acivicin, a specific γ-glutamyltransferase inhibitor. *Int. J. Cancer* **62,** 443–448.

Griffith, O. W., Bridges, R. J., and Meister, A. (1978). Evidence that the γ-glutamyl cycle functions *in vivo* using intracellular glutathione: Effects of amino acids and selective inhibition of enzymes. *Proc. Natl. Acad. Sci. USA* **75,** 5405–5408.

Griffith, O. W., and Meister, A. (1977). Selective inhibition of *gamma*-glutamyl-cycle enzymes by substrate analogs. *Proc. Natl. Acad. Sci. USA* **74,** 3330–3334.

Hanigan, M. H., and Pitot, H. C. (1985). *Gamma*-glutamyl transpeptidase—its role in hepatocarcinogenesis. *Carcinogenesis* **6,** 165–172.

Heisterkamp, N., Rajpert-De Meyts, E., Uribe, L., Forman, H. J., and Groffen, J. (1991). Identification of a human *gamma*-glutamyl cleaving enzyme related to, but distinct from *gamma*-glutamyl transpeptidase. *Proc. Natl. Acad. Sci. USA.* **88,** 6303–6307.

Holland, H. L., and Brown, F. M. (1998). Biocatalytic and chemical preparation of all diastereoisomers of methionine sulfoxide. *Tetrahedron: Asymmetry* **10,** 535–538.

Holland, H. L., Andreana, P. R., and Brown, F. M. (1999). Biocatalytic and chemical routes to all the stereoisomers of methionine and ethionine sulfoxide. *Tetrahedron: Asymmetry* **10,** 2833–2843.

Ikeda, Y., Fujii, J., Taniguchi, N., and Meister, A. (1995a). Human *gamma*-glutamyl transpeptidase mutants involving conserved aspartate residues and the unique cysteine residue of the light subunit. *J. Biol. Chem.* **270,** 12471–12475.

Ikeda, Y., Fujii, J., Anderson, M. E., Taniguchi, N., and Meister, A. (1995b). Involvement of Ser-451 and Ser-452 in the catalysis of human *gamma*-glutamyl transpeptidase. *J. Biol. Chem.* **270,** 22223–22228.

Jencks, W. P. (1971). Structure-reactivity correlations and general acid-base catalysis in enzymic transacylation reactions. *Cold Sp. Har. Sym. Quant. Biol.* **36,** 1–11.

Kasai, T., and Larsen, P. O. (1980). Chemistry and biochemistry of γ-glutamyl derivatives from plants including mushrooms (basidiomycetes). *Proc. Chem. Org. Nat. Prod.* **39,** 173–285.

Keillor, J. W., Ménard, A., Castonguay, R., Lherbet, C., and Rivard, C. (2004). Pre-steady state kinetic studies of rat kidney γ-glutamyl transpeptidase confirm its ping-pong mechanism. *J. Phys. Org. Chem.* **17,** 529–536.

Lee, D.-H., Ha, M.-H., Kim, J.-H., Christiani, D. C., Gross, M. D., Steffes, M., Blomkoff, R., and Jacobs, D. R. (2003). *Gamma*-glutamyltranspeptidase and diabetes—a 4 year follow-up study. *Diabetologia* **46,** 359364.

Lherbet, C., Morin, M., Castonguay, R., and Keillor, J. W. (2003). Synthesis of aza- and oxaglutamyl-*p*-nitroanilide derivatives and their kinetic studies with *gamma*-glutamyl-transpeptidase. *Bioorg. Med. Chem. Lett.* **13,** 997–1000.

Lherbet, C., and Keillor, J. W. (2004). Probing the stereochemistry of the active site of *gamma*-glutamyl transpeptidase using sulphur derivatives of L-glutamic acid. *Org. Biomol. Chem.* **2,** 238–245.

Lherbet, C., Gravel, C., and Keillor, J. W. (2004). Synthesis of S-alkyl L-homocysteine analogues of glutathione and their kinetics with L-γ-glutamyl transpeptidase. *Bioorg. Med. Chem. Lett.* **14,** 3451–3455.

Meister, A. (1973). On the enzymology of amino acid transport. *Science* **180,** 33–39.

Ménard, A., Castonguay, R., Lherbet, C., Rivard, C., Roupioz, Y., and Keillor, J. W. (2001). Nonlinear free energy relationship in the general-acid-catalyzed acylation of rat kidney γ-glutamyl transpeptidase by a series of γ-glutamyl anilide substrate analogues. *Biochemistry* **40**, 12678–12685.

Örning, L., Hammarström, S., and Samuelsson, B. (1980). Leukotriene D: A slow reacting substance from rat basophilic leukemia cells. *Proc. Natl. Acad. Sci. USA* **77**, 2014–2017.

Robertus, J. D., Kraut, J., Alden, R. A., and Birktoft, J. J. (1972). Subtilisin: A stereochemical mechanism involving transition-state stabilization. *Biochemistry* **11**, 4293–4303.

Schellenberger, V., Schellenberger, U., Mitin, Y. V., and Jakubke, H.-D. (1990). Characterization of the S'-subsite specificity of bovine pancreatic α-chymotrypsin via acyl transfer to added nucleophiles. *Eur. J. Biochem* **187**, 163–167.

Sian, J., Dexter, D. T., Lees, A. J., Daniel, S., Jenner, P., and Marsden, C. D. (1994). Glutathione-related enzymes in brain in Parkinson's disease. *Ann. Neurol.* **36**, 356–361.

Smith, T. K., and Meister, A. (1995). Chemical modification of active site residues in γ-glutamyl transpeptidase. *J. Biol. Chem.* **270**, 12476–12480.

Stole, E., and Meister, A. (1991). Interaction of *gamma*-glutamyl transpeptidase with glutathione involves specific arginine and lysine residues of the heavy subunit. *J. Biol. Chem.* **252**, 6042–6045.

Taniguchi, N., and Ikeda, Y. (1998). γ-Glutamyl transpeptidase: Catalytic mechanism and gene expression. *Adv. Enzymol. Relat. Areas Mol. Biol.* **72**, 239–278.

Tate, S. S., and Meister, A. (1974). Stimulation of the hydrolysis activity and decrease of the transpeptidase activity of γ-glutamyl transpeptidase by maleate; identity of a rat kidney maleate-stimulated glutaminase and γ-glutamyl transpeptidase. *Proc. Natl. Acad. Sci. USA* **71**, 3329–3333.

Tate, S. S., and Meister, A. (1985). γ-Glutamyl transpeptidase from kidney. *Methods Enzymol.* **113**, 400–419.

Thompson, G. A., and Meister, A. (1976). Hydrolysis and transfer reactions catalyzed by γ-glutamyl transpeptidase; evidence for separate substrate sites and for high affinity of L-cystine. *Biochem. Biophys. Res. Comm.* **71**, 32–36.

Thompson, G. A., and Meister, A. (1977). Interrelationships between the binding sites for amino acids, dipeptides, and γ-glutamyl donors in γ-glutamyl transpeptidase. *J. Biol. Chem.* **252**, 6792–6798.

Further Reading

Hiratake, J., Inoue, M., and Sakata, K. (2002). *Gamma*-glutamyltranspeptidase and *gamma*-glutamylpeptide ligases: Fluorophosphonate and phosphonodifluoromethylketone analogues as probes of tetrahedral state and *gamma*-glutamyl-phosphate intermediate. *Methods Enzymol.* **354**, 272–275.

Inoue, M., Hiratake, J., Suzuki, H., Kumagai, H., and Sakata, K. (2000). Identification of catalytic nucleophile of *Escherichia coli gamma*-glutamyltranspeptidase by *gamma*-monofluorophosphono derivative of glutamic acid: *N*-terminal Thr-391 in small subunit is the nucleophile. *Biochemistry* **39**, 7764–7771.

[28] γ-Glutamyl Transpeptidase in Glutathione Biosynthesis

By Hongqiao Zhang, Henry Jay Forman, and Jinah Choi

Abstract

Glutathione (GSH) is the most abundant nonprotein thiol in cells and has multiple biological functions. Glutathione biosynthesis by way of the γ-glutamyl cycle is important for maintaining GSH homeostasis and normal redox status. As the only enzyme of the cycle located on the outer surface of plasma membrane, γ-glutamyl transpeptidase (GGT) plays key roles in GSH homeostasis by breaking down extracellular GSH and providing cysteine, the rate-limiting substrate, for intracellular *de novo* synthesis of GSH. GGT also initiates the metabolism of glutathione S-conjugates to mercapturic acids by transferring the γ-glutamyl moiety to an acceptor amino acid and releasing cysteinylglycine. GGT is expressed in a tissue-, developmental phase-, and cell-specific manner that may be related to its complex gene structure. In rodents, there is a single GGT gene, and several promoters that generate different mRNA subtypes and regulate its expression. In contrast, several GGT genes have been found in humans. During oxidative stress, GGT gene expression is increased, and this is believed to constitute an adaptation to stress. Interestingly, only certain mRNA subtypes are increased, suggesting a specific mode of regulation of GGT gene expression by oxidants. Here, protocols to measure GGT activity, relative levels of total and specific GGT mRNA subtypes, and GSH concentration are described.

Introduction

Glutathione (GSH): Biological functions and metabolism

GSH(γ-glutamyl-L-cysteinyl-L-glycine) is the most abundant nonprotein thiol in cells, found ubiquitously at levels as high as 0.5–10 mM concentrations (Meister, 1988). GSH has multiple important biological functions, and these include conjugation of electrophiles, for example, in the detoxification of xenobiotic compounds and leukotriene metabolism, thiol-disulfide exchange reactions in the maintenance of normal cellular redox status, and antioxidant functions in the removal of both endogenous and exogenous oxidants (Meister, 1992). GSH has also been implicated in

Copyright 2005, Elsevier Inc. All rights reserved.
0076-6879/05 $35.00
DOI: 10.1016/S0076-6879(05)01028-1

cell signaling (Filomeni *et al.*, 2002; Sen, 2000) and been suggested to act as a reservoir of cysteine in the transport of amino acids.

Consequently, maintenance of the reduced GSH pool is important for the general welfare of cells, especially in situations of severe oxidative stress; moreover, inability to do so may be detrimental to cells and organisms as a whole. Indeed, an elevated GSH content generally confers protection against oxidative damage, whereas depletion of GSH tends to enhance such damage. GSH deficiency has been associated with various diseases (Townsend *et al.*, 2003; Wu *et al.*, 2004) such as cardiovascular disease (Giugliano *et al.*, 1995), neurodegenerative disorders (Cruz *et al.*, 2003; Dringen and Hirrlinger, 2003; Halliwell, 2001; Schulz *et al.*, 2000), aging (Richie, 1992), chronic lung disease (Rahman, 1999; Rahman and MacNee, 1999; Rahman *et al.*, 1999), and viral infections (Droge, 1993).

The reactions of GSH with peroxides are catalyzed by members of GSH peroxidase (GPx) family or peroxiredoxin 6 (Prx6, also called 1-cys Prx) and result in the production of GSH disulfide (GSSG):

$$2GSH + H_2O_2 \xrightarrow[Prx6]{GPx} GSSG + 2H_2O \tag{1}$$

GSH's reactions with electrophiles are catalyzed by GSH S-transferases (GST), resulting in the formation of glutathione conjugates. GSTs catalyze either reductive addition or exchange where X is usually a halide or hydroxyl moiety:

$$\begin{aligned} GSH + Electrophile \xrightarrow{GST} GS - Conjugate \\ GSH + Electrophile - X \xrightarrow{GST} GS - Conjugate + HX \end{aligned} \tag{2}$$

Physiologically, most GSH is found in the reduced form; less than 1% of GSH exists as GSSG, so that the ratio of GSH/GSSG is maintained at approximately 100:1. Cells partly maintain this high intracellular reduced GSH pool by recycling GSSG back to GSH with GSSG reductase (Eq. 3) (Meister and Anderson, 1983):

$$GSSG + NADPH + H^+ \xrightarrow{GSSGReductase} 2GSH + NADP^+ \tag{3}$$

Alternately, GSSG is extruded from cells by means of the multidrug resistance–associated protein that also transports GSH and GSH S conjugates out of cells (Lee *et al.*, 1997).

In multiorgan systems, GSH is also transported from one organ to help maintain intracellular GSH content in other organs (Anderson *et al.*, 1980; Lauterburg *et al.*, 1984). Exogenous GSH may be transported in its intact form by certain epithelial cells to help maintain their intracellular GSH content (Hagen *et al.*, 1988). However, extracellular GSH, GSSG, and glutathione S-conjugates are usually broken down to their constituent amino acids by γ-glutamyl transpeptidase (γ-glutamyl transferase, GGT; EC2.3.2.2). These amino acids are then taken up and used intracellularly for *de novo* synthesis of GSH (Meister, 1994; Suzuki *et al.*, 1993) through the γ-glutamyl cycle (Fig. 1). The *de novo* synthesis of GSH consists of two consecutive ATP-dependent reactions: in the first reaction, glutamate cysteine ligase (GCL, also known as γ-glutamylcysteine synthase) uses glutamic acid and cysteine as the substrates to produce γ-glutamylcysteine (γGC), which then reacts with glycine to form GSH in a reaction catalyzed by glutathione synthase. Detailed discussion of *de novo* synthesis of GSH and regulation of GCL can be found elsewhere (Deneke and Fanburg, 1989; Dickinson *et al.*, 2004; Griffith, 1999; Lu, 2000). The remaining chapter will focus on GGT, its roles in the metabolism of GSH and GSH S-conjugates, and the regulation of its gene expression.

Roles of GGT in γ-Glutamyl Cycle and Metabolism of GSH S Conjugates

GGT catalyzes the transfer of γ-glutamyl moiety from GSH, GSH S-conjugates, and other γ-glutamyl compounds to acceptors such as amino acids, dipeptides, and H_2O. The catalytic mechanism of GGT is well known (Taniguchi and Ikeda, 1998; Tate and Meister, 1981). As discussed previously, GGT plays a key role in γ-glutamyl cycle (Fig. 1) in the *de novo* synthesis of GSH (Meister, 1974; Stark *et al.*, 2003). Here, GGT

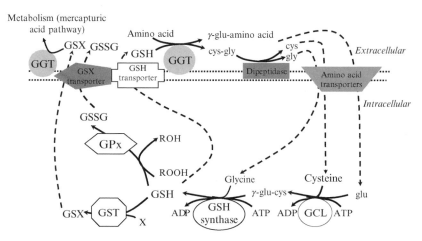

FIG. 1. γ-glutamyl cycle. GCL, glutamate cysteine ligase; GPx, glutathione peroxidase; GST, glutathione S-transferase. GGT, γ-glutamyl transpeptidase.

breaks down extracellular GSH to generate γ-glutamyl compounds and cysteinylglycine, which is further cleaved by membrane-bound dipeptidases. The constituent amino acids are then taken up and used by cells for intracellular resynthesis of GSH. One of the amino acids that are thus supplied by GGT is cysteine, the limiting substrate for GSH biosynthesis. Importantly, cysteine is a preferred acceptor for glutamate-amino acid conjugation reaction by GGT, and the product, γ-glutamylcystine, can be transported inside the cells and, after reduction to γ-GC, used directly for GSH biosynthesis (Anderson and Meister, 1983; Meister, 1984). This GSH salvage pathway bypasses the rate-limiting reaction catalyzed by GCL in *de novo* synthesis of GSH (Griffith *et al.*, 1981) and may play an important role in the maintenance of GSH in cells.

Therefore, GGT is critical in maintaining GSH and cysteine homeostasis (Hanigan and Ricketts, 1993) and for cellular antioxidant function (Forman and Skelton, 1990). GGT$^{-/-}$ mice showed glutathionemia, glutathionuria, cysteine deficiency, decreased tissue GSH levels, and increased susceptibility to oxidative stress, all of which could be reversed by *N*-acetylcysteine (NAC), a pharmacological source of cysteine (Barrios *et al.*, 2001; Griffith and Meister, 1980; Lieberman *et al.*, 1996; Jean *et al.*, 2002; Rojas *et al.*, 2000). These findings demonstrated the importance of GGT in GSH and cysteine homeostasis and suggested GGT as an integral component of endogenous antioxidant system in cells.

The cleavage of GSH S-conjugates and other γ-glutamyl compounds by GGT also plays an important role in the metabolism of natural biomolecules such as leukotriene C4, prostaglandins, and estrogen, as well as xenobiotic compounds such as carcinogens, mutagens, and drug compounds after their conjugation with GSH. In fact, the removal of γ-glutamyl moiety from these conjugates often constitutes an initial step in the conversion of these compounds to mercapturic acids in the detoxification reactions.

Besides these roles in the metabolism of GSH and GSH S-conjugates, other functions of GGT, such as the facilitation of amino acids transport (Hsu *et al.*, 1984; Vina *et al.*, 1983, 1990) and pro-oxidant effects, have also been described (del Bello *et al.*, 1999; Paolicchi *et al.*, 1997, 2002). The physiological significance of these findings, however, remains unclear.

Regulation of GGT Gene Expression

Expression Pattern and Gene Structures of GGT

GGT is mainly located at the external surface of plasma membrane but also exists in bodily fluids such as blood and epithelial lung-lining fluid. The expression pattern of this enzyme has been reviewed in detail elsewhere

(Chikhi *et al.*, 1999; Taniguchi and Ikeda, 1998; Tate and Meister, 1985; Whitfield, 2001). GGT is expressed in a tissue- and developmental phase–specific manner. For example, the protein is abundant in tissues with secretory or absorptive functions, such as kidney, pancreas, epididymis, and bile duct. Conversely, GGT activity in hepatocytes is prominent in fetal liver but disappears after birth. GGT is also expressed in a cell type–specific manner; even in tissues with low abundance of GGT, such as normal adult liver, the epithelial cells of bile duct show a relatively high level of GGT. In the brain, GGT is rich in microglial cells but poorly expressed in neurons.

GGT is one of the genes that show an extremely complex structure (Chikhi *et al.*, 1999; Pawlak *et al.*, 1988; Taniguchi and Ikeda, 1998). Rodents have a single copy of GGT gene, but several tandemly arranged promoters that generate multiple transcripts with different 5'-untranslated regions (5'-UTR) with the same protein-coding region to regulate its expression. For example, rat GGT is a single gene, and the transcription is controlled by five tandemly arranged promoters and, along with alternative splicing, generates seven different transcripts that share the same coding region for GGT protein but different 5'-UTR. These transcripts show different tissue-specific distribution that might be explained by the activation of different promoters in different tissues (Chikhi *et al.*, 1999).

Human GGT, on the other hand, is a multigene family consisting of at least seven GGT genes or pseudogenes (Pawlak *et al.*, 1988). Several GGT cDNAs have been cloned in human hepatoma cells, placenta, lung, and pancreas and, once again, GGT transcripts in these tissues displayed different 5'-UTRs but encoded the same protein (Visvikis *et al.*, 2001). The human GGT gene, like the rat GGT gene, may be regulated by several tandem promoters that, with alternative splicing, dictate the tissue-specific expression of GGT transcripts.

Increased Expression of GGT during Oxidative Stress

Acute exposure to oxidative stress is expected to decrease the GSH pool at least transiently, because GSH is used to remove oxidants. However, it is now well documented that cells can respond to low levels of oxidants and GSH-depleting agents by increasing GSH (Darley-Usmar *et al.*, 1991; Deneke *et al.*, 1989; Kugelman *et al.*, 1994; Ogino *et al.*, 1989; Poot *et al.*, 1987). GSH synthesis is up-regulated, and this seems to be a concerted process that involves modulation of GCL, GGT, and other enzymes in γ-glutamyl cycle (Dickinson *et al.*, 2002; Yamane *et al.*, 1998). Specifically, GGT expression can be up-regulated by redox-cycling quinones (Liu *et al.*, 1996a,b, 1998) in rat lung type II cells, hyperoxia

(Knickelbein *et al.*, 1996) and NO$_2$ in rat lung (Takahashi *et al.*, 1997), ethoxyquin- and aflatoxin B1 in rat liver (Griffiths *et al.*, 1995), and hypoxanthine and xanthine oxidase in rat epididymis (Markey *et al.*, 1998). The increased expression of GGT by oxidants has been demonstrated at both protein/activity and mRNA levels. Importantly, GGT has long been used as a clinical marker because its activity is elevated in many diseases (Arkkila *et al.*, 2001; Wannamethee *et al.*, 1995; Whitfield, 2001), including liver disease, cardiovascular disease, diabetes mellitus, and cancers (Hanigan, 1998; Vanderlaan and Phares, 1981), although the biological implication is still unclear. Many substances, including drugs, carcinogens, and alcohol, are also capable of increasing GGT gene expression in various cells and tissues (Brown *et al.*, 1998; Griffiths *et al.*, 1995). Increased expression of GGT during oxidative stress facilitates GSH turnover, *de novo* GSH synthesis, and metabolism and detoxification of GSH conjugates that increase cells' resistance to subsequent stress. Therefore, regulation of GGT gene expression is an important adaptive response and helps protect cells and tissues from oxidative injury.

Assays to Measure GGT Gene Regulation

GGT Activity Assay

GGT is a glycosylated protein that yields multiple protein bands when analyzed by Western blot (Kugelman *et al.*, 1994). Because of such difficulties, GGT is rarely measured at the protein level. Previously, we found that GGT activity correlated well with the deglycosylated protein level and have since used activity assays as a substitute for measuring GGT protein content (Kugelman *et al.*, 1994).

GGT has a broad specificity for γ-glutamyl compounds, and several spectrophotometric procedures have been described for analyzing its activity (Forman *et al.*, 1995; Meister *et al.*, 1981; Smith *et al.*, 1979; Tate and Meister, 1985). Most clinical laboratories use the IFCC (International Federation of Clinical Chemistry) method, described by Shaw *et al.* (1983). L-γ-glutamyl-*p*-nitroanilide is commonly used as a substrate for GGT activity assays, and the amount of *p*-nitroaniline generated is determined by monitoring absorbance at 410 nm. The use of L-γ-glutamyl-7-amino-4-methylcoumarin, a fluorogenic substrate, provides an alternative and more sensitive method of analyzing GGT activity. Methods can be selected on the basis of desired accuracy, abundance of GGT in the sample, amount of samples, and availability of laboratory equipment. Here, we describe a sensitive kinetic assay for quantifying GGT activity in cultured cells. The assay has been modified from the previously described method for use with standard fluorescence plate reader (see Eq. 4) (Forman *et al.*, 1995). The same method can be used to measure GGT activity in tissues or plasma samples (see the end of this section for details).

$$\gamma - glu - AMC \xrightarrow{GGT} glu + AMC \qquad (4)$$

γ-glu –AMC: L-γ-glutamyl-7-amino-4-methylcoumarin
AMC: amino-4-methylcoumarin

Reagents

Ammediol/HCl buffer, pH 8.6, 0.2 *M*
Glycylglycine, 0.2 *M*
γ–Glu-AMC in methoxyethanol, 10 m*M*
Bovine GGT standards (10 mU/ml; dilute to set up standards, Sigma Aldrich)
Triton X-100
1× PBS
0.1% triton-X100/PBS (1×)

Cells (from 100-mm dishes) are collected in 1 ml ice-cold 0.1 % triton-X100/PBS (1×) and then sonicated briefly in 1.5-ml tubes. Alternately, cell pellets can be stored at –80°. After sonication, samples are centrifuged at 4,000*g* for 5 min at 4°, and the supernatant is transferred to a new tube and stored at –80° until further analysis. GGT activity is stable for at least 2 months when stored this way. To measure GGT activity, 50 μl of supernatant (diluted in 0.1% triton-X100/PBS, as needed) is added to 150 μl of reaction mixture, containing 0.1 *M* ammediol/HCl buffer, pH 8.6, 20 m*M* glycylglycine, 200 μ*M* γ–glu-AMC, and 0.1% triton X-100, using 96-well plates. GGT activity is monitored for 15 min at 37° with excitation at 370 nm and emission at 440 nm. GGT activity is calculated against standard curve generated with known amounts of purified bovine GGT. Then, the activity is normalized by protein concentration, and the final activity is expressed as unit per milligram protein per minute. The specificity of GGT activity can be confirmed by preincubating samples with acivicin (AT-125; L-(γS,5S)-γ-amino-3-chloro-4,5-dihydro-5-isoxazoleacetic acid), a specific inhibitor of GGT. More than 95% of GGT activity should be inhibited. Acivicin interacts with GGT at a 1:1 molar ratio, and the inhibition is irreversible (Stole *et al.*, 1994). In many cells, 1 m*M* acivicin can completely inhibit GGT activity with 10 min incubation; 0.1 m*M* acivicin may require longer incubation (e.g., 30–60 min) to achieve the same degree of inhibition. Because the concentration of GGT varies in cells and tissues, the exact concentration and time of acivicin treatment should be experimentally determined.

To analyze GGT activity in tissues, tissues are first homogenized in 0.1% Triton-X 100/1× PBS and centrifuged to remove debris. Supernatant is then used to measure the tissue GGT activity, using the preceding procedure.

GGT mRNA Assay

GGT mRNA content can be determined with Northern blots or RNase Protection Assay (Liu et al., 1998) if mRNA is relatively abundant in the samples to be studied. However, if the mRNA cannot be readily detected by these approaches or if one needs to determine the relative abundance of GGT mRNA subtypes, real-time reverse transcription-polymerase chain reaction (RT-PCR) is recommended. Real-time PCR determines the amount of sequence-specific PCR products *real time* by quantifying the signal of fluorescent reporters that increases proportionally to the amount of RT-PCR product that is being generated. During the exponential phase of PCR, increases in the fluorescence signal (i.e., which is proportional to the amount of PCR product being generated) can be correlated to the initial amount of target template and, hence, the amount of mRNA in samples. Instead of comparing the accumulated signal at the end of the PCR reactions, the threshold cycle, which is the cycle number at which significant increase of fluorescence signal is first detected above baseline, is determined for each sample and compared between different treatment groups.

There are three kinds of real-time PCR fluorescence probes: hydrolysis probes (e.g., Taqman probes [Heid et al., 1996], molecular beacons [Abravaya et al., 2003; Mhlanga and Malmberg, 2001; Tan et al., 2004], and scorpions [Saha et al., 2001; Solinas et al., 2001]), hybridizing probes, and DNA binding agents. SYBR green I is a fluorogenic dye that displays little fluorescence while in solution but, on binding to the minor groove of double strand DNA, emits a strong fluorescent signal. SYBR green I provides a convenient and more cost-effective alternative to other approaches for quantifying products of real-time PCR reactions. The dye binds double-stranded DNAs in a non-sequence–specific manner; therefore, the major disadvantage to this method is nonspecific amplification of signals. With careful design of primers, optimization of PCR condition, and follow-up assays, including melting curve or dissociation analysis and DNA sequencing to identify amplicon, however, such problems can be overcome. The following section describes the analysis of GGT mRNAs in rat cells by real-time PCR using SYBR green I. To find protocols for measuring relative and absolute levels of mRNA by real-time PCR and a guide for the selection of internal controls, see Bustin (2000); Livak and Schmittgen (2001); Vandesompele *et al.* (2002); Schmid *et al.* (2003); Aerts *et al.* (2004); and Dheda *et al.* (2004).

Reagents

RNA extraction reagent (TRIzol reagent from Invitrogen)
DNase I (DNA-*free* reagent from Ambion)
Reverse transcription reagent (Taq-Man reverse transcription system from Applied Biosystems)
Primers, including forward and reverse primer
Real-time PCR reagent (SYBR green PCR master mix from Applied Biosystems)
PCR tubes (Smart Cycler tube, 25 μl)
Real-time PCR machine (Cepheid 1.2 real-time PCR machine)

Total RNA is extracted using TRIzol reagent (from Invitrogen) and treated with DNA-*free* DNaseI (from Ambion), according to manufacturers' recommendations. DNA-free RNA samples are then reverse transcribed using Taq-Man reverse transcription system (Applied Biosystems), and real-time PCR reactions are performed with a Cepheid 1.2 real-time PCR machine. Briefly, 5 μl of reverse transcription reaction product is added to reaction mixtures containing 12.5 μl SYBR green PCR master mix and the primer pair specific to total or different types of GGT mRNA. The total volume of PCR reaction is 25 μl. To measure total GGT mRNA content, the combined primer concentration in the final reaction is 0.25 μM, and the PCR condition is set as the following: 95° for 10 min (to activate DNA polymerase), 95° (to denature) for 2 min, 62° (for annealing) for 30 s, and 72° (for extension) for 30 s, for a total of 45 cycles. Fluorescence is detected during the annealing step. GAPDH mRNA content in each of the RNA samples is also quantified as an internal control. For GAPDH mRNA determination, the PCR reaction consisted of 2.5 μ RT reaction, 12.5 μl SYBR-green PCR mix, GAPDH primers, and water.

The comparative $\Delta\Delta$CT-method is used for the relative mRNA quantification. The relative quantification of target, which is normalized by an internal control (GAPDH) and untreated control, is calculated as the following:

$$Relative\ quantitation = 2^{-\Delta\Delta CT} \tag{5}$$

where $\Delta\Delta CT$ represents mean ΔCT of samples (treated or untreated) minus mean CT of control sample, and ΔCT represents mean CT of GGT minus mean CT of GAPDH. Threshold cycles (CT) are selected where all samples are in logarithmic phase.

Table I lists the sequences of specific primer pairs for total and type-specific GGT mRNA determination in rat tissues (Table I). Specificity of PCR products can be confirmed by DNA sequencing. To reduce nonspecific PCR amplification, high-performance liquid chromatography (HPLC) or gel-purified primers are preferred. Note that the primers can be changed to analyze GGT mRNAs from other species or any other mRNAs of known sequences

TABLE I

PRIMER PAIRS FOR RAT GGT MRNA REAL-TIME RT-PCR ASSAY: PRIMER PAIRS FOR TYPE IV AND V EACH GENERATE TWO PRODUCTS, RESPECTIVELY BECAUSE OF ALTERNATIVE SPLICING

GGT mRNA type	Primers
All GGTs	Forward 5'-ACCACTCACCCAACCGCCTAC-3' Reverse 5'-ATCCGAACCTTGCCGTCCTT-3'
I	Forward 5' -GCTCATCACATCAGGCACCC-3' Reverse 5' -AAACCCACAGCCAATCTTCC-3'
II	Forward 5' -CCACCAGTGTTGACCATCCTC-3' Reverse 5' -AAACCCACAGCCAATCTTCC-3'
III	Forward 5' -ATCCCAAGCCCTCCTCACC-3' Reverse 5' -AAACCCACAGCCAATCTTCC-3'
IV	Forward 5' -GCTTGTTGACCTTGGGCATCTG-3' Reverse 5' -AAACCCACAGCCAATCTTCC-3'
V	Forward 5'-GCTTGTTGACCTTGGGCATCTG-3' Reverse 5' -AAACCCACAGCCAATCTTCC-3'
GAPDH	Forward 5'-ACCCCCAATGTATCCGTTGT-3' Reverse 5'-TACTCCTTGGAGGCCATGT-3'

Support Protocol

Intracellular GSH and Cysteine Measurements by HPLC

Total intracellular GSH ([GSH] + 2 × [GSSG]) and total cysteine ([cysteine] + 2 × [cystine]) concentrations can be measured by HPLC according to the well-established method of Fariss and Reed (Fariss and Reed, 1987). Although many GSH assays are available, including the recycling assay (Tietze, 1969) and other assays, some of which are commercially available (e.g., Cayman Chemical, Chemicon, and OXIS, among others), this protocol offers one of the most sensitive method for simultaneous detection of GSH, cysteine, cystine, and GSSG in cells and tissues.

Reagents

10% perchloric acid/2 mM EDTA
10 mM iodoacetic acid in 0.2 mM m-creosol purple
1% 2,4 fluorodinitrobenzene (DNB, in HPLC-grade methanol)
1 M lysine
GSH, GSSG, cysteine, cystine (standards, prepared in 10% perchloric acid/2 mM EDTA)
γ-Glutamylglutamic acid (internal standard)
80% methanol (HPLC-grade, in HPLC-grade water, filtered)
0.5 M sodium acetate in 64% methanol (filtered)

Immediately after obtaining samples, samples are washed in 1× PBS and acidified in 10% perchloric acid/2 mM EDTA that contains 7.5 nmol of internal standard, γ-glutamylglutamic acid, per 0.45 ml (Bachem, Torrance, CA). Tissue samples should be sonicated and precipitated twice in this solution to ensure complete recovery of GSH. After centrifugation, 45 μl of 10 mM iodoacetic acid in 0.2 mM m-creosol purple is added to 0.45 ml of supernatant, and the pH is adjusted to 8–9 (i.e., purple) by adding 2 M KOH/2.4 M KHCO$_3$ drop wise. The mixture is incubated in the dark at room temperature for 15 min and then 0.45 ml of 1% DNB is added, and the reaction mixture is incubated at 4° overnight. Samples should be bright yellow at this step. Then, 45 μl of 1 M L-lysine was added to react with excess DNB. After 2 h incubation at 4° in the dark, salt is removed by centrifugation, and the supernatant is transferred to injection vials for analysis by HPLC. After derivatization, samples can be stored at 4° in the dark for ~2 weeks.

HPLC analysis of samples is then carried out with a Perkin Elmer HPLC system consisting of a Series 410 pump, LC-90 UV spectrophotometric detector, and LCI-100 integrator, using speri-5 amino column (Brownlee, from PerkinElmer). Elution solvents are 80% methanol (solvent A) and 0.5 M sodium acetate in 64% methanol (solvent B). After a 100-μl injection of the derivatized samples, the mobile phase is maintained at 70% solvent A and 20% solvent B for 5 min, followed by a 10-min linear gradient to 100% solvent B at a flow rate of 1.5 ml/min. The mobile phase is held at 100% solvent B for 40 min. DNB derivatives are detected at 365 nm. GSH, cysteine, cystine, and GSSG standards are run under the same conditions, and the peak areas are measured. GSH and GSSG concentrations are calculated based on the internal standard, as well as GSH or GSSG standard curves. GSH, cysteine, cystine, and GSSG concentrations are normalized by protein levels in samples and expressed as nmol/mg protein. For the determination of protein, protein pellets (after precipitation in 10% perchloric acid/2 mM EDTA) are dissolved in NaOH and analyzed by standard protein assays.

References

Abravaya, K., Huff, J., Marshall, R., Merchant, B., Mullen, C., Schneider, G., and Robinson, J. (2003). Molecular beacons as diagnostic tools: Technology and applications. *Clin. Chem. Lab. Med.* **41,** 468–474.

Aerts, J. L., Gonzales, M. I., and Topalian, S. L. (2004). Selection of appropriate control genes to assess expression of tumor antigens using real-time RT-PCR. *Biotechniques* **36,** 84–86, 88, 90–91.

Anderson, M. E., Bridges, R. J., and Meister, A. (1980). Direct evidence for inter-organ transport of glutathione and that the non-filtration renal mechanism for glutathione utilization involves gamma-glutamyl transpeptidase. *Biochem. Biophys. Res. Commun.* **96,** 848–853.

Anderson, M. E., and Meister, A. (1983). Transport and direct utilization of gamma-glutamylcyst(e)ine for glutathione synthesis. *Proc. Natl. Acad. Sci. USA* **80,** 707–711.

Arkkila, P. E., Koskinen, P. J., Kantola, I. M., Ronnemaa, T., Seppanen, E., and Viikari, J. S. (2001). Diabetic complications are associated with liver enzyme activities in people with type 1 diabetes. *Diabetes Res. Clin. Pract.* **52,** 113–118.

Barrios, R., Shi, Z. Z., Kala, S. V., Wiseman, A. L., Welty, S. E., Kala, G., Bahler, A. A., Ou, C. N., and Lieberman, M. W. (2001). Oxygen-induced pulmonary injury in gamma-glutamyl transpeptidase-deficient mice. *Lung* **179,** 319–330.

Brown, K. E., Kinter, M. T., Oberley, T. D., Freeman, M. L., Frierson, H. F., Ridnour, L. A., Tao, Y., Oberley, L. W., and Spitz, D. R. (1998). Enhanced gamma-glutamyl transpeptidase expression and selective loss of CuZn superoxide dismutase in hepatic iron overload. *Free Radic. Biol. Med.* **24,** 545–555.

Bustin, S. A. (2000). Absolute quantification of mRNA using real-time reverse transcription polymerase chain reaction assays. *J. Mol. Endocrinol.* **25,** 169–193.

Chikhi, N., Holic, N., Guellaen, G., and Laperche, Y. (1999). Gamma-glutamyl transpeptidase gene organization and expression: A comparative analysis in rat, mouse, pig and human species. *Comp. Biochem. Physiol. B. Biochem. Mol. Biol.* **122,** 367–380.

Cruz, R., Almaguer Melian, W., and Bergado Rosado, J. A. (2003). Glutathione in cognitive function and neurodegeneration. *Rev. Neurol.* **36,** 877–886.

Darley-Usmar, V. M., Severn, A., O'Leary, V. J., and Rogers, M. (1991). Treatment of macrophages with oxidized low-density lipoprotein increases their intracellular glutathione content. *Biochem. J.* **278**(Pt. 2), 429–434.

del Bello, B., Paolicchi, A., Comporti, M., Pompella, A., and Maellaro, E. (1999). Hydrogen peroxide produced during gamma-glutamyl transpeptidase activity is involved in prevention of apoptosis and maintenance of proliferation in U937 cells. *FASEB. J.* **13,** 69–79.

Deneke, S. M., Baxter, D. F., Phelps, D. T., and Fanburg, B. L. (1989). Increase in endothelial cell glutathione and precursor amino acid uptake by diethyl maleate and hyperoxia. *Am. J. Physiol.* **257,** L265–L271.

Deneke, S. M., and Fanburg, B. L. (1989). Regulation of cellular glutathione. *Am. J. Physiol.* **257,** L163–L173.

Dheda, K., Huggett, J. F., Bustin, S. A., Johnson, M. A., Rook, G., and Zumla, A. (2004). Validation of housekeeping genes for normalizing RNA expression in real-time PCR. *Biotechniques* **37,** 112–114, 116, 118–119.

Dickinson, D. A., Iles, K. E., Watanabe, N., Iwamoto, T., Zhang, H., Krzywanski, D. M., and Forman, H. J. (2002). 4-hydroxynonenal induces glutamate cysteine ligase through JNK in HBE1 cells. *Free Radic. Biol. Med.* **33,** 974.

Dickinson, D. A., Levonen, A. L., Moellering, D. R., Arnold, E. K., Zhang, H., Darley-Usmar, V. M., and Forman, H. J. (2004). Human glutamate cysteine ligase gene regulation through the electrophile response element. *Free Radic. Biol. Med.* **37,** 1152–1159.

Dringen, R., and Hirrlinger, J. (2003). Glutathione pathways in the brain. *Biol. Chem.* **384,** 505–516.

Droge, W. (1993). Cysteine and glutathione deficiency in AIDS patients: A rationale for the treatment with N-acetyl-cysteine. *Pharmacology* **46,** 61–65.

Fariss, M. W., and Reed, D. J. (1987). High-performance liquid chromatography of thiols and disulfides: Dinitrophenol derivatives. *Methods Enzymol.* **143,** 101–109.

Filomeni, G., Rotilio, G., and Ciriolo, M. R. (2002). Cell signalling and the glutathione redox system. *Biochem. Pharmacol.* **64,** 1057–1064.

Forman, H. J., Shi, M. M., Iwamoto, T., Liu, R. M., and Robison, T. W. (1995). Measurement of gamma-glutamyl transpeptidase and gamma-glutamylcysteine synthetase activities in cells. *Methods Enzymol.* **252,** 66–71.

Forman, H. J., and Skelton, D. C. (1990). Protection of alveolar macrophages from hyperoxia by gamma-glutamyl transpeptidase. *Am. J. Physiol.* **259,** L102–L107.

Giugliano, D., Ceriello, A., and Paolisso, G. (1995). Diabetes mellitus, hypertension, and cardiovascular disease: Which role for oxidative stress? *Metabolism* **44,** 363–368.

Griffith, O. W. (1999). Biologic and pharmacologic regulation of mammalian glutathione synthesis. *Free Radic. Biol. Med.* **27,** 922–935.

Griffith, O. W., Bridges, R. J., and Meister, A. (1981). Formation of gamma-glutamycyst(e)ine *in vivo* is catalyzed by gamma-glutamyl transpeptidase. *Proc. Natl. Acad. Sci. USA* **78,** 2777–2781.

Griffith, O. W., and Meister, A. (1980). Excretion of cysteine and gamma-glutamylcysteine moieties in human and experimental animal gamma-glutamyl transpeptidase deficiency. *Proc. Natl. Acad. Sci. USA* **77,** 3384–3387.

Griffiths, S. A., Good, V. M., Gordon, L. A., Hudson, E. A., Barrett, M. C., Munks, R. J., and Manson, M. M. (1995). Characterization of a promoter for gamma-glutamyl transpeptidase activated in rat liver in response to aflatoxin B1 and ethoxyquin. *Mol. Carcinog.* **14,** 251–262.

Hagen, T. M., Aw, T. Y., and Jones, D. P. (1988). Glutathione uptake and protection against oxidative injury in isolated kidney cells. *Kidney Int.* **34,** 74–81.

Halliwell, B. (2001). Role of free radicals in the neurodegenerative diseases: Therapeutic implications for antioxidant treatment. *Drugs Aging* **18,** 685–716.

Hanigan, M. H. (1998). gamma-Glutamyl transpeptidase, a glutathionase: Its expression and function in carcinogenesis. *Chem. Biol. Interact.* **111–112,** 333–342.

Hanigan, M. H., and Ricketts, W. A. (1993). Extracellular glutathione is a source of cysteine for cells that express gamma-glutamyl transpeptidase. *Biochemistry* **32**, 6302–6306.

Heid, C. A., Stevens, J., Livak, K. J., and Williams, P. M. (1996). Real time quantitative PCR. *Genome Res.* **6**, 986–994.

Hsu, B. Y., Foreman, J. W., Corcoran, S. M., Ginkinger, K., and Segal, S. (1984). Absence of a role of gamma-glutamyl transpeptidase in the transport of amino acids by rat renal brushborder membrane vesicles. *J. Membr. Biol.* **80**, 167–173.

Jean, J. C., Liu, Y., Brown, L. A., Marc, R. E., Klings, E., and Joyce-Brady, M. (2002). Gamma-glutamyl transferase deficiency results in lung oxidant stress in normoxia. *Am. J. Physiol. Lung Cell Mol. Physiol.* **283**, L766–L776.

Knickelbein, R. G., Ingbar, D. H., Seres, T., Snow, K., Johnston, R. B., Jr., Fayemi, O., Gumkowski, F., Jamieson, J. D., and Warshaw, J. B. (1996). Hyperoxia enhances expression of gamma-glutamyl transpeptidase and increases protein S-glutathiolation in rat lung. *Am. J. Physiol.* **270**, L115–L122.

Kugelman, A., Choy, H. A., Liu, R., Shi, M. M., Gozal, E., and Forman, H. J. (1994). gamma-Glutamyl transpeptidase is increased by oxidative stress in rat alveolar L2 epithelial cells. *Am. J. Respir. Cell Mol. Biol.* **11**, 586–592.

Lauterburg, B. H., Adams, J. D., and Mitchell, J. R. (1984). Hepatic glutathione homeostasis in the rat: Efflux accounts for glutathione turnover. *Hepatology* **4**, 586–590.

Lee, T. K., Li, L., and Ballatori, N. (1997). Hepatic glutathione and glutathione S-conjugate transport mechanisms. *Yale J. Biol. Med.* **70**, 287–300.

Lieberman, M. W., Wiseman, A. L., Shi, Z. Z., Carter, B. Z., Barrios, R., Ou, C. N., Chevez-Barrios, P., Wang, Y., Habib, G. M., Goodman, J. C., Huang, S. L., Lebovitz, R. M., and Matzuk, M. M. (1996). Growth retardation and cysteine deficiency in gamma-glutamyl transpeptidase-deficient mice. *Proc. Natl. Acad. Sci. USA* **93**, 7923–7926.

Liu, R. M., Hu, H., Robison, T. W., and Forman, H. J. (1996a). Differential enhancement of gamma-glutamyl transpeptidase and gamma-glutamylcysteine synthetase by tert-butylhydroquinone in rat lung epithelial L2 cells. *Am. J. Respir. Cell Mol. Biol.* **14**, 186–191.

Liu, R. M., Hu, H., Robison, T. W., and Forman, H. J. (1996b). Increased gamma-glutamylcysteine synthetase and gamma-glutamyl transpeptidase activities enhance resistance of rat lung epithelial L2 cells to quinone toxicity. *Am. J. Respir. Cell Mol. Biol.* **14**, 192–197.

Liu, R. M., Shi, M. M., Giulivi, C., and Forman, H. J. (1998). Quinones increase gamma-glutamyl transpeptidase expression by multiple mechanisms in rat lung epithelial cells. *Am. J. Physiol.* **274**, L330–L336.

Livak, K. J., and Schmittgen, T. D. (2001). Analysis of relative gene expression data using real-time quantitative PCR and the 2(-Delta Delta C(T)) Method. *Methods* **25**, 402–408.

Lu, S. C. (2000). Regulation of glutathione synthesis. *Curr. Top. Cell Regul.* **36**, 95–116.

Markey, C. M., Rudolph, D. B., Labus, J. C., and Hinton, B. T. (1998). Oxidative stress differentially regulates the expression of gamma-glutamyl transpeptidase mRNAs in the initial segment of the rat epididymis. *J. Androl.* **19**, 92–99.

Meister, A. (1974). The gamma-glutamyl cycle. Diseases associated with specific enzyme deficiencies. *Ann. Intern. Med.* **81**, 247–253.

Meister, A. (1984). New aspects of glutathione biochemistry and transport–selective alteration of glutathione metabolism. *Nutr. Rev.* **42**, 397–410.

Meister, A. (1988). Glutathione metabolism and its selective modification. *J. Biol. Chem.* **263**, 17205–17208.

Meister, A. (1992). Biosynthesis and functions of glutathione, an essential biofactor. *J. Nutr. Sci. Vitaminol. (Tokyo)* **Spec No,** 1–6.

Meister, A. (1994). Glutathione, ascorbate, and cellular protection. *Cancer Res.* **54,** 1969s–1975s.

Meister, A., and Anderson, M. E. (1983). Glutathione. *Annu. Rev. Biochem.* **52,** 711–760.

Meister, A., Tate, S. S., and Griffith, O. W. (1981). Gamma-glutamyl transpeptidase. *Methods Enzymol.* **77,** 237–253.

Mhlanga, M. M., and Malmberg, L. (2001). Using molecular beacons to detect single-nucleotide polymorphisms with real-time PCR. *Methods* **25,** 463–471.

Ogino, T., Kawabata, T., and Awai, M. (1989). Stimulation of glutathione synthesis in iron-loaded mice. *Biochim. Biophys. Acta* **1006,** 131–135.

Paolicchi, A., Dominici, S., Pieri, L., Maellaro, E., and Pompella, A. (2002). Glutathione catabolism as a signaling mechanism. *Biochem. Pharmacol.* **64,** 1027–1035.

Paolicchi, A., Tongiani, R., Tonarelli, P., Comporti, M., and Pompella, A. (1997). gamma-Glutamyl transpeptidase-dependent lipid peroxidation in isolated hepatocytes and HepG2 hepatoma cells. *Free Radic. Biol. Med.* **22,** 853–860.

Pawlak, A., Lahuna, O., Bulle, F., Suzuki, A., Ferry, N., Siegrist, S., Chikhi, N., Chobert, M. N., Guellaen, G., and Laperche, Y. (1988). gamma-Glutamyl transpeptidase: A single copy gene in the rat and a multigene family in the human genome. *J. Biol. Chem.* **263,** 9913–9916.

Poot, M., Verkerk, A., Koster, J. F., Esterbauer, H., and Jongkind, J. F. (1987). Influence of cumene hydroperoxide and 4-hydroxynonenal on the glutathione metabolism during *in vitro* ageing of human skin fibroblasts. *Eur. J. Biochem.* **162,** 287–291.

Rahman, I. (1999). Inflammation and the regulation of glutathione level in lung epithelial cells. *Antioxid. Redox Signal* **1,** 425–447.

Rahman, I., and Mac Nee, W. (1999). Lung glutathione and oxidative stress: Implications in cigarette smoke-induced airway disease. *Am. J. Physiol.* **277,** L1067–L1088.

Rahman, Q., Abidi, P., Afaq, F., Schiffmann, D., Mossman, B. T., Kamp, D. W., and Athar, M. (1999). Glutathione redox system in oxidative lung injury. *Crit. Rev. Toxicol.* **29,** 543–568.

Richie, J. P., Jr. (1992). The role of glutathione in aging and cancer. *Exp. Gerontol.* **27,** 615–626.

Rojas, E., Valverde, M., Kala, S. V., Kala, G., and Lieberman, M. W. (2000). Accumulation of DNA damage in the organs of mice deficient in gamma-glutamyltranspeptidase. *Mutat. Res.* **447,** 305–316.

Saha, B. K., Tian, B., and Bucy, R. P. (2001). Quantitation of HIV-1 by real-time PCR with a unique fluorogenic probe. *J. Virol. Methods* **93,** 33–42.

Schmid, H., Cohen, C. D., Henger, A., Irrgang, S., Schlondorff, D., and Kretzler, M. (2003). Validation of endogenous controls for gene expression analysis in microdissected human renal biopsies. *Kidney Int.* **64,** 356–360.

Schulz, J. B., Lindenau, J., Seyfried, J., and Dichgans, J. (2000). Glutathione, oxidative stress and neurodegeneration. *Eur. J. Biochem.* **267,** 4904–4911.

Sen, C. K. (2000). Cellular thiols and redox-regulated signal transduction. *Curr. Top. Cell Regul.* **36,** 1–30.

Shaw, L. M., Stromme, J. H., London, J. L., and Theodorsen, L. (1983). International Federation of Clinical Chemistry. Scientific Committee, Analytical Section. Expert Panel on Enzymes. IFCC methods for measurement of enzymes. Part 4. IFCC methods for gamma-glutamyltransferase [(gamma-glutamyl)-peptide: Amino acid gamma-glutamyl-transferase, EC 2.3.2.2]. IFCC Document, Stage 2, Draft 2, 1983–01 with a view to an IFCC Recommendation. *Clin. Chim. Acta* **135,** 315F–338F.

Smith, G. D., Ding, J. L., and Peters, T. J. (1979). A sensitive fluorimetric assay for gamma-glutamyl transferase. *Anal. Biochem.* **100,** 136–139.

Solinas, A., Brown, L. J., McKeen, C., Mellor, J. M., Nicol, J., Thelwell, N., and Brown, T. (2001). Duplex Scorpion primers in SNP analysis and FRET applications. *Nucleic Acids Res.* **29**, E96.

Stark, A. A., Porat, N., Volohonsky, G., Komlosh, A., Bluvshtein, E., Tubi, C., and Steinberg, P. (2003). The role of gamma-glutamyl transpeptidase in the biosynthesis of glutathione. *Biofactors* **17**, 139–149.

Stole, E., Smith, T. K., Manning, J. M., and Meister, A. (1994). Interaction of gamma-glutamyl transpeptidase with acivicin. *J. Biol. Chem.* **269**, 21435–21439.

Suzuki, H., Hashimoto, W., and Kumagai, H. (1993). *Escherichia coli* K-12 can utilize an exogenous gamma-glutamyl peptide as an amino acid source, for which gamma-glutamyltranspeptidase is essential. *J. Bacteriol.* **175**, 6038–6040.

Takahashi, Y., Oakes, S. M., Williams, M. C., Takahashi, S., Miura, T., and Joyce-Brady, M. (1997). Nitrogen dioxide exposure activates gamma-glutamyl transferase gene expression in rat lung. *Toxicol. Appl. Pharmacol.* **143**, 388–396.

Tan, W., Wang, K., and Drake, T. J. (2004). Molecular beacons. *Curr. Opin. Chem. Biol.* **8**, 547–553.

Taniguchi, N., and Ikeda, Y. (1998). gamma-Glutamyl transpeptidase: Catalytic mechanism and gene expression. *Adv. Enzymol. Relat. Areas Mol. Biol.* **72**, 239–278.

Tate, S. S., and Meister, A. (1981). Gamma-glutamyl transpeptidase: Catalytic, structural and functional aspects. *Mol. Cell Biochem.* **39**, 357–368.

Tate, S. S., and Meister, A. (1985). Gamma-glutamyl transpeptidase from kidney. *Methods Enzymol.* **113**, 400–419.

Tietze, F. (1969). Enzymic method for quantitative determination of nanogram amounts of total and oxidized glutathione: Applications to mammalian blood and other tissues. *Anal. Biochem.* **27**, 502–522.

Townsend, D. M., Tew, K. D., and Tapiero, H. (2003). The importance of glutathione in human disease. *Biomed. Pharmacother.* **57**, 145–155.

Vanderlaan, M., and Phares, W. (1981). Gamma-glutamyltranspeptidase: A tumour cell marker with a pharmacological function. *Histochem. J.* **13**, 865–877.

Vandesompele, J., De Preter, K., Pattyn, F., Poppe, B., Van Roy, N., De Paepe, A., and Speleman, F. (2002). Accurate normalization of real-time quantitative RT-PCR data by geometric averaging of multiple internal control genes. *Genome Biol.* **3**, research0034.1–0034.11.

Vina, J., Puertes, I. R., Montoro, J. B., and Vina, J. R. (1983). Effect of specific inhibition of gamma-glutamyl transpeptidase on amino acid uptake by mammary gland of the lactating rat. *FEBS Lett.* **159**, 119–122.

Vina, J. R., Blay, P., Ramirez, A., Castells, A., and Vina, J. (1990). Inhibition of gamma-glutamyl transpeptidase decreases amino acid uptake in human keratinocytes in culture. *FEBS Lett.* **269**, 86–88.

Visvikis, A., Pawlak, A., Accaoui, M. J., Ichino, K., Leh, H., Guellaen, G., and Wellman, M. (2001). Structure of the 5' sequences of the human gamma-glutamyltransferase gene. *Eur. J. Biochem.* **268**, 317–325.

Wannamethee, G., Ebrahim, S., and Shaper, A. G. (1995). Gamma-glutamyltransferase: Determinants and association with mortality from ischemic heart disease and all causes. *Am. J. Epidemiol.* **142**, 699–708.

Whitfield, J. B. (2001). Gamma glutamyl transferase. *Crit. Rev. Clin. Lab. Sci.* **38**, 263–355.

Wu, G., Fang, Y. Z., Yang, S., Lupton, J. R., and Turner, N. D. (2004). Glutathione metabolism and its implications for health. *J. Nutr.* **134**, 489–492.

Yamane, Y., Furuichi, M., Song, R., Van, N. T., Mulcahy, R. T., Ishikawa, T., and Kuo, M. T. (1998). Expression of multidrug resistance protein/GS-X pump and gamma-glutamylcysteine synthetase genes is regulated by oxidative stress. *J. Biol. Chem* **273**, 31075–31085.

[29] Prooxidant Reactions Promoted by Soluble and Cell-Bound γ-Glutamyltransferase Activity

By Silvia Dominici, Aldo Paolicchi, Alessandro Corti, Emilia Maellaro, and Alfonso Pompella

Abstract

Recent studies have provided evidence for the prooxidant roles played by molecular species originating during the catabolism of glutathione (GSH) effected by γ-glutamyltransferase (GGT), an enzyme normally present in serum and on the outer surface of numerous cell types. The reduction of metal ions by GSH catabolites is capable of inducing the redox cycling processes, leading to the production of reactive oxygen species and other free radicals. Through the action of these reactive compounds, cell membrane GGT activity can ultimately produce oxidative modifications on a variety of molecular targets, involving oxidation and/or S-thiolation of protein thiol groups in the first place. This chapter is a survey of the procedures most suitable to reveal GGT-dependent prooxidant reactions and their effects at the cellular and extracellular level, including methods in histochemistry, cytochemistry, and biochemistry, with special reference to methods for the evaluation of protein thiol redox status.

Introduction

Gamma-glutamyltransferase activity (GGT; EC 2.3.2.2), normally found in serum as well as in the plasma membrane of virtually all cells, catalyzes the first step in the degradation of extracellular GSH (i.e., the hydrolysis of the γ-glutamyl bond between glutamate and cysteine). In so doing, GGT releases cysteinyl-glycine, which is subsequently cleaved to cysteine and glycine by plasma membrane dipeptidase activities. Stark *et al.* (1993) first proposed that the catabolism of GSH could play a prooxidant role in selected conditions. These authors suggested that the GGT-mediated generation of the more reactive thiol cysteinyl-glycine could cause the reduction of ferric iron Fe(III) to ferrous Fe(II), thus starting a redox-cycling process liable to result in the production of reactive oxygen species (ROS). GGT was shown to stimulate a GSH-dependent lipid peroxidation in model systems involving Fe(III) complexes as redox catalysts and purified linoleic acid peroxidizable substrate. The possible involvement of other metal cations (e.g., Cu^{++}) was also demonstrated (Glass and Stark, 1997).

METHODS IN ENZYMOLOGY, VOL. 401
Copyright 2005, Elsevier Inc. All rights reserved.
0076-6879/05 $35.00
DOI: 10.1016/S0076-6879(05)01029-3

The prooxidant effect of GGT observed in these systems was attributed to the formation of cysteinyl-glycine and cysteine, which reduce Fe(III) more efficiently than does GSH (Spear et al., 1997; Tien et al., 1982). Thiol-induced iron reduction and redox cycling lead to the production of ROS (Aust et al., 1985), as well as of thiyl (–S•) radicals; the latter can easily react to form disulfides. The following overall sequence can be envisaged as set into motion by GGT-mediated catabolism of GSH (Dominici et al., 1999) (GC-SH = cysteinyl-glycine):

$$GSH \text{--------} GGT \text{--------} > \text{glutamic acid} + \text{GC-SH}; \tag{1}$$

$$\text{GC-SH----} pH > 7.0 \text{---} > \text{GC-S}^- + \text{H}^+; \tag{2}$$

$$\text{GC-S}^- + \text{Fe}^{3+} \text{--------} > \text{GC-S}^{\cdot} + \text{Fe}^{2+}; \tag{3}$$

$$\text{Fe}^{2+} + \text{O}_2 \text{---------} > \text{Fe}^{3+} + \text{O}_2^-; \tag{4}$$

$$\text{O}_2^- + 2\text{H}_2\text{O} \text{-------} > 1/2\text{O}_2 + 2\text{H}^+ + \text{H}_2\text{O}_2 \tag{5}$$

In a pathophysiological perspective, the possibility thus exists that sites of GGT activity may act as sites of promotion of metal ion reduction and redox cycling, with consequent stimulation of oxidative processes. Our own studies have verified this hypothesis in a series of conditions, including liver carcinogenesis, human atherosclerosis, kidney ischemia, and GGT-expressing cancer cells (reviewed by Dominici et al., 2003). Importantly, it was also established that GGT/GSH-mediated prooxidant reactions can participate to redox modulation of both proliferative and apoptotic signals, which prompted the definition of "GSH catabolism as a signaling mechanism" (Paolicchi et al., 2002).

Determination of GGT Activity in Biological Samples

Soluble and Cell-Membrane-Bound Enzyme

Although the physiological substrate for GGT is GSH, the activity of GGT is usually determined by the hydrolysis of a synthetic substrate, most commonly γ-glutamyl-p-nitroanilide (GPNA; Sigma). On enzyme activity, yellow-colored p-nitroaniline is released, with a broad peak absorbance around 380 nm. The procedure described for the kinetic determination of

GGT in serum (Huseby and Stromme, 1979) is suitable for soluble GGT and for GGT solubilized from tissue samples or cultured cells.

For soluble GGT, the sample (50 μl) is added to 1 ml of prewarmed 0.1 M Tris-HCl buffer, pH 7.8, containing 4.6 mM GPNA and 10 mM MgCl$_2$, and the reaction is started by adding the acceptor for the transpeptidation reaction: 100 μl of a 575 mM solution of glycyl-glycine (pH adjusted to 7.8). The reaction is monitored spectrophotometrically at 405 nm (thermostatted at 37°) against an appropriate blank to subtract spontaneous hydrolysis of the substrate. The activity is calculated using a molar extinction coefficient of 9200, and expressed as U/l, a unit of enzyme activity being the amount of enzyme that will catalyze the hydrolysis of 1 μmol of substrate/min.

When cell-membrane-bound GGT is to be determined, cells are washed three times with Dulbecco's phosphate-buffered solution, resuspended in ice-cold Tris-HCl (10 mM, pH 7.8), and disrupted in a glass–glass Dounce homogenizer (20 strokes). After centrifuging at 400g (5 min) to separate intact cells and nuclei, the supernatant is centrifuged at 22,500g (60 min, 4°). Aliquots (50 μl) of the resuspended cell membrane pellet are used for the kinetic determination.

The same procedure can be used for determination of GGT activity in crude tissue homogenates, where readings, however, can be complicated by excessive turbidity of samples. The reaction can be performed at 37° in test tubes and stopped with acid protein precipitants (e.g., acetic acid), and the mixture is then centrifuged before reading against appropriate blanks. Because the absorbance spectrum of p-nitroaniline depends on ionic strength and pH of the medium and decreases dramatically below pH 5.0, the millimolar extinction coefficient of p-nitroaniline should be redetermined in the exact conditions of the assay.

Histochemical and Cytochemical Demonstration of GGT Activity

In histochemical procedures, the visualization of GGT activity is obtained through the spontaneous reaction between a diazonium salt and the naphthylamine released by hydrolysis of a synthetic GGT substrate, γ-glutamyl-4-methoxy-2-naphtylamide (GMNA; Sigma). Such "azo-coupling" reaction causes the precipitation of an insoluble dye at tissue sites containing enzyme activity. The diazonium salt Fast Blue BB (Sigma) (Rutenburg et al., 1969) offers the advantage of a well-focused precipitation of the red stain and low unspecific stain of proteins. The poor stability of Fast Blue BB in watery solutions, however, does not allow incubations longer than 10 min, thus reducing the sensitivity of the procedure. Unfixed cryostat sections (10–15 μm) are dried in air and incubated at room temperature in freshly prepared reagent mixture. For the latter, the substrate GMNA (10 mg) is dissolved in 200 μl DMSO and 200 μl 1 N NaOH; 3.6 ml of distilled water are then added (this solution can be stored at 4° up to 3 days). Immediately before use, the preceding solution (4 ml) is added to 76 ml of a freshly prepared solution containing 40 mg Fast Blue BB and 60 mg glycyl-glycine in Dulbecco's PBS, pH 7.8. After 10 min, the reaction medium is discarded, sections are washed in saline, and the stain is fixed by a 2-min incubation in 0.1 M CuSO$_4$. Slides are then washed in distilled water and mounted with Kaiser's glycerol gelatin (Merck).

For cytochemical procedures (or when higher sensitivity is required) the 40 mg Fast Blue BB used in the previous procedure can be substituted with 28 mg Fast Garnet GBC (Sigma). Despite causing a more intense yellow unspecific staining of proteins, this stain is more stable in solution, allowing incubation of samples up to 100 min (Khalaf and Hayhoe, 1987). To prevent damage caused by long incubations, cell smears or tissue imprints are briefly fixed (15 s) in ice-cold acetone, or in phosphate-buffered acetone/formaldehyde (Yam *et al.*, 1971) and immediately rinsed in ice-cold bidistilled water (×3) before staining.

GGT-Dependent Lipid Peroxidation (LPO)

Detection of LPO in Tissue Sections

GGT is expressed in cells of chemically induced hyperplastic nodules of rat liver, as well as in other model preneoplastic lesions observed in experimental carcinogenesis (Warren *et al.*, 1993). GGT-dependent prooxidant processes can occur in such preneoplastic lesions, and it has been suggested that they may affect their evolution by interfering with redox regulation of proliferative or apoptotic processes (Pompella *et al.*, 1996a).

A convenient experimental model to study these phenomena is the N-nitroso-diethylamine (DEN)/2-acetylaminofluorene (2-AAF)–induced hepatocarcinogenesis in the rat, as described by Solt and Farber (1976). Male Sprague-Dawley rats (250–300 g) fed a standard laboratory chow (Nossan, Italy) are used. The animals are given a single dose (100 mg/kg body wt) of DEN, intraperitoneally, followed 2 weeks later by the administration of 2-AAF, 0.02% (w/w) in diet, for 2 further weeks. At the end of the first week of 2-AAF treatment, partial hepatectomy is performed. Within 4 weeks from the end of the 2-AAF treatment, the animals are sacrificed, and the livers are removed. The appearance of preneoplastic foci in tissue can be assessed by the usual histological stains (toluidine blue,

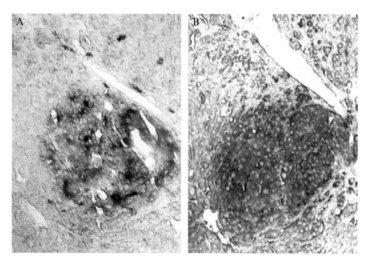

FIG. 1. Selective involvement of hepatocytes of chemically induced liver preneoplastic lesions in GGT-dependent lipid peroxidation. Serial cryostat sections (unfixed) obtained from the liver of a DEN/2-AAF-treated rat, sacrificed at the end of the initiation-promotion schedule. (A) GGT activity; (B) NAH-FBB reaction after induction of LPO by incubation in the presence of GSH, glycyl-glycine, ADP, and iron (60 min) × 50.

hematoxylin–eosin) and by the histochemical determination of GGT activity (Rutenburg *et al.*, 1969).

For the determination of GGT-mediated LPO, fresh cryostat sections (thickness, 10 μm; unfixed) are allowed to dry in the cryostat chamber and are then incubated (37°, 45–120 min) in a pre-warmed medium as follows: 0.15 M KCl, 50 mM Tris-maleate buffer (pH 7.4), 0.05–2.0 mM glutathione, 20 mM glycyl-glycine, 50 μM FeCl$_3$, and 1 mM ADP as iron chelator. At the end of incubation times, sections are rinsed in several changes of saline and processed for the histochemical and microphotometric determination of LPO-derived carbonyl functions by means of the 3-hydroxy-naphthoic acid hydrazide–fast blue B (NAH-FBB) procedure, as reported by Pompella and Comporti (1991). Figure 1 shows an example of the results obtainable with the procedure described.

Detection of LPO in Isolated Cells

In GGT-expressing isolated cells, lipid peroxidation can be induced by simple activation of enzyme activity, by addition to cell suspensions of its substrate and of a suitable glutamyl acceptor. Incubations are thus performed in Hanks' buffer, pH 7.2, containing GSH, 2 mM and glycyl-glycine, 20 mM. A source of redox-active iron can be added (e.g., ADP-chelated ferric iron; ADP-Fe, 1.5 mM–150 μM, respectively). For control, GGT inhibition can be conveniently obtained using acivicin (AT125; 1 mM). Lipid peroxidation can be determined as the content of malonaldehyde (MDA). Medium is discarded, and cells are yielded in 5% TCA. After centrifugation, TCA extracts are reacted with 1 vol 0.67% thiobarbituric acid (TBA) at 100° for 10 min; after cooling of samples, absorption is directly determined at 532 nm (Pompella et al., 1987). Alternately, as a means for sensitizing the assay, the so-called thiobarbituric acid reacting substances (TBARS) can be determined instead; 2 vols TBA reagent are added then to each culture flask, and the mixtures are reacted at 95° for 15 min, essentially as described (Buege and Aust, 1978). Before spectrophotometry, extraction of the chromophore with n-butanol (1:1, v/v; stir thoroughly; centrifuge for separation of phases) is necessary in this case.

GGT-Dependent Peroxidation of Isolated Low-Density Lipoprotein (LDL)

Catalytically active GGT is detectable in human atherosclerotic plaques (Paolicchi et al., 2004). In addition, clinical studies have consistently documented the association of increased plasma GGT levels with accelerated evolution of atherosclerotic heart disease (Emdin et al., 2001; Pompella et al., 2004). It is conceivable that GGT-dependent prooxidant reactions may be involved at some stage in pathogenesis of the disease, possibly by participating in production of peroxidized LDL.

Convenient studies can be carried out with isolated LDL (Paolicchi et al., 1999). LDL is isolated by ultracentrifugation from the plasma of healthy volunteers. After extensive dialysis against PBS, pH 7.4, at 4°, LDL (dissolved in PBS, 0.1 mg protein/ml) is incubated (37°) with purified GGT (25–100 mU/ml). Alternately, LDL is incubated with GGT-expressing cells (e.g., U937 histiocytic lymphoma) at a concentration giving comparable GGT activities per unit volume. ADP-chelated iron (150 μM chelator: 15 μM FeCl$_3$) is also added. GGT activity is then stimulated by addition of GSH (0.5–2 mM) and glycyl-glycine (20 mM). At incubation intervals, aliquots are withdrawn, and LDL peroxidation is determined by assaying for TBARS as described previously. The stimulating effects of GGT-mediated GSH metabolism on LDL oxidation in vitro are illustrated in Fig. 2.

GGT-Dependent Production of Reactive Oxygen Species (ROS)

The production of ROS after iron reduction induced by the GGT-mediated catabolism of GSH has been repeatedly documented (Del Bello et al., 1999; Dominici et al., 1999; Drodzd et al., 1998; Maellaro et al., 2000) and can be achieved using one or the other of several procedures.

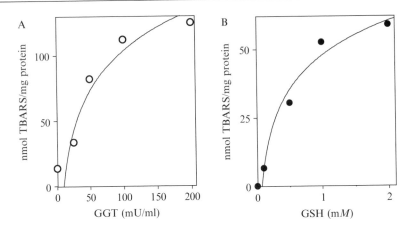

Fig. 2. GSH- and GGT-dependence of LDL oxidation. Incubations (0.5–1 ml) contained LDL (0.1 mg protein/ml), glycyl-glycine (20 mM), and ADP-Fe (150 μM chelator: 15 μM FeCl₃) in PBS, pH 7.4, 37°. In (A) the reactions were started by adding GSH (2 mM), plus increasing amounts of GGT. In (B) GGT was held constant (50 mU/ml), and reactions were started by adding increasing concentrations of GSH. Values determined at 30 min of incubation.

NBT Reduction by Isolated Cells

A simple and rapid procedure for determination of cellular production of superoxide anion is the detection of blue precipitates formed by nitro blue tetrazolium on reduction. Cells are cultured in chamber slides or equivalent support. Incubations (37°, 30–45 min) are performed in Krebs–Ringer buffer (pH 7.4) containing 1.25 mg/ml NBT (Sigma), 2 mM GSH, 20 mM glycyl-glycine, and ADP-chelated iron (1.5 mM chelator: 150 μM FeCl₃). Slides are then gently washed and mounted with an aqueous medium (e.g., Kaiser's glycerol gelatin; Merck).

Fluorimetric Determination of Hydrogen Peroxide Production

Of the numerous existing biochemical procedures, fluorimetric methods are the best suited for determination of H_2O_2 produced by living cells in culture. A number of published studies have used 2′,7′-dichlorofluorescin diacetate (DCF-DA), whose fluorescence intensity is increased on oxidation. However, the actual specificity of DCF in detecting individual ROS has never been explained; rather, serious objections have been raised (Kalinich et al., 1996; Rota et al., 1999). In our hands, good results have been obtained with fluorimetric methods based on horseradish peroxidase (HRP), in which the H_2O_2 present in the sample is specifically reduced by

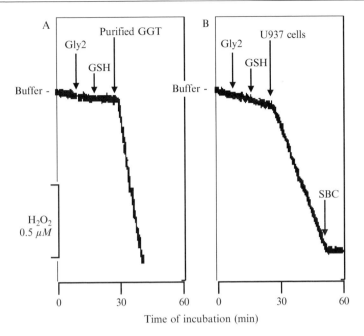

FIG. 3. GSH- and GGT-dependent production of hydrogen peroxide. Decrease of scopoletin fluorescence in the presence of horseradish peroxidase. Vertical bars indicate the fluorescence decrease corresponding to a concentration of 0.5 μM H_2O_2, as established in preliminary calibration experiments using diluted standards. (A) Production of H_2O_2 after the addition of purified GGT protein (corresponding to 18 mU/ml enzyme activity) to Hanks' buffered solution containing the substrate GSH (100 μM) and the $\hat{\gamma}$ glutamyl acceptor glycyl-glycine (gly2, 1 mM); (B) production of H_2O_2 after the addition of an equivalent amount of cell membrane-bound GGT activity, expressed at the surface of U937 histiocytic lymphoma cells (3 × 10^6/ml); the phenomenon is suppressed on addition of the competitive GGT inhibitor, serine/boric acid complex (SBC, 10/20 mM final conc.).

the enzyme, whereas a fluorescent hydrogen donor is concomitantly oxidized (Dominici et al., 1999; Maellaro et al., 2000).

One such method is based on monitoring the decrease in fluorescence of scopoletin during its oxidation catalyzed by type II horseradish peroxidase (HRP; EC 1.11.1.7) (Root et al., 1975). Calibration curves are obtained with additions of standard H_2O_2 in the range 0.1–5 μM to cell-free incubation medium. For real-time determinations, cultured cells are harvested and resuspended (3 × 10^6/ml) in HBSS without Ca^{2+} and Mg^{2+}, in the presence of 2 μM scopoletin (Sigma) and 50 nM HRP (Sigma). Cells are then incubated at 37° under continuous gentle magnetic stirring in the cuvette of a Perkin-Elmer 650-10S fluorimeter (settings: excitation/emission wavelengths, 350/460 nm; slits, 4 nm; sensitivity, 0.3). Incubations are performed in the presence of GGT substrates (GSH and glycyl-glycine, e.g., 0.1 and 1 mM, respectively) and can include specific inhibitors (AT125; serine/boric acid complex), as well as antioxidant agents. Figure 3 shows the results obtainable with this technique.

Scopoletin can be substituted with other suitable probes (e.g., N-acetyl-3, 7-dihydroxyphenoxazine, A6550; Molecular probes, Eugene, OR, USA; 50 mM final conc.; Mohanty et al. [1997]). A6550 becomes highly fluorescent only after oxidation by HRP; thus, in this case, the relative increase in fluorescence intensity is measured (fluorimeter settings: sensitivity, 1; slits, 4 nm; excitation/emission wavelengths, 590/645 nm). When using adhering cells, all reagents can be added to the incubation medium, and fluorimetric determinations can be performed in medium aliquots withdrawn at time intervals.

Prooxidant Effects of GGT Activity on Protein Thiol Redox Status

GGT-Induced Protein Thiol Oxidation

ELISA Determination of Cell Surface Protein SH Groups Using Maleimide-Peroxidase. GGT active site is oriented toward the outer cell surface; prooxidant reactions originating from GGT-mediated GSH metabolism thus primarily involve the cell microenvironment and the cell surface in the first place. With adhering cells, a quantitative evaluation of the redox status of –SH groups of proteins located at the cell surface can be easily accomplished by ELISA assay (Maellaro et al., 2000). Cells are seeded in 96-well microtiter plates (5×10^4 cells/well) and incubated for 24 h at 37°. The experimental treatments (GGT stimulation and inhibition, etc.) are performed in the same plates at 37° for increasing time intervals. At the end of incubations, cells are washed twice with Tris-buffered saline and immediately exposed to polymerized maleimide-activated peroxidase (Sigma) at 37° for 10 min. After washing, the peroxidase substrate o-phenylene diamine dichloride (SigmaFast tablet sets) is added; 15 min later the development of color is determined in an ELISA microplate reader at 405 nm.

Determination of Cell Surface Protein Thiols by Western blot. A more detailed evaluation of protein thiol redox status at the cell surface can be obtained using a combined SH-biotinylation/immunoblot/ECL procedure (Dominici et al., 1999). After experimental treatments, cells are washed once, resuspended (1×10^6/ml) in fresh medium without FCS, and exposed for 15 min at 37° to the non-cell–permeant thiol label

N-(Biotinoyl)-N'-(iodoacetyl)ethylenediamine (BIA; Molecular Probes). A final concentration of 100 μM is obtained adding small aliquots (10 μl/ ml) of a 10 mM stock solution of BIA in DMSO. Cells are then washed in fresh medium, pelleted, and lysed by incubating for 45 min at 4° with a small volume (approximately 10 $\mu l/10^6$ cells) of a lysis buffer composed as follows: 0.5% (v/v) Triton X-100, 5 mM Tris, 20 mM EDTA, 50 mM NEM (pH 8.0). Cell lysates are then centrifuged (15 min, 17,000g, 4°), and aliquots of the supernatants are used for protein content determination. Additional lysate aliquots are used for separation of cellular proteins by 10% SDS-PAGE, using biotinylated high molecular mass standards in the range 66.2–116.0 kDa as molecular weight markers. Such separating conditions are suitable for obtaining information on the redox status of cell surface proteins over a relatively wide mass range. Gels are subsequently blotted onto nitrocellulose filter paper (Sartorius, FRG) soaked in transfer buffer (48 mM Tris, 39 mM glycine, 0.04%, w/v, SDS and 20%, v/v, methanol). The labeling of protein bands with BIA, indicating the presence of reduced protein thiols, is revealed by exposing blots to streptavidin– POD conjugate (Roche), followed by development by enhanced chemiluminescence, using Kodak X-Omat AR films (exposure times: 10 s to 5 min). Alternately, chemiluminescence can be analyzed with a BioRad ChemiDoc apparatus and the BioRad QuantityOne software. The same apparatus can be used for densitometric analysis of developed films. Figure 4 reports an example of densitometric profiles obtained with U937 histiocytoma cells before and after stimulation of membrane GGT activity.

Cytochemical Visualization of Cell Surface Protein Thiols. Several maleimide-based, cell-impermeable, thiol-specific probes are available for this purpose, such as BIA mentioned previously. Another useful compound is 3-(N-maleimidylpropionyl) biocytin (MPB; Molecular probes) (Pompella *et al.*, 1996b). After treatment, cells are reacted with MPB, 25 μg/ml in phosphate-buffered saline, for 30 min at room temperature. Thoroughly washed cells are smeared on clean glass slides; in the case of adhering cells, these can be seeded in Lab Tek chamber slides (Promega, Madison, WI, USA), or alternatively in six-well plates containing (autoclaved) coverslips, which can be subsequently removed and handled separately. Cell smears or monolayers are then postfixed with cold anhydrous acetone (10 min), rehydrated, and treated with a suitable revealing system. When ABComplex/AP (Dako) is used, for example, the results in light microscopy can be as reported in Fig. 5. Using fluoresceinated extravidin (Sigma) instead, specimens become suitable for examination with confocal fluorescence microscopy. Figure 6 shows the results obtained with a BioRad MRC-500 confocal imaging system, equipped with a 20 mW argon ion laser, using a Nikon Plan-Apo 60 (NA 1.4) oil immersion objective. Confocal

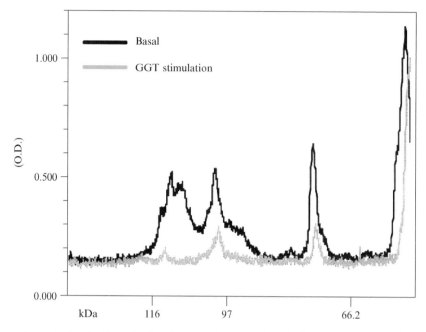

FIG. 4. Determination of cell surface protein thiols after labeling of U937 cells with the non-cell permeant thiol label N-(biotinoyl)-N'-(iodoacetyl) ethylenediamine (BIA). Densitometric profiles obtained from immunoblots revealed by an enhanced chemoluminescence system. *Black profile,* controls; *gray profile,* effects of stimulation of GGT activity (= addition of GSH, 200 μM and glycyl-glycine, 2 mM, final concs.).

microscopy allows one to obtain optical sections just a few micrometers thick, which may enable the discrimination of cell surface from the rest of the cell body. Detailed machine settings were previously described (Pompella *et al.,* 1996b).

Assay of Thiol Redox Status in Individual Proteins by Mal-PEG. When the thiol redox status has to be analyzed in individual proteins, the procedure using maleimide-polyethylene glycol (MalPEG; Shearwaters Polymers Inc. [Huntsville, AL, USA]) for tagging of reduced protein thiols can be used, as recently described (Wu *et al.,* 2000). MalPEG (m.r. $\approx$5500) reacts with available reduced thiols on protein; the molecular weight of cellular proteins will be thus increased in proportion with the number of SH groups available for the reaction. After treatments for modulation of GGT activity, cells are lysed with a buffer containing 1% Triton X-100, MalPEG is added (100 μM final conc.), and reaction is allowed for 30 min at room temperature under gentle mixing. Loading buffer is then added to

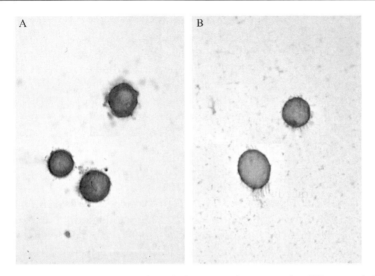

Fig. 5. Cytochemical demonstration of plasma membrane protein –SH groups in U937 histiocytoma cells using 3-(N-maleimidylpropionyl) biocytin (MPB) and an AB-Complex revelation system. (A) Controls; (B) effects of GGT stimulation (= addition of GSH, 200 μM and glycyl-glycine, 2 mM, final concs.; 15 min).

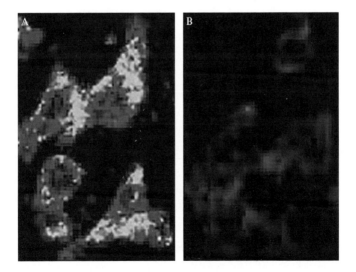

Fig. 6. GGT-dependent loss of cell surface protein thiols in melanoma Me665/2/60 cells. (A) Controls; (B) effects of GGT stimulation after addition of GSH and glycyl-glycine (0.1 and 1 mM, respectively; 45 min). Confocal laser scanning fluorescence microscopy of unfixed cell monolayers reacted with the cell-impermeable thiol-specific probe MPB, followed by fluoresceinated extravidin.

reaction mixtures, and samples are separated by SDS-PAGE (8% polyacryl-amide). Negative controls should be obtained by treating cells with N-ethyl maleimide (NEM) to prevent the binding of MalPEG. After blotting, individual protein bands are identified using specific antibodies and an enhanced chemoluminescence detection system (Roche). Densitometric analysis of band volumes can be performed after imaging of immunoblots with a BioRad ChemiDoc apparatus using the QuantityOne software. The results of a typical experiment, aiming to evaluate redox effects on the TNFR-1 receptor of melanoma cells, are reported in Fig. 7.

GGT-Induced Protein S-Cysteyl-Glycylation

Protein S-thiolation is one of the prooxidant effects ensuing from activation of plasma membrane GGT, as documented in experiments using [^{35}S]-labeled GSH (Dominici *et al.*, 1999). We have subsequently shown that this S-thiolation results, at least in part, from the binding to cellular and extracellular proteins of cysteinyl-glycine, the reactive thiol produced during GGT-mediated GSH metabolism; the process may be

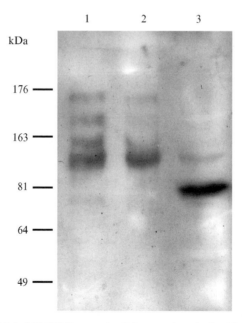

FIG. 7. Immunoblot of MalPEG-reacted cellular proteins, obtained from Me665/2 human melanoma cells. Anti-TNFR-1 antibodies reveal several protein bonds, depending on the availability of reduced SH groups in protein, lower molecule weights correspond to more oxidized forms of the receptor protein. *Lane 1*, controls; *lane 2*, effects of GGT stimulation (addition of GSH, 0.2 m*M* and glycyl-glycine, 2 m*M*, 45 min before cell harvesting and MalPEG treatment; *lane 3*, negative controls (NEM-treated).

thus termed "protein S-cysteyl-glycylation" (Corti et al., 2005). Cysteinyl-glycine–containing mixed disulfides are formed as a result of activation of GGT, obtained by administration to cells of its substrates GSH and γ-glutamyl acceptor glycyl-glycine but are also detectable in resting, untreated GGT-expressing cells, suggesting that the process is occurring in basal conditions during routine GGT activity. Interestingly, the increased binding of cysteinyl-glycine to proteins after GGT activation was found to correspond to a decreased protein S-glutathionylation, allegedly the result of direct oxidation of protein SH groups induced by GGT-derived ROS (Corti et al., 2005).

For evaluation of GGT-induced protein cysteyl-glycylation, procedures for reduction of protein mixed disulfides and HPLC analysis of released non-protein thiols are to be used. After GGT-modulating treatments, aliquots of incubation mixtures are diluted in 0.2 M Tris-HCl buffer, pH 8.2, containing 20 mM EDTA and 1% (w/v) SDS. For measurements on cellular proteins, cells are washed twice with PBS and treated with 10% TCA (20 min, 4°). For measurements on proteins of culture media, aliquots are acidified with 10% TCA and centrifuged at 15,000g (10 min at 4°). In all cases, protein precipitates are washed twice with diethyl ether and resuspended in fresh Tris-HCl buffer. Samples and standards are then incubated (30 min, room temp) with Tris-(2-carboxyethyl)-phosphine (TCEP; Molecular Probes), 5 g/l final conc., to achieve disulfide reduction. Proteins are then precipitated with 5% TCA (final conc.); samples are vortexed immediately and centrifuged (10 min, 15,000g).

Soluble thiols are then derivatized for HPLC analysis. After centrifugation, aliquots (100 μl) of the supernatants are added to test tubes prepared with 250 μl of 125 mM borate buffer, pH 9.5, containing 4 mM EDTA, 110 μl of a 1 g/l solution of the fluorescent probe 7-fluorobenzo-2-oxa-1,3-diazole-4-sulfonate (SBD-F; Fluka), and 20 μl of 1.55 M NaOH. Derivatization is carried out for 60 min at 60° in the dark. Samples are then acidified with 20 μl of 0.4 M H_3PO_4 and kept at –20° until HPLC analysis. The latter is carried out as described (Frick et al., 2003; Pfeiffer et al., 1999). Reverse-phase HPLC is performed using a Resolve C18 column (Waters, Milford, MA, USA), and a Beckman 125P pump equipped with a Rheodyne 7125 injection valve, fitted with a 200-μl sample loop. Mobile phase (5% methanol in 0.2 M KH_2PO_4, adjusted to pH 2.7; filtered, and degassed before use) is delivered at a flow rate of 0.6 ml/min. Calibration is obtained with external standards using 10 mM stock solutions of cysteine, cysteinyl-glycine, and GSH prepared in 0.12 M perchloric acid and stored at –20°. Runs are analyzed by fluorimetric detection (filter settings; 385 nm excitation; 515 nm emission).

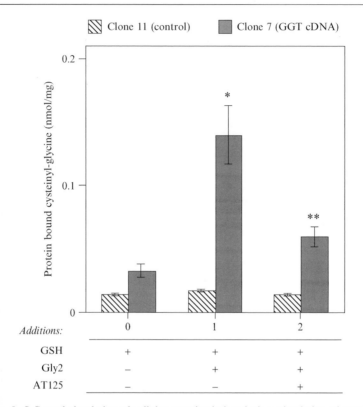

FIG. 8. S-Cysteyl-glycylation of cellular proteins induced after stimulation of membrane GGT in Me665/2 melanoma cell clones. "Clone 11" cells do not express membrane GGT activity, which is instead expressed by "Clone 7" cells after stable transfection with a vector containing GGT cDNA. Determinations were made after 30 min of incubation of cells in HBSS, to which GSH (2 mM), glycyl-glycine (gly2; 20 mM) and the irreversible GGT inhibitor AT125 (1 mM) were added, as indicated. Data are from Corti et al. (2005).

Figure 8 shows the GGT-dependence of cysteinyl-glycine binding to cellular proteins in GGT-expressing compared with non-expressing melanoma cells.

Concluding Remarks

The generation of ROS and reactive thiol metabolites during the GGT-mediated metabolism of extracellular glutathione can primarily affect the redox status of protein thiols, thus producing regulatory effects on the cell itself and its microenvironment. Effects of this kind have been described in several distinct experimental models, suggesting that GGT prooxidant function may represent a general pathophysiological mechanism. The selection of suitable methods, some of which have been described in this chapter, is critical for the assessment of roles possibly played by GGT-dependent prooxidant reactions in pathogenesis of human disease conditions.

Acknowledgments

The studies reported in this chapter were supported by the Italian Ministry for Education, University and Research (MIUR/FIRB and COFIN funds), as well as by the Associazione Italiana Ricerca sul Cancro (A. I. R. C., Milan, Italy).

References

Aust, S. D., Morehouse, L. A., and Thomas, C. E. (1985). Role of metals in oxygen radical reactions. *Free Rad. Biol. Med.* **1,** 3–25.

Buege, J., and Aust, S. D. (1978). Microsomal lipid peroxidation. *Meth. Enzymol.* **52,** 302–310.

Corti, A., Paolicchi, A., Franzini, M., Dominici, S., Casini, A. F., and Pompella, A. (2005). The S-thiolating activity of membrane γ-glutamyltransferase: Formation of cysteinyl-glycine mixed disulfides with cellular proteins and in the cell microenvironment. *Antioxid. Redox Signal.* **7,** 911–918.

Del Bello, B., Paolicchi, A., Comporti, M., Pompella, A., and Maellaro, E. (1999). Hydrogen peroxide produced during γ-glutamyl transpeptidase activity is involved in prevention of apoptosis and maintainance of proliferation in U937 cells. *FASEB J.* **13,** 69–79.

Dominici, S., Paolicchi, A., Lorenzini, E., Maellaro, E., Comporti, M., Pieri, L., Minotti, G., and Pompella, A. (2003). γ-Glutamyltransferase-dependent prooxidant reactions: A factor in multiple processes. *BioFactors* **17,** 187–198.

Dominici, S., Valentini, M., Maellaro, E., Del Bello, B., Paolicchi, A., Lorenzini, E., Tongiani, R., Comporti, M., and Pompella, A. (1999). Redox modulation of cell surface protein thiols in U937 lymphoma cells: The role of γ-glutamyl transpeptidase-dependent H_2O_2 production and S-thiolation. *Free Rad. Biol. Med.* **27,** 623–635.

Emdin, M., Passino, C., Michelassi, C., Titta, F., L'abbate, A., Donato, L., Pompella, A., and Paolicchi, A. (2001). Prognostic value of serum γ-glutamyl transferase activity after myocardial infarction. *Eur. Heart J.* **22,** 1802–1807.

Frick, B., Schrocksnadel, K., Neurauter, G., Wirleitner, B., Artner-Dworzak, E., and Fuchs, D. (2003). Rapid measurement of total plasma homocysteine by HPLC. *Clin. Chim. Acta* **331,** 19–23.

Glass, G. A., and Stark, A. A. (1997). Promotion of glutathione-γ-glutamyl transpeptidase-dependent lipid peroxidation by copper and ceruloplasmin: The requirement for iron and the effects of antioxidants and antioxidant enzymes. *Environ. Mol. Mutagen.* **29,** 73–80.

Huseby, N. E., and Strömme, J. H. (1979). Practical points regarding routine determination of gamma-glutamyltransferase (γ-GT) in serum with a kinetic method at 37°. *Scand. J. Clin. Lab. Invest.* **34,** 357–363.

Kalinich, J. F., Ramakrishnan, N., and McClain, D. E. (1996). The antioxidant Trolox enhances the oxidation of 2',7'-dichlorofluorescin to 2',7'-dichlorofluorescein. *Free Rad. Res.* **26**, 37–47.

Khalaf, M. R., and Hayhoe, G. J. (1987). Cytochemistry of γ-glutamyltransferase in haemic cells and malignancies. *Histochem. J.* **19**, 385–395.

Maellaro, E., Dominici, S., Del Bello, B., Valentini, M. A., Pieri, L., Perego, P., Supino, R., Zunino, F., Lorenzini, E., Paolicchi, A., Comporti, M., and Pompella, A. (2000). Membrane γ-glutamyl transpeptidase activity of melanoma cells: Effects on cellular H_2O_2 production, cell surface protein thiol oxidation and NF-kB activation status. *J. Cell Sci.* **113**, 2671–2678.

Mohanty, J. G., Jaffe, J. S., Shulman, E. S., and Raible, D. G. (1997). A highly sensitive fluorescent micro-assay of H_2O_2 release from activated human leukocytes using a dihydroxyphenoxazine derivative. *J. Immunol. Methods* **202**, 133–141.

Paolicchi, A., Dominici, S., Pieri, L., Maellaro, E., and Pompella, A. (2002). Glutathione catabolism as a signalling mechanism. *Biochem. Pharmacol.* **64**, 1029–1037.

Paolicchi, A., Emdin, E., Ghliozeni, E., Ciancia, E., Passino, C., Popoff, G., and Pompella, A. (2004). Atherosclerotic plaques contain γ-glutamyl transpeptidase activity. *Circulation* **109**, 1440.

Paolicchi, A., Minotti, G., Tonarelli, P., Tongiani, R., De Cesare, D., Mezzetti, A., Dominici, S., Comporti, M., and Pompella, A. (1999). Gamma-glutamyl transpeptidase-dependent iron reduction and low density lipoprotein oxidation—a potential mechanism in atherosclerosis. *J. Invest. Med.* **47**, 151–160.

Pfeiffer, C. M., Huff, D. L., and Gunter, E. W. (1999). Rapid and accurate HPLC assay for plasma total homocysteine and cysteine in a clinical laboratory setting. *Clin. Chem.* **45**, 290–292.

Pompella, A., Paolicchi, A., Dominici, S., Comporti, M., and Tongiani, R. (1996a). Selective colocalization of lipid peroxidation and protein thiol loss in chemically induced hepatic preneoplastic lesions: The role of γ-glutamyl transpeptidase activity. *Histochem. Cell Biol.* **106**, 275–282.

Pompella, A., and Comporti, M. (1991). The use of 3-hydroxy-2-naphthoic acid hydrazide and Fast Blue B for the histochemical detection of lipid peroxidation in animal tissues—a microphotometric study. *Histochemistry* **95**, 255–262.

Pompella, A., Cambiaggi, C., Dominici, S., Paolicchi, A., Tongiani, R., and Comporti, M. (1996b). Single-cell investigation by laser scanning confocal microscopy of cytochemical alterations resulting from extracellular oxidant challenge. *Histochem. Cell Biol.* **105**, 173–178.

Pompella, A., Emdin, M., Passino, C., and Paolicchi, A. (2004). The significance of serum γ-glutamyltransferase in cardiovascular diseases. *Clin. Chem. Lab. Med.* **42**, 1085–1091.

Pompella, A., Maellaro, E., Casini, A. F., Ferrali, M., Ciccoli, L., and Comporti, M. (1987). Measurement of lipid peroxidation *in vivo*: A comparison of different procedures. *Lipids* **22**, 206–211.

Root, R. K., Metcalf, J., Oshino, N., and Chance, B. (1975). H_2O_2 release from human granulocytes during phagocytosis. *J. Clin. Invest.* **55**, 945–955.

Rota, C., Chignell, C. F., and Mason, R. P. (1999). Evidence fro free radical formation during the oxidation of 2',7'-dichlorofluorescin to the fluorescent dye 2',7'-dichlorofluorescein by horseradish peroxidase: Possible implications for oxidative stress measurements. *Free Rad. Biol. Med.* **27**, 873–881.

Rutenburg, A. M., Kim, H., Fischbein, J. W., Hanker, J. S., Wasserkrug, H. L., and Seligman, A. M. (1969). Histochemical and ultrastructural demonstration of γ̄-glutamyl transpeptidase activity. *J. Histochem. Cytochem.* **17,** 517–526.

Solt, D., and Farber, E. (1976). New principle for the analysis of chemical carcinogenesis. *Nature* **263,** 701–706.

Stark, A-A., Zeiger, E., and Pagano, D. A. (1993). Glutathione metabolism by γ-glutamyl transpeptidase leads to lipid peroxidation: Characterization of the system and relevance to hepatocarcinogenesis. *Carcinogenesis* **14,** 183–189.

Tien, M., Bucher, J. R., and Aust, S. D. (1982). Thiol-dependent lipid peroxidation. *Biochem. Biophys. Res. Commun.* **107,** 279–285.

Warren, B. S., Naylor, M. F., Winberg, L. D., Yoshimi, N., Volpe, J. P. G., Gimenez-Conti, I., and Slaga, T. J. (1993). Induction and inhibition of tumor progression. *Proc. Soc. Exp. Biol. Med.* **202,** 9–15.

Wu, H.-H., Thomas, J. A., and Momand, J. (2000). p53 protein oxidation in cultured cells in response to pyrrolidine dithiocarbamate: A novel method for relating the amount of p53 oxidation *in vivo* to the regulation of p53-responsive genes. *Biochem. J.* **351,** 87–93.

Yam, L. T., Li, C. Y., and Crosby, W. H. (1971). Cytochemical identification of monocytes and granulocytes. *Am. J. Clin. Pathol.* **55,** 283–290.

Further Reading

Drozdz, R., Parmentier, C., Hachad, H., Leroy, P., Siest, G., and Wellman, M. (1998). γ-Glutamyltransferase dependent generation of reactive oxygen species from a glutathione/transferrin system. *Free Rad. Biol. Med.* **25,** 786–792.

Paolicchi, A., Tongiani, R., Tonarelli, P., Comporti, M., and Pompella, A. (1997). Gamma-glutamyl transpeptidase-dependent lipid peroxidation in isolated hepatocytes and HepG2 hepatoma cells. *Free Rad. Biol. Med.* **22,** 853–860.

Spear, N., and Aust, S. D. (1994). Thiol-mediated NTA-Fe(III) reduction and lipid peroxidation. *Arch. Biochem. Biophys.* **312,** 198–202.

Author Index

Bennett, B. M., 143
Bennett, C., 199
Benson, A. M., 2, 29, 191, 267
Bentz, B. G., 119
Benz, W. K., 217
Benzekri, A., 79, 83, 88, 89, 91
Berendsen, C. L., 118, 119
Bergado Rosado, J. A., 469
Berge, J. B., 210, 212, 227
Bergevoet, S. M., 22
Bergman, B., 265
Bergman, K., 125
Bergman, T., 137, 140, 179, 263
Bergmann, B., 243, 244, 245, 249
Berhane, K., 24, 380
Berks, M., 79
Bernardini, S., 127
Bernassola, F., 127
Bernat, B. A., 139, 144, 199, 368, 369, 371, 374
Bernat, B. L., 368, 370, 374
Bernaudin, J. F., 414
Bernstein, P. S., 3
Bernström, K., 450
Berry, D. A., 23
Bertini, L., 430, 442, 443
Bertolino, A. P., 83
Bertuzzi, M., 310, 311
Betzel, C., 245, 249
Beuckmann, C. T., 355
Beussman, D. J., 171
Bexter, A., 311
Bezencon, C., 312, 318, 321
Bhakat, P., 137
Bhat, M. B., 94
Bhat, N. K., 101
Bhatnagar, A., 383, 385, 389, 392, 396
Bhatnagar, R. S., 101
Bhosale, P., 3
Biagini, A., 25
Bianco, A., 24, 120
Biasi, F., 391
Bichler, J., 324
Bickers, D. R., 314, 324
Bidoli, E., 311
Bieseler, B., 170
Bignon, J., 308
Bik, D. P., 314
Billiar, T. R., 298
Billon, M. C., 412
Birktoft, J. J., 454

Birney, E., 103, 227, 228
Bitsch, A., 314
Björnestedt, R., 257, 271, 357, 361
Blacher, R. W., 430
Black, M., 125
Black, S. M., 118
Blackburn, A. C., 63, 64, 67, 68, 69, 70, 71, 72, 73, 74, 79, 82, 89, 91, 92, 94, 355
Blakesley, R. W., 101
Blanch, G. P., 321
Blass, C., 227, 228
Blaszkewicz, M., 309
Blay, P., 471
Blochberger, T. C., 428
Blocki, F. A., 348
Blomkoff, R., 450
Blomqvist, A., 151
Bluvshtein, E., 470
Board, P., 212, 279, 281
Board, P. G., 4, 6, 7, 12, 15, 16, 19, 25, 31, 33, 62, 63, 64, 66, 67, 68, 69, 70, 71, 72, 73, 74, 79, 80, 82, 83, 85, 86, 87, 88, 89, 91, 92, 93, 94, 95, 100, 101, 117, 170, 179, 180, 193, 210, 212, 227, 228, 231, 232, 242, 255, 257, 268, 269, 271, 279, 342, 345, 350, 355, 409
Bode, A. M., 127, 326
Bode, C., 392, 396
Boehlert, C. C., 250
Boerwinkle, E., 25
Boffetta, P., 308, 310
Bogaards, J. J. P., 27
Böger, P., 173
Bogers, J., 119
Bogoyevitch, M. A., 111, 112
Bogwitz, M. R., 212, 214, 217
Bohle, D. S., 298
Bohnenstengel, F., 29
Bolanos, R., 227, 228
Bolt, H. M., 63, 308, 309
Bolton-Grob, R., 256, 261, 263, 357, 361
Bonaldo, M. F., 101
Bonfield, J., 79
Bonifant, C. L., 298
Bonvalet, J. P., 414
Boonchauy, C., 103, 104, 105, 110, 233
Boonyaphiphat, P., 315
Boor, P. J., 380, 391
Booth, H. S., 63, 79
Booth, J., 2, 8, 227, 354, 379
Booth, M. N., 298

Subject Index

A

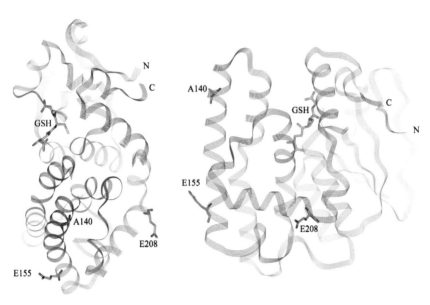

WHITBREAD *ET AL.*, CHAPTER 5, FIG. 4. A structural representation of hGSTO1. Two views of hGSTO1 derived from the protein database coordinates 1eem (Board *et al.*, 2000). The position of genetically variable residues and GSH are indicated.

HEBERT *ET AL.*, CHAPTER 10, FIG. 3. Three-dimensional map of the MGST1 trimer truncated at 4.5 Å resolution obtained by merging data from the p6 and $p22_12_1$ crystals forms. The view is from the lumen side at a slight inclination from the membrane plane. A mask was applied to omit parts of the cytoplasmic domain to clearly visualize the orientations of the transmembrane helices. The most likely subdivision into a monomer is illustrated.

```
DmgstD25     --MDLYNMSQSPSTRAVMMTAKAVGVEFNSI-QVNTFVGEQLEPWFVKINPQHTIPTLV  56
DmgstD26     MPNLDLYNFPMAPASRAIQMVAKAIGLELNSK-LINTMEGDQLKPEFVRINPQHTIPTLV  59
DmgstD21     --MDFYYMPGGGGCRTVIMVAKAIGLELNKK-LINTMEGEQLKPEFVKLNPQHTIPTLV  56
DmgstD24     --MDFYYSPRGSGCRTVIMVAKAIGVKLNMK-LINTLEKDQLKPEFVKLNPQHTIPTLV  56
DmgstD23     --MDFYYSPRSSGSRTIIMVAKAIGLELNKK-QLRITEGEHLKPEFLKLNPQHTIPTLV  56
DmgstD22     --------------MVGKAIGLEFNKK-INTLKGEQMNPDFIKINPQHSIPTLV  40
DmgstD1      -MVDFYYLPGSSPCRSVIMTAKAVGVELNKK-LINLQAGEHLKPEFLKINPQHTIPTLV  57
DmgstD10     --MDLYYRPGSAPCRSVLMTAKAIGVEFDKKTIINTRAREQFTPEYLKINPQHTIPTLH  57
DmgstD27     --MDFYYHPCSAPCRSVIMTAKAIGVDLNMK-LLKVMDGEQLKPEFVKLNPQHCIPTLV  56
DmgstD9      -MLDFYYMLYSAPCRSILMTARAIGLELNKK-QVDLDAGEHLKPEFVKINPQHTIPTLV  57
                  .    .:.:*:*:.::        :    ::.*  :::::*.***. ****
```

```
DmgstD25     DNLFVIWETRAIVVYLVEQYGKDDS-LYPKDPQKQALINQRLYFDMGTLYDGIAKYFFPL  115
DmgstD26     DNGFVIWESRAIAVYLVEKYGKPDSPLYPNDPQKRALINQRLYFDMGTLYDALTKYFFLI  119
DmgstD21     DNGFSIWESRAIAVYLVEKYGKDDY-LLRNDPKKRAVINQRLYFDMGTLYESFAKYYYPL  115
DmgstD24     DNGFSIWESRAIAVYLVEKYGKDDT-LFFKDPKKQALVNQRLYFDMGTLYDSFAKYYYPL  115
DmgstD23     DNGFAIWESRAIAVYLVEKYGKDDS-LFFNDPQKRALINQRLYFDMGTLHDSFMKYYYPF  115
DmgstD22     DNGFTIWESRAILVYLVEKYGKDDA-LYPKDIQKQAVINQRLYFDMALMYPTLANYYYKA  99
DmgstD1      DNGFAIWESRAIQVYLVEKYGKTDS-LYPKCPKKRAVINQRLYFDMGTLYQSFANYYYPQ  116
DmgstD10     DHGFAIWESRAIMVYLVEKYGKDDK-LFFKDVQKQALINQRLYFDMGTLYKSFSEYYYPQ  116
DmgstD27     DDGFSIWESRAILIYLVEKYGADDS-LYPSDPQKKAVVNQRLYFDMGTLFQSFVEAIYPQ  115
DmgstD9      DDGFAIWESRAILYLAEKYDKDGS-LYPKDPQQRAVINQRLFEDLSTLYQSYVYYYPQ  116
                *.  *  :**:*** :**.*:*.   .   * *.  :::*::*****:**:. :.       :
```

```
DmgstD25     LRT--GKPGTQENLEKLNAAFDLINNFLDGQDYVAGNQLSVADIVILATVSTTEMVDFDL  173
DmgstD26     FRT--GKFGDQEALDKVNSAFGFINTFIEGQDFVAGSQLTVADIVILATVSTVE-----  171
DmgstD21     FRT--GKPGSDEDLKRIETAFGFLDTFIEGQEYVAGDQLTVADIAILSTVSTFEVSEFDF  173
DmgstD24     FHT--GKPGSDEDFKKIESSFEYLNIFLEGQNYVAGDHLTVADIAILSTVSTFEIFDFDL  173
DmgstD23     IRT--GQLGNAENYKKVEAAEEFLDIFLVGQDYVAGSQLTVADIAILSSVSTFEVVEFDI  173
DmgstD22     FTT--GQFGSEEDYKKVQETFDFINTFIEGQDYVAGDQYTVADIAILANVSNFDVVGFDI  157
DmgstD1      VFA--KAPADPEAFKKIEAAEEFINTFIEGQDYAAGDSLTVADIAIVATVSTFEVAEFDF  174
DmgstD10     IFL--KKPANEENYKKIEVAEEFINTFIEGQTYSAGGDYSLADIAFLATVSTFDVAGFDF  174
DmgstD27     IRN--NHPADPEAMQKVDSAFGHLDTFIEDQEYVAGDCLTIADIAILASVSTFEVVDFDI  173
DmgstD9      LFEDVKKPADPDNLKKIDDAFAMENTLIKGQQYAAINKLTLADFAILATVSTFEISEYDF  176
                .     :      .::: :*   ::: *  .* *.    ::**:::*:**: .       :
```

```
DmgstD25     KKFPNVDRWYKNAQKVTPGWDENLARIQSAKKFLAENLIEKL-  215
DmgstD26     ----------------------------------
DmgstD21     SKYSNVSRWYDNAKKVTPGWDENWEGLMAMKALEDARKLAAK-  215
DmgstD24     NKYPNVARWYANAKKVTPGWEENWKGAVELKGVFDARQAAAKQ  216
DmgstD23     SKYPNVARWYANAKKITPGWDENWKGLIQMKTMYEAQKASLK-  215
DmgstD22     SKYPNVARWYDHVKKITPGWEENWAGALDVKRIEEKQNAAK-  199
DmgstD1      SKYANVNRWYENAKKYTPGWENWAGCLEFKKYFE-------  209
DmgstD10     KRYANVARWYENAKKLTPGWEENWAGCQEFRKYFDN------  210
DmgstD27     AQYPNVASWYENAKEVTPGWEENWDGVQLIKKLVQ--ERNE--  212
DmgstD9      GKYPEVVRWYDNAKKVIPGWEENWEGCEYYKKLYLGAIINKQ-  218
                ..          ..  . ...
```

TU AND AKGÜL, CHAPTER 13, FIG. 2. Amino acid alignment of *delta* class GSTs. The amino acid sequences were obtained from the flybase (www.flybase.org). The alignment was performed using ClustalW version 1.8 (Chenna *et al.*, 2003). Dm, *Drosophila melanogaster*; red, small and hydrophobic residues (AVFPMILW); blue, acidic residues (DE); magenta, basic residues (RHK); green, hydroxyl, amine and basic residues (STYHCNGQ). The columns identical in all sequences are labeled with a (*) sign. The columns that contain conserved and semiconserved substitutions are marked with a (:) and (.) sign, respectively. The Ser residues of the putative catalytic site are labeled by a (↑) sign.

```
DmgstE1    MSSSG VLYGTDLSPCVRTVKLTLKVLNLDYEYKEVNLQAGEHLSEEYVKKNPQHTVPML  60
DmgstE2    -MSDKLVLYGMDISPPVRACKLTLRALNLDYEYKEMDLLAGDHFKDAFIKKNPQHTVPLL  59
DmgstE10   -MANLILYGTESSPPVRAVILTLRAIQLDHEFHTLDMQAGDHLKPDMLRKNPQHTVPML  58
DmgstE9    -MGKLVLYGVEASPPVRACKLTLDAIGLQYEYRLVNLLAGEHKTKEFSLKNPQHTVPVL  58
DmgstE5    -MVKLTLYGVNPSPPVRACKLTLAALQLPYEFVNVNISGCEQLSEEYLKKNPEHTVPTL  58
DmgstE6    -MVKLTLYGLDPSPPVRAVKLTLAALNLTYEYVNVDIVARAQLSPEYLEKNPQHTVPTL  58
DmgstE8    -MSKLILYGTEASPPVRAAKLTLAALGIPYEYVKINTLAKETLSPEFIRKNPQHTVPTL  58
DmgstE7    -MPKLILYGLEASPPVRAVKLTLAALEVPYEFVEVNTRAKENFSEEFIKKNPQHTVPTL  58
DmgstE4    -MGKISLYGLDASPPTRACLLTLKALDLPEEFVFWNLFEKENFSEDFSKKNPQHTVPLL  58
DmgstE3    -MGKLTLYGIDGSPPVRSVLLTLRALNLDFDYKIVNLMEKEHLKPEFIKINPLHTVPAL  58
            :*** :** .*: ***.* :.:: ::         .  ** ***** *

DmgstE1    DDNGTFIWDSHAIAAYLVDKYAKSDELYPKDLAKRAIVNQRLFEDASVIYASIAN-VSRP  119
DmgstE2    EDNGALIWDSHAIVCYLVDKYANSDELYPRDLVIRAQVDQRLFEDASILFMSLRN-VSIP  118
DmgstE10   EDGESCIWDSHAIIGYLVNKYAQSDELYPKDPLKRAVVDQRLHFETGVLFHGIFKQLQRA  118
DmgstE9    EDDGKFIWESHAICAYLVRRYAKSDDLYPKDYFKRALVDQRLHFESGVLFQGCIRNIAIP  118
DmgstE5    EDDGNYIWDSHAIIAYLVSKYADSDALYPRDLLQRAVVDQRLHFETGVVFANGIKAITKP  118
DmgstE6    EDDGHYIWDSHAIIAYLVSKYADSDALYPKDPLKRAVVDQRLHFESGVVFANGIRSISKS  118
DmgstE8    EDDGHFIWDSHAISAYLVSKYGQSDTLYPKDLLQRAVVDQRLHFESGVVFWNGLRGITKP  118
DmgstE7    EDDGHYIWDSHAIIAYLVSKYGKTDSLYPKDLLQRAVVDQRLHFESGVIFANALRSITKP  118
DmgstE4    QDDDACIWDSHAIMAYLVEKYAPSDELYPKDLLQRAKVDQLMHFESGVIEESALRRLTRP  118
DmgstE3    DDNGFYLADSHAINSYLVSKYGRNDSLYPKDLKKRAIVDQRLHYDSSVVTSTG-RAITFP  117
            *.   :  :  :****  *** :*.  .* ***:*   ** *:* :.::.:.:    . :  .

DmgstE1    FWINGVTEVPQEKLDAVHQGLKLLETFIGNSPYLAGDSLTLADLSTGPTVSAVP-AAVDI  178
DmgstE2    YFLRQVSLVPKEKVDNIKDAYGHLENFLGDNPYLTGSQLTIADLCCGATASSLA-AVLDL  177
DmgstE10   LFKENATEVPKDRLAELKDAYALIEQFLAENPYVAGPQLTIADFSIVATVSTLHLSYCPV  178
DmgstE9    LFYKNITEVPRSQIDAIYEAYDFLEAFIGNQAYLCGPVITIADYSVVSSVSSLV-GLAAI  177
DmgstE5    LFFNGLNRIPKERYDAIVEIYDFVETFLAGHDYIAGDQLTIADFSLISSITSLV-AFVEI  177
DmgstE6    VLFQGQTKVPKERYDAIIEIYDFVETFLKGQDYIAGNQLTIADFSLVSSVASLE-AFVAL  177
DmgstE8    LFATGQTTIPKERYDAVIEIYDFVETFLTGHDFIAGDQLTIADFSLITSITALA-VFVVI  177
DmgstE7    LFAGKQTMIPKERYDAIIEVYDFLEKFLAGNDYVAGNQLTIADFSIISTVSSLE-VFVKV  177
DmgstE4    VLFFGEPTLPRNQVDHILQVYDFVETFLDDHDFVAGDQLTIADFSIVSTITSIG-VFLEL  177
DmgstE3    LFWENKTEIPQARIDALEGVYKSLNLFLENGNYLAGDNLTIADFHVIAGLTGFF-VFLPV  176
            :*:  :    :       ::  *:     ::  *    :*: **     . .   .      :

DmgstE1    DPATYPKVTAWLDRLNKLPYYKEINEAPAQSYVAFIRSKWTKLGDK-------------  224
DmgstE2    DELKPKVAAWFERLSKLPHYEEINLRGLKKYINLLKPVIN-LEQ----------------  221
DmgstE10   DATKYPKLSAWLARISALPRYEEINLRGARLIADKIRSKLPKQFDKLWQKAFEDIKSGAG  238
DmgstE9    DAKRYPKLNGWLDRMAAQPNYQSLNGNGAQMLIDMFSSKITKIV---------------  221
DmgstE5    DRLKYPRIIEWVRRLEKLPYYEEANAKGARELETILKSTNFTFAT--------------  222
DmgstE6    DTTKYPRIGAWIKKLEQLPYYEEANGKGVRQLVAIRKTNFTFEA---------------  222
DmgstE8    DTVKYANITAWIKRIEELPYYEEACGKGARDLVTLLKKFNFTFST--------------  222
DmgstE7    DTTKYPRIAAWFKRLQKLPYYEEANGNGARTFESFIREYNFTFASN------------  223
DmgstE4    DPAKYPKIAAWLERLKELPYYEEANGKGAAQFVELIRSKNFTIVS-------------  222
DmgstE3    DATKYPELAAWIKRIKELPYYEEANGSRAAQIIEFIKSKKFTIV-------------  220
            *    *..: *. :: * *:.           :

DmgstE1    —
DmgstE2    —
DmgstE10   KQ 240
DmgstE9    —
DmgstE5    —
DmgstE6    —
DmgstE8    —
DmgstE7    —
DmgstE4    —
DmgstE3    —
```

Tu and Akgül, Chapter 13, Fig. 3. Amino acid alignment of *epsilon* class GSTs. The amino acid sequences were obtained from the flybase (www.flybase.org). The alignment was performed using ClustalW version 1.8 (Chenna *et al.*, 2003). *Dm, Drosophila melanogaster*; red, small and hydrophobic residues (AVFPMILW); blue, acidic residues (DE); magenta, basic residues (RHK); green, hydroxyl, amine and basic residues (STYHCNGQ). The columns identical in all sequences are labeled with a (*) sign. The columns that contain conserved and semi-conserved substitutions are marked with a (:) and (.) sign, respectively.

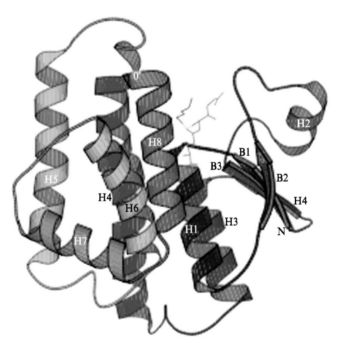

RANSON AND HEMINGWAY, CHAPTER 14, FIG. 2. The 3-D structure of AgGSTD1-6 bound to the inhibitor, S-hexylglutathione (shown in stick representation). Reproduced with permission from *Acta Crystallographica* with permission from Dr. Liqing Chen.

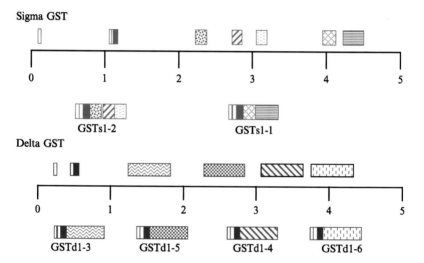

RANSON AND HEMINGWAY, CHAPTER 14, FIG. 3. Schematic diagram showing the alternative transcripts produced by alternative splicing of the *Anopheles gambiae* GSTd1 and GSTs1 genes. The genomic sequence is shown above the scale bar, transcripts are shown below. Empty rectangles indicate 5'UTR regions, solid or shaded rectangles represent coding regions.

3 5282 00599 1396